Community Development Series

The papers gathered in this volume support the notion that *information* may be as important a generator of design in the years ahead as building technology, symbolism, and aesthetic impulse were in the past. The concepts of form and function are not new to those who are designing our environment, but insights based on research can now give new meanings to those words. For practitioner, client, and general reader, these volumes illuminate and detail a new dimension in the environmental arts.

Series Editor: Richard P. Dober, AIP

ENCLOSING BEHAVIOR / Robert Bechtel

URBAN ENVIRONMENTS AND HUMAN BEHAVIOR: An Annotated Bibliography / edited by Gwen Bell, Edwina Randall, and Judith Roeder

EDUCATIVE ENVIRONMENTS FOR CHILDREN: Implications for Design and Research / edited by Gary Coates

COMPUTER GRAPHICS FOR COMMUNITY DEVELOPMENT / William Fetter

DESIGNING FOR HUMAN BEHAVIOR / edited by Jon T. Lang, Charles H. Burnette, and David A. Vachon

APPLYING THE SYSTEMS APPROACH TO URBAN DEVELOPMENT / Jack LaPatra

ENVIRONMENTAL DESIGN: The Role of Preference, Perception and Satisfaction / George L. Peterson

ENVIRONMENTAL DESIGN RESEARCH, VOL. I: Selected Papers / edited by Wolfgang F. E. Preiser

ENVIRONMENTAL DESIGN RESEARCH, VOL. II: Symposia and Workshops / edited by Wolfgang F. E. Preiser

TERRAIN ANALYSIS: A Guide to Site Selection Using Aerial Photographic Interpretation / Douglas Way

Environmental Design Research Association

Community Development Series

ENVIRONMENTAL DESIGN RESEARCH

volume one
selected papers

editor   Wolfgang F E Preiser

fourth international
edra conference

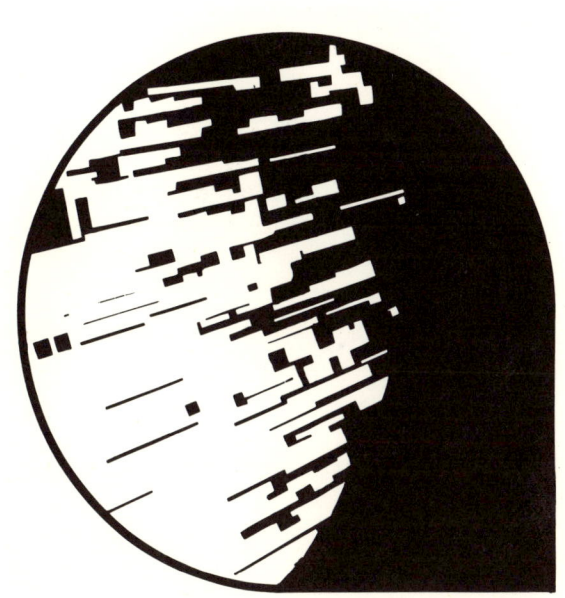

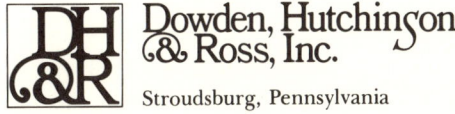

Dowden, Hutchinson & Ross, Inc.
Stroudsburg, Pennsylvania

Copyright © 1973 by Dowden, Hutchinson & Ross, Inc.
Library of Congress Catalog Number: 73-2010
ISBN: 0-87933-029-5

All rights reserved. No part of this book covered by the copyrights hereon may be reproduced or transmitted in any form or by any means—graphic, electronic, or mechanical, including photocopying, recording, taping or information storage and retrieval systems—without written permission of the publisher.

Manufactured in the United States of America

Executive distributor outside the United States and Canada: John Wiley & Sons, Inc.

ERRATUM

Environmental Design Research Vol. 1
    Ed. W. F. E. Preiser

A critical portion of the first sentence on page 21 has been omitted. The sentence should read:
"Wilkie found that at the poles of upper and lower class, people tend to be conservative for different reasons, in the upper because they are oriented to peer group expectation, and in the lower to community tradition through lack of confidence in their ability to cope with the environment."

PREFACE

This is the first of two volumes of ENVIRONMENTAL DESIGN RESEARCH, the proceedings of the Fourth International EDRA Conference which was held on April 15-18, 1973. The College of Architecture at Virginia Polytechnic Institute and State University sponsored the event. Contained in this volume are the proceedings of the paper sessions of EDRA 4 which were organized in response to the "Call For Papers." The second volume will include the symposia and workshops. In these volumes an attempt was made to indicate some structure in these infinite fields for contributions by researchers in disciplines ranging from the physical to the behavioral sciences.

Compared with the organization of the past three EDRA conferences, a number of differences in content and structure of this conference and its proceedings will be noted. They not only reflect shifts of concerns but they may also be indicative of a gradual maturing of the field of environmental design research. While there are a number of manifestations of this process-exemplified by the publication of journals, books, bibliographies and the establishment of specialized research and academic programs, to our knowledge, no successful attempt has yet been made to define the concern and boundaries of environmental design research. It is the purpose of the present two volumes to make a modest beginning in this direction.

If environmental design research is to be recognized as an independent research domain, the question of what differentiates it from other established fields of endeavor needs to be answered. It could be hypothesized that in the past technology was in congruence with the socio-cultural conditions and the "Zeitgeist" of the times. Recent technological developments have occurred independently of the social system and also faster than the necessary social change. The consequence of this trend has been an ever increasing number of conflict situations in man-environment interactions. To reduce conflicts a holistic view of man-environment interaction is needed. It is also necessary to bring up to the conscious level the consequences of decision making in the field of environmental design. In addition, multidiciplinary research rather than research in a single discipline must be undertaken to establish synchrony between user needs and choices provided by new technological developments.

The goal of this research must be to further the understanding of man-environment interaction, and to devise methodologies for the analysis and synthesis of the man-constructed environment, as well as for the management of resources and influences governing human and environmental systems. As methodologies for environmental analysis are developed to higher degrees of sophistication, the question of applicability of research findings to environmental programming and design must be raised. More and more emphasis will be placed on the need for information feedback from the researcher to the design practitioner. One can foresee the possible emergence of a new type of professional, the environmental analyst and programmer, who will be mediator and consultant to both designers and their clients.

Conference Structure and Content

The organization of this conference was set up to operate at two levels: First, a local planning committee was established at VPI & SU which was responsible for basic concepts and structure. Second, an external, interdisciplinary planning committee tried to determine content and quality standards of the material to be included.

A conceptual framework which differentiates between the analysis, synthesis and management aspects of the field, was devised to organize the subject matter. A second three-part structure which contains general ways of dealing with research issues, namely theory, methodology, and empirical findings is expressed by the division of the conference presentations into three kinds of events:

Volume One

> PAPER SESSIONS summarize and criticize 43 selected paper submissions which were solicited through the EDRA "Call for Papers." Authors, panelists and the session participants discuss these papers which are meant to communicate current research findings.

Volume Two

> SYMPOSIA, WITH INVITED PAPERS, are theory oriented. They are intended to comprehensively assess the status of current knowledge and prospects of future directions in the respective fields contributing to environmental design research.

> WORKSHOPS are methodology oriented. Intended as learning experiences, they deal with the demonstration of problem solving processes, with methodological applications to environmental analysis and generally with topics of current concern.

The common pattern of presentation at past conferences involved the full presentation of each paper in sessions organized to reflect the papers to be presented. This approach seemed unsatisfactory since the mere communication of methods and data without some framework of reference made understanding difficult. Therefore, symposia with invited papers were introduced with the goal of synthesising and critically assessing existing knowledge in what appeared to be the major areas of concern in environmental design research. Symposia are intended to convey to the newcomer, both a general overview and an idea of the frontier of knowledge in a given topic area, whereas workshops are intended to focus on current issues and the demonstration of methods. Workshops do not attempt to cover the field topically as do the symposia. They are organized in response to participant interest.

The selection of topics addressed at this conference reflects a greater emphasis on behavioral subjects than was observed at previous environmental design research conferences. Thus, environmental psychologists, linguists, cultural anthropologists and researchers from disciplines earlier considered only remotely related to environmental design are now primarily involved in it. Grouped into nine sessions papers addressed topics like: issues in man-environment theory; environmental cognition; methods of environmental analysis and design languages and methods. Symposia and workshop topics included future oriented technology and other hardware systems. A "new" and current domain "Environmental Design Research in the Social and Political Context" was introduced as a means of bringing together environmental design researchers in the universities and government agencies as well as professionals, and financial and legal groups involved in environmental problem solving for minority groups in the United States are relevant to this concern. A special effort is being made to move the field of environmental design research out of the academic vacuum into the problems of the "real world."

Criteria of Selection

The various types of sessions and events involved different review procedures. The proceedings of the paper sessions (Volume One) contain 43 out of 250 originally submitted contributions. They were reviewed twice anonymously by three independent reviewers. The classification of papers and the selection of appropriate reviewers presented some difficulties. In the case of submissions for symposia and workshops it was mainly the task of their organizers to select and invite the contributors. The papers included in these events represent specialists' viewpoints on their respective fields of competence. The criteria for the selection of material for presentation here included the paper's potential contribution to theory and the soundness of methodology employed, and originality of the paper's concept. Previously published papers or mere proposals for future projects were discouraged. Discussion of relevance and implications for environmental design and achievement of objectives as stated in the paper as well as clarity of content and presentation were considered important criteria.

Certain papers lacking the perfection in presentation were accepted because they showed promise for future elaboration, especially, to encourage students to compete with established researchers, to increase international communication and participation and to provide learning experiences. In order to further the cause of creating an international network of communication among environmental design researchers the publication of an "INTERNATIONAL DIRECTORY OF BEHAVIOR AND DESIGN RESEARCH" was initiated by the conference organizers, the American Institute of Architects (AIA) and the Association for the Study of Man-Environment Relations (ASMER, Inc.). The directory is scheduled to appear in the fall of 1973.

Acknowledgements

A great number of dedicated individuals have helped in the laborious process of preparing this conference. We would like to give credit to all of the contributors who are listed below. We owe thanks to the host of the conference, the College of Architecture at Virginia Polytechnic Institute and State University, especially to Dean Charles H. Burchard and his faculty members for making available their resources of time, funds and goodwill. We thank the publishers of these volumes, Dowden, Hutchinson and Ross, Inc. for their cooperation and understanding regarding the problem of getting countless contributors to produce voluntarily, and on time. Special recognition is given to the conceptual and editorial advice provided by Charles H. Burnette and Gary T. Moore.

Blacksburg, Virginia
December, 1972
Wolfgang F. E. Preiser

EDRA 4 ORGANIZATION

The following compilation includes the names of individuals involved in the preparation of the EDRA 4 Conference. A complete listing is found in the appendix and in Volume II of the proceedings.

Conference Planning Committee

Wolfgang F. E. Preiser, Conference Chairman
Omer Akin, Assistant Chairman
C. Craig Frazier, Planning Assistant
    Division of Environmental and Urban Systems, College of Architecture,
    Virginia Polytechnic Institute & State University, Blacksburg, Virginia 24061

Irwin Altman, Department of Psychology, University of Utah, Salt Lake City, Utah

Charles H. Burnette, AIA, Philadelphia, Pennsylvania

Daniel H. Carson, Architecture, University of Wisconsin, Milwaukee

Aristide H. Esser, Research Center, Rockland State Hospital, Orangeburg, New York

Michael Kennedy, Architecture, University of Kentucky, Lexington, Kentucky

William Michelson, Department of Sociology, University of Toronto, Canada

Special Assistance

Michael Harrell, Graphic Design/Alan Shinder, Photography/Shirley Miles, Sharon Vaught, Secretaries/Clark Jones, Program Specialist/Katie Ahern and Mark Lapping, Editorial Assistance/Joseph Wang, Public Relations/David Pasquale, Transportation/Meral Akin, Inci Ankara, Kamal Choudhury, Sal Choudhury, James Johnson, Sam Reifsynder, Clerical Assistance.

Advisory Committee

Charles H. Burchard/Olivio Ferrari/Alan W. Steiss
    College of Architecture, Virginia Polytechnic Institute & State University,
    Blacksburg, Virginia

Environmental Design Research Association (EDRA, Inc.): Steering Committee

Henry Sanoff, Chairman, North Carolina State University, Raleigh
Daniel H. Carson, Vice-Chairman, University of Wisconsin, Milwaukee
Sidney Cohn, Treasurer, University of North Carolina, Chapel Hill

CONTENTS

PREFACE, Wolfgang F.E. Preiser v

CONFERENCE ORGANIZATION viii

1. THEORETICAL ISSUES IN MAN-ENVIRONMENT RELATIONS
    1.0 Introduction
        Richard A. Chase, Session Chairman 2
    1.1 Aspectual And Hierarchical Characteristics Of Environmental Codes
        Alton J. DeLong. 5
    1.2 An Ethological Approach To Community Design
        Barrie B. Greenbie 14
    1.3 Toward An Ecological Theory of Adaptation And Aging
        Lucille Nahemow and M. Powell Lawton 24
    1.4 Archetypal Place
        Mayer Spivack 33
    1.5 Environmental Stress And Flexibility In The Housing Process
        Mete H. Turan 47

2. VISUAL ATTRIBUTES OF ENVIRONMENTS
    2.0 Introduction
        Gary H. Winkel, Session Chairman 60
    2.1 Toward A Perceptual Tool In Urban Design: A Street Simulation Study
        Michael C. Cunningham, John A. Carter, Carter P. Reese, Bruce C. Webb. 62
    2.2 Presentation And Judgement Of Planned Environment And The Hypothesis
        Of Arousal
        Carl-Axel Acking and Rikard Küller. 72
    2.3 Youth's Perception And Categorizations Of Residential Cues
        Henry Sanoff 84
    2.4 Dynamics Of Preference For Visual Attributes Of Housing Environments
        Peter G. Flachsbart and George L. Peterson 98

3. HUMAN RESPONSES TO THE NATURAL AND MAN-MADE PHYSICAL ENVIRONMENT
    3.0 Introduction
        Harvey M. Choldin, Session Chairman 108
    3.1 A Housing Analysis Of The Cheyenne And Arapaho Of Oklahoma
        Arn Henderson and Richard D. Bauman 111
    3.2 Contextual Fittingness Of Everyday Activity Encounters
        Omer Akin. 123
    3.3 Behavioral Evaluation Of A Juvenile Treatment Center: Case Study Of
        A Planning Methodology
        Robert R. Hahn 138
    3.4 Types Of Cities: The Subnational Urban Environment And Some Design
        Implications
        Robert B. Bechtel 150
    3.5 Psychology And Environmental Management For Outdoor Recreation
        George L. Peterson 161

CONTENTS

4. ENVIRONMENTAL RESEARCH AND DESIGN FOR DIFFERENT AGE GROUPS
   - 4.0 Introduction
     Jerome T. Durlak, Session Chairman 176
   - 4.1 Child's Conception Of Places To Live In
     Joseph Muntanola Thornberg 178
   - 4.2 Age Related Differences In The Use Of Space
     Leanne G. Rivlin, Maxine Wolfe and Marian Beyda 191
   - 4.3 Physical Planning For Increased Cross-Generation Contact
     Edward H. Steinfeld 204
   - 4.4 Measuring Environmental Dispositions Of Elderly Females
     Paul G. Windley 217
   - 4.5 Researching The Behavioral Implications Of Residential Design: The Case Of Elderly Housing
     Sandra C. Howell 229

5. ENVIRONMENTAL COGNITION
   - 5.0 Introduction
     Roger M. Downs, Session Chairman 240
   - 5.1 Personal Construct Theory And Environmental Evaluation
     Basil Honikman 242
   - 5.2 An Analysis Of Four Measures Of Cognitive Maps
     Roger B. Howard, Sara D. Chase, Mark Rothman 254
   - 5.3 Predictors Of Environmental Preference: Designers and "Clients"
     Rachel Kaplan 265
   - 5.4 Cognitive Maps, Human Needs And The Designed Environment
     Stephen Kaplan 275
   - 5.5 Construing The Physical Environment: Differences Between Environmental Professionals And Lay Persons
     H. Stephen Leff and Paul S. Deutsch 284

6. QUANTITATIVE TECHNIQUES IN ENVIRONMENTAL ANALYSIS
   - 6.0 Introduction
     Daniel H. Carson, Session Chairman 300
   - 6.1 Studying Community Renewal Data Relationships With Multiple Factor Analysis
     David Seader 302
   - 6.2 A Bi-Racial Comparison Of Density Preferences in Housing in Two Cities
     Eric Schweitzer, Gwen Bell, John Daily 312
   - 6.3 Types Of User Building Evaluation
     Trevor Denton, J. McCollum, C. Peter Ind, Richard Stutsman 324
   - 6.4 Demodynamics - A Statistical Theory Of Demographic Equilibrium
     Francisco N. Arumi 334

7. DECISION MAKING TOOLS
   - 7.0 Introduction
     Donald P. Grant, Session Chairman 346
   - 7.1 The Value Of User Evaluation Studies To The Design Process And Formula Financing
     Joan C. Simon 348

7.2 An Economy Model For Generating, Evaluating, And Selecting
    Architectural Design Alternatives
    Volker Hartkopf 359
7.3 Proposal For A Diagrammatic Language For Design
    Robert E. David 371
7.4 The Simulation Of Age Related Sensory Losses: A New Approach To
    The Study Of Environmental Barriers
    Leon A. Pastalan, Robert K. Mautz II, John Merrill 383

8. SPACE PLANNING TECHNIQUES
   8.0 Introduction
       Weldon E. Clark 394
   8.1 Requirements For Man-Machine Collaboration In Design
       Charles M. Eastman 396
   8.2 An Evaluation Of Space Planning Methodologies
       Elliott E. Dudnik and Robert Krawczyk 414
   8.3 Syntactic Structures For Site Planning
       Christos I. Yessios 428
   8.4 Utility Ratings And 0-1 Programming In Housing Design
       Richard Hobson and Imre Kohn 440
   8.5 An Integrated Methodology For Office Building Elevator Design
       Margaret A. Frederking and Alton J. Penz 448

9. DESIGN LANGUAGES AND METHODS:
   9.0 Introduction
       Charles H. Burnette, Session Chairman 462
   9.1 Modular Programming
       David Anstrand, Richard Dagenhart, John Moore, Val Thomas 463
   9.2 A Role-Oriented Approach To Problem Solving By Groups
       Charles H. Burnette, Gary T. Moore, Lynn Simek 481
   9.3 Piecemeal Social Engineering: A Case Study
       Patrick J. Quinn and J. MacGregor Smith 493
   9.4 Synthesis In Design - An Interdisciplinary Essay
       Pattabi G. Raman 506
   9.5 U Graph: A Decision Making Aid
       Arie P. Schinnar 519
   9.6 Designing Off The Street
       Michael Kreski and David Rejeski 529

APPENDIX Reviewers of Abstracts and Papers 537
   Author index 541
   Subject index 549

ONE THEORETICAL ISSUES IN MAN-ENVIRONMENT RELATIONS

Chairman: Richard A. Chase, John Hopkins Hospital, Baltimore, Md.

Panelists: Sidney N. Brower, Dept. of Planning, City of Baltimore, Md.
C. F. Graumann, Dept. of Psychology, Univ. of Heidelberg, Germany
Sven Hesselgren, Royal Inst. of Technology, Stockholm, Sweden
Albert Mehrabian, Dept. of Psychology, U.C.L.A., Los Angeles

Authors: Alton J. DeLong, "Aspectual and Hierarchical Characteristics of Environmental Codes"
Barrie B. Greenbie, "An Ethological Approach to Community Design"
Lucille Nahemow and M. Powell Lawton, "Toward an Ecological Theory of Adaptation and Aging"
Mayer Spivack, "Archetypal Place"
Mete H. Turan, Environmental Stress and Flexibility in the Housing Process"

THEORETICAL ISSUES IN MAN-ENVIRONMENT RELATIONS: INTRODUCTION   1.0

Richard Allen Chase, Session Chairman

Associate Professor
Department of Psychiatry and Behavioral Sciences
The Johns Hopkins University
School of Medicine
Baltimore, Maryland 21205

The papers in this section touch on a number of important common themes. They are concerned with ways of conceptualizing those aspects of environment, and those aspects of human behavior that offer greatest promise of illuminating a general systems theory of man-environment relations that will allow prediction and control of the human consequences of physical design. Some of the contributions focus on relevant dimensions of human behavior, such as territoriality, and basic life-support behaviors such as eating, sleeping, defense and reproduction. Others focus more on ways of describing aspects of environmental structure that seem to influence human behavior in important ways, such as flexibility with respect to the amounts of space available and the ways in which the same space may be arranged under changing circumstances. All of the papers contrast adaptive and maladaptive patterns of behavior, and perceive that patterns of relationships between the structure of environments and the structure of human behavior determine whether the human behavioral repertoire grows or diminishes in amount and competence. There is not, as yet, any general agreement about the languages that are most suitable to the description of these patterns of relationships. Each paper offers suggestions in this matter, but the suggestions are quite diverse. The technical languages used by the various authors seem to be more diverse than their underlying concerns and emerging conceptual insights. This is characteristic of the state of conversation in this field at this point in time. As concern continues to shape new concepts, new concepts will begin to shape plans for experiences -- experiences organized and controlled so that they will ultimately constitute experiments. As this evolution proceeds, words and concepts will be increasingly viewed as tools, to be shaped by the consequences of their use. In the meantime, the provocative and diverse statements of our authors constitutes an essential step in the evolution of a new field of enquiry.

The paper by Alton DeLong is concerned with ways of conceptualizing the environment that will facilitate productive exchange of information between social scientists and designers. It is suggested that the environment be conceived of as a system of communication, characterized by objective complexity (etics), perceptual simplicity (emics) and events related to each other in orderly ways over time (tactics). Complexity of environmental structure varies heirarchically. Study of the environment, whether analytically or synthetically oriented, requires specification of the level of complexity (heirarchical characteristics) as well as aspects of the environment the investigator is concerned with (e.g., fine-grained objective events, or economizing generalizations based upon recurrent patterns of user behaviors). In natural environments, all levels of complexity, and all features communicate simultaneously. Our understanding of environments must comprehend all levels of complexity, and all features. To accomplish this objective, conceptual frameworks must be utilized that accommodate analytic and synthetic modes of enquiry in a complementary fashion. The view of environments as systems of communication is thought to satisfy this criterion.

The paper by Barrie Greenbie is organized around interest in "territory" and "territorial behaviors". Territory is viewed as the space surrounding discrete sets of social activities. Human social behavior is determined by phylogenetically old parts of the brain, subserving the basic sensory, perceptual and emotional capabilities of the organism. In addition, phylogenetically newer parts of the brain allow anticipation, planning, and the abstract, objective processes of analytic intellectual behavior that prediction and planning require. As a result, the human organism organizes space in terms of concepts as well as traditional physical (geographical) design. In primitive societies geographical and conceptual space tend to coincide; customs and spaces are neatly overlapped. In complex societies, the organization of conceptual space and the organization of physical space are not as interdependent. Cultural segregation of physical space is viewed as facilitating social identity and cross-cultural communication and trust.

The paper by Lucille Nahemow and M. Powell Lawton begins with a series of generalizations concerning man-environment relations: 1) man-environment systems are transactional (interdependent) in character, 2) behaving organisms seek homeostasis, 3) behavior change may be initiated by the individual or by changes made in the environment, 4) the competence of an individual may be so poor that an adaptive relationship will be difficult in any environment, and 5) unsatisfactory environments can limit the adaptive behaviors of even the most competent individuals. For each individual, there is a range of environmental stimulation, challenge, and responsivity within which behavior is generally adaptive, and accompanied by a sense of well-being. Exceeding the limits of this range, in either direction, results in subjective discomfort, and objective evidence of maladaptive behavior. Moderately challenging environments support individual competence, whereas inadequately-challenging environments produce extinction behavior, and unduly-challenging environments produce escape behavior. In the case of aging individuals, there is a common tendency to provide an environment that is too simplified and stereotyped, thereby resulting in a reduction of competence, continued personal growth and self-actualization.

The places that support behaviors most important to the survival and continued growth of individual and social life are labeled "archetypal" places in Mayer Spivack's paper. The total set of behaviors accommodated by archetypal places for human populations is listed as follows: nesting, sleeping, mating, childbirth, nursery, healing, grooming, nourishment, excretion, storing, looking out, playing, locomotion, meeting, working, competing, learning, and worshipping. Healthy individuals make connections between all of the archetypal places necessary for complete expression of their needs and interests. Adaptive cultures provide the necessary number and kind of archetypal places to allow individuals and groups to design and re-design their own networks of places to accommodate the changing spectrum of needs and interests that are shaped by the continuing evolution of the individual and the culture. Omissions of archetypal places result in disordered behavior. The extent to which the needs of individuals and groups are accommodated by available archetypal places provides a measure of environmental adequacy, and thereby, a way of predicting patterns of individual and group behavior.

The paper by Mete Turan is concerned with the structure of the physical environment, and the interaction between individuals and environments. Subjective appraisal of input from the environment as a threat to the adaptive equilibrium of an individual thereby transforms a previously "neutral" component of environmental input into a "load", and produces a state of "environmental stress". Environmental stress is accompanied by subjective discomfort, and initiates efforts at coping intended to re-establish a more adaptive equilibrium. If the coping efforts are ineffective, "environmental strain" results as a function of "load" and "stress", and is manifested by deformation in the structure of man-environment interaction. The more limited the ability of an environment to change as a function of changing conditions (low environmental flexibility), the higher the strain will be from a constant load. An environment that can readily be modified as a function of changing needs of its inhabitants is said to possess high environmental flexibility. Flexibility can be analyzed in terms of the number of ways in which a given space can be utilized as well as the quantity of space available. The ability to change space utilization patterns as well as the amounts of space available for particular uses allows households to adapt to evolutionary shifts in the number and ages of occupants, as well as periodic changes in the social functions that must be accommodated by the family group. Increased environmental flexibility results in increased ability of individuals and groups to cope with environmental stress.

ASPECTUAL AND HIERARCHICAL CHARACTERISTICS
OF ENVIRONMENTAL CODES

Alton J. De Long

School of Architecture & Planning
University of Texas
Austin, Texas 78712

Abstract

The environment is considered as a code. Codes are systems characterized by aspectual integration and hierarchical organization. These properties are discussed and considered as a frame of reference for relating diverse findings acquired through different analytical emphases and as providing a basis by which synthetic and analytic orientations can be seen as complementary.

Introduction

This paper represents an attempt to briefly identify the environment as a system of communication which is hierarchically organized, with each level of hierarchical complexity characterized by an aspectual integration.

The major theorectical issues concerning hierarchically organized and aspectually integrated systems will not be taken up in this paper, as they have been carefully outlined in detail elsewhere (1,2,3,4,5). The usefulness as well as the necessity of such an organizational framework will be assumed well understood. The reason is simple: we want to begin probing a type of thinking that will highlight the complementary nature of synthetic and analytic orientations and which will recognize the necessity of each viewpoint. A basic requirement is the development of a way of specifying environmental features such that synthetic and analytic points of view can be simultaneously identified.

The designer who has traditionally coveted his ability to synthesize diffuse data into a crystallized concrete expression is finding it increasingly more difficult to cope with research data relevant to the improvement of his designs. The analytically oriented social scientist, on the other hand, whose job it is to generate relevant data is experiencing considerable difficulty in discovering cogent criteria for timely research. The dilemma seems partially related to the fact that within the context of environmental design, analytic and synthetic orientations are no longer free to determine their own criteria of relevance. They are constraining each other as never before.

The concepts contained in this paper are rooted in a discipline which is as much art as it is science, structural linguistics. As a discipline, linguistics is naturally oriented toward both synthetic and analytic processes because communication, by definition, simultaneously involves encoding and decoding. And the similarities between design and encoding and science and decoding do not seem fortuitous.

Aspectual and Hierarchical Characteristics

Over the past several years, we have elaborated the environment as a code. It is a code very much like language in a variety of ways. The environment constitutes an extended medium capable of conveying information through messages which require encoding and decoding. It has discrete properies which are sequentially ordered and simultaneously configured according to specific rules. And it is hierarchically organized. Aspectually, the environment is simultaneously characterized by objective complexity, relative perceptual simplicity, and constraints of order. Hierarchically, the environment constitutes a multilayered, integrated totality which impinges upon users in a systematic manner.

Etic, Emic and Tactic Aspects

Etics, emics and tactics can best be introduced through a simple illustration. Consider a line with an indefinite number of points. If we suppose this line to represent reality, we can assume it to be a relatively objective representation since each point along the line is a _different_ point, and we can postulate as many points as we wish, depending upon how fine grained we wish to observe. This line, then, represents a considerable amount of complexity -- no two events are equivalent; all are unique. We will refer to this line as the _etic_. Next, suppose that we decide it is not necessary to accurately distinguish between every one of our indefinite number of points. We might feel that the difference between p34 and p35 is irrelevant. In fact, we arbitrarily determine that the only differences which are relevant is whether any given point lies within one of several ranges (say p1-p99, p100-p199, etc.). We can now transform our original line into a series of ranges, or classes. In other words, we can collapse our original continuum and adequately represent it in its entirety by several points, much as we can quite adequately represent a triangle by three points rather than by three intersecting lines. We can refer

1. MAN-ENVIRONMENT RELATIONS: THEORY / 7

to these points which represent ranges of irrelevant variation as emes, and the method of representing our original etic line by these points as emic. Emic representation, then, is a shorthand way of accounting for the objective complexity of the etic, and it is important to note that an emic description and an etic description are both representations of the same thing. The emic description merely eliminates all irrelevant and nonfunctional variation from the descriptive process.

The classes, or emes, constitute the units of the system under study. If we have ten such units and rely upon them to send messages, we can send eleven different messages -- ten messages corresponding to our ten different units and "no message" corresponding to a lack of unit transmission. Obviously the system is impoverished, and not very flexible. By combining our previous units, we can greatly expand the number of potential messages, so that if we agree to combine our units until we have a message length of three units, the number of possible messages is 10^3, or 1,000. Such a system, however, requires that we accurately receive every unit in the sequence if the appropriate message is to be understood. Any "noise" in the environment, or any failure in transmission or reception results in a communication breakdown. As a result, communication becomes an exacting, tedious and time-consuming operation. It would be vastly more efficient if we limited the sequential organization of our messages so that enough redundancy were present to allow us to "fill in" with reasonable accuracy any missed unit in the message flow. We could not communicate as many different messages, but of those permissable, we would have a greater guarantee of accurate reception. The incorporation of rules which restrict the ordering of emic units so that sequential predictability is present, is know as tactics.

Every communication system, whether language, kinesics, proxemics or the environment, is characterized by these three aspects, and all must be accounted for before the system can be adequately described and understood. From a slightly different perspective, we can speak of the etic as the given complexity of the stream of behavior, and the emic (classes) and tactic (order) can be seen as error-reducing mechanisms which make behavior not only possible, but intelligible as well.

The etic, emic and tactic aspects of behavior (and the environment constitutes a medium of behavior just as does language or body motion) interact with one another to yeild strata, or levels of behavior. If we think of the etic as the continuum of behavior (the stream) and the emic as how that continuum is segmented, then tactics constitute the ways in which segments are ordered. But once we apply rules (tactics) to units (emics) we have, in effect, generated the basis for a new continuum. This new continuum, in turn, requries segmemtation and ordering; and we have another level in our system. In short, emics + tactics at level x, provide the etic baseline for level y. Emics + tacitcs at level y, in turn, provide the etic baseline for level z, etc.

Two additional factors should be mentioned before considering hierarchical levels. First, aspectual analysis requies a logic different from that normally employed in social science. The definition of classes (emes) is based upon <u>distributional logic</u>: it is where things occur rather than how often they occur that assumes overriding importance (6,7,8). Units (emes) are thus relationally defined, and it is not possible to <u>reduce</u> one level to another in any meaningful manner.

Second, the tripartite aspects of each level as here defined, are commensurate with a variety of different models including those of the transactionalists (9), the developmental paradigm of Piaget for both sensori-motor development and the acquisition of thought and logic (10), the phylogenetic and functional development of the central nervous system (11,12,13), and the various components of emotional processes (14,15).

<u>Hierarchical Organization</u>

One of the functions of hierarchical organization is to severely limit what is related to what. In so doing, hierarchical organization effectively limits awarness to one level of complexity at any given point in time. A simple example may serve to clarify this point. Consider two investigators studying a medium we will assume has just become prominent and is still not fully understood; a situation environmentalists currently find themselves in. The first argues that /s/ and /z/ are distinctly different in English and constitute separate units. He cites, for example, that the only difference between the forms "sing" and "zing" and "bus" and "buzz" (when spoken) is the presence of either /z/ or /s/. Therefore they must be different units or the native speaker of English would never know the forms were different. The second investigator, however, argues the opposite: /s/ and /z/ are perceptually equivalent. He offers as evidence the observation that /s/ is systematically added to forms ending in voiceless consonants to make them plural (cat + s) whereas /z/ is added to forms ending in vowels or voiced consonants (bay + z, dog + z) for pluralization. The second investigator insists that the users of the system treat /s/ and /z/ as being perceptually equivalent, and further, that they inform him they have always <u>heard</u> the plural endings on "cat" and "dog" as being the same.

Who is correct? Both, of course. But the apparently contradictory results cannot be adjudicated until they explicitly recognize that the phenomenon they are dealing with occur on different levels of hierarchical complexity. If they continue to fail to recognize the hierarchical aspects of the system they are studying, their only alternative would be to decide that people are inconsistent and unpredictable in their behavior. Meanwhile, speakers of English will continue to use English without confusion, knowing quite well when /s/ and /z/ are functioning as distinct units (phonemes) and when they are functioning as members of the same unit, namely pluralization (morphemes).

This hypothetical illustration has value from several points of view. First, it should be obvious that there are limitations on what is related to what, and that these limitations follow the boundaries imposed by hierarchical levels. If everything were indeed related to everything else, the contrasting relationships between /s/ and /z/ at one level would interfere with the establishment of a relationship of mutual equivalence at another. Relational analyses within the context of systems, then, are relative and must be exhaustively conducted only between events which occur on the same level of complexity.

Second, because relationships established at one strata are not carried over to the next, awareness is limited to a single level at any point in time. Popular arguments to the contrary, the notion of hierarchical levels in systems, does not imply reductionism, since relationships at any given level are characteristic of event distribution on that level and cannot, by definition, be reduced to any other level. Since reduction is not possible, awareness must be limited to a single level. The practical advantage of the <u>relative</u> autonomy of each level resides in the fact that external factors impinging on the system cannot disrupt the entire system. The system has intrinsic stability. Without such characteristics, systems could not learn(16).

A simplified example of the aspectual and hierarchical characteristics of the environment as a system of communication is shown in schematic form in Figure 1.

<u>Implications of Aspectual and Hierarchical Characteristics
of Systems for the Study of Man-Environment Relations</u>

The concern for aspectual and hierarchical characteristics of systems in this paper highlights the fact that a genuine concern for the man-environment relationship requires a specification of at least two things: the level of complexity one is dealing with, and the aspect being considered. For it is essential to have at least a rough idea of these two characteristics before any intelligent assessment of relative significance can be made. It is quite pointless and a waste of precious time to debate what features, or levels of the environment have the most impact on behavior: <u>the environment communicates on numerous levels in equally numerous ways simultaneously</u>.

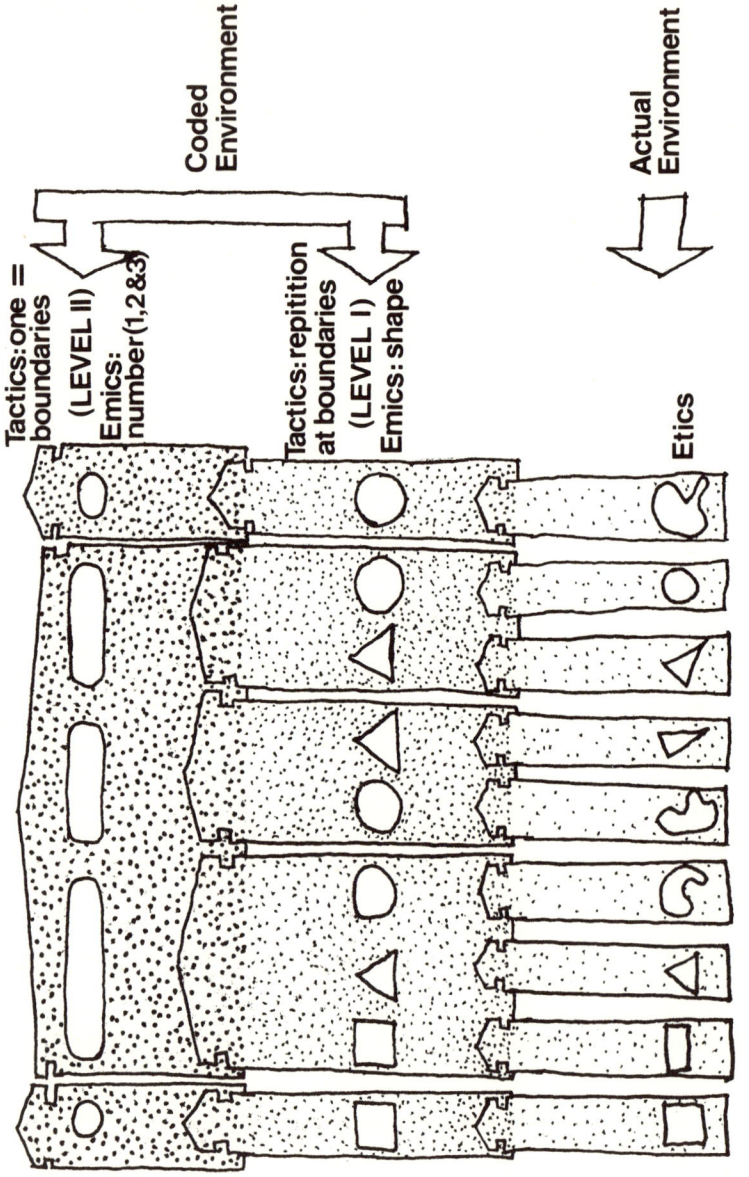

Figure 1: Aspectual and hierarchical characteristics of the environment

1. MAN-ENVIRONMENT RELATIONS: THEORY / 11

We desperately need good researchers operating at as many levels as possible. But the productive coordination of their efforts demands an understanding of how their separate endeavors are ultimately related to one another. Unless findings at all levels can be productively related and viewed as fundamentally complementary, integration and synthesis become nearly impossible, promoting unnecessary debate over which findings are the most relevant. All findings are relevant and can be taken advantage of once a framework for their organizational integration is available.

A significant portion of contradictory findings and specific differences would seem related to the fact that researchers operate at quite different levels and in different ways. Investigators often tend to concern themselves with only one aspect. Those who are concerned with etics are always the most detailed, and presumed the most "rigorous" and "precise". In short, popular stereotype attributes them with being the most scientific. But even the etically oriented must simplify their data, and more often than not, such simplification consists of statistical formulation based principally upon counting exercises. The value of this approach is its fine grain data base. Its weakness lies in the fact that the structural classes of behavior (emes) are not a function of how <u>often</u> particular events occur, but rather <u>where</u> (distribution) they occur. Those emically oriented are principally concerned with condensed, short-hand versions of reality. They characteristically cover much more ground, but do so only by virtue of the fact that they employ numerous unstated assumptions about the nature of their data. The strength of these investigators is that they provide a conceptual framework within which much data can be handled. Their weakness is that often their frameworks embody unstated assumptions which may or may not be pertinent to the nature of their data. Popular stereotypes which continually recur usually cast these individuals as being "soft", "subjective" and as being overly concerned with anecdotal evidence. Finally, we have those interested primarily in tactics, or how events are related to one another over time. Individuals with this orientation are often concerned with the manipulation of units relative to one another employing mathematical models which emphasize sequential processes. Their relative strength resides in acceptance of the sequential (or configurational) properties of behavior. Their implicit weakness lies in the need to assume the nature of the units to be sequentially ordered as well as the occasional neglect of the irrational aspects of behavior (the mere fact that something is logically possible does not imply it is empirically permissable).

The tendency for different investigators to be oriented in these different ways most often leads to intolerable misunderstanding and the feeling that these different points of view are fundamentally incompatible. Yet, all of these points of view mutually complement one another. In isolation, each is both laudable and ludicrous.

Overriding all of this we have the tendency for analytic and synthetic endeavor, which appears to sharply divide and segregate individuals who often are concerned about precisely the same sets of issues. The historical predilection to treat these two approaches as mutually exclusive rather than as being complementary is partially historical accident: a function of the fact that design and science have never had a medium in common. The consideration of the environment as a medium of communication, in which encoding and decoding of information and messages is as essential as it is continuous, would seem to provide an adequate common ground.

An often ignored and neglected concomittant of the synthetic and analytic orientations having a medium in common is the unavoidable necessity of each approach having to accept constraints imposed by the other; and, as a result, of having to undergo alterations. This does not mean that designers should strive to become scientists or that scientists should attempt to become designers. The designer who attempts to shroud himself in quasimathematical models or accrue statistical accoutrements for the sake of achieving scientific respectability may well be missing the point. Design should no longer be conducted for the designer and science no longer fabricated solely according to criteria supplied by the scientist. The willingness of the scientist and the designer to genuinely accommodate the requirements of one another is what is truly exciting about coming to grips with the man-environment relationship. The notion of the environment as a system of communication implicitly fosters such accommodation; and the hierarchical and aspectual characteristics of the environment provide initial sets of criteria which can partially stabilize the context of the dialogue.

References

1. De Long, A. J. Coding behavior and levels of cultural integration: synchronic and diachronic adaptive mechanisms in human organization. In: J. Archea & C. Eastman (Eds.) EDRA II, Pittsburgh: Environmental Design Research Association, 1970.

2. De Long, A. J. Review of E. T. Hall, The hidden dimension. (paperback edition), Man-Environment Systems, 1971, 1, R-9.

3. De Long, A. J. A context for the concept of culture. Preprint, School of Architecture, University of Texas, 1972.

4. De Long, A. J. Variability, complexity and movement in environmental communication. Architectural Design, 1972, XLII, 641-645.

5. De Long, A. J. The communication process: a generic model for man-environment relations. Man-Environment Systems, 1972, September.

6. Austin, W. M. Criteria for phonetic similarity. Language, 1957, 33, 538-544.

7. Austin, W. M. Phonotactics and the identity theorem. Studies in Linguistics, 1960, 15, 14-18.

8. De Long, A. J. Environment as code. Presented at the 139th Meeting of the American Association for the Advancement of Science, Washington, D.C., December, 1972.

9. Tibbetts, P. The transactional theory of human knowledge and action: notes toward a "behavioral ecology." Man-Environment Systems, 1972, 2, 37-59.

10. Piaget, J. & Inhelder, B. The psychology of the child. New York: Basic Books, 1970.

11. MacLean, P. Phylogenesis. In: P. Knapp (Ed.) Expression of the emotions in man. New York: International Universities Press, 1963.

12. MacLean, P. The brain in relation to empathy and medical education. Journal of Nervous and Mental Disease, 1967, 144, 374-382.

13. Esser, A. H. Environmental design needs empathy to combat pollution. In: W. F. E. Preiser (Ed.) Environmental design perspectives. Orangeburg, N.Y.: The Association for the Study of Man-Environment Relations, 1972.

14. Pribram, K. The new neurology and the biology of emotion. American Psychologist, 1967, 22, 830-838.

15. Spitz, R. Ontogenesis: the proleptic function of emotion. In: P. Knapp (Ed.) Expression of the emotions in man. New York: International Universities Press, 1963.

16. Beck, H. Minimal requirements for a biobehavioral paradigm. Behavioral Science, 1971, 16, 442-455.

AN ETHOLOGICAL APPROACH TO COMMUNITY DESIGN

Barrie B. Greenbie

Associate Professor, Regional Planning
University of Massachusetts, Amherst

Abstract

Ethological concepts of territory are related to ethnic and other social groupings. MacLean's theory of the "triune brain" is used as a metaphor to define two imperfectly integrated psychological environments, one sensual and territorial, the other abstract and symbolic. Those with the narrowest "conceptual worlds" will be most dependent on geographical boundaries; thus it is the poorest and least educated members of a society who will be most likely to seek security in physical territory and least likely to be able to obtain it. A hierarchy of private/public spaces is proposed as a model for community design.

The report of the Commission on Population Growth and the American Future (1) once again confirms what is well known: the effects of over-crowding, both on human societies and the non-human environment, have become the most crucial problems of our time. To meet them a variety of proposals for redistribution and resettlement have emerged, raising such controversial issues as urban renewal, exclusionary zoning, and new town development.

Both the problems and the proposed cures lend great importance to two parallel developments within the behavioral sciences. One of these is the renewed interest of sociologists in ethnicity (2)(3)(4). The other is a growing body of knowledge generated by ethologists regarding territorial behavior of men and other animals. The seminal works of Edward T. Hall (5) and Robert Sommer (6) have faced head-on the complexities of human culture in terms of the infra-structure of man's evolutionary inheritance. Much fresh thinking and creative research has followed, even if a widely accepted body of theory has yet to emerge.

The egalitarian myths of our own culture have tended to ignore cultural distinctions, without (quite obviously) eliminating those distinctions. Territory tends to be considered only as personal space, and is too often implicitly looked upon as being "anti-social". A close look at the work of ethologists reveals that among all social animals, including man, territorial spacing is not anti-social but the means by which social organization becomes possible. If looked at in the broadest sense it can be considered the space surrounding social activity which distinguishes one social entity from another. I have therefore redefined "territory" for planning purposes as the expression of social organization in spatial terms.

The question of instinct in man remains a bugaboo to traditional humanists, although the "nature-nurture" controversy is getting to be pretty much a straw man among serious researchers. Because so much confusion remains in the minds of scientific laymen, I should like to preface my discussion with a summary of my own conclusions, based on personal interviews with a number of prominent scientists who have used ethological methods in the investigation of both human and non-human behavior, which were conducted over a two year period with the help of the National Endowment for the Arts:

Most, if not all, animals maintain around themselves at least one of three kinds of space: (a) personal space, (b) small group space, and (c) range. The first two of these spaces may move with the animals, or they may move within them. In either case <u>collective</u> movement takes place within the third.

Most animals, and apparently all social animals, maintain some sort of hierarchy or ranking order within their group. Most social animals will repel strangers who intrude into their social group space, i.e., territory, whether fixed or mobile. Aggressive competition, invloving some sort of symbolic display, will be used to maintain both status and territory.

Aggression within non-human species rarely involves physical destruction (7) unless conditions are pathological (e.g., over-crowding). However, the psychological, or adrenocortical results of symbolic defeat may often be harmful to the losing animal, resulting in inability to breed and often to survive (8).

What is often overlooked is that the tendency to maintain social hierarchies and territory will not only vary from species to species but will also vary within species with seasonal and circadian (daily) rhythms, and may only appear, or be especially prominent, at breeding times. Territorial behavior has a subjective perceptual component, even in very simple animals, as well as an objective geographical one, and is part of a system of what Wynne-Edwards calls "conventional tokens" (9), which enable animals to organize themselves socially. In human beings the phenomenon is complicated by our complex symbolic world made possible by the capacity for abstract thought.

MacLean has postulated three anatomical levels of the brain which correspond to the stages of evolution from vertebrates through mammals to man (10). He notes that what he calls the "neural chassis" is essentially the same for all vertebrates; it governs behavior patterns required for survival of the individual and the species, e.g. feeding, flight, and reproductive activity. In mammals there has developed around this primitive brain what is called the limbic lobe or "old" cortex. In higher animals, according to MacLean, it is this part which gives rise to practically all forms of emotional feeling. He calls this the "lower mammalian" brain. The neocortex increases in size and complexity as we come up the evolutionary scale and is most thoroughly developed only in man. In MacLean's theory the neocortex is primarily the area which contains the circuits pertaining to anticipation and planning.

MacLean believes that the three levels of brain function together imperfectly and he calls man's brain the "triune brain". There is some debate concerning his concept among medical researchers, and I must hasten to disclaim any qualifications for entering into it. But whatever the physiological facts may be, MacLean's concept does offer a useful metaphor for viewing certain readily observable contradictions in the behavior of our species. We may assume that each of us lives, not in one environment, but two environments simultaneously. We may further suppose that it is the interaction between these environments that has made culture, and its correlates in language and technology, possible, and consequent opportunities for altering the external environment.

However, Esser has noted that the clash of images which derive from these different ways of coding experience, result in what he calls "social pollution" (11). This describes very well what unfortunately so often happens in human relationships. I think that we may improve our ability to function in the "real world" harmoniously, by building into our habitats provision for two mental environments. I propose that we view one of our internal environments as being organized very much as it is for other mammals, in terms of familiar and alien conspecifics, and that we recognize that we are prompted to respond within it by fundamentally similar urgings and constraints. This is the environment that is "territorial" in an animal sense. For exampel, Esser's studies of psychiatric wards indicate that it is the most mentally debilitated who are the most reliant on territorial space (12). Our other environment is organized by the neo-cortex. This is an environment made up entirely of abstractions, that is, of patterned relationships between perceived and anticipated objects and events in the environment without references to our emotional individuality. While normal human beings dwell in both environments, and both must be synthesized adequately in the "real world", these internal worlds have very different spatial parameters. The problem is that an external or "real world" environment structured only in limbic terms will not work for technological man. A "real world" environment structured only in neocortical terms will in one way or another be repudiated, either passively by ignoring it, or actively by destructive assaults. What is now happening, it seems to me, in all industrial societies is that we are rejecting, like a transplanted heart, the abstract mechanisms that are keeping us, socially speaking, alive.

Our troubles arise, as Esser suggests, when we fail to recognize the different realities of our two inner environments, and organize both behavior and space appropriate to one in the context of the other. Except for very primitive people, and relatively simple sub-cultures within complex societies, the purely limbic aspects of territoriality cannot explain all of our behavior. In this sense the charges of over-simplified extrapolations regarding territory are often justified. But this does not mean that we humans have surrendered all our territorial propensities, but rather that we have merely transformed them. On the limbic level we relate to specific individuals and places in finite terms on a relatively small scale. On the neocortical level we organize experience with our conceptual coding system which transcends physical time and space. But we nevertheless then tend to defend and seek status within conceptual space as if it were its physical counterpart.

In primitive societies geographical space and "conceptual territory" will tend to coincide, i.e., certain customs and traditions will take place in a certain space, and the space will be defined by where the customs take place. This is the most stable but limited situation. As we extend our conceptual worlds through language, and the range of our physical world through technology, there is an increasing disparity between geographical and conceptual space. We may then give priority in our behavior to one or the other, and may exhibit considerable ambiguity in the process.

As with most social animals, we will seek security at the center of our territory when frightened and vulnerable. We will exhibit more exploratory or aggressive behavior when motivated by need or excess or free energy. We may do so in terms of physical space <u>or</u> in what Calhoun calls "conceptual space" (13) or both in what I will call "conceptual territory".

<u>The larger the conceptual world of the individual, the less dependent he is likely to be on geographical territory</u>, although, ironically, the more likely he will be able to obtain it. Conversely, the more limited the individual's conceptual world, the more he will rely on possession of geographical space. Thus it is the lease educated (usually, but not always, the most poor) who are most likely to need clearly defined home boundaries, who are least likely to have them and are most likely to suffer from being dispossessed.

The need for emotional security and protection of both the individual and the small group can best be provided by <u>small scale</u> physical space protected in some manner from intrusion (limbic territory). On this level there is both an optimum size and an optimum density for human societies. <u>This kind of territory is most satisfactory when culturally (not racially) segregated; heterogeneous social interaction should occur by permission only, not by right</u>, but in democratic societies this pre-supposes a reasonably equitable allocation of space and resources. <u>Conceptual space</u> is the arena in which the cultural interactions and large scale cooperative activities so necessary for complex civilizations take place. On this level the impulse for exploration can best be fulfilled, for conceptual space must be fully integrated in fact and in law. The physical correlate of conceptual space is public space, e.g., parks, markets, civic centers, industries, and universities. These must be kept open to all and at the same time secure from any form of physical aggression.

I have defined territory as the expression of social organization in spatial terms. Since social organization takes place in physical space, and is inevitably constrained by it, it involves not merely a two-way transaction between individuals or groups, but a three-way one between them and the environment. Any change at one point in this triad will alter relationships between the other two. Each of us identifies himself (a) as an individual within a group, (b) as a member of a group vis-a-vis other groups, and (c) in terms of a particular geographical area. "Where are you from?" is often synonymous with "Who are you?" and "Who are your relatives or associates?". In the model I am advancing we may expect that on the limbic level, intergroup relations will frequently involve some sort of competition and often xenophobia and hostility. But on the <u>intellectual</u> level, the individual is a detail. It is the ability of conceptual man to leave himself out of the picture that enables him to cope with objective complexity. A "real world" for complex technological man must therefore be one which enables us to integrate these perceptual worlds. It will be an hierarchy of groups, within which individuals reside more or less securely, and between which the more adventuresome can move more or less freely without xenophobia or territorial defensiveness.

We may relate this concept to Fred Fischer's "cardinal distance values" (14), security contrasted with freedom of movement as positive perceptions of external space, and confinement in contrast with the sense of being lost as negative responses. I have arranged these on a continuum (Fig. 1), with "security" and "confinement" at one pole, and "freedom of movement" and "sense of being lost" at the other, giving the upper set a positive sign and the lower set a negative sign. We may assume each one of us will view our position in such terms, depending on our state of being at the moment, (e.g., energetic or tired, healthy or sick, young or old, etc.)

CARDINAL DISTANCE CONTINUUM (after Fischer)

Fig. 1

Any satisfactory scheme for organizing "real world" space in social terms must take into account not only conceptual space and limbic space, but also provide <u>relative security and freedom</u> within those spaces. Limbic space, if my hypotheses are correct, will be concerned with security-freedom relationship within the small group, whereas abstract conceptual space will consist primarily of those relationships between groups (e.g. ideologies) on a rising scale of complexity.

I suggest that our egalitarian philosophy, and the legal and physical structures which have been built on it in western democracies, as well as in communist societies, have set up an oversimplified polarity between "private" and "public". In capitalist democracies the right to private property is a fundamental tenet of the culture. The psychological importance of "limbic territory" on the level of the family is also recognized; it is protected, at least in principle, against search and seizure without due process and in social attitudes such as "a man's home is his castle". The private home is also implicitly recognized in communist coutries because ignoring it would be virtually impossible. But all philosophies based on an ideology of human brotherhood tend to set private space at one end of what I shall facetiously call a "discontinuous continuum", with "public space" and "public good" at the other (Fig. 2). I think we need to organize urban physical space socially in terms of three categories corresponding to the ethologists' definitions of animal spaces: (a) personal space (b) small group space, i.e., defended territory, and (c) home range (Fig. 3). The first and second of these are essentially the limbic environment for man, and represent two kinds of privacy, while the links between second and third constitute conceptual space and are the real business of what planners call "the public sector". The private limbic space may be considered to be relatively fixed and stable. It is apparently related to certain innate propensities within all human beings, and it has an optimum size and workable limits. As Hall has shown us, it differs in important details from one culture, to another, and he believes this is a fundamental aspect of human communication within cultures as well as an obstacle to communication between them. In essential form it is what zoologists call species specific. Hall calls this "infra-culture". The public conceptual space

that I am adding to the equation is a
hierarchial arrangement, or orders,
capable of infinite complexity and infinite complexity and infinite variation
and is reasonably independent of culture.
For example, it will be easier to get
agreement between an airplane pilot
from America and one from Japan on the
spatial requirements of an airfield than
for a home. To a great extent what goes
on in limbic space is only the business
of those who occupy it, whether on the
personal or group level. It is <u>private property</u> in a <u>social sense</u>. But what
goes on in <u>conceptual space</u> is the business of anyone who cares to enter one
or more of its hierarchial orders at any given place and time and all those who
manage technology. It is the exclusive property of no one; it cannot and should
not be "defended" but only protected, like air and water, from aggressive misuse or excessive monopoly by one group to the detriment of another. This conceptual space might be called "public range".

Fig. 2

Fig. 3

At the birth of the twentieth century Ebenezer Howard devised a way of structuring Homo sapiens' habitat which provided a hierarchy of discreet centers buffered
from each other by greenbelts. I propose that we add to this model a social-
psychological one composed of small self-defined groups, or limbic territories,
buffered from each other by areas of public space which serves as the social
correlate of physical open space. Figure 4 illustrates this paradigm. The
social buffers may also be physical greenbelts or parks, but they may as well
be urban artifacts, or combinations of these, such as markets, civic centers,
and institutional spaces. The physical characteristics of what I conceive of
as "public space" will vary with the image systems and cultural norms of the
groups that interact within them. The only requirement is that such space be
freely accessible, which is not always advisable in natural spaces. As densities
build up more and more, "conceptual space" will increasingly be represented by
social structures rather than physical ones, such as trades, professions, clubs,
etc. For instance, an inter-disciplinary association like EDRA may serve as
"public conceptual space" between "professional territories".

The task of the physical planner in delineating public space in each case is to
provide the widest range of opportunities and minimize sources of conflict between
them, while protecting the security of the private and group territories.

Cassel and his associates (15) (16) (17) have concluded that the destructive
consequences of crowding do not derive from either densities or the stress of
interaction <u>per se</u>, but rather from the particular kinds of stress that arises
when familiar customs, status hierarchies, and emotional group support systems
based on specific cultures are disorganized. Such disorganization may occur
when unfamiliar conspecifics are forced into restricted spaces which do not
permit retreat to a private "territory" of the familiar group or family, or
where the social coding processes that make possible interpersonal communication

are disrupted through migration. He
believes that this does not necessarily
lead to social unrest and violence; it
may merely lead to disease!

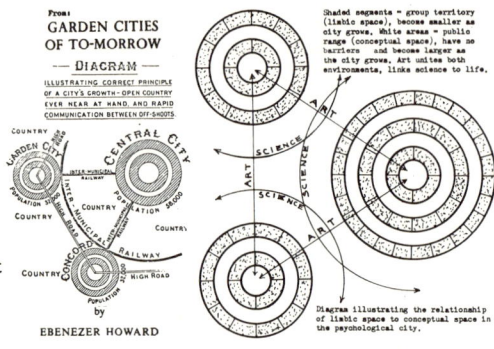

There is also reason to believe that
social identity is facilitated by clearly
perceived physical boundaries, and that
where cultural identity ceases to become associated with physical territory, it
is replaced by "conceptual territories",
e.g. ideologies, which may be considerably
more rigid than the physical ones and
therefore prevent, rather than facilitate,
cross-cultural communication, cooperation,
and trust.

Obviously, if these conclusions are correct, they have the most serious implications for design and planning of human habitations. For many years sociologists
and other observers have noted that the mass relocations of poor people and the
destruction of urban enclaves in our larger cities, which have accompanied large
scale urban renewal and public housing projects, have been contributing to the
social breakdown of our cities (18) (19) (20) (21). It is crucial that proposals
to relocate the poor and minorities in the suburbs and in new towns do not repeat
these errors. Ethological studies of animals support sociological and anthropological research which suggests that protection of ethnic and other cultural
constellations are an indispensible condition for human cooperation and well-
being.

Two generations ago McKenzie developed the concept of "human ecology" in which
he drew analogies from plant and animal studies (22). Most recently Suttles (23)
has reexamined the idea of community, suggesting that the conception of community
is a much more complex phenomenon then is suggested by the traditional concept
of "neighborhood", and that individuals actually will inhabit a variety of
communities, depending on their activities and circumstances. I think that by
accepting a distinction between "limbic" and "conceptual" space, based on MacLean's
"triune brain", we can allow for this complexity and at the same time account
for the peculiarly tenacious tendency to form closed social-physical units.

But there is important reason to believe that there is considerable spread between various individuals within any group as to their position on the limbic-
conceptual space continuum. In a study of migration patterns among Argentine
peasants, Wilkie (24) found that a relatively conservative peasant community,
which in a typical sociological view would be considered "lower class", actually
was stratified not only with a lower, middle and upper class, but with two kinds
of middle class. He turned up this information by combining the anthropologist's
"participant'observer" method, which corresponds closely to the ethological one,
with a multivariate statistical technique. These two methods reveal complexities
that a monolithic view of class structure and linear analysis often covers up.

Wilkie found that at the poles of upper and lower class, people tend to be confident in their ability to cope with the environment. In addition, a portion of the middle class also resembled these groups. One of his conclusions is that dynamic behavior correlates with trust in the environment, and that this presumably comes from the security of feeling that one can cope with it successfully. By contrast, dependence on the authority of tradition corresponds to mistrust of the environment. But Wilkie's studies suggest that both types of individuals will appear in any "class", including what may appear by normative standards to be the "lowest".

One immediate lesson from all this for planners and designers is that we ourselves are likely to come out of this conceptually adventuresome group with migratory inclinations, and that will be as true for black advocates as for "WASP" establishmentarians. We will tend to model community designs on our own demands for diversity and stimulation, projecting as universal good what is in fact a special minority interest, even if a socially most important one. We will continually come up with environments that are quite intolerable for others, who, if only because they are not planners and decision makers, have less reason to trust the environment. The real challenge is to devise environments which will truly protect both kinds of personalities in all social strata, providing both optimum opportunity and optimum security under varying conditions for diverse populations at different stages of the life cycle. The goal should be to provide both of Leyhausen's hierarchies, the absolute, based on individual energy levels, and the relative one, based on territorial identity (25).

On the other hand to recognize the importance of group identity is not to suggest that the structuring of human communities primarily in terms of enclaves of ethnic and other homogeneous groups will by itself produce an ideal society. Homogeneous small group territories must not be isolated villages, but fully accessible to public spaces where people can interact cross-culturally as their interests, needs and energies permit. There is certainly no room in this concept for economic, cultural, or racial "Berlin walls" that divide one sort of ghetto from another. The right to travel is an inalienable right in any society that can be called a civilization.

But planners must recognize two psychological limitations. One is that the capacity for cross-cultural interaction is a function of security; only people with highly developed self-images and the sense of belonging either to a physical or conceptual territory of their own can comfortably face the stress of uncertainty that comes with continuing change in milieu. It is one thing to travel away from home; it is another to be a displaced person with no way to return home, or no home to return to. The other limitation is statistical; probably only a minority of individuals either seek or will tolerate a high amount of individual social or cultural mobility. The proportion will be highest with young adults and decline steadily with advances in the life cycle. Thus no theory or formula, can relieve the planner/designer from the obligation to examine carefully and in detail, with all the skill, objectivity, intuition, and empathy at his command, the cultural and temperamental characteristics of every population for whom he or she proposes to organize space.

References

1. U.S. President's Commission on Population Growth and the American Future. Population and the American Future. Washington, D.C.: Government Printing Office, 1972.

2. Gans, H.J., The Urban villagers. New York: MacMillan, 1967.

3. Glazer, N., & Moynihan, D., Beyond The Melting Pot. Cambridge, Mass.: The MIT Press, 1963.

4. Novak, M., The Rise of the Unmeltable Ethnics: Politics and Culture in the Seventies. New York: MacMillan, 1972.

5. Hall, E.T., The Hidden Dimension, New York: Doubleday, 1966.

6. Sommer, R., Personal Space, Englewood, N.J.,: Prentice-Hall, 1969.

7. Lorenz, K.Z., On Aggression, New York: Harcourt, Brace & World, 1966.

8. Watson, A., & Moss, R., Spacing As Affected by Territorial Behavior, Habitat and Nutrition in Red Grouse (Lagopus I. Scoticus). Behavior and Environment, Esser, A.H. (ED.) New York: Plenum Press, 1971, 92-111.

9. Wyne-Edwards, V.C., Animal Dispersion in Relation to Social Behavior, New York: Hafner Publication Co., 1962.

10. MacLean, P.D., The Triune Brain, Emotion, and Scientific Bias. The Neuroscince: Second Study Program, Schmitt, F.O. (Ed.) New York: Rockefeller University, 1970.

11. Esser, A.H., Social Pollution. Social Education, 1971, 35, 10-18.

12. Esser, A.H. & Others. Territoriality of Patients on a Research Ward. Recent Advance in Biological Psychiatry, Wortis, J. (Ed.) New York: Plenum Press, 1965, v. VII.

13. Calhoun, J.B., Space and Strategy of Life. Behavior and Environment, Esser, A.H. (Ed.) New York: Plenum Press, 1971, 329-387.

14. Fischer, F., Der Wohnraum. Zurich: Verlag fur Architektur im Artemis, 1965. English translation, The Living Space, unpublished ms.

15. Cassel, J., & Tyroler, H.A., Epidemiological Studies of Culture Change. I Health Status and Recency of Industrialization. Archives of Environmental Health, 1961, 3, 25.

16. Cassel, J., Health Consequences of Population Density and Crowding. Rapid Population Growth, National Academy of Sciences, Baltimore: Johns Hopkins Press, 1971, ch. 12.

17. Cassel, J., Physical Illness in Response to Stress. Social Stress, Levine, S. & Scotch, N.A. (Eds.) Chicago: Aldine-Atherton Press, 1970, ch. 7.

18. Anderson, M., The Federal Bulldozer. Cambridge, Mass.: The MIT Press, 1964.

19. Jacobs, J., The Death and Life of Great American Cities. New York: Random House, 1961.

20. Yancy, W.L., Architecture, Interaction, and Social Control; The Case of a Large-Scale Public Housing Project. Environment and Behavior, 1971, 3, 3-21.

21. Rosenthal, J., Housing Study: High Rise-High Crime. The New York Times, October 26, 1972, sect. C, 45.

22. Hawley, A.N. (Ed.) Roderick D. McKenzie on Human Ecology; Selected Readings. Chicago: University of Chicago Press, 1968.

23. Suttles, G.D., The Social Construction of Communities. Chicago: University of Chicago Press, 1972.

24. Wilkie, R.W., Toward a Behavioral Model of Peasant Migration: An Argentine Case Study of Spatial Behavior by Social Class Level. Population Dynamics of Latin America: A Review Bibliography, Thomas, R. (Ed.) Muncie: Ball State University Press, 1972.

25. Leyhausen, P., The Communal Organization of Solitary Mammals. Symposium, Zoological Society, London, 1965, 14, 249-63. Also in Proshansky, I.M. & Others (Eds.). Environmental Psychology, New York: Holt, Rinehart & Winston, 1970.

Author's Note

The research on which this paper is based was supported in part by a grant from the National Endowment for the Arts.

TOWARD AN ECOLOGICAL THEORY OF ADAPTATION AND AGING

Lucille Nahemow and M. Powell Lawton

Philadelphia Geriatric Center
5301 Old York Road
Philadelphia, Pa. 19141

Abstract

The environmental docility hypothesis suggests that environmental stimuli ("press", in Murray's terms) have a greater demand quality as the competence of the individual decreases. The dynamics of ecological transactions are considered as a function of personal competence, strength of environmental press, the dual characteristics of the individual's response (affective quality and adaptiveness of behavior), adaptation level, and the optimization function. Behavioral homeostasis is maintained by the individual as both respondent and initiator in interaction with his environment. Hypotheses are suggested to account for striving vs. relaxation and for changes in the individual's level of personal competence. Four transactional types discussed are environmental engineering, rehabilitation and therapy, individual growth, and active change of the environment.

Recent work in the psychology of stimulation (1) has led to theoretical advances in the area of social ecology. We propose an elaboration in this area that is middle-range, in the sense of attempting to account for a limited aspect of human behavior. This contribution to the theory of man-environment relationships deals with the aspects of human responses that can be viewed in evaluative terms, that is, behavior that can be rated on the continuum of adaptiveness, and inner states that can be rated on the continuum of positive to negative. This is, perhaps, a limited view of the human response repertory, but it stems from the traditional concern of the psychologist with mental health and mental illness. Similarly, our view of environment for this purpose is limited to the "demand quality" of the environment, an abstraction that represents only one of many ways of dimensionalizing the environment. We shall use our knowledge from the area of gerontology to provide content for the theoretical structure, but suggest that the constructs are more generally applicable to any area involving the understanding of mental or social pathology [see Lawton and Nahemow, (2) for a more complete discussion].

One way to begin is to look at the old ecological equation

$$B = \underline{f} \ (P, E)$$

to acknowledge its veracity and familiarity, but linger on a few of its implications:

1. All behavior is transactional, that is, not explainable solely on the basis of knowledge about either the person behaving or the environment in which it occurs.

2. Multiple antecedents may lead to the same behavior. Different personal qualities in different contexts may behave similarly, but the "meaning" of the behavior is not comprehensive unless both person and situation are analyzed.

3. The homeostatic principle is illustrated, in that the same behavior can be maintained in the face of a change in either the behaving individual or the environment, providing an appropriate change occurs in the second of the pair of determinants.

4. Behavior change may be instigated at either the personal or the environmental level.

If we add the evaluative element to behavior (adaptive vs. non-adaptive behavior or positive vs. negative affect), the equation also implies that

5. Even in the "best" environments, some individuals will be unable to behave in an adaptive manner.

6. Even the most capable individuals may not behave in an adaptive manner in the most malign environments.

The above implications are concerned with the prediction of the outcome of various person-environment transactions. Very early in our association with social gerontology it became plain that environmental solutions, as opposed to personality-change solutions, were prescribed for the problems of older people. It was clear that social planners, designers, and people in the helping professions were operating on the basis of the "environmental docility hypothesis". This hypothesis states: ".......the more competent the organism--in terms of health, intelligence, ego strength, social role performance, or cultural evolution--the less will be the proportion of variance in behavior attributable to physical objects or conditions around him....With high degrees of competence he will, in common parlance, rise above his environment. However, reduction of competence, or deprived status, heightens his behavioral dependence on external conditions" (3).

This hypothesis was formulated on the basis of their finding that elderly apartment dwellers who were female, or foreign-born, or in poorer health, were more likely to choose physically proximate neighbors as friends than were males, native-born, or healthy tenants. Other research has provided findings consistent with this notion. Rosow (4), for example, found that working-class (low status) elderly were more dependent upon their local neighborhoods for social interaction and help than were middle-class elderly. Mangum (5) used multiple regression analysis to determine the relative contributions of environmental and personal factors in predicting adjustment to planned housing. For the low-income tenants, environmental factors were most predictive while for the higher-income tenants, personal factors were more predictive.

The model that we are proposing requires the following definitions:

    1. <u>Individual</u> <u>competence</u> is the enduring ability that enables an individual

to function--the analogue of "personality trait" as the inner aspect of behavior. Actually, competences are many, depending on the area, such as intelligence, motor and perceptual ability, social tact, and so on. The designation of degree of competence should specify the particular area of competence, and is meant to refer to intraindividual, enduring characteristics that vary within minimum and maximum limits.

2. <u>Environmental</u> <u>press</u> are used in Murray's (6) sense to refer to aspects of the environment that act in concert with a personal need to evoke behavior by the subject. At this level we refer to external aspects of the environment that are presumed to have some motivating force for the individual whether he is aware of them or not ("alpha press," in Murray's terms). Aspects of the environment defined as those that are <u>perceived</u> as important to the individual ("beta press") are not included here, nor are the infinite number of aspects of the environment that do not impinge on the subject in any way. "Demand character" is the index of the total magnitude of the environment's effect on the individual, whether he is aware of the effect or not. The demand character may sometimes be estimated statistically in terms of the proportion of variance accounted for by environmental factors. The demand quality in extreme form may be termed "stress", though by no means are all press stressful.

3. <u>Adaptive behavior</u> is the externally observable behavior of the individual evaluated either in terms of social norms or of an a priori value system based on the assumption that pleasure to others, fulfillment of one's own potential, and the performance of complex tasks are separate but equally important bases for the establishment of norms.

4. <u>Affective</u> <u>response</u> is the self-evaluated quality of experience, ranging from positive through neutral to negative. Every person-environment transaction may be evaluated in terms of either and sometimes both the adaptive quality of the behavior involved and the quality of affect.

Further elaboration of the model requires the use of Helson's (7) concept of adaptation level (AL) and Wohlwill's optimization function (1).

5. <u>Adaptation level</u> is the perceiver's receptor status when the value of a stimulus is perceived as neutral, that is, as neither warm nor cool, loud nor soft, pleasant nor unpleasant. Much of the time we are at adaptation level with respect to our environment. A major aspect of our capacity to cope with the tasks of living involves our being able to screen out awareness of our proximate visual, auditory, thermal, and other environments, in order to concentrate attention and effort on focal tasks.

6. The "<u>optimization function</u>" suggests that for moderate levels of stimulation positive affect is engendered by stimuli that depart in either direction from AL (8, 1). As stimuli proceed further toward either higher or lower levels of intensity, they may begin to evoke a negative inner response.

1. MAN-ENVIRONMENT RELATIONS: THEORY / 27

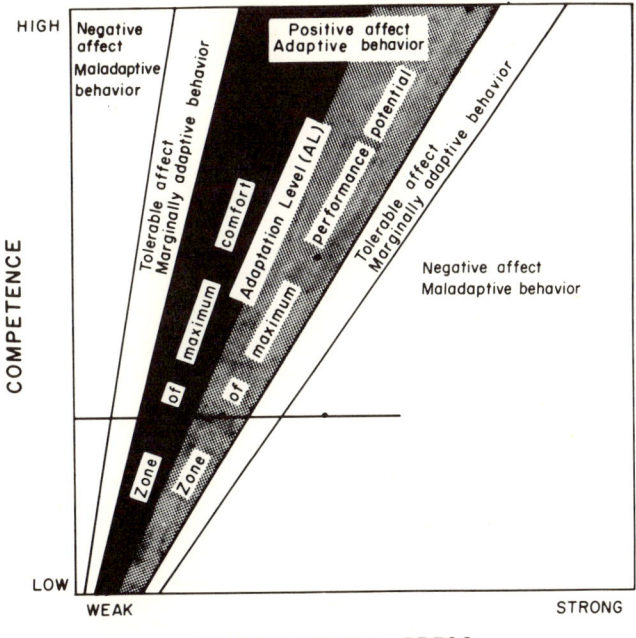

Figure 1

Graphic representation of an ecological theory of adaptation and aging

The Model

    The theoretical model, represented graphically in Figure 1, shows individual competence, which is represented on the ordinate, and environmental press, shown on the abscissa. The diagonal line labeled AL represents a theoretical mean adaptation level for individuals of differing competence interacting with their environments. For an individual in a particular environment, the ebb and flow of environmental press remain within a constant range, resulting in the establishment of the adaptation level for that individual. Individuals of a given level of competence would be distributed normally to the right and left of the AL point for that level. Their AL fluctuates at different points in time and with respect to specific stimuli in accordance with the conditions elaborated by Helson (7).

There is a range of environmental press adjacent to the individual's adaptation level where he experiences an inner sense of wellbeing vis-a-vis his environment and his behavior is adaptive (the shaded area of Figure 1). To either side of this positive outcome area is an area where higher and lower levels of press may test the limits of affective and behavioral adaptation. When the environmental press are either much greater or much less than those to which the individual has grown accustomed, he will experience a sense of discomfort and his behavior will become maladaptive. The region of positive affect and adaptive behavior is wider for persons of high competence. The dynamics of adaptation level are based upon homeostatic principles. Constant temporal variation occurs in both individual competence and environmental press. The beta component of environmental press (the perceived environment) will typically vary within what seems to be an objectively constant environment. Adaptation level is thus a theoretical point around which both personal competence and environmental press vary. For individuals who are both high and low in competence, the impact is vastly different for the competent and incompetent. For a highly competent person the zone of positive affect is sufficiently large so that normal oscillation in strength of environmental press will very rarely throw the individual beyond the shaded central zone. As we travel down the scale in competence, the central zone shrinks and the buffer zone, the area of tolerable affect, similarly shrinks. This means that for an individual of very low competence random variation in press will take him beyond the central zone where he has a sense of environmental mastery. This differential impact of environment press is stated by the environmental docility hypothesis.

The low point of environmental press might occur in sensory deprivation situations, the high point in many types of stressful or overloading situations. Individuals of high competence have a wide latitude of capacity to interact with the environment in ways that maximize positive affect. Note that as the individual's competence increases, the variability in environmental press which he can comfortably tolerate increases. Consequently, few high-competence individuals will show the breakdowns in behavior or affect that occur beyond adaptation range. Within that range, the following homeostatic reactions occur. When the environmental press are high, the person will become increasingly sensitized to his environment and try to make sense of it. When he is successful the perceived environmental complexity will be reduced. In Murray's terms, the beta component will be simplified. When occupying a region where environmental press are at a minimum, exploration and sensation-seeking will occur so as to increase the beta component of environmental press.

There are two different kinds of outcomes which concern us in the transaction between individual and environment: an affective and a behavioral response. Wohlwill's (1) concept of optimization deals with the affective response to a stimulus as a function of its deviation from adaptation level. Slight variations from AL produce positive affect, but large variations produce negative affect. The gradient is the same for both positive and negative discrepancies from AL. The present theory incorporates this aspect of Wohlwill's theory. When we consider the variable of performance, however, we find that the zone of maximum performance is found to be at an environmental press level which is above adaptation level. For example, when environmental press increase slightly, the problems the individual faces are increased but still remain within the individual's capacity to solve. He is therefore put on his mettle. Frustrations may increase but the satisfactions derived from achievement increase as well and the person experiences positive affect. If the individual receives less challenge than usual from his environment, frustrations diminish and he relaxes. This comfort is more immediately rewarding, but there is less of the delayed gratification that comes from achievement, or increasing one's competence. Thus, the zone of positive affect includes levels of environmental press that are both higher and lower than AL. However, the subjective nature of the feeling is quite different in the two directions. We have therefore labeled one side the region of maximum performance, the other the region of maximum comfort.

This ecological theory of aging posits that the individual is operating at his best when the environmental press are moderately challenging. If the environment offers too little challenge, the individual adapts by becoming lethargic and thus operates below his capacity. On the other hand, it may be that the environment is too stressful and he has adapted by turning off. Actually, it is possible for both conditions to occur sequentially. Consider the example of a widower who had become used to having many services performed by his spouse. Overwhelmed by the unaccustomed demands of the new life, he retreats psychologically and as a consequence is placed in a nursing home. At first the simplified environment is appropriate to his psychologically weakened condition. Press level and personal competence are well matched. However, as the person gains confidence in coping with his simplified and benign environment, he gradually discovers that it is excessively simple. In fact, he is operating near the low-press borderline of his own positive adaptation area. Within his affective comfort zone he relaxes, but unchallenged he becomes dull.

When the environmental press are very strong the individual may panic and attempt to escape from the field either physically or psychologically. This escape-oriented behavior is not geared to dealing with the situation represented by the immed-environment, but it is adaptive in the sense that it may remove the person from an intolerable situation. In our model, this would mean a reduction in competence as well, which would result in his being diagnosed as "disturbed", but still might represent the best possible temporary solution to the stress situation.

The following illustration exemplifies a situation in which the dynamics of the system diminish an individual's competence. An eighty year-old woman who is showing signs of declining competence, but who is still able to function in accustomed surroundings is relocated. The environmental press increase beyond her tolerance level and her functioning is markedly impaired. A danger for the person of declining competence is that of being removed to a too-supportive environment. The person of moderate competence who, with some help, could continue to function in the community, might also adapt downward to the limited environment of an institution, that is, become too content with functioning in the zone of comfort. In this case, she finds herself in a situation where it is difficult either to seek or to find stimulation. However, the environmental press are not so constantly low that she is really driven to the extremity of "maladaptive behavior." She seemingly adjusts very well to the home environment, but her powers decline markedly. What has happened is that in a situation with consistently low environmental press, the transactional balance adjusts with an adaptation level at lowered press, which leads to diminution of competence.

It is also possible for the dynamics of the system to improve the individual's level of functioning. Sivadon (9) attempted to design a mental hospital in which the person was introduced to challenging environments in small doses so that he gradually built up a tolerance for greater environmental press, and ultimately changed his adaptation level. An increase in competence would theoretically follow.

An important element in the theory is the relationship of time to the dynamics of adaptation. Typically, environmental press diminish over time as a natural consequence of the adaptation process. Thus, there tends to be a drift of AL from the right to the left in the diagram. The individual forms a cognitive map of his surrounds, develops concepts to reduce the environment to manipulable chunks, adjusts his coping mechanisms to the outer world and engages in other behaviors consistent with the process of adaptation. When entering a new environment, just the right amount of stimulation may be forthcoming for an individual <u>at</u> <u>that</u> <u>moment</u> <u>in</u> <u>time</u>, With increased familarity, however, the environmental press gradually diminish. At first it was challenging, then comfortable, and finally dull.

The environment of many older people is reduced in complexity, in terms of lowered role demands, less economic freedom, dwindling inter-personal worlds, and in some cases deprived physical surroundings -- a weakening of environmental press. Concomitantly, their competence may be reduced by comparison to younger people. It is our thesis that reduced press extended over a long time may lead to adaptation levels for **sensory** and affective experience that are significantly lower than those of younger people and ultimately lead to reduced personal competence.

For every individual there is an area in which changing environmental press are associated with positive outcome and an area in which self-initiated or externally initiated action may have a considerable impact upon the level of environmental demand in his behavioral world. As individual competence decreases, the area where maladaptive behavior and negative affect are risked becomes enlarged. Small changes in level of environmental press in people of low competence may evoke gross changes in quality of affect or behavior. Consequently, environmental intervention for therapeutic goals may be most fruitfully applied to this population.

Applications of these concepts to intervention schemes may be illustrated by the ecological change model implied by the basis ecological equation. Intervention may be applied either to the environment or to the individual. The individual's role may be as initiator or as a respondent to an external change. This transaction may be represented by the following dichotomies which suggest the fourfold prescriptive model for change shown in Figure 2:

1. MAN-ENVIRONMENT RELATIONS: THEORY

Point of Application	The individual's role	
	Respondent	Initiator
The environment	A Social & environmental engineering	B The individual redesigns his environment
The individual	C Rehabilitation, pscho-therapy	D Growth, self-therapy

Figure 2 - Ecological change model

The traditional concerns of the helping professions are represented in change attempts applied to the individual by trained therapists (cell C). While one may argue that effective change requires much active initiatory behavior from the subject, his independence is much more complete in the processes of growth and self-actualization (cell D). Our large body of design professionals operates primarily in cell A, where we presumably take into account not only the needs of individuals of low competence but the fact that life is made easy for most of us by programming a substantial amount of our everyday behavior. That is, we are at AL most of the time for very good reasons, and it is therefore appropriate to attempt to design in such a way as to free our attention for what really counts. Presumably in cell B the individual is initiating environmental change which may be either rewarding in itself (stimulus-seeking, tension creation) or in its outcome, whether that be ultimately an outcome that raises his competence, re-establishes an adaptation level that is within the bounds of positive affective experience, or enables him to shift his AL toward the "comfort" side.

Thus change may be approached from four points of view, any of which may, given the right person and the right situation, optimize the functioning level of the individual. It is clear that for any given level of supportive service, there will be some people who are too well and some who are too sick to enable a positive outcome. Our institutionalized widower may find that his socially engineered environment has made too many pre-digested decisions for him, and that there is practically no way for him to construct actively any part of his environment. Another institutional resident of lower competence may attempt more environmental change than he can handle (e.g., hoarding, fecal decoration, or occupation of another's territory) and downward environmental programming--e.g., increased support or institutional control --may be necessary. Other areas of application of these principles may be found in housing, the operation of neighborhood services, or the design of parks and recreation areas. In any of these pursuits, the important principles are (a) that support and demand are equally important in maintaining behavior and (b) that both tension reduction and tension creation are personally satisfying, depending on the person and the situation.

References

1. Wohlwill, J.F. Behavioral response and adaptation to environmental stimulation. In Damon, A. (Ed.) Physiological anthropology. Cambridge, Mass.: Harvard University Press, forthcoming.

2. Lawton, M.P. & Nahemow, L. Ecology and adaptation in the aging process. In Eisdorfer, C. & Lawton, M.P. (Eds.) Psychology of the aging process. Washington, D.C.: American Psychological Association, 1973.

3. Lawton, M.P. & Simon, B. The ecology of social relationships in housing for the elderly. The Gerontologist, 1968, 8, 108-115.

4. Rosow, I. Social integration of the aged. New York: Free Press, 1967.

5. Mangum, W. Adjustment in special settings for the aged. Unpublished Ph.D. dissertation, University of Southern California, 1971.

6. Murray, H.A. Explorations in personality. New York: Oxford, 1938.

7. Helson, H. Adaptation level theory. New York: Harper & Row, 1964.

8. McClelland, D.C., Atkinson, J.W., Clark, R.A., & Lowell, E.L. The achievement motive. New York: Appleton-Century, 1963.

9. Sivadon, P. Space as experienced: Therapeutic implications. In Proshansky, H.M., Ittelson, W. H., & Rivlin, L.G. (Eds.) Environmental psychology. New York: Holt, Rinehart, & Winston, 1970.

ARCHETYPAL PLACE   1.4

Mayer Spivack

Harvard Medical School, Laboratory of Community Psychiatry

Abstract

    The theory of Archetypal Place perhaps should be called the theory of whole environments. It is an attempt to identify the meaningful parts of the human environment. When this environment does not provide all settings necessary for the total human behavior spectrum, individual functioning and the quality of society may be impaired. Such a population exists in a state of <u>setting deprivation</u>. Thirteen settings, an irreducible group, are designated as <u>archetypal places</u>. Each is associated with a significant whole behavior, which is in turn keyed to developmental time or period in the life cycle, with a need or drive, and with the drive's object. The combination of the drive, object, the time, and archetypal place, forms the <u>critical confluence</u>. Setting deprivation results when behaviors at the critical confluence are blocked--because environments are archetypally inadequate.

*not even the smallest, the most tender**

not even the smallest, the most tender
of the animals has forgotten
the small place, the walls
of his touch.

the field mouse remembers
the hole under the piled hay;
always and forever
he will be earth-aching
smelling the warm manure of spring.

even the shy earthworm
returns. The webbed frog
and all secretly winged
beasts. desire is
like an old onion
at their hearts. they return.
who knows what longings of the mole
go unfulfilled?

...Besides the home, in many animals another differentiation is often met with, namely, places where special metabolic functions are performed. These may be places for the dismembering of prey among the predators, disgorging of pellets among raptorial birds, drinking places, and places for defecation and urination, as well as spots at which certain gland secretions are deposited (scent marks) and larders. Bathing places and wallows are important, too, and so are rubbing posts, etc. All these localities are connected in typical cases by well known paths and are normally visited at definite times. (1)

---

* Copyright 1968 Kathleen Spivack (from Peace Feelers)

Work leading to the generation of this theory was funded by NIMH (Grant No. MH 15314-03).

Our existence as city building and city dwelling men is marked by a tragic paradox. While we aspire to build a world which is the realization of our dreams, we grope to escape from the physical tangle and social wreckage of our urban nightmare like dreamers unable to wake.

The designers and developers of our physical environment have seen their task, as if schooled in noblesse oblige, as that of designing an aesthetic system within which other men <u>should</u> be content to live. Their buildings and cities have evolved most often from idiosyncratic, intuitive fantasies in which spaces and forms are moulded by a priori aesthetic principles. Architects are encouraged to conceive individual buildings in terms of their visual qualities--almost as sculpture. Buildings which meet sculptural criteria may be good--and necessary--if the environment is not to become even uglier than it is.

But they are not good enough.

Aesthetics must not be our greatest urban concern. Buildings, and the cities which they in turn build, must successfully establish an environment which is capable of structuring and supporting human behavior patterns--family life, meaningful working life, education settings--at their optimum levels.

Even that will not be enough. As our urban cultures and social structures evolve and develop, our behavior patterns evolve and grow in complexity. The new environment of the city must be able to sustain the load of the old ways and the new ways together. It must be as adaptable as man himself, and capable of rapid, sensitive adjustment.

One other factor threatens the holistic nature of the human habitat: our environments are for the most part designed and built by a few for use by the many. As absentee ownership and large scale development increase, fewer people have the opportunity or the "power" to significantly influence or even modify the form of their shelter. This practice guarantees that, in the absence of evaluation procedures, whatever omissions or mistakes are made by the designer will be repeated, and will become the burden of all to live in. This amplification of error has for many years continued unquestioned and unchecked. No doubt some of our contemporary urban crises, social and physical, are in part the legacy of this practice.

While change is one of the more obvious features of the urban scene, change frequently only adjusts the form of the environment to reflect rising land values. In order to argue more convincingly for adjustments to accommodate the humanist issues most often violated in house and city building, designers and social scientists require a coherent, common, perspective and theory. Thus, this paper presents a theoretical framework within which may be integrated the efforts of both social science and the design professions. There are three parts to the theory, all three of which are conceptually and dynamically linked, yet each retains its value when considered independently, as well. First is the concept of <u>setting deprivation</u>; second, the system of <u>archetypal places</u> and human life cycle requirements; and third, the concept of the <u>critical confluence</u>.

When houses, neighborhoods, towns and cities do not adequately provide all of the components or behavior places necessary for the fullest kind of human existence, the population can be said to be in a state of setting deprivation. I should like to suggest that setting deprivation is responsible for a considerable part of the social disorganization, some of the mental and physical illness, and much of the general human misery which surrounds the lives of people in contemporary society.

When people live in environments restricted to a severely limited range of settings (2) in which to carry out all the behavior that constitutes the human repertoire, their ability to function as individuals and family groups, and the integrity and quality of their society, may be impaired. People fail to maintain deep, lasting interpersonal relationships, they may suffer in their ability to work, provide or eat food, to sleep in deep renewing comfort, play, raise children, explore and protect territory, to meet with their peers, and make decisions which control the shape and quality of life. Each of the foregoing functions, and others, are associated with thirteen characteristic settings in the physical environment, with the rooms and furniture which focus and support behavior patterns in specific and appropriate ways. Such settings, taken together, in their smallest irreducible group, are herein identified as archetypal places.

In this paper I will describe how each archetypal place is associated with a significant whole behavior, which is in turn keyed to developmental time or period in the life cycle, with a need or drive, and with the object of that drive. The combination of the drive, the drive's object, the time and the archetypal place in which all are brought together form what I will call the critical confluence. Thus setting deprivation results when full behaviors at the critical confluence are blocked--within the lives of individuals and populations--because their environments are archetypally inadequate.

Setting Deprivation

The concept of archetypal setting deprivation derives from two sources; the first is the work of Roger Barker, whose conception of the behavior setting (3) has been borrowed, albeit rather loosely. It is too early to assign to the archetypes a set of rigid parameters; we do not yet know which of the many possible criteria that could be included in a definition of the term, and the phenomenon of a setting, will turn out to be most essential, nor do we understand their combination or relative proportion. Barker's definition should therefore stand as the temporary expedient. Since setting deprivation is a testable proposition, any definition will be only as good as it is useful in the field. While Barker's work is wonderfully precise, it is also detailed to the point of being cumbersome. The potential researcher of setting deprivation is urged to attempt his own definition of what criteria uniquely constitute a setting, while retaining the basic list of archetypal categories if they prove adequate and exhaustive.

The second source of the concept was suggested by the author's work in mental hospital settings. It became apparent that institutionalized residents lacked opportunities to be active. That the inmates' discomfort was not only a result of their various mental illnesses became quite clear when they could occasionally be observed in a different setting (swimming, for example) which offered some relief to their boredom, and a chance to test and momentarily regain their sense of personal competence, worth and identity. In a new setting, inmates behaved more like healthier people, and looked--and probably felt--less ill. A review of the settings normally available to these patients revealed that many settings normally open to non-inmates living normal lives were missing from their everyday environment. The sicker the patients seemed to the staff, the more restricted they were, until ultimately they were confined to the "day room", and not even allowed access to their own bed. At that point the reciprocal nature of their torment became apparent: the sicker one seemed, the more one would be confined; the more one's confinement, the sicker one seemed. Eventually the only variation in the inward spiral of experience occurred when the patient was transferred permanently to a back ward reserved for intractable patients.

In some of the more recent mental health centers (4) which are gradually replacing the old mental hospitals, one may observe similarly ill patients who look to be much more engaged, less "chronic", and far more active. Most such centers feature strong activities programs and provide interesting and varied environments to house them. Since that time, the same kinds of observations have been made--and questions asked--in non-institutional settings, e.g., cities and towns, when the author has observed what appeared to be significant amounts of inappropriate, random-appearing or delinquent behavior. So far these inquiries have been informal, and are unpublished. Perhaps as a result of this theoretical effort, more serious studies may be initiated.

Setting deprivation can result from a spatial distribution of functional places within the community which is in conflict with--or incongruent with--the desires and capabilities of the population. Opportunities may be too far away for walking, and walking may be the most desirable way to get there. Mothers with small children want to be able to meet outside of their homes and go shopping at the spur of the moment, on foot. Access to some settings may be restricted to the wealthy, by virtue of their unprofitable nature, resulting in scarcity and privileged use patterns (5). Reduced access can occur if facilities are removed to distant specialized parts of the metropolitan area, such as the medical care areas or entertainment areas, shopping areas, etc. For people without a car the trip length, the time and expense may render such trips either infrequent or impossible.

Ultimately the most significant and frequent deprivation results when planners, developers and architects build large scale new environments such as housing projects, suburbs, downtowns, industrial parks, new towns, hospitals and schools. At this scale major archetypal elements are often simply left out--forgotten. Our finished architecture and urban renewal efforts often resemble an unfinished jigsaw puzzle--important pieces of the whole image are missing. In this fashion, whole populations may be deprived of opportunities to develop their complete, rich human repertory. Narrow, invariant environments may develop grotesque societies and stunted lives.

While our informed behavior can, obviously, modify even the worst environment, narrow environments reduce our experience and expectations, and they must in turn modify our behavior. This is a kind of feedback loop which in good circumstances is responsible for the co-evolution of the species' behavior and its environment, together.

1. MAN-ENVIRONMENT RELATIONS: THEORY / 37

In the best circumstances, this loop of environment-behavior interaction and influence works for us: we reorganize, build and rebuild, we adapt, grow and expand our abilities and horizons. In crowded conditions, in poverty, illness or oppression, it often works against us.

If individuals are under stress or in a condition of poverty or illness, it will be much harder for them to change their environment; they will probably lack access to suitable political power and authority, to actual tools, money and time. Under just such conditions lives are most vulnerable to being distorted by outside forces--the inconvenient arrangement of the city or house, the sensory poverty and social sterility of public housing or hospital, the imposition of limited housing opportunities by political decree.

Archetypal Places (6)

The following system of archetypal places generically describes the fundamental collection of functional places used by man and other animals in the continuing business of daily life. Hediger has described the subdivisions of territory in animal habitats (7) in similar terms, and must be credited with inspiring and anticipating these spatial-behavioral categories. It is hoped this new systematic schema will organize and give additional dimensions to emerging--as well as older--information. [See Figure 1.]

Archetypal places, and the configurations they describe, denote space with highly specific--and for some species dimensionally exact--sets of specifications. Take for example the underground sleeping place of a prairie dog (sleeping place is an archetype). Specific to the prairie dog is a sleeping space in the form of a blind hollow cul de sac, perpendicular to an underground passage or route (another archetype) having approximately the same dimensions as the animal's body form, usually constructed within about three feet of the ground surface, slightly elevated from the main tunnel floor level, or at an angle to a vertical shaft. (8)

The general archetype, sleeping place, is possibly universal in that nearly all species sleep (with the exception of reptiles and perhaps some antelope (9)), and usually sleep in a characteristically constructed or selected sleeping place, in a typical position. (10) The species-specific nature of the sleeping place, and in the case of man its culturally specific character as well, is not universal. Thus, we must also be concerned with the qualities of the species-specific archetypal elements in a given animal's habitat. From this kind of information one can generate a prescriptive system of environmental design specifications which will match the complex building blocks of a species' drives, behavior patterns and social organization.

Archetypal places fall into three classes:

I   The total set of behaviorally defined archetypal places. This inclusive group is so far comprised of 13 place types. (11) This is the smallest mutually exclusive set of all possible spaces associated with needs, drives, and their realization, social life, psychological life motifs, biological existence, and maintenance of species population levels. They are conceived of as the minimum group of settings which together are necessary for support of the healthy life of a human family and the larger community. **They** are modified by culture.

II The species-specific set of archetypal places. As the behavior of more species becomes precisely known, differences will appear in their use of space to satisfy behavioral requirements. For instance, some animals sleep wherever nightfall finds them. Most birds excrete wherever they are, but will not excrete in the nest. Some animals don't use shelter at all.

III The culturally specific set of archetypal places. In man, cultural variation further refines and shapes the archetypal place and its emphasis in use.

The Spatial Web of Behavior

A connected web of archetypal places is woven by the animal and the human alike who, as the shuttle in a loom, run over their daily routes the continuous thread-stream of their behavior, connecting all the significant (archetypal) settings or places in their life experience.

For each genus and its ecological niche, there are probably characteristic kinds of social and spatial organization, and particular behavior patterns associated with archetypal places. The pattern which evolves when the behaviors in space are laid out in the home range of the species, however, may differ across species and even for local groups or "cultures". These species-specific patterns have been studied by ethologists, for some species in particular detail, (12) with reference to where each component of the total behavioral repertoire (in non-archetypal terms) of a species is placed in its home range. Records approaching a high level of detail also exist in the accounts of archeologists. (13) A detailed list and description of archetypal settings for man are delineated in Figure 1.

In any animal's natural habitat the separate functional places or archetypes may not be artificially divided, as by walls, in any way. Rather, it is the connections or routes between parts of the range that are most apparent. For any given species, some archetypes may be compressed and contained within others, or appear together in constellations. The raccoon, for example, excretes where—and often while—he drinks, "washes" and eats his food, a combination of three archetypes. The raccoon is in little danger of living in a wrong or incomplete environment. However, men, unlike raccoons, build walls, and walls are used to divide and isolate functions by subdividing space within the home range. Once these walls are up, the use of the space is relatively fixed and unchanging. If the space was poorly apportioned and designed at the outset, the shortcomings remain ever afterward.

We have lived so long in large cities and houses, that the earliest integration with our natural habitat has been overwhelmed and destroyed. The use of houses as shelters evolved in response to climatic factors, and economic and social evolution. With the development of megalopolitan scale city growth, the integration of the house on the land or the village in the countryside, and the ecological balance in which they once stood, was shattered. We have come a long way from Eden. Nor would most of us recognize the place, let alone be able to live there even as well (or poorly) as we do in our contemporary chaos. Unfortunately, neither do we live particularly well or healthily in our predominant options--houses in cities. We have lost the skills and opportunities, but not the drives of primitive men. We have, to borrow from Rene Dubos, overadapted. (14) We are trapped, behaviorally, physically and conceptually, in our houses.

The behavioral counterpart of archetypal place, what people do in these settings, constitutes the "meaning" in our environment. It is what makes a <u>place</u> out of a space. Living overlong in an environment composed of too few or improperly organized archetypal possibilities drains from our lives the meaning, the social and psychological contexts, and opportunities to act in meaningful ways. The desirable, even traditional, behavior patterns of communication, mutual government, peace-keeping and child care, recreation, courtship and family life disintegrate or disappear without the support of appropriate archetypal settings.

Mutation of social behavior, sometimes maladaptive, will result as old behavior patterns disappear in the wake of sudden changes in the environment. Populations adapt even to the most barren of surroundings. They may also maladapt, evolving to the character of uncontrolled and unecological forces.

If we neglect to provide the complete range of archetypal places within our communities, if we do not compensate in the larger community for those we no longer can contain within our homes, we may expect new social behavior patterns to arise suddenly. These, lacking the stable support of a strong archetypal setting, will change rapidly according to fad or to demogoguery.

Communities in a state of setting deprivation may produce feelings of rootlessness, disorientation and a dissolution of the cohesive bonds present in more healthy states of social organization. This is perhaps a major source of our contemporary anomic urban life with all of its attendant danger and discontent.

If an archetype is lost to a community, we should see consequent changes in the structure and location of behavior in its population, echoing the pattern of those behaviors whose accommodations have been lost, disturbed or distorted. We cannot be sure if the behavior will be displaced to another setting, mutate into a new kind of behavior in the same mode, or seem to disappear altogether, only to turn up transmuted into emotional pathology, higher divorce rates, or crime.

In the best circumstances, a healthy individual will have access to each of the archetypal places all the time, and show increased use intensity at the "proper" time at the dictates of life cycle requirements. The ideal, however, is the rare experience of the middle and upper classes in relatively industrialized nations, and perhaps also of those living in traditional cultures or pre-industrial areas.

Archetypal settings are the containers of culture. In them the spirit of a society--the identity, unity and vitality of a people--are initially and continuously moulded. A society's customs, rituals and mores are shaped and supported by the spaces which house them. The form of these places slowly evolves under the influence of changes in the culture. If, as a result of population growth, economic change, urbanization, war or legislation, the environment is reorganized in a way which shortchanges us by eliminating or reducing access to archetypes, the areas in which we may expect to see emergent "mutant" behavior patterns should be predictable.

Fig. 1

	Characteristic Tasks	1 SHELTER	2 SLEEP	3 MATE	4 GROOM	5 FEED	6 EX-CRETE	7 STORE	8 TERR-ITORY	9 PLAY	10 ROUTE	11 MEET	12 COMPETE	13 WORK
		Elemental protection, protection for nesting activities; retreat from stimulation, aggression, threat; social contact; emotional recuperation...	Neurophysiological processes; recuperation; rest; reduced stimulation; labor and birth; postnatal care of mother and child; death.	Courting rituals; pair-bonding; copulation; affectional behavior; communication.	Washing; mutual grooming.	Eating, slaking thirst; communication; social gathering; feeding others.	Excreting; territorial marking.	Hiding of food and other property; storage; hoarding.	Spying; contemplating; meditating; planning; waiting; territorial sentry; defending; observing.	Motor satisfactions; role testing; rule breaking; fantasy; exorcising; creation; discover; dominance testing; synthesis.	Perimeter checking; territorial confirmation; motor satisfactions; social and community control.	Communication; dominance testing; governing; education; worship; socialization; meditation; cosmic awe; moral concerns.	Agonistic ritual; dominance testing; ecological competition; inter-species defense; intra-species defense & aggression; mating; chauvinistic conflict.	Hunting, gathering, earning; building; making.
A	INFANCY Reflex control; orientation; communicate w/siblings & parents.	…	…	…	…	…	…	…	…	…	…	…	…	…
B	CHILDHOOD Gain motor, social, verbal, intellectual, emotional competence.													
C	ADOLESCENCE Forge identity; establish peer group relations; social/sexual exploration.													
D	COURTING-MATING Group w/peers; pair-bond; obtain sexual privacy.													
E	REPRODUCTION, CHILD CARE Nesting/nurturing; symbiosis; socialization.													
F	MIDDLE LIFE Care of aging parents; re-emphasis on worldly affairs; redefine identity.													
G	AGING MATURITY Maintain identity, contact, health; accept care by others, mortality.													

GENERATIONALLY RELATED LIFE CYCLE STAGES

THE TOTAL SET OF BEHAVIORALLY DEFINED ARCHETYPAL PLACES

Fig. 1 (cont.) **A**:1 Protection from elemental extremes; explore dwelling. A:2 Recognize bed; learn daily rhythms. A:3 XX A:4 Lose fear of wet face, sudden temperature change; regular grooming as primary contact ritual. A:5 Regulate feeding satisfactions. A:6 Discover excretion as separate from self; associate with setting and time. A:7 Acquire confidence in food abundance. A:8 Identify bed as primary secure place. A:9 Explore close environment; develop manipulative, cognitive skills. A:10 Route connects parts of shelter structures, provides orientation & change; motor satisfaction. A:11 XX A:12 Master frustration in competition w/siblings for attention & toys. A:13 (see A:9)/**B**:1 Differentiate subsettings; retreat from overstimulation, threat; emotional recuperation. B:2 Associated bed w/fatigue; learn volitional control of sleep; illness and recuperation. B:3 XX B:4 Learn to bathe, dress oneself. B:5 Coordinate feeding tools; communication; differentiate food from symbiotic source in mother. B:6 Autonomously control excretion. B:7 Learn to prepare food. B:8 Establish play "turfs"; orient to neighborhood; play protect territory from lookout; plan, wait. B:9 Role modeling; interact w/peers; fantasy, exercise, exorcism, creation, discovery, dominance testing. B:10 Enlarge route maps; differentiate settings, provide social encounters; learn safe wandering limits. B:11 Regular play/meeting rituals & places; elaborate functions; dominance testing. B:12 Games; fight; agonistic ritual; dominance testing. B:13 Acquire intellectual, motor skills./**C**:1 Find alternate private shelter: auto, attic, stairwell. C:2 XX C:3 Meet w/opposite sex in private, public settings; obtain sexual privacy anywhere: autos, barns, etc. C:4 Groom for mating encounters. C:5 Communicate w/peers over food & drink. C:6 Privacy in excretion. C:7 XX C:8 Expand territory into intellectual domains. C:9 Learn autonomous hobbies. C:10 Provides social contact w/opposite sex. C:11 Meet w/peers, both sexes; establish new rituals. C:12 Sexual display: cars, sports, clothes (see C:3). C:13 Refine, test skills in chosen work./**D**:1 Find new shelter. D:2 Share bed w/mate. D:3 Select mate; achieve couple privacy. D:4 XX D:5 Share food w/mate; increase food abundance. D:6 XX D:7 Enlarge larder for family. D:8 Expand territory to include mate. D:9 (see D:12) D:10 Maintain community of contacts. D:11 Meet w/couples. D:12 Personal display; ecological, mating competition. D:13 Apply skills toward life support./**E**:1 Expand shelter for offspring (see E:5). E:2 Maintain sexual privacy against invasion by new young family. E:3 XX E:4 XX E:5 Increase abundance; feed family; gather, communicate w/family. E:6 XX E:7 Increase capacity & variety of food. E:8 Expand territory to include young & check frequently. E:9 XX E:10 XX E:11 Expand functions, contacts; governing, educating, mystical awe. E:12 Display in common values; conspicuous consumption. E:13 Improve capacities, performance.//**F**:1 Shelter contracts as young leave. F:2 through F:7 XX F:8 Territorial needs contract as young leave shelter. F:9 through F:13 XX /**G**:1 Maintain location or adjust to imposed change; adapt surroundings to needs. G:2 More time in bed, sleep less; possible confinement, compression of world to bedside. G:3 Adjust sexuality to changing libido; possible illness or loss of mate (see G:2) G:4 Possible inability to care for self. G:5 Arrange special diet; reduction of taste, smell spectra. G:6 Possibly require aid and equipment; lowered mobility may reduce functional dependability. G:7 Possibly require assistance gathering & preparing food. G:8 Passive observation of archetypal activities performed by others. G:9 New leisure activities to fit changing capacities. G:10 Reduction in home range scale; fear of exposure to attack. G:11 Need for contact w/ & support from peers. G:12 Probable withdrawal from competition/defeat by young; defensive, evasive postures. G:13 Less active roles w/in former context; fend off retirement.

Thus, archetypal places as herein conceived are associated with, and resonate to, the deepest needs of the human organism. These places are the ones with which we identify strongly, as "my bedroom", "my study" (work place). Obviously, the concept of territory and territorial behavior overlaps with the concept of archetypal place in significant ways.

Each of the archetypes is in a sense a subtype of territory. When, as in the mental hospital or jail, only one or two archetypes are available at a given time, the "required but missing settings are functionally imposed upon whatever settings there are, and compressed within them. When settings are thus forced to handle an overload of functions, each orphaned from its own archetypal place, the resulting incongruence between behavior and place may appear bizarre and chaotic. The individual may become disoriented, and may respond with a variety of adaptive or maladaptive measures, most of which probably make him look "sicker" to his caretakers.

The systems of personal space, territory, and archetypal place are dynamically related: if two individuals are forced by hospital crowding to set up parts of their territory, their sleeping place for instance, within each other's personal or intimate social spaces, a strongly felt interference results which may affect the whole territory or its shape (change rooms, move the bed); personal and social behavior (fight with roommate); or the function (lie tensely awake at night, vigilant, sleep daytimes in a dayroom chair). This same circumstance can so threaten the integrity of the personal space and body boundary that withdrawal and depression symptoms may result. Alternatively, if neither function, behavior or location can be altered successfully, the "need" will be blocked and may be displaced and imposed onto some other functional place or archetype in a disguised form, for instance, the loss of a secure sleeping place may transform as feelings of general insecurity about all one's own places and possessions. To pursue the example, in the feeding place one might perhaps observe defensive postures and retentive hoarding of food, overemphasis of "my place at table", and even greediness. Thus, the effects of a deprivation in one mode may have far-reaching effects on other archetypal places, following a typical pattern, in this case boundary-defining and hoarding.

By referring to the archetypes and the functions they support, it should be possible to evaluate the adequacy of any habitat from the scale of a city to the elephant house at the zoo. Further, by using the archetypes as a program support or a checklist of environmental adequacy, an architect should be better able to design structures that do no violence to their users' needs and ways. Used thoughtfully, the archetypes may aid in the design of a higher architecture which aspires to enrich and satisfy the enormous repertoire of human behavior. Finally, by generating and examining a matrix of archetypal requirements for a house or community, one should be able to discover and then predict the range and ratios of environmental resources which will allow and support full realization of both individual and community life.

The Critical Confluence

The span of human life may be resolved into developmental phases. In order to relate spatial requirements in the archetypes to the lives of individual men and communities, we must attend to these developmental stages and their distinct spatial requirements. Figure 1 presents a simplified life cycle schema for the human family, and relates life cycle phase to the rise and evolution of each archetype. Various other life cycle schemata have been proposed--in psychology (15), ethology (16), literature (17)--none of which conveniently adapt to our purposes.

Each phase of the human life cycle has not only a central, drive-related task--such as child rearing--but also an appropriate (archetypal) physical environment for the proper support and resolution of behaviors related to these tasks. Thus, in the context of the right archetypal surround, we are free to engage in a critical set of actions--such as cradling and nursing an infant. In order to successfully engage in these movement patterns, and to experience the events fully and to the ultimate satisfaction of the drive, particular temporal and physical criteria must be met.

The appropriateness of the total setting, or environment, can be specifically described in terms of four essential boundary conditions: 1) having experienced or being in the grip of a motivating need or drive; 2) having that urge occur within an appropriate time context (developmental time, "cultural time", life cycle phase, seasonal time, circadian time); 3) having access to an appropriate archetypal space or place; and 4) having the object available--as in the case of a nursing mother, the infant. Figure 2 is a visual representation of the foregoing conditions and their interrelatedness.

The Whole Action occupies the center of the diagram for it is immersed within, and dependent upon, each of the four components (Drive, Time, Place, Object). Should access be prevented to one of these four key elements, it can be predicted that the whole act will be impaired or prevented. The individual will experience a double frustration: first, by not having access to the necessary element (D,T,P or O), and second, being denied the satisfaction of a completed Drive-to-Object cycle.

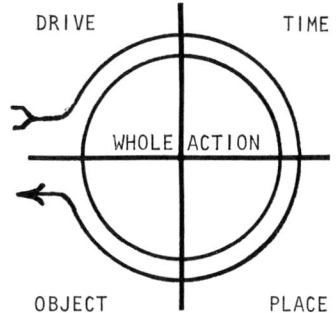

Fig. 2 THE CRITICAL CONFLUENCE

Critical confluence crises are proposed as typical of, and co-emergent with, the central events of our lives. Each is the central them of a life cycle phase.

The critical nature of the confluence derives from the linking of the Time and Drive criteria carrying the implications of a biological and behavioral developmental timetable and critical period phenomena.

Erikson (18) has postulated a multistage epigenetic system, in which each stage follows in regular sequence upon its precursor. There is explicit in his theory the notion that it is necessary to "satisfactorily complete" or "live through" the crucial events and life experiences which are the foci of each stage, before one may advance through to the succeeding ones. The availability of archetypal environmental situations which fit and fulfill the culturally adapted archetypal place expectations for each stage of development in the life cycle may partake of the same sequential, "critical" quality. Spatial misfit or unavailability may retard or prevent psycho-social advances. In the earliest years, especially in infancy, there is evidence (19) that stimulus variety and intensity of the infant's immediate spatial surround (relevant qualities on a continuum of environmental appropriateness to that stage) may play a critical part in the physiological and intellectual development of the child. Incompletion or deprivation in these critical areas might result in retarded emotional or even physical development, or in other emotional difficulties.

The successful resolution of a developmentally based physiological drive or psycho-social need is dependent upon the availability of a fitting archetypal place or its approximation in the terms of the culture. The behavior, biologically signalled, is intimately wedded to and threaded through the place. The place is supportive of the behavior to such a degree that in the absence of the appropriate place type, a drive may be severely or completely frustrated. Therefore place, like all of the other confluence components, can be called critical in the same sense.

The behavioral differentiation of places is characteristic of nearly all species. Habitats are divided into a spectrum of specific functional places (20) where the whole behavioral repertoire of the animal (man included) can be performed within species-specific contexts. In these specially designated functional places, the Drive-Time-Place-Object conjunctions occur with dependable regularity, i.e., daily and seasonally throughout the life span. As the life cycles advance with growth and age, new drives (and their objects) supervene over old ones, with concomitant new place requirements. The components of the older conjunctions may continue to be used, but they will no longer be critical to the further growth and health of the organism. In this way the individual's life can be described or represented as a continuing and overlapping series of critical confluence crises. Each of these, when frustrated or blocked by, say, too early emergence of a drive (adolescent mating), the premature appearance of an object (lunch, off schedule at 10 a.m.) or an inadequately organized or maladaptive spatial environment (restrictive public housing, slum conditions, mental hospital or other total institutional living conditions), can prevent or retard further maturational or psychological advances and otherwise impair maintenance of physical, psychological or social well-being.

If the boundary conditions described by a confluence crisis are met, then it can be expected that, all else being equal, the animal or man will thrive. Such a fortuitous condition can be described as a complete Drive-Object cycle, which is in the end self-extinguishing. The pressures and energy of the drive are relieved temporarily or permanently by its resolution and satisfaction under optimum environmental conditions.

Because an individual may have for the moment gained control over the forces of nature or society as they impinge upon his life, does not mean that he will keep his advantage. Walls crumble: anthills must be continuously rebuilt, the rent may be raised. The particular state of crisis, or crisis potential, may be expected to continue throughout the period of each cycle phase. The maintenance of control, and

1. MAN-ENVIRONMENT RELATIONS: THEORY / 45

the ability of the organism to cope with changes and demands from its environment, will constantly be at issue, and will constitute its central and continuing effort.

We thus face first the necessity of recognizing, and then achieving a qualitatively and quantitatively close match between the timing and development of human needs and drives, and the features, capacities, richness and poverty of the environment. The concept of ecological balance will gain new dimensions as we discover in depth how an ecologically balanced environment produces and supports health.

Summary

The total range of thirteen archetypal places must be available to the population at all times, even if their use is periodic. They will be used by individuals and families in ways which are predictable. In the course of the human life cycle, an individual's behavior patterns may change many times at the urging of his drives, as a function of his family status, biological maturity or social condition. In accordance with these changes, one uses the space available in rather specialized ways. The species-specific (or for man, culturally-specific) archetypal place forms must vary to meet the requirements of men at different ages and life stages. For example, a meeting place for three and four-year-olds would only in a few general ways resemble a faculty club or town meeting.

As the need for a particular archetypal setting becomes more pronounced due to development and maturation, or to changes in family composition, the availability of that place becomes more important. At some point in time, the space will become crucial to successful performance of life tasks. This time will occur when the object of the most pressing drive (or current life motif) occurs in conjunction with the felt need. This convergence of the dimensions of time and place with the motivating force of a drive in the presence of the proper object of the drive constitutes what is herein termed the critical confluence.

In the longer perspective, the lives of individuals intermesh in the structure of a family: we must then understand how the family unit, however large or small, generates a life cycle of its own. The life cycle of an individual is linear and sequential, resembling a song for solo voice: the life cycle of the family resembles a canon or round sung by many voices. Both perspectives, the individual and the family, are necessary if we are to project the implications of the archetypal system and the critical confluence theory into the requirements for design of houses and communities. It should be the task of the community to provide an appropriately designed richness and variety of spatial types for individuals and families.

Within the system of the archetypes and the critical confluences, there may emerge a new predictive theory of individual and social behavior with respect to space, maturational level, and life cycle position. This theory should prove useful at any scale of study, from the level of the individual through to the society, and from his single room flat to the receding boundaries of the megalopolis: it suggests a decision-making hierarchy and strategy for investing the physical plans of cities and houses with greater relevance to human life. While the exact content and shape of the archetypes will vary from culture to culture, changing as the culture changes, the archetypes will be seen to have remarkable consistency within a given culture, and probably great correspondence even across cultures within reasonable limits. Where strong differences in archetypal form or content exist across cultures, then we will have developed an interesting and useful tool for cross-cultural comparison and analysis.

Notes

1. H. Hediger. Wild Animals in Captivity. NY, Dover Publications, Inc. 1964.
2. Roger Barker. Ecological Psychology. Stanford, Stanford Univ. Press. 1968.
3. Ibid., pp.18-28.
4. Community Mental Health Centers Act (1963), Title II, Public Law 88-164.
5. Playing spaces such as ski resorts, summer resorts at the ocean or lakeside, horseback riding, golf, boating, theater and tourism, all are restricted to the use of high and middle-income families because of their relatively great expense. The children of the poor suffer more in this respect, having little or no access to safe and interesting play space.
6. The term "archetypal place" has no meaning beyond those specified in the functional definition given here, and in the beginning of the section so entitled. The term "archetypal" is not intended to carry any conceptual overlay from the work of C. Jung. The qualities of place and behavior that are called archetypal are modeled upon and developed from observed common and fundamental human behaviors which appear to be, but are not necessarily, innate to humans. Thus the use of the term archetypal as it is employed here might be synonymous with "prototype' or "fundamental" places which fit these (innate?) needs precisely.
7. Hediger, op.cit.
8. John A. King. "The Social Behavior of Prairie Dogs", Scient. Amer., Oct. 1959.
9. H. Hediger. "Comparative Observations on Sleep", Proceedings of the Royal Soc. of Med., Feb. 1969.
10. Ibid.
11. Presumably more such categories will be described by other investigators, and will enrich the potential of the system as a predictive tool.
12. E.g., Darling. A Herd of Red Deer. London, Oxford U. Press. 1937. George B. Schaller. The Mountain Gorilla: Ecology and Behavior. U. of Chicago Press. 1963. John B. Calhoun. The Ecology and Sociology of the Norway Rat. HEW, Bethesda. 1962. Alison Jolly. Lemur Behavior: A Madagascar Field Study. Chicago. 1966.
13. Douglas Frazer. Village Planning in the Primitive World. NY, Braziller. 1968.
14. Rene Dubos. "Man Overadapting", Psych. Today, Feb. 1971.
15. Erik Erikson. Childhood and Society. NY, Norton. 1963.
16. Irvin DeVore. Primate Behavior. NY, Holt, Rinehart. 1965.
17. E.g., the riddle of the Sphinx. See Edith Hamilton. Mythology. NY, NAL. 1942.
18. Erikson, op.cit. Erikson clearly links his own epigenetic theory to the concept of criticality: "...ego qualities...emerge from critical periods of development..." In "Eight Ages of Man": "...psycho-social development proceeds by critical steps. 'Critical' being a characteristic of turning points, of moments of decision between progress and regression, integration and retardation."
19. Rene A. Spitz. The First Year of Life: A Psychoanalytic Study of Normal and Deviant Development of Object Relations. NY, Int. Univ. Press. 1965.
20. H. Hediger. Wild Animals in Captivity.

ENVIRONMENTAL STRESS AND FLEXIBILITY IN THE HOUSING PROCESS   1.5

Mete H. Turan

Assistant Professor
College of Architecture
University of North Carolina at Charlotte

Abstract
A theoretical framework for the environmental stress and environmental flexibility concepts has been presented. The physical stress model has been employed to draw an analogy and to construct the environmental stress-strain model within the "internal ecological process." The implication of this constructed model has been focused specifically on the housing process and on the design of housing. Possibile indicators of strain in the housing process have been stated. Propositions which will aid the design process and which will also help to shed light on some of the hard pressing problems of housing have been set forth.

Introduction
The true nature of space is not absorbed or conceived through the sensory experiences alone. Sensory experiences, for example, relative to color, shape, size, texture, tone, and so forth, merely scratch the surface of spatial consciousness. For any environment and its spatial characteristics to be perceived and become part of an individual's activity, a sensory experience is interpreted in the light of associations (with symbolic meanings) and eventually registered as a conscious effort of observation. Man's values and attitudes, which are products of his symbolic creative capacity, are also the result of his awareness of his surroundings.

Behavior or action is a result of the basic needs and the personality structure of the individual combined with series of processes of adjustment to different situations and physical conditions. These intervening processes take place between the cognitive processes of appraisal of external stimuli and the activities that man performs which are influenced by that appraisal.

Human behavior is not merely a passive adjustment process. Subjective appraisal of the external environment often interferes with our psychological state. A _stress_ may occur as a function of this subjective perception of the environment. This is analogous to the physiological forms of stress in which there are some traceable changes within the organism such as autonomic disturbances or microbehavioral reactions due to a discrepancy between the demands of the external environment and the individual's ability to respond to these demands.(1)   If our cognitive functioning is disturbed, our behavior pattern will be modified. The physical conditions of the environment are felt socially and psychologically and their final influence has both psychological and physiological manifestations.

To give a wider perspective to the issues that we will be focusing on in this paper, a schematic "general behavior model" is presented. (FIG. 1). Although an elaboration on this model is not included here, we will elaborate on the "internal ecological process" of this model. Our concern is both with the structure and the exchange process, the structure being the actual physical setting and the exchange process being the interaction between the environment and the individual. We will deal with the structure in process by means of stress analysis. The reason to adopt a stress syndrome is twofold: 1. conceptually: stress is an integrating concept which can fundamentally connect the interrelated but practically isolated disciplines of physiology, psychology, sociology, architecture, etc.; 2. operationally: stress analysis may shed light on some of the hard pressing problems in the housing process, whether or not solutions are generated.

A Conceptual Model of Environmental Stress (2)
As indicated in our general behavior model, external environmental forces, or the input from the external environment, are the major portion of the stimuli for an individual's actions. During the intervening process, before the action is taken or executed, a filtering process will qualify the characteristics of the input and consequently the behavior in response to that input.

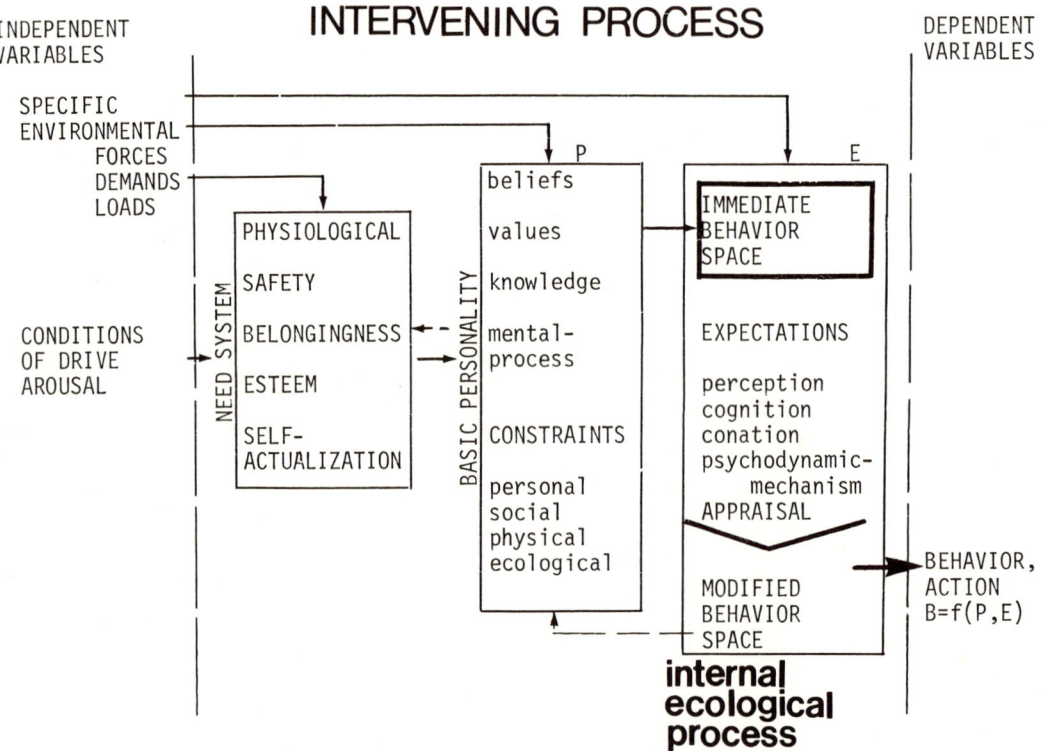

Figure 1.
GENERAL BEHAVIOR MODEL
Adopted and modified from Tolman(1951).

1. MAN-ENVIRONMENT RELATIONS: THEORY / 49

Let us now elaborate on each stage of the internal ecological process. We shall adopt the term "immediate behavior space" from Tolman, who defines it as a "particularized complex of perceptions (memories and inferences) as to objects and relations and the 'behaving self', evoked by the given environmental stimulus situation and by a controlling and activated belief-value matrix." (3)

To be more specific, we can define "immediate behavior space" as a group of objects that is percieved by an individual and takes into account their relative orientations and distances. A behavior space includes not only the particular objects but also their specific spatial, temporal, aesthetic, functional and other relations to one another.

An input entering the immediate behavior space may have originated in the external environment or it may have been generated in the internal environment. In the first case, the input is more independent of the psychological system than in the second, where the input is more dependent on the state of the system (it may be the product of a feed-back). For example, a certain noise coming from outside, from the street perhaps, is an illustration of the first case, and the desire to lie down and relax is an example of the second.

Inputs are neutral and merely informative until they are transformed into loads. This transformation occurs due to a subjective appraisal of the input as a threat situation. Broadbent's (4) filter theory, in which a biased selection of the nervous system acts upon the input, supports the concept of this transformation. To illustrate this input-load transformation, we can expand our noise from the street example: One is sitting in a room and reading, not being particularly bothered by a sound (e.g. low frequency, continuous) coming from the street outside. The sound changes suddenly from low to high frequency, reaching a disturbing level. (5) At this point, the new sound (noise) causes the filter to select the auditory information in the book as input to the perceptual system. A neutral input (the initial sound) has been transformed into a load (noise), causing distraction and perhaps even annoyance.

This appraisal process occurs after the input has reached the immediate behavior space. To understand this process clearly and to aid us in describing our stress model, the definitions of certain concepts will be stated. In order to construct our model, we will draw upon the physical stress concept both for analogies and terminology.

Since we have differentiated between an input and a load, let us define precisely what an environmental load is: any input that has the capacity to be a threat to the maintenance of a system, to its range of stable equilibrium, harmony or adaptiveness, and which threatens the achievement of a state to which some process is directed. The word "threat" has been used by most stress researchers in its more generally accepted meaning, suggesting damage, destruction, injury, menace, etc. Due to reasons that will be apparent when we define the concept of 'environmental stress', we need to expand the idea of threat so that it will include less dramatic and traumatic events such as annoyance, discontent, disruption, distraction, general uneasiness, and so forth. Some examples of environmental loads in the housing situation are: the inability of a mother to be in visual or verbal contact with her child while she is in the house and the child plays out of doors (6); inadequate acoustical insulation of the sleeping areas of a house (7); the lack of visual privacy from the outside (8); inappropriate layout of different areas of a house in terms of their function; a lack or excess of natural light coming into the house at a specific time of day; a lack of privacy within a dwelling (9); and inadequate space for large-scale socializing (10).

Environmental stress is a state which occurs when the individual is subjected to environmental load(s). Environmental stress arises from the appraisal process as a result of the perception, cognition and conation of the individual. All the information (the environment at that moment to the individual) is utilized or modified depending upon the qualities of the personality structure of the individual.

The intensity of the environmental load is generally measured against a hierarchy of needs.(11) Alternative paths to goals can be evaluated against these needs. Of course, this is not always a good measure of reality nor does it always lead to a rational choice, as is often assumed by architects in creating a "rational" design. "The irrationality and maladaptiveness does not come primarily from the intervention of emotions in thought processes, but rather from the fact that threat places the psychological system in jeopardy and that alternatives for coping with threat are tied to motives, beliefs and expectations concerning the situation, which differ from person to person."(12)

We can say that stress is partially a subjective phenomenon. Its sources lie in the physical, physiological, psychic, and social events or demands. As was mentioned earlier, it is derived either externally or internally. It can have a specific state as well as a diffuse state. Since stress underlies both adaptive and maladaptive behavior (13), we can conclude that it is not always harmful but may even at certain times provide a positive support in reaching a particular goal.

Environmental strain is the result of stress and the ineffective coping with a particular load. It is not independent of the load or the stress. It is the indicator of the joint result of the load and stress which manifests itself by the deformation of a structure, or in the displacement or deflection of the boundary of stability. Some examples of environmental strain are: discontentment with the spatial arrangement of a house; distraction while reading in a room; annoyance caused by noise coming from neighbors; the emotional (in a negative way) attitude toward the texture and color of a wall; worrying about one's child who cannot be seen from the window inside the house, and so forth. (See FIGURES 2. and 3. for schematic representation of stress-strain model.)

Environmental Flexibility
The relationship between stress and strain in a physical substance is a characteristic of that particular material. The ratio of stress to strain equals the modulus of elasticity of the material. According to this constant, the behavior of the material in terms of deformation, elasticity, plasticity, fatigue, etc. under various loads, will differ. This constant of proportionality determines the nature of the curve between stress and strain in the elastic range.

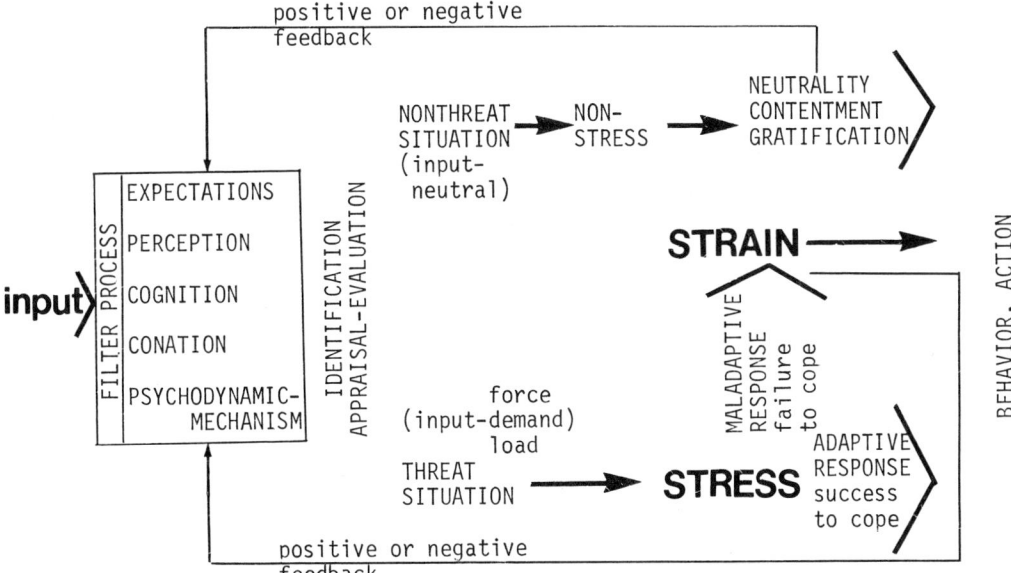

Figure 2.
PSYCHOLOGICAL SYSTEM - INTERNAL ECOLOGICAL PROCESS

Without further delineating the physical stress model (14), let us state our analogous stress-strain model considered in environmental terms: we shall introduce the concept of <u>environmental flexibility</u> analogous to the modulus of elasticity. Environmental flexibility (EF) is a variable related to a particular environmental-spatial order. It affects the magnitude of the environmental strain pertaining to the environmental aspects of a particular load which may put the psychological system into a stress state. Essentially it is the ratio of environmental stress to environmental strain, pertaining both to the environmental qualities and its physical and socio-psychological attributes. Thus, strain, the outcome of the stress syndrome, is directly dependent upon the stress and indirectly dependent on the EF. In other words, the lower the flexibility of an environment under a constant load or stress, the higher the strain due to that load (the converse being, the higher the flexibility, the lower the strain).

A high EF will exist in an environment that is more responsive to change, that is more capable of variation and modification. The flexibility of an environment is its capacity and readiness to yield to the influence of the inhabitants. Degrees of

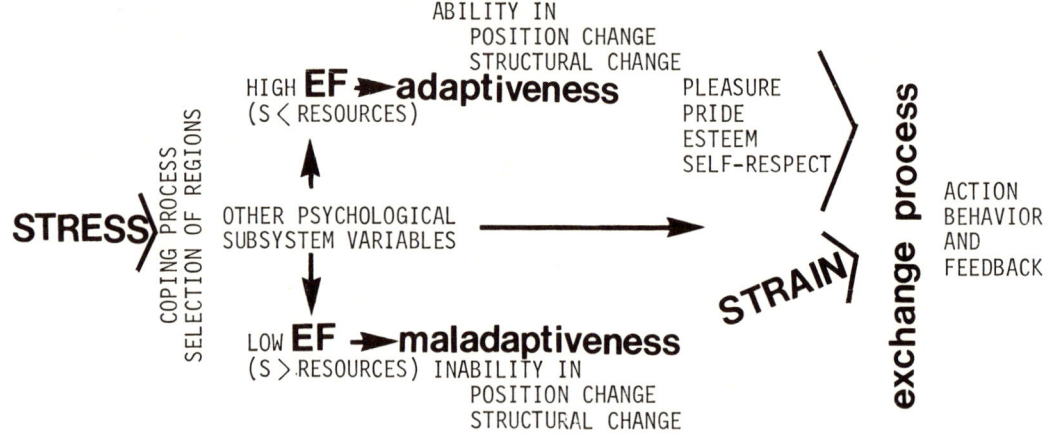

Figure 3.
STRESS - STRAIN MODEL

flexibility can be expressed by attributes such as, adaptibility (the capacity to be modified or altered), adjustability (the willingness for correspondence or harmony without being radically changed), and the faculty of accomodation (specific or momentary adaptation). <u>It is our central hypothesis that those strains which are results of environmental attributes of a specific spatial order, can be reduced in proportion to the degree of flexibility of that environment.</u>

Flexibility can be analyzed from two different but related points of view. One view is the utilization of a given space, wherein the action systems (at a higher level) and activity patterns (at a lower level) are important. Behavior patterns or episodes (15) in their aggregate can make up an activity and their distributions and smoothness in time and space can affect the individual's internal ecological process. The inability to utilize a given space, according to a particular need at a specific time, will affect the series of behavior patterns that follow.

The second aspect of flexibility pertains to the quantity of available space. This is concerned mainly with the changing population of a household and changing needs with respect to the spatial requirements. The growth and shrinkage of a household can be felt by the changing requirements for physical space. If the first aspect of flexibility, the utilization of a given space, is considered in the short run, in which immediate alterations and accomodations are necessary, one can consider the quantitative aspect to be in the long run, where the alterations and accomodations are neither immediate nor urgent, but ultimately necessary.

Short-term requirements may include the following: the rearrangement of furniture for a temporary social gathering; the utilization of the dining room or area for the temporary functioning of different household activities (e.g. ironing, studying, etc.); using the living room as a sleep area; using the bathroom as a dark room; and so forth. Most of these examples illustrate multi-functional usage of certain defined areas. Some may require major structural changes (removal of a wall or addition of a partition) while some may require only minor rearrangements (moving of furniture or changing the position of decorative artifacts on walls, tables, etc.).

The long term aspect of flexibility already mentioned can pertain to the changing population over time in household, which presents significant problems for house design. Different stages of a household present new and different demands in the realm of space arrangements. People normally do not move successively to a different home each time there is a major change in spatial requirements. Therefore, flexibility, which would allow <u>free movement</u> and alternative uses of space in the house, becomes a priority. These changes can occur not only when the size of the family becomes greater or smaller but also when certain necessities demand rearrangement of living space. An example of the latter is that at first it is essential to have the baby's room near that of his parents', but when the child gets older it often becomes desireable to offer him more privacy with a room further away from the parents'. While the toddler may need supervision during play and sleep activities, the school-age child may require a certain amount of seclusion from the rest of the household. Flexibility is of great importance in such instances and the potential for it must be provided if rearrangement is to take place.

Our emphasis on environmental flexibility is due in part to its importance in the coping process. If the stress state has been established, the coping process starts functioning in the situation and attempts to reduce or eliminate the anticipated threat. The coping processes are dependent upon the cognitive processes of appraisal. It is also very likely that they are affected by spatial influences. The faculty to have control, to manipulate, to manage or to rearrange space, can influence the "effectance"(16) of the coping process. Similar to Lewin's concept of "locomotion" (17) (change of position; change of structure), the environmental aspects of the coping process may reduce or eliminate or on the contrary promote the achievement of a certain state toward which a process is directed.

The coping process may involve either a change in position or a change in the structure. As was mentioned earlier, constituents of behavior space, orientations and distances may be other than spatial. This gives the coping process temporal, social, mechanical and other qualities to manipulate. This hypothetical construct, which is an intervening variable within the internal ecological process, acts as a bridge between the behavior of an individual from one region of behavior space to another. (18). A psychological region is a component of a life space, according to Lewin's definition. Therefore, everything that characterizes the situation which affects the psychological system can be represented as a physical or psychological region. The possible behaviors that can occur in a situation are dependent on the past experiences of the individual and his personality structure, as well as upon the presence of objects, i.e. the immediate environment. The process of coping is selection from several perceived and available regions which will result in the particular behavior or action. It is not the behavior or the action itself.

In addition to the utilitarian demands for new regions, there are also personal, idiosyncratic reasons or desires to change the position of or to alter a physical region. For instance, the desire to study in the dining area rather than in one's room, or the housewife's desire to rearrange the living room furniture, or the wish to reorganize the decorative artifacts in the display area, or addition of some more flowers to the windowsill, or the novel arrangement of lighting fixtures, and so forth will all require some degree of flexibility from the spatial-order.

There is a field of forces (or "valences")(19) in the environment that act upon the individual. This field acts between the individual and any psychological object, be it a physical object, an episode, an activity, a situation, or a goal. Valences are the manifestations of different values; they are the subjective perception resulting from the interaction between the individual's need system and the environment at the moment.

As the needs and desires of the individual change, the collection of all the forces in the field and the environment change too. The state of the person may modify the valences. Environmental flexibility is not only a function of the spatial order, but also of the valences of the situation.

Our main interest is in the ability to alter or manipulate the structure of a given environment when confronted with a stress situation. We must recognize that the environmental flexibility plays a most important role in the selection (in order to reduce stress) of regions. On a more general level, environmental flexibility plays a significant role in all exchange processes that occur between man and his environment and it is one of the dynamic properties of a situation.

EF also is closely connected with the momentary state of the individual. A person can behave differently at different times in the same environment as a result of the cognitive process of appraisal. This is not only a property of the structure but also of the exchange process between man and his environment. In other words, flexibility is a linkage between the cognitive process of appraisal and the structure of a situation.

In a broader scope Lewin (20) makes use of a similar concept: "degree of fluidity", covers the whole of the psychological system. To him a "situation is more fluid the smaller the forces which are necessary, other conditions being equal, to produce a given change in the situation."(21). Since our argument for environmental flexibility in the stress-strain relationship is parallel to Lewin's concept of structural changes, but only at a lower level, we can say that EF is included in the "degree of fluidity" or it is a subset of fluidity.

Outside the limits of ergonometric and anthropometric requirements, and beyond the ranges of "personal space"(22) and "territoriality"(23), a freedom of movement must be provided for the individual. That is, a freedom to move within an environmental order according to the choices and selections of the individual. This is another component of environmental flexibility. In other words, environmental flexibility affects the quality and the quantity of operations necessary to move freely and to have a choice in the spatial arrangement and to modify the spatial-order. The elimination or reduction of the "excessive adjustive burdens"(24) is provided by the environmental flexibility through this movement.

The inability to adapt to situations is a highly influential factor on the attitudinal attributes (cognitive, evaluative and affective). On the other hand, this aspect of the exchange process definitely is a design controllable element that can be furnished through the environmental flexibility. Our intention is not to conclude that environmental flexibility is the only factor in alleviating such a complex problem, nor would we wish to assert that man-built environment is the sole determinant in people's attitudes and behavior. But at least our observations and assessments of real life situations (especially of architect designed housings) incline us to believe that much more consideration and attention must be given to this aspect in the design process if we want the living environments to assist the inhabitants in their coping process with the environmental loads that create stress states. Promotion of this type of flexibility in the housing process may aid the inhabitants in their coping process.

Summary
- Within a general behavior model, the internal ecological process consists of filtering and appraisal of the environmental input which is then either perceived as neutral or transformed into a load. This perception depends upon both the structure of the environment and the exchange process between man and environment.
- Environmental stress is a state resulting from both the physiological and psychological loads of the environment.
- Stress is rationally or irrationally coped with and often a hierarchy of needs is employed during this intervening process; failure in coping leads to strain.
- Environmental stress-strain model developed is analogous to the physical stress-strain relationship.
- The main thesis, that the environmental flexibility is a most important factor affecting the coping process,is presented.
- Environmental flexibility is the variable facility to change the utilization of given space or the quantity of available space.

Propositions
1. Environmental stress is a state intervening between the constraints of the environment and the resulting efforts to reduce these constraints.
2. The intensity of environmental stress varies directly with the external and the internal environmental loads associated with that particular stress.
3. Strain, the indicator of environmental loads and the stress state, manifests itself through annoyance, discontent, distraction, general uneasiness as well as in more dramatic ways such as harm, destruction, total or partial dysfunction, etc.
4. The probability of maladaptive responses (leading to strain) due to environmental stresses varies directly with the intensity and duration of those stresses.
5. The intensity of environmental strain varies inversely with the "environmental flexibility".
6. Greater "environmental flexibility" increases the resources of the individual to cope with the environmental load(s) in that situation.

Notes

1. Some studies dealing specifically with the "environmental stress" focused on the

"physical environmental stressors" with an emphasis more on the physiological consequences on man. For a further discussion see D.H. Carson and B.L. Driver "An Ecological Approach to Environmental Stress,"The American Behavioral Scientist, September 1966, pp. 8-11 and also by the same authors, "An Environmental Approach to Human Stress and Well Being: With Implications for Planning," Mental Health Research Institute, University of Michigan, Preprint 194, Ann Arbor, June 1970.

2. Space limitations have forced us to refrain from giving even a brief review of stress models that other researchers have developed. Different conceptualizations in both physiological and psychological domains are quite common in most of the works that we could study. Our work, needless to say, has its common origin in the definitions and the terms that have been attempted to be developed in these earlier works. It is on these preceding studies that we tried to advance the concept of stress, specifically in the area of man-environment relationship. A representative selection is contained in the following references:
   -Bosowitz,H., and Others, Anxiety and Stress. New York: McGraw Hill, 1955.
   -Dohrenwend,B.P., "The Social Psychological Nature of Stress," Journal of Abnormal and Social Psychology, v. 62, n. 2, March 1961, pp. 203-230.
   -Howard,A. and Scott,R.A., "A Proposed Framework for the Analysis of Stress in the Human Organism,"Behavioral Science, v. 10, n. 2, April 1965, pp. 141-160.
   -Janis,I.L., "Problems of Theory in the Analysis of Stress in the Human Organism," Journal of Social Issues, v. 10, n. 3, 1954, pp. 12-25.
   -Lazarus,R.S., Psychological Stress and the Coping Process. New York: McGraw Hill, 1966.
   -Mechanic,D., Students Under Stress. Glencoe, Illinois: Free Press, 1962.
   -Selye, H., The Stress of Life. New York: McGraw Hill, 1956.

3. Tolman,E.C., "A Psychological Model," in T.Parsons and E.A. Shils (Eds.), Toward A General Theory of Action. New York: Harper and Row, 1962, (1951), pp. 279-361.

4. Broadbent,D.E., Perception and Communication. New York: Pergamon, 1958.

5. Broadbent,D.E., "Effects of Noises of High and Low Frequency on Behavior," Ergonomics, v. 1, 1957, pp. 21-29.

6. Three of the many studies that emphasize the visual and verbal contact and the children's play habit in housing complexes are:
   -Reynolds,I. and Nicholson,C., "Living Off the Ground," The Architects' Journal August 20, 1969, pp. 459-470.
   -Stevenson,A., Martin,E. and O'Neill,J., High Living. New York: Melbourne University Press, 1967.
   -White, L.E., "The Outdoor Play of Children Living in Flats: An Enquiry into the Use of Courtyards as Playgrounds," in H.M.Proshansky, W.H.Ittelson, and L.G. Rivlin (Eds.), Environmental Psychology. New York: Holt, Reinhart & Winston, 1970, pp. 37-382.

7. Most of the studies on acoustics range from physiological effects to social and psychological consequences and there are also those which treat the sound from

1. MAN-ENVIRONMENT RELATIONS: THEORY / 57

purely physical point of view. For the social and psychological effects, see L. Kuper, "Neighbor on the Hearth," in H.M. Proshansky, W.H.Ittelson,and L.G. Rivlin (Eds.), Environmental Psychology. New York: Holt, Rinehart & Winston, 1970, pp. 246-255.

8. Kuper,L., "Neighbor on the Hearth," in H.M. Proshansky, W.H.Ittelson,and L.G. Rivlin (Eds.), Environmental Psychology, New York: Holt, Rinehart & Winston, 1970, pp. 246-255.

9. Hole,V., "Social Effects of Planned Housing," The Town Planning Review, v. 30, n. 2, 1959, pp. 161-173.
Plant,J., "Some Psychiatric Aspects of Crowded Living Conditions," American Journal of Psychiatry, v. 9, n. 5, 1930, pp. 849-860.

10. Fried,M. and Gleicher,P., "Some Sources of Residential Satisfaction in an Urban Slum," Journal of American Institute of Planners, v. 27, n. 4, November 1961, pp. 305-315.

11. The <u>need system</u> can be conceptualized as a set of interconnecting domains, each domain corresponding to a different order of hierarchy of needs. In his hierarchy of needs, Maslow assumes that unless the basic "lower" needs are satisfied (a state of proper gratification), the "higher" needs can not develop. The ordering of basic needs from lower to higher runs as follows: physiological needs, safety needs, belongingness and love needs, esteem needs, and need for self-actualization. For more detailed explanation, see A.H. Maslow, "'Higher' and 'Lower' Needs," Journal of Psychology, v. 15, 1948, pp. 433-436. For further discussion of the need system and its role in the development of personality see also Maslow's Motivation and Personality. New York: Harper and Row, 1970, (1954).

12. Lazarus,R.S., "Cognitive and Personality Factors Underlying Threat and Coping," in S.Levine and N.A. Scotch (Eds.), Social Stress. Chicago: Aldine, 1970, pp. 143-164.

13. Dohrenwend,B.P., "The Social Psychological Nature of Stress," Journal of Abnormal and Social Psychology, v. 62, n. 2, March 1961, pp. 203-230.

14.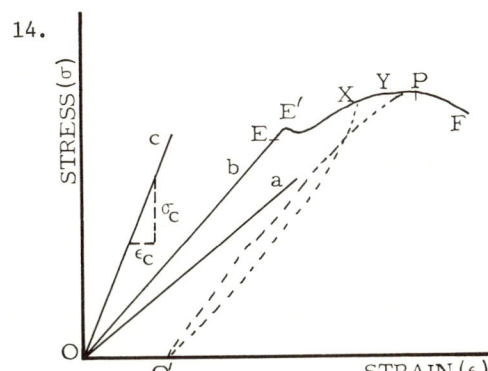

A very brief explanation of the physical stress model is given below:

$$\text{Stress } (\sigma) = \frac{\Delta \text{Force } (F)}{\Delta \text{Area } (A)} \text{ ; } F \text{ being normal to}$$

the Area as ΔA approaches zero.

$$\text{Modulus of Elasticity } (E) = \frac{\sigma}{\text{Strain } (\epsilon)} \text{ ;}$$

ϵ being the amount of deformation the body undergoes due to the action of F.

Ranges given on the diagram:
OE - Elastic range
EE'- Elastic limit

E'P - Plastic Limit
- if the body is loaded at O and unloaded at X: permanant deformation and strain hardening (OO' is the deformation);
- whereas in the elastic range (OE), after every loading, the body will regain its original shape without being deformed.

XO'Y - hysteresis loop
P   - ultimate strength
PF   - necking
F   - failure

- material "c" has greater Modulus of Elasticity (E) than materials "a" and "b";

<u>Theory</u> of <u>Elasticity</u> (Hooke's Law): The strain (deformation of an elastic body) is proportional to stress, within the elastic range. The proportionality constant (E) is the Modulus of Elasticity (or Young's Modulus of Elasticity).

15. In their study, where "episode" (or behavior pattern) is accepted as the basic unit of analysis, F.S. Chapin and H.C. Hightower refer to it as "a reasonably homogenous interval in the lifetime of an individual, an interval of time which is devoted to a single dominant purpose." (p. 8). See their study: "Household Activity Systems - A Pilot Study," Center for Urban and Regional Studies, Institute for Research in Social Science, U.N.C. Chapel Hill, May 1966.

16. In his critique of Freud's psychosexual theory and in his analysis, White suggests that the "effectance" concept (the desire to have a direct effect on the environment) be added to the development model. This concept in the personality theory emphasizes the importance of the environment in the growth process. For the elaboration of the concept, see R.W. White, "Competence and the Psychosexual Stages of Development," in M.R. Jones (Ed.), Nebraska Symposium on Motivation. University of Nebraska, 1960.; In her book, With Man in Mind. Cambridge: MIT, 1970, Constance Perin makes use of the same concept in the explanation of interaction of man and environment and of adaptation in terms of ego strength. (see pp. 45-48).

17.&18. Lewin,K., Principles of Topological Psychology. New York: McGraw Hill, 1936.
19. Lewin,K., A Dynamic Theory of Personality. New York: McGraw Hill, 1935.
   Lewin,K., "The Conceptual Representation and the Measurement of Psychological Forces," Contributions to the Psychologic Theory, v. 1, n. 4, 1938.
20.&21. Lewin,K., Principles of Topological Psychology. New York: McGraw Hill, 1936.
22. Sommer,R., Personal Space. Englewood Cliffs, N.J.: Prentice-Hall, 1969.
23. Hall,E.T., The Hidden Dimension. Garden City, N.Y.: Anchor-Doubleday, 1969.
24. Schorr,A.L., "Housing and Its Effects," in H.M. Proshansky, W.H. Ittelson, and L.G. Rivlin (Eds.), Environmental Psychology. New York: Holt, Rinehart & Winston 1970, pp. 319-333.

TWO VISUAL ATTRIBUTES OF ENVIRONMENTS

Chairman: Gary H. Winkel, Environmental Psychology, C.U.N.Y., N. Y.

Panelists: John Betak, Dept. of Geography, McMaster Univ., Hamilton, Canada
Florence Ladd, Grad. School of Design, Harvard Univ., Cambridge
James Wise, Dept. of Psychology, Ohio State Univ., Columbus
Donald Appleyard, Dept. of City & Reg. Planning, Univ. of California, Berkeley

Authors: Michael C. Cunningham, John A. Carter, Carter P. Reese, Bruce C. Webb, "Toward a Perceptual Tool in Urban Design: A Street Simulation Study"
Carl-Axel Acking and Rikard Kuller, "Presentation and Judgement of Planned Environment and the Hypothesis of Arousal"
Henry Sanoff, "Youth's Perception and Categorizations of Residential Cues"
Peter G. Flachsbart and George L. Peterson, "Dynamics of Preference for Visual Attributes of Housing Environments"

VISUAL ATTRIBUTES OF THE ENVIRONMENT: INTRODUCTION   2.0

Gary H. Winkel, Session Chairman

Environmental Psychology Program
City University of New York

The papers comprising this section represent a mix of substantive and methodologically oriented approaches to the problems of human response to visual attributes of the environment. Before I attempt to place these reports in some perspective and comment upon the problems and promises which they have raised, I would like to summarize briefly the papers comprising this session.

Cunningham et. al. were concerned with the development of a technique which could be used in urban design problems to predict user response to various design proposals. The focus of their research was the urban street. Using a combination of video tapes, photographs and movies, Cunningham and his colleagues have argued that responses to black and white video tapes of "real" streets are created by pasting photos of the actual streets on a "black face". The response measures employed were a short Semantic Differential scale and a test measuring the amount of interest generated by the "trip" down the street. In this study color was an important cue since using color movies resulted in greater expressed interest in the street. The authors concluded that it would be possible to make use of television or movies to simulate proposed changes in urban street design.

Acking and Kuller have summarized an extensive body of research which they and their associates have undertaken at Lund's Institute of Technology. These authors have insisted upon the development of a stable and reliable dependent measure which can be employed in a wider variety of environmental settings and under many different conditions. They now have a set of Semantic Differential scales relevant to judgments of the built environment. Using these scales Acking and Kuller compared a number of presentational techniques employed by designers to communicate their prospective clients. Illustrational plans, schematic and naturalistic models, perspective drawings, colored slides and color movies taken in a naturalistic model were all evaluated in terms of their effectiveness in recreating the adjective factor structure taken from ratings of an actual environment. Thus each of the presentational devices listed above represented a way of simulating the actual environment. The research question involved ranking the effectiveness of each of these presentational formats.

Analysis of their data indicate that movie films taken in the naturalistic model was the most successful way of recapturing the ratings given to the real environment. These investigators are now expanding their simulation capacity by building a television system which will allow a potential user of the environment both to move through it and to express his or her opinion about the qualities of the simulated environment.

Sanoff used single photographs of actual houses to determine which cues urban and rural teenagers use to make judgments of similarity and differences in built forms.

Once having understood which cues are used to make judgements of equivalence Sanoff attempted to understand how environmental personality preferences, past housing experience and choices between architect and non-architect designed housing contributed to expressed housing preferences. Sanoff's results indicate that the factor accounting for preference was the residential background of the judge (urban or rural). Personality variables such as "Preference for Modernism", or "Need to Explore" were not very helpful in predicting preference. Perceived status cues for homes also influence preference judgements in the positive direction.

Flachsbart and Peterson employed single photographs of actual houses and apartments and asked observers who differed on the basis of race and income to indicate those design characteristics which accounted for their housing preferences. These authors concluded that an individual's preference is influenced by those design characteristics of which the person is deprived and that observers tend to inflate the value of that factor. This generalization holds only if two further conditions are met: (1) the observer must not perceive any more pressing needs and (2) he must not have adapted to that of which he has been deprived.

As Cunningham et. al. suggest another of the original intentions regarding simulation was that it would allow the designer to consider a range of different design options and determine which combinations of elements were most effective in meeting design objectives. We could then hopefully identify which design parameters seemed most crucial to user perceptions. This step is essential for a better understanding of how the complex cues of any environment are coded and weighted by the perceiver.

Sanoff's efforts to determine the physical cues which are employed in making judgments of housing similarity certainly is a way to begin this process of noting which elements of that complex visual situation are most important in relating to judged preference. To be truly effective however, such procedures must allow the designer to manipulate elements of the designed environment through a visual medium. The television procedure described by Acking and Kuller and currently being employed by Appleyard and Craik at Berkeley allow the possibility of very rapid changes in the design models which comprise the presentation format.

Flachsbart and Peterson have used an alternative procedure to get at the information. These authors select pictures, pair them with one another on the basis of selected variables, and then ask respondents to indicate their preferences. I think that this approach is much less desirable because there is no guarantee that the criteria employed by the observers are necessarily the same as those reported verbally by them.

In summary the issues which have been raised in my comments should not be taken to mean that I am pessimistic about the future of simulation procedures. On the contrary, I think that they hold promise for dealing with some aspects of person-environment transactions. More careful attention to the issues discussed will, I hope, lead to the fulfillment of the promise.

TOWARD A PERCEPTUAL TOOL IN URBAN DESIGN: A STREET SIMULATION PILOT STUDY   2.1

Michael C. Cunningham, Assistant Professor of Urban Design, John A. Carter, Carter P. Reese, Bruce C. Webb

Division of Environmental and Urban Systems, College of Architecture, Virginia Polytechnic Institute and State University

Abstract

This paper describes a research-related studio experience of a group of graduate urban design students and faculty at the Virginia Polytechnic Institute and State University. The study involves video tape simulations of urban streetscapes and is directed toward the eventual development of an urban design tool to be used in testing aspects of human perceptual response to physical design projects before they are implemented. Results of this study substantiate the existence of a "cognitive gap" between designers and lay persons in terms of their interest response to black and white, color, and video tape films of both real and simulated urban street environments. (Interest response to visual stimuli is taken as a measure of environmental complexity.) Non-designers registered about one-half the number of "interest peaks" as did lay persons in tests involving the same streetscape. Semantic differential and interest cue test results to video tapes of "real" streets were not found to be significantly different from those of simulated streets. Color appears to be an important variable in affecting interest responses.

Introduction

Those of us who teach or study urban design are sensitive to the criticism that our discipline still is not a discipline--that it lacks well-defined boundaries, that it has few tools or techniques to call its own, that the results of urban design efforts are often only bigger and not necessarily better architecture. For this reason we have engaged in scavenger hunts among other disciplines such as operations research, economics, sociology, anthropology, political science, and psychology in search of concepts, methods, and even hand-me-down equipment.

The general goal of the study described here is to explore the field of perceptual psychology as a generative link between physical design of the urban environment and human behavior. The setting for the study is a design laboratory which is vertically integrated to include ten fourth-year architecture students and five graduate urban design students.

The specific objectives of this series of pilot studies are as follows: (a) to devise techniques for simulating the urban street environment which are simple and inexpensive enough to be used during the design formulation stage of a project by students and practitioners, (b) to develop testing procedures for eliciting direct responses from sample groups of potential users to the simulated environment which can be compared with the designer's behavioral objectives for a proposed project, and (c) to determine which of a subject's biographical variables, if any, exhibit a high correlation with his pattern of responses.

The investigation rests upon the basic assumptions that (a) the environmental planning professional's response to the physical environment is, in many respects, different from those of a lay person (it is assumed that such differences are particularly significant in those cases where the professional is personally involved in environmental intervention through planning or design), (b) real environments can be simulated effectively through the use of cinematography (films taken at eye level provide a comparatively satisfactory means for simulating the real visual environment), and (c) responses to interest cues displayed in films reflect the relative degree of complexity and unity in the visual environment and are also dependent upon the biographic composition of the perceiver.

Some questions concerning the use of simulations which this study addresses are: (a) How do levels of visual interest perceived in films of simulated environments compare with those of the real environment? (b) What is the effect, if any, of the presentation medium? and (3) Do correlations exist between selected biographical characteristics of the subjects (such as age, field of study and size of home town) and their responses to urban streets? However, due to the open-ended and exploratory nature of this study, the reader is cautioned not to accept the conclusions drawn here to be definitive.

Background

Literature is plentiful regarding aspects of the differences with which designers and lay persons relate to environments. Alexander (1) posited that designers are socialized by their education to think about environments with a different frame of reference than do lay persons. Herschberger (2) concluded that the professional education of architects caused them to attach greater importance to building aspects perceived to be potent. Leff and Deutsch (3) found that environmental planning professionals tend to be more concerned about concrete aspects of environments and referred most often to the socio-psychological and emotional dimensions of meaning. On the other hand lay persons tended to think about the environment in physical-spatial and ethno-demographic terms. Non-designers generally perceived environments to be less predictable than did their professional counterparts.

The still inexplicit nature of the cognitive differences between designers and lay persons is problematic in several ways. First, the differences create ambiguities in meaning which hamper communication and sometimes cause counter-productive interrelationships between the professional and his client. Such situations, which are common in urban planning and increasingly so in architecture, are aggravated by the dichotomy between very real clients and statistically imagined users. Second, the inexplicitness of the differences forces us to treat them as detrimental to prosthetic design, as a gap to be filled in and leveled out rather than as strengths which in reality could be exploited.

Promising methods used to uncover cognitive differences and make them more explicit include mapping, semantic differential and grid techniques, and observational studies. These methods share an inherent impracticability in terms of potential application by professionals to specific design development projects. As tools they are too specialized in terms of the skills required to operate them and too general in terms of the information they output. They are enormously tedious and time consuming. In short, most of the techniques evolved for cognitive study are of use to theory builders and cannot be easily applied by decision-makers who are in a position to commit vast human and economic resources to projects and programs which are virtually irrevocable.

The Case for Simulation

As a tool space simulation offers a reasonably simple and inexpensive means of evaluating some aspects of a project while it is in the design-development phase. It is generative and predictive and utilizes skills such as scale modeling and photography which environmental planners and designers commonly acquire as part of their professional education.

Films of models of an environment appear to provide the most satisfactory simulation results. Acking and Küller (4) compared six different methods of presentation: illustrational plans, perspective drawings, white schematic models, color slides from a naturalistic model, colored naturalistic models and color films made from a naturalistic model. These were related to five semantic dimensions including pleasantness, enclosedness, complexity, social status, and unity. Color films made from eye level in a "naturalistic model" proved to be the least faulty means of representation.

Although video tape, film and computer graphics have been used in simulation studies, their use is most often restricted to presentations (5,6) and only occasionally are they used as in-house design tools. Attention (undue in our opinion) is usually given to laborious "realistic" modeling techniques not withstanding the fact that a relatively high degree of abstraction can be interpreted easily by the perceiver (7). Video tape has the important advantage of immediacy over conventional film. It requires no development time and can be monitored while the recording is being made.

Spatial Settings, Equipment and Samples

The space chosen for the simulation is the section of Wisconsin Avenue in Washington, D. C., between "M" and "P" Streets. A photomontage scale model (1:50) was made by photographing each building facade and mounting the strips to styrofoam panels on both sides of the road adjacent to paper sidewalks. A Sony portable video tape camera fitted with a HCI model-scope, Model 98, was mounted on a wheeled dolly to allow manual adjustment in both the horizontal and verticle planes. Video-tapes were made with the lens of the modelscope at approximately six feet (to scale) above ground level. The equipment was pushed along guide rails at various rates of speed in order to simulate pedestrian and automobile sequences ranging from 0 to 30 miles-per-hour.

The black and white video tape simulation of the model was compared with three films of real streets: (a) a super-8 mm. color film of Wisconsin Avenue, (b) a black and white video tape reproduction of the preceding, and (c) a super-8 mm. color film of Jefferson Street in Roanoke, Virginia. The video tapes were presented on 14" monitors and the films were shown on white screens placed approximately eight feet from the subjects.

Research Design

The research design held constant the observer groups and the response devices while varying the presentation medium in an attempt to determine the effect which the medium might have on the response patterns of the subjects. Biographical data was collected from the subjects prior to the viewing of the first film and was later compared with response patterns.

Two response devices were used. In the first of these each subject was given a three position toggle switch wired to a Sanborn pen recorder, Model 67-1200. They were instructed to register interest responses to visual "excitment" perceived on either the right or left side of the screen by pushing the toggle in the corresponding direction. These responses were recorded on heat sensitive graph paper to be correlated later with the films.

Following the viewing of each film subjects were given twelve word pairs on opposite ends of a scale (+3 to -3) and were asked to determine the most appropriate rating for the film sequence which was just viewed.

Responses to the three films for both tests were compared and correlations were determined between response patterns and various biographical data for each subject. Subjects for the experiments were students and faculty at Virginia Tech. The total sample was comprised of two separate groups of thirty-four and twenty-one persons each, only three of whom were female.

Interpretation of Data

The mean interest responses as well as their positions on the recording sheet were different for each film shown. The responses to the video tape of the model of Wisconsin Avenue were not markedly different (in trend) from those of the video tape of the actual street. The color film of Wisconsin Avenue was found to be somewhat more interesting than the black and white video tape of the same street.

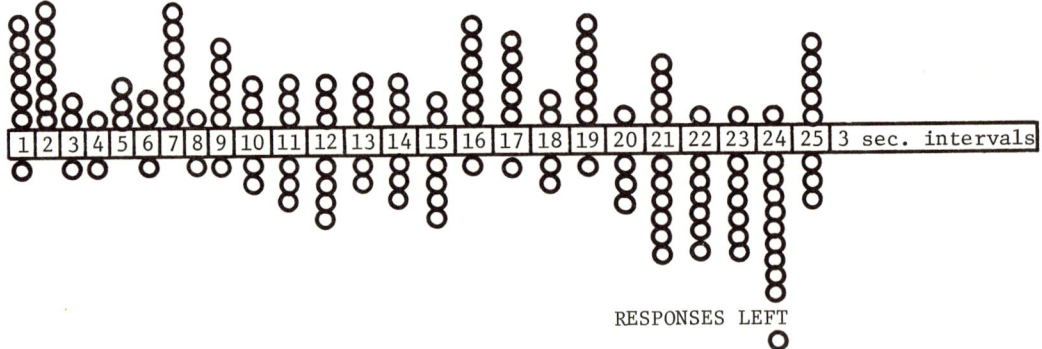

Figure 1. INTEREST RESPONSES—BLACK & WHITE FILM

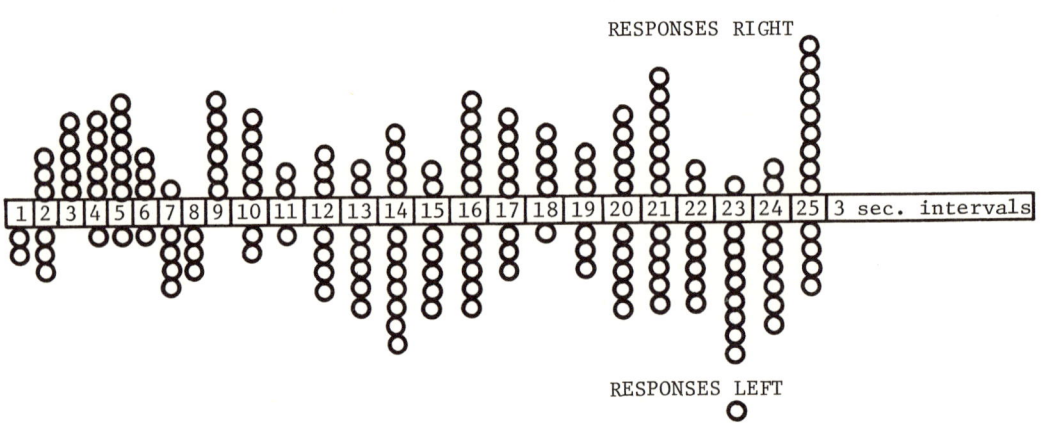

Figure 2. INTEREST RESPONSES—COLOR FILM

2. VISUAL ATTRIBUTES OF ENVIRONMENTS / 67

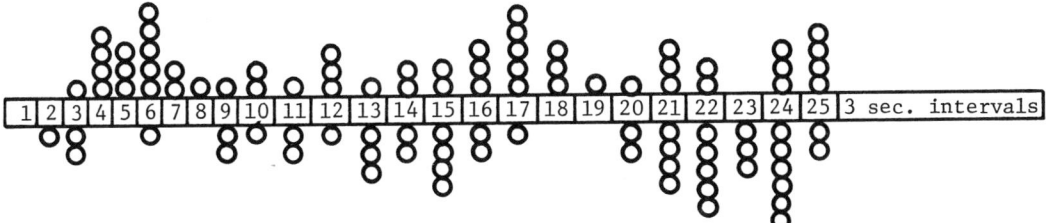

Figure 3. INTEREST RESPONSES--VIDEO TAPE OF MODEL

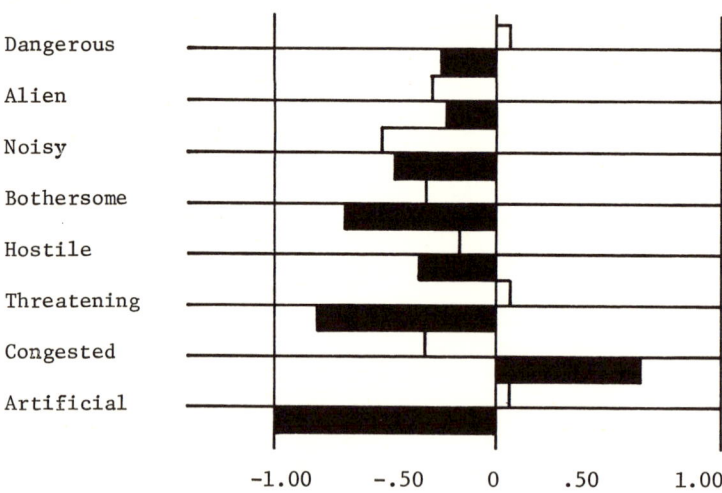

Figure 4. BLACK & WHITE FILM AND MODEL FILM

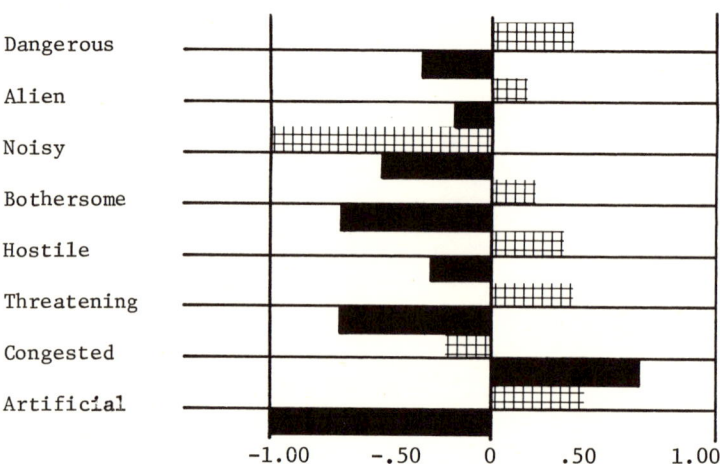

Figure 5. COLOR FILM AND MODEL FILM

2. VISUAL ATTRIBUTES OF ENVIRONMENTS / 69

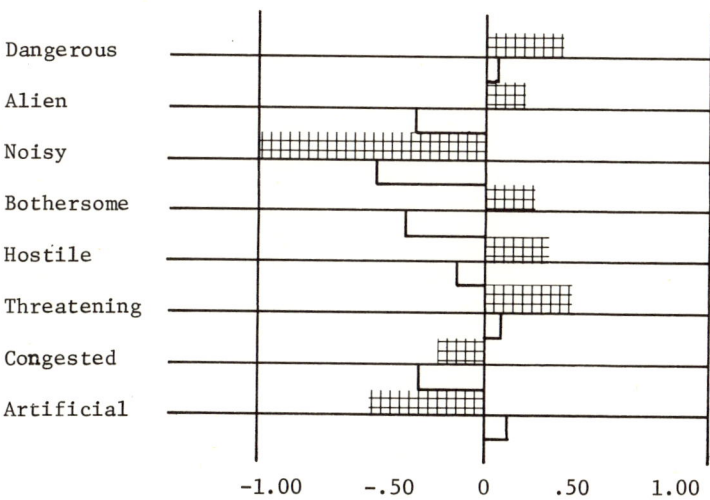

Figure 6. BLACK & WHITE FILM
AND COLOR FILM

Key:

Video Tape ■
Black & White Film ☐
Color Film ▦

The compared mean responses were 7.9 for the video-tape versus 9.1 for the color film.

A stepwise multiple regression was run in an attempt to isolate the most significant biographical variables influencing response patterns. Age, design versus non-design background, and life style history were all correlated with responses for each subject. Results indicate that only the design versus non-design variable had any significant effect on the response patterns elicited from the subjects. Designers registered nearly twice the number of interest responses as did those persons with non-design backgrounds. No attempt has yet been made to correlate designer or lay responses with categories of visual cues such as natural or man-made.

In the semantic differential test the responses to the color film of the Georgetown street were significantly different from the responses to either the black and white video tape of the street or the model. However, similar responses were recorded for the black and white tape of the actual Georgetown street and the black and white tape of the model of the same street.

Although color is an important variable influencing perception, it appears that designers can elicit useful approximations of user responses to their projects through the use of simple models displayed on black and white video tape.

Conclusions and Discussion

The results of the pilot studies described above substantiate the suspicion of the experiementers that designers see things with a special perceptual frame of reference. They tend to be considerably more aware of (and perhaps susceptible to) visual stimuli than are the non-designers. The experimenters view this as support for the continued development of simulation techniques which can provide useful feedback information to the designer during his design activities.

Responses to the video tapes of the simulation model were judged to be sufficiently correlated with responses to films of the real environment to encourage continuation of the study. However, an experimental difficulty was caused by the noisy toggle switches and primitive projection equipment used. Convenience dictated that groups of four subjects viewed the films at the same time in a common space. In some cases there appeared to be a "triggering" effect when one subject's response initiated the responses of other subjects seated in close proximity. Although the exact affect of this phenomenon is undetermined, it could be minimized by using silent switches and eliminated through individual monitoring of the films.

Further exploration should be directed towards two specific problems. First, the interest responses registered by the means described here are dependent upon the subject's intellectual manipulation of information. It would be considered desirable to elicit direct responses by measuring physiological changes not consciously produced by the subject. Measurement of brain wave emission, particularly alpha waves, is suggested as one possible means of compiling complementary data.

The second issue is considerably more "messy." Studies to date have compared responses to _films_ of streets without comparing them to behavior in the _actual_ street itself. Before simulations of this kind will become a reliable descriptive and evaluative tool in urban design, they must either approximate the real world experience and include some means of accounting for haptic, olfactory, tactile, and auditory sensory stimuli, or means must be devised to compensate for the lack of such stimuli.

Simple simulation techniques seem to offer one means of augmenting the traditional tools of the designer. They promise cheap (the cost of equipment used in this study is approximately $3,500) and effective means for evaluating certain aspects of human response to design projects before they are implemented as a part of the public environment. It is anticipated that the results of testing simulated spaces can be fed back into the design development phase of planning/design projects in order to avoid at least some predictable errors.

Notes

1. C. Alexander, Notes on the Synthesis of Form, Cambridge, Mass.: Harvard University Press, 1964.

2. R. Herschberger, "A Study of Meaning and Architecture," Paper presented at the First Annual Conference of the Environmental Design Research Association, Chapel Hill, June, 1969.

3. H. Leff and P. Deutsch, "Construing the Physical Environment: Differences Between Environmental Planning Professionals and Lay Persons," Unpublished manuscript, Harvard University, 1972.

4. C. Acking and R. Kuller, "Presentation and Judgement of Planned Environment and the Hypothesis of Arousal," Unpublished manuscript, Lund Institute of Technology, 1972.

5. See J. M. Anderson, "A Television Aid to Design Presentation," Architectural Research and Teaching, 1970, 1(2), 20-24.

6. See S. Rose and M. Pierce, "Television as a Design Tool," American Institute of Architects Journal, 1967, 47 (3), 76-80.

7. See R. Gregory, "On How Little Information Controls So Much Behavior," Ergonomics, 1970, 13(1), 25-35.

PRESENTATION AND JUDGEMENT OF PLANNED ENVIRONMENT AND THE HYPOTHESIS OF AROUSAL 2.2

Carl-Axel Acking and Rikard Küller

Lund Institute of Technology
Section of Architecture
Department of theoretical and applied Aesthetics
Fack 725, 220 07 Lund, Sweden

Abstract

To find a systematic method for measuring and describing perception of human environment groups of subjects rated various environments with descriptive words, which were given the form of rating scales. Data was processed by means of factor analysis, through which eight main dimensions were obtained. Some of these dimensions were used in an attempt to compare different methods for presentation of architectural projects. Movie filmen from eye-level in naturalistic models turned out to be most promising, which led to the construction of an advanced equipment for model presentations by means of television technique in combination with continuous response-registration. Some of the semantic dimensions are probably related to physiological arousal. This is being investigated at present in two different experiments.

Part 1

A semantic model for describing perceived environment

When this research project started six years ago the basic hypothesis was as follows: The total perception of architectural environment can be described in a limited number of meaningful dimensions. Those dimensions can be separately defined and measured and they are valid within determinable limits.

This hypothesis was tested by letting groups of experimental subjects judge different environments with semantic rating scales. Osgood, Suci and Tannenbaum had shown that it was possible to obtain a fairly simple semantic descriptive model by treating data with factor analysis. His research group mainly worked with measuring meaning of different concepts, but also made some experiments in the field of aesthetics. They obtained three main dimensions with great validity and another four which seemed to be more difficult to interpret. They had however some difficulty in measuring the dimensions in a unitary way for different research areas.

By limiting the area to be perception of architectural environment we hoped to obtain dimensions which would be more easily interpretable and meaningful for an architectural situation, and maybe also possible to measure through a standardized procedure. Just as Osgood and his group we used seven-graded semantic scales, but

2. VISUAL ATTRIBUTES OF ENVIRONMENTS / 73

instead of defining those by opposite words as a rule we used one descriptive word for every scale.

Figure 1.

The Osgood scale:

BAD : ___ : ___ : ___ : ___ : ___ : ___ : ___ :   GOOD

The type of scale used by us:

                              GOOD

Slightly ☐ ☐ ☐ ☐ ☐ ☐ ☐ Very

Our experiments were carried through and the data treated in two different ways. According to the first alternative a group of people judged several environments in many different scales. For every environment the group average was calculated and used as basic statistic for estimating correlations and for the factor analysis. This procedure will result in factors of dimensions which may be said to describe average differences between environments. Usually we presented these environments in the form of colour slides and the treatment will be named design 1.

According to the second alternative an experimental group judged only on environment, also in this case with many different scales. In this case the individual ratings were used for estimating correlations and for factor analysis. Factors or dimensions obtained in this way can be said to describe individual differences in judging the environment in case. This alternative was used in full scale studies and will be named design 2.

In the first experiment a group of students judged colour slides of 15 interiors in 78 different rating scales. The descriptive words were a representative sample from a Swedish Academy glossary. The analysis of this material was made according to design 1 and resulted in eight factors, which could be given a fairly meaningful interpretation. Those factors were named:

      PLEASANTNESS
      SOCIAL STATUS
      ENCLOSEDNESS
      ORIGINALITY
      COMPLEXITY
      AFFECTION
      UNITY

In the next experiment a group of students of technology judged one interior room in the same 78 variables. In this experiment, which was treated according to design 2, three factors were obtained and could be interpreted as PLEASANTNESS, COMPLEXITY and ENCLOSEDNESS. This was taken as a confimation of the fact that at least some of the factors earlier obtained were valid also for interiors in full scale.

In the third experiment a group of students judged a housing area and data was treated according to design 2. The response variables were about the same as earlier. In this case four factors came out, which could be interpreted as PLEASANTNESS, ORIGINALITY, COMPLEXITY and ENCLOSEDNESS. The factors earlier obtained could now also be said to be valid to at least some extent for judgement of exteriors in full scale.

At this point Acking and Sorte started to investigate landscapes in colour slides and Gärling investigated working environment in full scale. These authors showed that the above mentioned factor structure was valid also in those cases. In the study on landscapes the following factors turned out: PLEASANTNESS, COMPLEXITY, UNITY and ENCLOSEDNESS. In judging working environments PLEASANTNESS, UNITY, ENCLOSEDNESS, SOCIAL STATUS and COMPLEXITY came out.

Finally we made a new experiment according to design 1; a study where both the experimental group as well as the environments had a more heterogeneous character. Of the earlier used response variables 33 were kept and another 33 new ones were added. The factor analysis resulted in eight factors and those could be given the same interpretation as those obtained in the very first experiment. We therefore concluded that the semantic rating of environment with advantage can be described in these very eight dimensions.

Now we went on to study the group stability for the different dimensions. Comparisons were made between architects, students, house wifes, students of architecture, interior architecture, landscapes architecture as well as more general groups. Very experienced architects seemed to deviate very much in the factor of pleasantness. It also seemed likely that the frame of reference made up by the experimental situation had an influence on the ratings. If the same subject was allowed to rate several environments those were often better differentiated than if every environment was judged by different groups.

The different dimensions could to a certain extent be validated by using the semantic method paralell with other methods. The complexity turned out to depend on the number of units that made up the environments and also the colouring especially saturation of the colours in the environment. Unity could be operationally defined as a lack of disturbing units, especially of permanent character, in the environment. Enclosedness showed a negative correlation with the lightness of the walls of the room. Potency and social status also showed certain correlations with the colouring.

From the collected results we have interpreted the eight dimensions in the following way.

2. VISUAL ATTRIBUTES OF ENVIRONMENTS / 75

PLEASANTNESS generally accounts for the largest part of the variance. The dimension can be regarded as a measure of the amount of pleasantness and security that an individual perceives or feels in the environment. There is no doubt that this dimension is the same as the one Osgood´s group called Evaluation factor. Examples on positively and negatively loaded words in this dimension are: UGLY, STIMULATING, SECURE, BORING, IDYLLIC, GOOD, PLEASANT, BRUTAL.

COMPLEXITY is a measure of the liveliness and complexity of the environment. It seems to be very much the same as Osgood´s et al Activity factor. Examples: MOTLEY, SUBDUED, LIVELY, COMPOSITE.

UNITY can be interpreted as a measure of how well the different parts in the environment fit together. Examples: FUNCTIONAL, OF PURE STYLE, CONSISTENT, WHOLE.

ENCLOSEDNESS seems to describe the closedness or on the other hand the spaciousness and lightness of an interior or exterior room or the spacing of a group of buildings. Examples: CLOSED, OPEN, DEMARCATED, AIRY.

POTENCY describes a kind of potential powerfulness of the environment. It also shows a fairly clear aspect of sex, which means that an environment is more or less associated with one sex or the other. It seems to be the same as Osgood´s et al Potency factor. Examples: MASCULINE, FRAGILE, POTENT, FEMININE.

SOCIAL STATUS can be said to be an economical and social measurement. Examples: EXPENSIVE, WELL-KEPT, SIMPLE, LAVISH.

AFFECTION describes the age of an environment but also a feeling for the old and genuine. Examples: MODERN, TIMELESS, AGED, NEW.

ORIGINALITY, finally, is a measure of the unusual and surprising in an environment. What has not earlier been seen gets a high value in this dimension. It might correspond to Osgood´s et al Novelty factor. Examples: CURIOUS, ORDINARY, SURPRISING, SPECIAL.

These eight dimensions are in princip independent of each other. Pleasantness, Complexity, Unity, Enclosedness etc are measures of different aspects and can vary independently of each other. It seems likely that for every environment there exists neutral zones in the different dimensions, zones, which are determined by the individual´s expectations. A danc-hall and a sick-room, for example, may be expected to have different complexity. It seems possible that different environments must have different semantic profiles to make it possible for people to find themselves at home and feel secure. We are at present working along these lines.

It is also possible that the subject´s personality is important for how he perceives and judges his surroundings. By testing a group of teachers with the Eysench Personality Inventory they could be devided into more or less extrovert and more or less neurotic. They were then asked to judge how they believed themselves to be influenced by different kinds of environments. They believed themselves to be significantly more extroverted in environments that were high in pleasantness than in environments that were low in pleasantness. They also believed themselves to become more stable in environments with high pleasantness. The less neurotic half of the subjects also believed themselves to be favourably influenced in environments which were low in complexity.

We have put together the best of the rating scales that we have used into a test, where pleasantness is measured with 8 different scales and the other dimensions with 4 scales each. To try the reliability of the test it was given to a group of students of technology, who rated a lecture room two times with a week in between. The reliability of the re-test-situation was satisfying for all the eight factors and the correlations between different factors were very low. We have come to the conclusion that the semantic descriptive model is meaningful for architectural environment and that the different dimensions have high face validity and can be measured with a standardized procedure. We work at present with the validity of the different dimensions, which you can see in the last part of this paper and also with using the semantic model in practical work on architectural environment.

Part 2

Presentation of planned environment

When judging architectural projects many different kinds of information may be used. Faulty information results in bad decisions. This is very often the case when one is going to judge the visual character of an architectural project. If, for instance, a perspective drawing of a housing area shows big green trees, the consumer may get an incorrect view about what the area will be like to live in. In a discussion about alternative plans for a housing area it is possible that a certain method of presentation favours one alternative to another. One-family houses may for instance look very boring on a ground plan while an area with high houses may seem exciting. In reality it might be the other way around. The better the information about the project coincides with the perceived reality, the higher are the possibilities that correct decisions will be taken. In this study different ways of presenting two housing areas have been compared. The housing areas existed in reality so that the judgement of reality could be used as criterium. The following ways of presentation were compared:

1 <u>Surveys</u>
Illustrational plan in black and white, scale 1:400
White schematic model, scale 1:400
Coloured naturalistic model, scale 1:400

2 <u>Pictures</u>
Perspective drawings in black and white
Colour slides form eye-level in the naturalistic model

2. VISUAL ATTRIBUTES OF ENVIRONMENTS / 77

3   Colour <u>movie</u> filmed from eye-level in the naturalistic model

The two housing areas were about ten years old and could be considered to have fairly determinable limits.
Area 1. This area consisted of one-family houses, situated in the town of Malmö. Although the different houses were alike and situated in a rather regular form their inhabitants had given them rather individualistic looks by using different material in the fences, by planting different types of trees, bushes etc.
Area 2. This area consisted of three-storeyed family houses situated in Lund. The families had no possiblity to influence the exterior of the area. The buildings were grouped around a large open place with abundant vegetation. In this green area there was a play ground.

The areas were judged in reality under two different conditions, nice and bad weather, each by a group consisting of 20 subjetcs. Comparable groups each of 20 subjects judged the two areas as presented by each of the methods of presentation mentioned above. The judgements were made with semantic rating scales and measures were taken on the following five semantic dimensions.

    Pleasantness
    Enclosedness
    Complexity
    Social status
    Unity

Comparisons between every method of presentation and judgements in reality were made by analysis of variance. The results of this statistical treatment will now be presented. When judged in reality the two areas got a certain score in each of the five dimensions and they could accordingly be related to each other. Suppose that the one-family area gets a higher point than the three-storeyed area in reality in one of the dimensions. For a method of presentation to be good it must show the same relationship. If not, the information from this method of presentation is useless.

For three of the five dimensions the illustrational plan and the white model did not give the correct relationship. For the illustrational plan the faults turned up in the following dimensions; Pleasantness, Complexity and Unity. For the white model the faults came out in Enclosedness, Complexity and Unity. The perspective drawings did not give the correct relationship in Enclosedness and Unity. For the naturalistic model and the movie the relationship was incorrect for one dimension only. This was for the naturalistic model Unity, while for the movie it was Social status. The only method of presentation that kept the correct relationsip in all the five dimensions was colour slides from eye level in the naturalistic model. The analyses of variance also gave the possibility to make several other evaluations of the different methods of presentation. This made it possible to rank the methods in the following way.

Table 1. Amount of faulty information for different presentation methods in five different dimensions.

	Pleasantness	Enclosedness	Complexity	Social status	Unity	Total
Illustrational plan	3			1	2	6
Perspective drawings	1	2	1	1	1	6
White schematic model	2	1	1		1	5
Colourslides from naturalistic model	1		2	1		4
Coloured naturalistic model	2				1	3
Colour movie from naturalistic model	1			1	1	3
Total	10	3	4	4	6	

It might be seen that illustrational plan, perspective drawing and white model contained most faulty information. This might have been expected, but it is quite serious as those are the methods most often used in practise. The analyses also show that pleasantness is the most difficult dimension to predict in a correct manner. This is also very serious as pleasantness might be expected to be the most important dimension. It seems quite clear that the naturalistic model especially when presented filmed from eye-level is the best way of presenting a project. It contains the most valuable and least faulty information, when one wants to predict perception of reality.

This conclusion has led us to construct an equipment which makes possible continuous rating with semantic scales, in combination with continuous presentation of planned environment. Similar equipment exists in the Netherlands and at Berkeley, California, but has been elaborated by us and will now be presented.

2. VISUAL ATTRIBUTES OF ENVIRONMENTS / 79

Figure 21. Equipment for presentation and judgement of architectural projects.

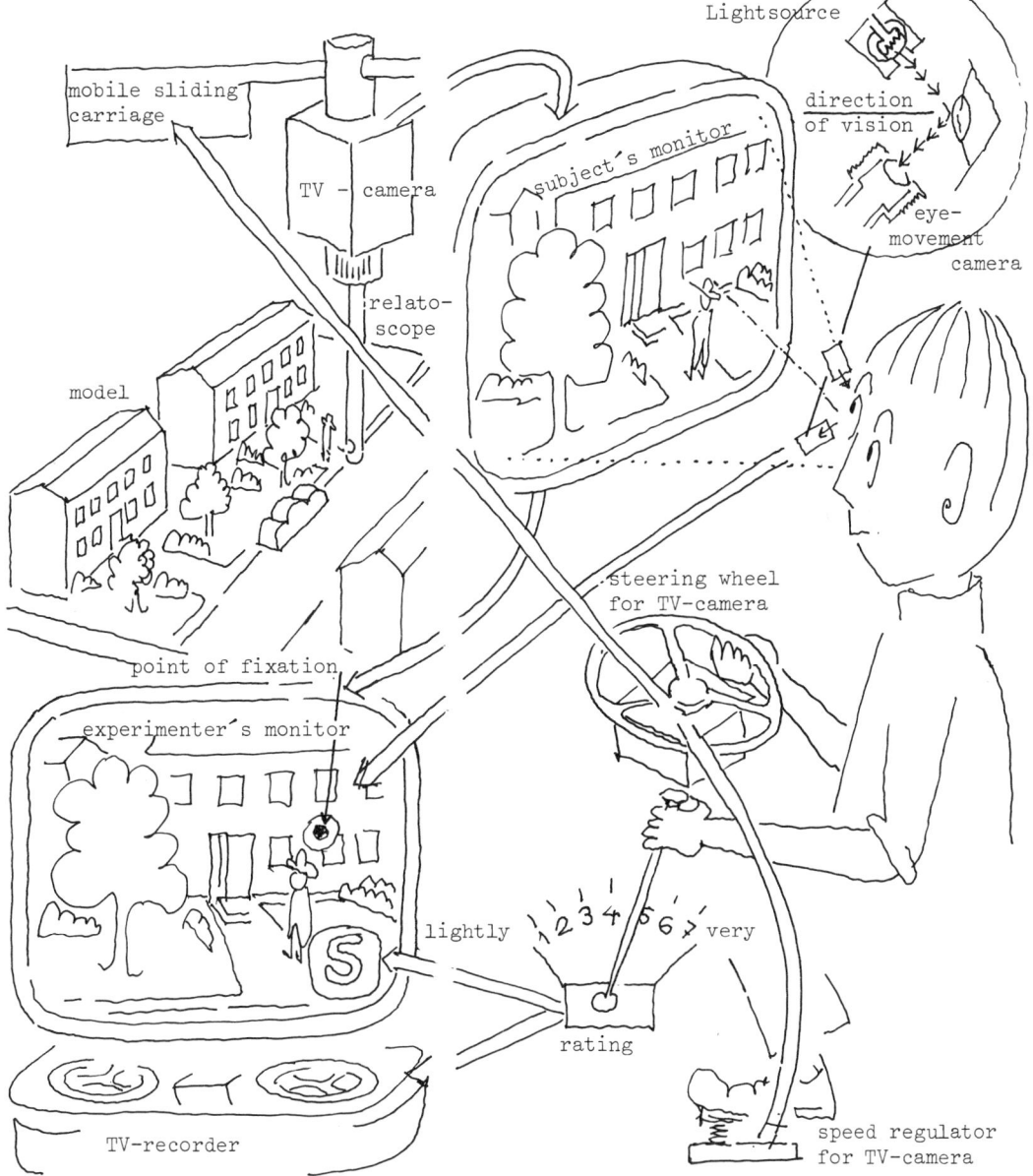

A mobile television camera is situated over a naturalistic model. A relatoscope, that is a long pencilshaped tube with a lens, extends from the camera down into the model and gives a view from eye level. This picture goes to two different monitors, one monitor for the experimental subject and one for the experimenter. We will later return to the experimenter and instead focus on the experimental subject.

He is sitting in front of his monitor and sees the view from the naturalistic model. In front of him is a foot pedal and a steering wheel. As he steps on the pedal impulses go to the mobile television camera, which is hanging over the model in a mobile sliding carriage. The camera starts to move. The harder he steps the faster the camera moves over the model. The relatoscope dipping down into the model moves along with the camera, which means that the experimental subject when looking on his television screen seems to move around in the model. If he turns his wheel the camera starts moving in a new direction and also the relatoscope rotates around its axes to a corresponding degree. The feed-back is immediate on the monitor so that the subject can drive from a main street into a smaller street or he can turn on the street and start driving the other way. He can also stop for instance on a square and look around etc. By using different speeds this movement might simulate walking, biking or driving a car.

From an experimental point of view it might hereby be important to know what the subject is looking at at a certain moment, that is, his spot of fixation on the television monitor in front of him. To know this we light one of his eyes with a narrow lightbeam which is reflected by the cornea and recorded by another television camera. Thanks to the eye not being completely spherical the reflected lightbeam will move along with the eye. In this way the point of fixation is registered and shows as a white moving spot on the monitor of the experimenter.

It is also important to measure how the experimental subject perceives the space he is travelling in. How does he continually perceive its size, complexity, pleasantness, etc. To report this he has at his disposal a lever that can be manipulated with one hand. The lever has seven distinct steps and works as a kind of seven-graded rating scale. He might for instance have got the instruction to judge the widths of the street. Suppose he found it unnecessarily wide. In that case he moves the lever to the right and if he finds it too narrow he moves it to the left. If the street seems to have the adequate width he puts the lever some place in between. This information too goes to the experimenter's monitor in digital form, and he can continually study the response and presumably the perception of the subject, when the latter is travelling through the model.

Finally all this information, the view from the model itself, the fixation spot and the semantic rating can be recorded on tape. In addition to the visual stimulation we also have the possibility to play back-ground noise, for instance street sounds, recorded in a realistic situation. Such back-ground noise seems to contribute to give the subject an experience of reality.

Part 3

The semantic model and the hypothesis of arousal

Will it be possible to predict with the semantic method the neurophysiological arousal effect of architectural environment. Will it be possible to determine the highest and lowest acceptable limit of this arousal level for a specific environmental situation. And finally is there any systematic relationship between personality and such limits.

The following situation can be used to illustrate the meaning of arousal. Suppose a person is walking in the wood during the night and that he suddenly hears a gun firing close to him. He suddenly becomes wide awake (cortical arousal reaction), he gets afraid (affective arousal reaction), his heart starts beating very fast (vegetative arousal reaction) and his muscles get tense (spinal arousal reaction). This means that the firing has led to a general arousal reaction, that is the whole organism has reacted on the stimulation. This reaction can be studied as a desynchronization of the electroencephalic rythm.

We will now turn to a discussion of the environment. We have found that every environment is perceived as having a certain complexity which means that it contains a certain amount of easily distinguishable units. These units are also perceived to belong together to a higher or lesser degree so that the environment becomes more or less easy to grasp and survey. This aspect can be described as unity. We are now working along the hypothesis that the degrees of complexity and unity in the environment are of vital importance for the arousal level that this environment will give rise to. When a subject stays in an environment which he perceives as very complex and of very low unity he is overflowed by a stream of impulses which leeds to constant activation and he seems to have no possibility to rest or recover. Examples of such environments are crowded trafic situations or working environments with a lot of machinery, control panels and moving machine parts. Even if the subject stays for a long time in such a situation he will not get used to it and the arousal level will remain to high. The organism is forced to constant readiness as in situations of emergency and this will in turn result in stress and psychosomatic symptoms.

In environments with very low complexity and very high unity on the other hand the individual will face too low activation. He will experience monotony and as has been shown in experiments on sensory deprivation he might start to day dream, hallucinate or even fall asleep. Environments of this type might be highways during nights and also certain types of monotonous working places. It would be invaluable to be able to predict already when planning an environment the average activating influence it will have. If this influence shows to be too high or too low it may then be easily changed.

It is well known that the arousal system of the human organism has its center in the reticular formation in the brain stem, from which pulses are sent out to other parts of the central nervous system. This makes it possible to study activation or arousal by means of electroencophalogram. It is also possible to use other physiological measurements as heart-beat, blood-pressure, etc, but also pupillary reflexes and eye movements. We are at present doing several studies using such methods, where subjects are presented to an environment either through the relatovision described above or in full scale. The results of these methods are then related to semantic measurements of the environmental situation. The results are to be followed up in field studies in environments with varying complexity and unity. For this purpose we are planning to use medical data about the subjects.

At the present time two different studies are taken place.
1. The experimental subject seats in front of the eye movement recorder. On the screen of the recorder slides of environments are shown to the subject. Each slide is shown for a fixed time during which the subjetc is to watch carefully the environment and at the same time judge what he sees with the semantic rating lever. Ratings are done of complexity, unity and pleasantness. The hypothesis is that number of eye movements are positively correlated to complexity. At the same time the pattern of eye movement is studied. Data is also treated from the combined knowledge of pleasantness, unity and complexity. We actually believe that there might be some fairly complicated interactions.

This experiment has certain very interesting features. It is obvious that the subject fairly soon decides on what rating he wants to give to a certain environment. But after having looked at the slide for some time he will tend to change his mind and will move the lever. We are interested to know whether this changing of mind, i.e. moving the lever, is coinciding with a change of fixation. At the present time we have not yet analysed the results but the data will be treated before the end of 1972.

2. The experimental subject is placed in a full scale room and supplied with a semantic rating lever. The room is presented during different conditions. In some of the conditions the room is fairly empty of furniture and have weak colours on the walls and details. In other cases the room is highly furnished and has strong colours. During his visit in the room the subject is furnished with electrodes, which make it possible to record his eletroencophalogram. The hypothesis is that the desynchronization of the electric activity of the brain will be directly related to the degreee of complexity but also inversely to the degree of unity of the environment and that the subject´s attitude, whether he feels himself in a pleasant or unpleasant mood, will influence the arousal pattern.

It is well known that stress, for instance, can easily result, if the person dislikes the situation, if he has no control over the immediate environment. Let us take noise as an example. A situation with a high noise-level will accordingly be stressing to the person who dislikes noise but can not control its level. On the other hand he will stand much noise as long as he makes it himself or choose it deliberately as when listening to a symphony or even modern pop music.

To test the influence of the environment on the cortical activity the electroencophalogram i registered eletronically on tape and then analysed frequency by frequency. It is known that the fairly slow alfa-waves, around 8 per second, are normal for a relaxed situation and that every type of stimulation causes a desynchronization and a higher percentage of fast waves in the brain. It has recently become possible to analyse this type of data in a fairly exact manner.

We expect these studies to be finished in June 1973. If it turns out, and we have many indications already, that there is a definite relationship between the semantic judgement and the physiological status of the experimental subject, this will open wide and important ways for getting at and designing better architectural environment. Such work will proceed along two different lines. The one and most obvious line is perhaps jurys or by using the presumtive consumers themselves. The other line of advance will be to elaborate a theoretical model for describing how the different dimensions of perceived environment interact and also to relate those expectations that every person has on his environment to what he actually perceives.

References

1. Osgood,C.E., Suci,G.J., Tannenbaum,P.H., 1957. The Measurement of Meaning, University of Illinois Press, Urbana.
2. Osgood,C.E., Suci,G.J., Tannenbaum,P.H., 1957. See reference 1.
3. Acking,C.A., Sorte,G.J., 1972:2. Metoder för presentation av planerad miljö. Lund Institute of Technology. Department of theoretical and applied Aesthetics. Lund.
4. Gärling,T., 1970. Upplevelsen av en arbetsmiljös arkitekturala utformning. En undersökning med semantiska skalor. Psykologiska Institutionen Stockholms Universitet. Stockholm.
5. Osgood,C.E., Suci,G.J., Tannenbaum,P.H., 1957. See reference 1.
6. Osgood,C.E., Suci,G.J., Tannenbaum,P.H., 1957. See reference 1.
7. Osgood,C.E., Suci,G.J., Tannenbaum,P.H., 1957. See reference 1.
8. Osgood,C.E., Suci,G.J., Tannenbaum,P.H., 1957. See reference 1.

Literature

Acking,C.A. and Küller,R., 1972:1. Miljöupplevelsens dimensioner. En semantisk beskrivningsmodell. Lund Institute of Technology. Department of Theoretical and applied Aesthetics. Lund.

Birkmayer,W. and Pilleri, G., 1965. Hjärnstammens reticulära formation och dess betydelse för de vegetativa funktionerna och affektlivet. E Hoffman - La Roche & Co., Basel

Küller,R., 1972. A semantic model for describing perceived environment. National Swedish Institute for Building Research. Document D 12 1972. Stockholm.

YOUTH'S PERCEPTION & CATEGORIZATIONS OF RESIDENTIAL CUES

By Henry Sanoff, Associate Professor of Architecture
School of Design, North Carolina State University

Abstract

This study attempts to explore ways in which the social learning takes place through classification schemes youths use to describe and evaluate residential settings, as well as the meaning they are likely to draw from the categories. The ways in which people discriminately group different stimuli and treat them alike, or equivalence making, is a part of their cognitive development, and a process rarely approached in formal education. This research attempts to explore teenagers' perceptions of the social meaning of the environment. The research design included 150 high school students, residing in urban and rural settings, who were asked to describe similarities and differences between 12 selective photographs of residential settings. A content analysis of the sorting techniques revealed descriptive and affective categories as well as hierarchial preference patterns corresponding to their environmental experiences.

Introduction

The residential environment that is so much a part of everyday life has embedded within it cues about the social system which contain symbolic connotations that transcent shelter alone. While the way in which this type of social learning takes place has been rarely studied in detail, it is important to understand the classification scheme that children use to comprehend the environment as well as the meaning they are likely to draw from the categories. Little is known about the kinds of differences that young people are apt to notice yet it is clear that the "house" transmits cues which they use to build social categories about the nature and desirability of their environments. It also appears that everyday environmental experiences of urban and rural children may have a very different quality such that it can be a pervasive influence upon their perceptions as well as aspirations.

Kelly (1) views perceptions of sameness and differences as the basic elements of cognitive structure. He suggests that if we discover the numbers and kinds of dimensions that are used to compare and contrast stimuli, then we shall have a framework from which to derive preferences. This exploratory study, then, began with the following questions as a basis for investigation: What kinds of perceptual cues are children most likely to notice? Do children from different environments differ in the criteria they use for comparison? Do they use the same criteria to compare and evaluate visual stimuli? (2)

The ability for youths to discriminately group different things and treat them as alike permits them the ability to cope with the diversity of the environment. This equivalence making is a part of cognitive development and particularly through ikonic representation can descriptions be made of different features of the environments. Equivalence with ikonic representation might more likely be accomplished by grouping items according to perceptual likeliness, however, the basis for judging similarities may be more extensive. Olver and Hornsby (3), suggest five main modes for distinguishing equivalence: perceptible, functional, affective, nominal, and fiat.

1. Perceptible: The individual describes the items equivalent on the basis of immediate phenomena qualified such as size, shape or on the basis of location.
2. Functional: The individual may base equivalence on the use of function of items.
3. Affective: Items may be described as equivalent on the basis of an emotion they arouse or of an evaluation of them.
4. Nominal: Items may be grouped by a name that exists for them in the language.
5. Fiat: Items may be described as the same or alike without providing any further information.

While these modes are primarily based on verbal materials, it can be shown that when using ikonic representation of identical functional equivalence (a house), judgements of similarity are more extensive than "perceptible".

It is also evident that in order for this knowledge to have some predictive utility in forecasting the nature of environmental preferences, the individual's disposition to the environment, which may be personalogically meaningful, needs to be understood. Research in the theory of personality (4) suggests that people relate to the everyday physical environment in stable, characteristics ways just as they relate to themselves and to others according to enduring patterns.

A measurable dimension of environmental disposition can be described as sensation seeking(5) which has been linked to preferences for complex and novel stimuli (6). Research in complexity and novelty (7) showed that complex stimuli elicit the investigatory reflex more strongly than others, with conditions of diversity, complexity, novelty and ambiguity in a composition, more readily leading to arousal and attention. An individual's disposition to the environment may also include preference for novelty as well as the importance of the environment on his daily behavior. This research, then, is an attempt to describe classes of environmental descriptions and their association with designed and non-designed house types as well as personological relationships to residential preferences.

Research Design

It has been suggested by Brunswik (8), that the generality of statistical results, or representativeness may be more important than proper sampling of subjects. Representativity of the multiple-cue residential settings in this study was attempted by observations of house types in the coastal plains of North Carolina of towns less than 2,500 population. Eight major categories of house types were established accounting for 84.5% of the rural housing stock, while the remaining 15.5% explained all the variations of house types over .3% per category. From the eight categories of major house types, five categories are of houses generally classified as substandard or were constructed prior to 1930. Pictures J and F represent 18.8% of the house types while pictures A, D, & C, account for 1.3% or the two remaining categories. Picture H did not represent a substantial proportion of rural housing, but is increasingly more common in suburban development. The remaining six photographs are not representative of existing stock, but were carefully selected for their unique nature, relative cost and size to the more representative houses and availability in the study locale. Controls were established for lighting conditions, scale and the miminization of supplementary cues such as furniture toys, etc. The research design held constant observer groups, the media of presentation and the response formats but varied the environmental displays (ED).

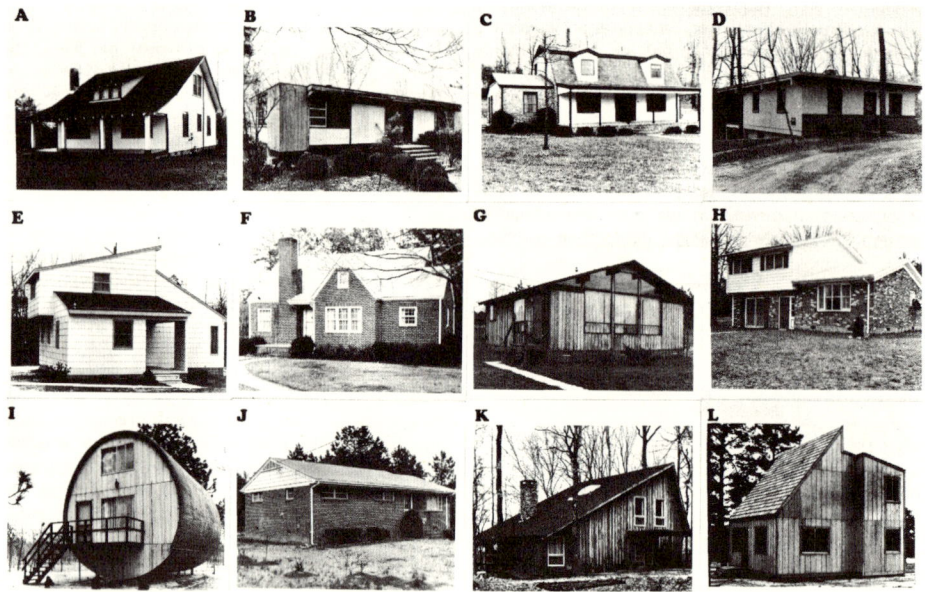

The sample of 153 respondents, consisting of eleventh and twelfth grade high school students were drawn from urban and rural settings. Place of residence was chosen to be a major variable in order to discern any differences in perception between urban and rural teenagers. In an interview, 87 urban and 66 rural teenagers were asked to respond to the following:

1. Sorting for similarities and ranking of all house types.
2. Questions about their present house type.
3. A set of twenty-eight true-false statements about their disposition to the environment.

Descriptive Attributes from Sort: Similarity Groupings

The teenagers were asked to select from an array of 12 pictures, those which go together and then describe how they are alike. The descriptions, which numbered over 60, were then classified into seven major categories: house form, house detail, house quality, house context, house style, size and socio-economic status. It was also evident that these descriptions are value laden and have positive or negative connotations which varied with the respondents and their disposition toward the house types. House style, quality, and socio-economic status were assumed to be qualitative descriptions of social status.

This classificatory scheme, then, was utilized in the context analysis of the teenagers' similarity structure in an attempt to describe both social and visual criteria suggested by the inferential cues in the house pictures.

Figure 2: CLASSIFICATION OF DESCRIPTIVE ATTRIBUTES (Similarity Criteria: SORT)

House Form	Social Status	House Details	House Style	House Context	House Size	House Quality
Flat	Low	Brick	Modern	Landscaped	One story	Comfortable
Gable	Middle	Combined materials	Ordinary	Pines	Two story	Spacious
Gambril	High	Wood	Traditional	Shrubbery	Small	
Round	Status	Painted	Old-fashioned	City	Large	
Angular	Subdivision	Dark trim	Plain	Isolated		
Symmetrical	Vacation	Vertical	Barn	Bleak		
High Pitch		Horizontal	Rustic			
Low Pitch		Dormers	Country			
Rectilinear		Columns	Commercial			
Horizontal		Porch	Unique			
		Chimney	Strange			
		Outside stairs	Unusual			
		Windows	Split level			
		Shutters	Simple			
			Complex			

The free sort test was intended to elicit the most salient categories of similarities. From the data, it can be observed that the highest degree of similarity was found between house types K and L, J and F, and B and D (Table 1). The major factors accounting for the similarity between house types K and C were described as shape, roof line, and surface materials, while for houses F and J, the same descriptions were used with the addition of plain and ordinary. Pictures B and D were described as "alike" primarily because of roof line and shape. A relatively larger number of urban than rural children grouped house A and E with K. The rural children primarily associated house A with C. House types A, E, and K were similar to the urban group because of roof line, shape, materials, country look, and low-income image. The rural residents described houses A and C as similar because they are both "old fashioned", "two stories", have "long front porches", "columns in the front", and similar "front windows". They also suggested a strong association between G and I because of the "large windows" and use of "wood", while the urban group linked houses G and K because of the use of "wood" and that they looked "modern"(Table 1).

The paired similarities from the free sort suggest some differences between urban and rural perceptions. The most important distinction between the groups were with house types B and G, H and J, and houses E and H. The urban sample associated types BG and HJ. The associations were based upon "house style" (BG) and "house details"

(EH). The rural sample associated pictures B and D on the basis of "house form" as the dominant factor. There was also a significant variation (.01) between the rural group's association with houses E and H on the basis of "form" as well as pictures E and L related by "style".

The urban students showed some ability in associating groups of three pictures to a greater degree than those residing in rural settings. The greatest differences occurred in the grouping between pictures BG and K made by the urban youths, as well as the triad groups HJF and GKL.

It is evident from the data in Table 2 that physical criteria (form, details, and size) were the predominant criteria used as a basis for grouping houses. Clearly, however, a majority of the teenagers did select nonphysical criteria such as house style at least once to group houses. Explicit social status criteria were recorded by less than half of each group and physical setting had very little influence in recording similarity. The majority of teenagers in the sample who used the smallest number of different criteria (three or less) for sorting selected "house details" most often. In fact, over half of this group of students also selected "house form" criteria. There was no substantial difference in place of residence along this dimension. Though "house details" and "form" are obvious criteria for similarity, those students who generated four or more criteria tended to describe similarities on a social status basis. Thus, the major first sort criterion appeared to be at a "perceptible" level where comparisons were made between physical characteristics of the house types. The second criterion used for similarity could be described as "affective", where groupings were made along quality dimensions. The third criterion in the free sort was described as "nominal" where similarities were distinguished by their name only.

House Triads

In the triad task, the students were presented with three house photographs and asked to select the two which were alike and explain their choice. The twelve house types were grouped into four sets of triads. The triads were intended to test the relative strength of the different kinds of categories. Therefore, the content of the triad reasons were classified into the same categories as the free sort responses.

Triad No. 1 consisted of house types A, E, and H where most of the urban students (50.0%) associated A and E, the rural linked E and H (56.3%) (Table 4). The predominant reason for the similarities between A and E was described as "house details" (48.9%) while 83.3% of the rural sample equated E and H because of "house form". Triad No. 2 consisted of houses BD and G where both urban and rural students described B and D alike, 65.1% and 81.2% respectively, with the major defining characteristics described as "house form". Triad No. 3 included types CF and J where an almost equal amount from both urban and rural groups (65.1% and 62.5% respectively) (Table 4) associated EDF and J. The predominant explanation for the similarities was described by both groups as "house details". Triad No. 4, house types IL and K received a homogeneous response where both groups linked L and K where the predominant criterion was "house form", while the respondents used many descriptions for distinguishing similarities and difference, almost 50% of all categories for both groups are explained as "house form".

The predominant criteria for triad association for both groups was "house form" with "house detail" almost as salient. This further substantiates the similarity from the free sort task described previously. The major variation between groups was the category of "house style" where the urban students described similarities from a quality basis more so than the rural group (Table 5). "Social status" was less frequently cited in the triad task than in the free sort.

Table 1: PAIRED SIMILARITIES FROM THE FREE SORT

House Pairs	Urban (N=87)		Rural (N=66)		Total (N=153)	Chi-Square Level of Significance
	No.	%	No.	%		
AC	36	41.4	38	57.6	74	
AF	15	17.2	12	18.2	27	
AE	26	29.8	11	16.7	37	
AK	27	31.0	13	19.7	30	
BD	56	64.4	53	80.3	109	
BG	26	29.8	6	9.1	32	.01
BI	14	16.1	5	7.6	19	
CH	40	46.0	24	36.4	64	
CF	24	27.6	11	16.7	35	
DJ	12	13.8	7	10.6	19	
EL	34	39.1	32	48.5	66	
EH	12	13.8	25	37.9	37	.01
EK	21	24.1	15	22.8	36	
FJ	54	62.1	50	75.8	104	
FH	37	42.5	19	28.8	56	
GL	39	44.8	22	33.3	61	
GK	52	59.8	26	39.4	78	
GI	38	43.7	38	57.6	76	
HJ	39	44.8	16	24.2	55	.05
IL	43	49.4	28	42.4	71	
KL	59	67.8	43	65.2	102	
House Triads						
GKL	36	41.4	18	27.3		
HFJ	30	34.5	13	19.7		
BGK	17	19.5	1	1.5		.01

Percents represent the proportion in each sub-group judging the pictures to be similar.

When comparing the total number of criteria used by both groups, it was evident that the vast majority of both groups cited three or less which varied considerably from the free sort. The single most important criterion for those selecting three or less was "house details". None of the students chose more than five criteria for describing similarities.

Table 2: HOUSE SORT CRITERIA BY PLACE OF RESIDENCE

Residence	House Form No.	%	Social Status No.	%	House Details No.	%	House Style No.	%	House Context No.	%	House Size No.	%	House Quality No.	%
Urban (N=87)	62	71.3	42	48.2	73	83.8	70	80.4	14	16.1	39	40.2	11	12.7
Rural (N=66)	54	81.8	25	37.8	63	95.5	59	89.3	10	15.1	37	56.0	4	6.0

The percents in each cell represent the proportion of the sub-group who used this kind of similarity as criterion for sorting at least once.

Table 3: NUMBER OF DIFFERENT HOUSE SORT CRITERIA BY PLACE OF RESIDENCE

	Total Number of Criteria					
	0-3		4-5		6 or more	
Residence	No.	%	No.	%	No.	%
Urban	37	42.5	44	50.5	6	7.0
Rural	22	33.3	42	63.6	1	3.1

Table 4: PAIRED ASSOCIATIONS FROM THE TRIAD GROUPS BY RESIDENCE

	House Types	No.	%	House Types	No.	%	House Types	No.	%	House Types	No.	%
Urban	AE	43	50.0	BD	56	65.1	CF	26	30.2	IL	25	29.1
	AH	12	14.0	BG	28	32.6	FJ	56	65.1	IK	3	3.5
	EH	31	36.0	DG	2	2.3	CJ	4	4.7	LK	58	67.4
Total		86	100.0		86	100.0		86	100.0		86	100.0
Rural	AE	23	35.9	BD	52	81.2	CF	24	37.5	IL	8	12.5
	AH	5	7.8	BD	8	12.5	FJ	40	62.5	IK	0	0.0
	EH	36	56.3	DG	4	6.3	CJ	0	0.0	LK	56	87.5
Total		64	100.0		64	100.0		64	100.0		64	100.0

Table 5: HOUSE TRIAD CRITERIA BY PLACE OF RESIDENCE

Residence	House Form No.	%	Social Status No.	%	House Details No.	%	House Style No.	%	House Context No.	%	House Size No.	%	House Quality No.	%
Urban (N-87)	69	79.1	20	22.9	67	77.0	67	77.0	10	11.5	31	35.6	4	4.6
Rural (N-66)	58	87.8	8	12.1	53	80.3	34	51.5	5	7.9	35	53.0	0	.0

The percents in each cell represent the portion of the sub-group who used this kind of similarity as criterion for sorting at least once.

House Preferences

The students were asked which of the house types they preferred and why. When comparing the preference from both groups, it can be clearly seen that for the rural students house type H is preferred by a majority (59.1%) while for the urban group, less than one-half cumulatively prefer houses C and H. There is also a greater variety of preferences articulated by the urban sample. The major reasons given by the rural group for their preference of house H was its "bigness" and that it was a "split-level" type of house. The urban students also described their preferences for H as "big" and "split-level".

When the house pictures were collapsed into designed and non-designed categories there appears to be further distinctions made distinguishing rural and urban students. While 56.5% of the urban group preferred non-designed houses, 80.3% of the rural students preferred non-designed houses. The predominant reasons given by urban group for their preference for designed houses was "modern", the "setting" and "different" while the rural group described their preferences for non-designed houses as "comfortable", "big", and "split-level".

It is also evident from the data that most of the positive responses from the urban residents describing their preferences were "modern", "style", "setting", and "comfortable looking". House type B had the highest consistency of response, all indicating "modern" while some suggested "the setting". While house type K had the highest number of respondents indicating a preference for "the setting", house H had the greatest variety of descriptions, the most prominent of which was "split-level". The negative responses, suggesting least preference, were "ugly", "common", "shape", and "old fashioned". House type A was disliked most because it was described as "old fashioned", and in some instances "ugly". Houses E and L were disliked primarily for their "shape" while J was least preferred because it was "common". House type K was ranked low by many because it was "run down", while some liked its "rustic" appearance.

The rural respondents described the positive characteristics of what they preferred as "big", "comfortable", "split-level" and "beautiful". House H scored highest in each of the categories, with house C being preferred because it is "beautiful" and for its "style". The predominant negative characteristics perceived by the rural group was "shape", "ugly", "old fashioned", and "style", clearly more similar to the urban group. House K was disliked most because it was described as "old-fashioned" while houses I and L were disliked because of their shape. It is also important that pictures I and L are relatively unique in the area and are not readily associated with particular social class characteristics. It appeared from these findings that there is greater agreement between groups about what is disliked, both in the descriptions and in actual house types than what is preferred. When comparing the reasons for similarity groupings of the houses and the reasons for preferences, the major difference appears to be in the frequency of use of the "affective" descriptions such as style and status (Table 6) compared to former groupings on the basis of "perceptibility". It is clear that social quality indicators were used primarily to explain preferences and are far more meaningful in making social evaluations than the sorting tasks.

House Preference and Present House Type

In an attempt to explain certain preferences, an analysis was made of the student's present house type and his aspirations in order to detect any relationship. A comparison was made between the predominant first preference, house type H, which accounted for a total of 38.0% (58) of the total sample. From this group only 8.6% currently reside in a similar house type, while 37.8% selecting picture H as a first preference, currently reside in a house similar to type C. It is also important to note that 56.0% of those students residing in J preferred a house like H. The second largest group preferring house H resided in a house similar to A (17.3%). This group represents 48.0% of all those living in a home similar to A (Table 8). These findings suggest a pattern adhered to by one-half of the sample which is clearly of a hierarchical nature.

Table 6: PREFERRED HOUSE TYPE AND CRITERIA FOR CHOICE BY RESIDENCE GROUPS

Residence	House Form		Social Status		House Detail		House Style		House Context		House Size		House Quality	
	No.	%	No.	%	No.	%	No.	%	No.	%	No.	%	No.	%
Urban (N=77)	1	.9	1	.9	7	9.9	45	58.4	6	7.8	6	7.8	11	14.3
Rural (N=57)	1	1.8	2	3.4	3	5.3	29	50.9	3	5.3	4	7.0	15	26.3

Table 7: HOUSE PREFERENCE (LIKE VERY MUCH) BY RESIDENCE GROUPS

Status Rank	House Type	Urban	Rural
3	A	6.9	6.0
2	B	32.2	15.2
1	C	54.0	59.0
3	D	6.9	7.6
2	E	5.7	6.0
2	F	33.3	51.5
1	G	40.2	24.2
1	H	66.7	94.0
3	I	32.2	27.2
3	J	18.4	48.5
1	K	48.2	16.7
2	L	20.6	6.0

The proportions represent the number of people in each cell who had selected the house type for the "like very much" category. Status ranks were developed by judges.

Since the visual displays are characteristic of the region, it is not unlikely to find both urban and rural respondents residing in similar house types. From the entire sample, approximately 9.2% (14) lived in a house similar to type H. From that group, five students selected house H as their first choice. Most of the respondents lived in housing similar to types F and J, yet 56.4% of those living in J selected H as their first choice as well as 47.6% of those living in house type A compared to 22.6% of those living in houses similar to F.

Only 5.1% of those preferring house H to all other displays presently reside in a designed house. Over one-third (38.6%) of the total sample selected H as their first choice compared to 14.6% who selected K, the highest preferences of the designed houses. From this group, 13.6% reported living in dwellings similar to B and L, which are both examples of designed houses.

When the twelve house types were aggregated into classifications of "designed" and "non-designed", interesting differences between groups emerged. Most of the rural students (80.3%) selected non-designed houses as their first choice compared to 43.5% from the urban group who selected "designed houses" (Table 9). Similarly, there was almost no difference between designed and non-designed houses as the last choice for the urban sample (50.6% and 49.4% respectively) compared to 68.8% of the rural respondents who rated designed houses as their last choice from the twelve house photographs.

Environmental Disposition and House Preference

A series of true-false statements related to the students' disposition toward the physical environment were asked of the respondents. The statements included aspects of "exploratory behavior", "sensation seeking," and a general disposition towards "new and unique environments". The factor analytic method was employed to group responses into definable sets. The four categories were described as "modernism", "stimulus seeking", and "territoriality" as shown in Table 11.

Factor 1's attributes are predominantly descriptive of "modernism". Since the twelve pictures were equally divided between designed or modern and traditional house types, it was assumed that there would be some relationship between house preference and Factor 1 scores. Therefore, the nine attributes with the highest factor loading (.40 or greater) were selected for scoring. Since the responses to each of the nine questions had a "true" answer, each was designated with a "1" while the "false" answers were given a "0" score. The respondents' answers were then summed, yielding a range of scores from 0-9. The individual scores were aggregated into an index of low (0-3), medium (4-6), and high (7-9). From the 150 respondents, 25.3% were low, 44.0% medium, and 30.3% in the high index.

Comparisons were then made between each of the house pictures (A-L) and the "modernism" index. When the sample group was disaggregated into rural and urban, the differences were evident. From the data, it was evident that 26.9% of those in the high index and selecting house H were from urban areas compared to 60% of the rural students who selected H. In fact, 42.3% of the urban sample in the high index selected "modern" house types as their first choice compared to 10% of those from the rural areas, while 30.0% of each group scored in the high Factor 1 category (Table 11).

When comparing true-false responses to the question, "I prefer buildings that are new and completely different from all other buildings", and first preference of the house types, particularly when the designed and non-designed houses are aggregated, it was found that 48.2% of the urban students replying "true", ranked designed houses first while 33.3% of those replying false preferred designed houses. (Table 12) The rural respondents replying "true" to the same statement accounted for 25.7% of the designed houses while those replying "false" were 11.1%. When comparing preferences for "new and different" with the selection of designed house types, it appeared that a majority (69.4%) of the urban students who replied "true" selected designed types. The proportion of rural students was similar (76.9%)

The response "true" to the question, "I would like to tell my residence apart from all others", compared with preferences for designed houses indicated 44.7% of the urban students were in this category compared to 31.3% who indicated "false". Respondents from the rural group indicated that 25.7% responding "true" preferred designed houses. (Table 13) The question, "I like modern architectural styles better than most of the older styles", was more ambiguous than the previous and, as a result, the distinctions were not as evident. It was clear that of the 56.9% urban youths responding "true", who preferred non-designed houses, 26.4% preferred house H. The rural respondents replying "true", yet selecting houses, were 18.9%. (Table 14) House type H accounted for 62.2% of the "true" responses to this statement and had been classified as non-designed house. It is apparent that "modern architectural style" was an ambiguous concept and varied with the respondents' experience, historical perspective, and level of perceptual awareness.

While professional architects have agreed that pictures E, H, and J, particularly, are traditional house types, non-professionals would be responding from a different frame of reference. House H represents the type presently being constructed and therefore associated with "modern", but the reference is temporal rather than stylistic.

Table 8: HOUSE PREFERENCE H and RESIDENCE TYPE

	House You Live In		Total No. Living in House Type		Rural		Urban	
	No.	%	No.	%	No.	%	No.	%
A	10	17.3	21	13.7	13	19.7	8	9.2
B	0	0.0	2	1.3	0	0.0	2	2.3
C	4	6.9	14	9.2	4	6.1	10	11.5
D	0	0.0	1	.6	1	1.5	0	0.0
E	1	1.7	1	.6	1	1.5	0	0.0
F	7	12.1	31	20.3	6	9.1	25	28.7
G	2	3.4	4	2.6	3	4.5	1	1.2
H	5	8.6	14	9.2	4	6.1	10	11.5
I	0	0.0	0	0.0	0	0.0	0	0.0
J	22	37.8	39	25.5	24	36.4	15	17.2
K	0	0.0	0	0.0	0	0.0	0	0.0
L	0	0.0	1	.6	0	0.0	1	1.2
NONE	7	12.1	25	16.4	10	15.1	15	17.2
TOTAL	58	100.0	153	100.0	66	100.0	87	100.0

One of the more interesting and general conclusions is that house preference has its roots embedded in social and cultural values, rather than in physical or perceptible factors, per se. This is best described by the strong similarities the respondents observed between house types E and H, particularly those from rural areas and the location of E and H in the rank ordering scales. Again, type H received the first preference from both residence groups, with a particularly higher proportion of responses from the rural students. House type E received a consistently lower preference from both groups, yet the predominant reason advanced for the strong preference to H was two story or split level, a feature that is equally apparent in house E. It appears then, that there are intervening variables such as perceived cost and status implications that account for preferential judgements.

The other interesting aspects of this study is that one-half of those youths living in houses similar to type A and J preferred house H to any other alternative, which clearly suggest a hierarchy of preference, or aspiration from either old housing or inexpensive new housing (particularly for low-moderate income) to a preference for clearly middle-income status represented by house type H. Since the associations and connotations of house E are different than that of H (though it does display the vernacular forms of the region and represents a contemporary "designed" house) it was not regarded as "modern" by the respondents, particularly those who indicated their strong preference for "modern styles". It is evident that "modern" is a value laden concept with different meanings to different individuals. To the design professions, the classification of house types may be more sensitive than to the everyday user. "Modern" to an architect represents a current philosophy as well as a form which exemplifies the ideology, where the user may describe "modern" as what is being constructed on his street or in his neighborhood, a temporal frame of reference.

Table 9: HOUSE TYPES COMPARED TO RANK ORDER BY RESIDENCE GROUPS

	Urban				Rural			
	First Choice		Last Choice		First Choice		Last Choice	
	No.	%	No.	%	No.	%	No.	%
Designed	37	43.5	44	50.6	13	19.7	44	68.8
Non-Designed	48	56.5	43	49.4	53	80.3	20	31.2
Total	85	100.0	87	100.0	66	100.0	64	100.0

Table 10: FACTOR LOADINGS

	Factor Loading
Factor 1 - Modernism	
Modern architectural styles	-0.50
Living in a modern high-rise apartment	-0.54
Wild color variations	-0.49
People adapt rapidly to new physical surroundings	-0.43
Physical environment to be responsive to my wishes and intentions	-0.48
Relaxed in most modern buildings	-0.45
Modern, planned community	-0.69
Buildings that are new and completely different from other buildings	-0.43
Convenience of movable walls in my residence	-0.58
Factor 2 - Humanism	
Little affected by physical surroundings	0.43
Live in a small town where I know everyone by name	0.58
Modern buildings seem cold and austere	0.65
Buildings made of metal and glass human nature	0.56
Modern furniture always looks uncomfortable	0.52
Modern buildings lack place for people to meet	0.67
Factor 3 - Stimulus Seeking Exploratory Behavior	
Explore unfamiliar places	-0.75
Adventuresome person	-0.72
Visiting unfamiliar places	-0.74
Factor 4 - Territoriality	
Physical surroundings often play an important role in my dreams	0.67
My self-identity is a function of my physical surroundings	0.48

Table 11: FACTOR "MODERNISM" AND HOUSE PREFERENCE BY RESIDENCE GROUPS

House Type	Urban Low No.	Urban Low %	Urban Medium No.	Urban Medium %	Urban High No.	Urban High %	Rural Low No.	Rural Low %	Rural Medium No.	Rural Medium %	Rural High No.	Rural High %
Designed	9	36.0	14	41.0	11	42.3	4	30.8	5	15.6	2	10.0
Non-Designed	16	64.0	20	59.0	15	57.7	9	69.2	27	84.4	18	90.0
Total	25	100.0	34	100.0	26	100.0	13	100.0	32	100.0	20	100.0

Table 12: I WOULD PREFER BUILDINGS THAT ARE NEW AND DIFFERENT FROM ALL OTHER BUILDINGS COMPARED TO FIRST PREFERENCE

First Preference	Urban (N=84) True No.	Urban (N=84) True %	Urban (N=84) False No.	Urban (N=84) False %	Rural (N=66) True No.	Rural (N=66) True %	Rural (N=66) False No.	Rural (N=66) False %
Designed Houses	25	48.2	11	33.3	10	25.7	3	11.1
Non-Designed Houses	26	51.8	22	66.7	29	74.3	24	88.9
Total	51	100.0	33	100.0	39	100.0	27	100.0

Table 13: I WOULD LIKE TO TELL MY RESIDENCE APART FROM ALL OTHERS COMPARED TO FIRST PREFERENCE

First Preference	Urban (N=84) True No.	Urban (N=84) True %	Urban (N=84) False No.	Urban (N=84) False %	Rural (N=66) True No.	Rural (N=66) True %	Rural (N=66) False No.	Rural (N=66) False %
Designed Houses	30	44.7	5	31.3	13	23.2	0	0.0
Non-Designed Houses	37	55.3	11	68.7	43	76.8	10	100.0
Total	64	100.0	16	100.0	56	100.0	10	100.0

Table 14: I LIKE THE MODERN ARCHITECTURAL STYLES BETTER THAN MOST OF THE OLDER STYLES COMPARED TO FIRST PREFERENCE

First Preference	Urban (N=84) True No.	Urban (N=84) True %	Urban (N=84) False No.	Urban (N=84) False %	Rural (N=66) True No.	Rural (N=66) True %	Rural (N=66) False No.	Rural (N=66) False %
Designed Houses	25	43.1	12	44.4	10	18.9	3	23.0
Non-Designed Houses	33	56.9	15	55.6	43	81.1	10	77.0
Total	58	100.0	27	100.0	53	100.0	13	100.0

This suggests that there is some evidence to reinforce the notation that the physical environment transmits certain clues about the social system. These clues are imbedded in individual backgrounds and experiences which may vary according to the patterns and styles of residential development. It certainly refutes arguments of normative perceptions or economic or regional homogeniety. Future research into environmental cues, particularly residential environments, will require tighter controls of the visual material in order to discern the extent to which categorical descriptions differentiate judgments of similarity or preference. Though representivity of object types was attempted, it is important that future research utilizing environmental displays incorporate more rigorous representative measures.

2. VISUAL ATTRIBUTES OF ENVIRONMENTS

NOTES

1. Kelly, G., The Psychology of Personal Constructs, New York, Norton, 1955.

2. Steinitz, Victoria, How Children Categorize Social Stimuli, HEW, National Center for Educational Research, 1971.

3. Olver, R., and Hornsby, I., "On Equivalence" in Bruner, J.S., Olver, R.R., Greenfield, P.M., et.al., Studies in Cognitive Growth, New York, Wiley, 1966.

4. McKechnie, G., "Measuring Environmental Disposition with the Environmental Response Inventory," in Eastman, C., and Archea, J., EDRA 2, Pittsburgh, 1970.

5. Zuckerman, M., Kolin, E.A., Price, I., Zoob, I., "Development of a Sensation Seeking Scale", Journal of Consulting Psychology, 28, pages 477-482, 1964.

6. Markman, R., Sensation Seeking and Environmental Preference (Unpublished dissertation), North Carolina State University, 1971.

7. Berlyne, D.E., Conflict and Arousal and Curiosity, McGraw Hill, New York, 1960.

8. Brunswik, E., Perception and the Representative Design of Psychological Experiments, University of California, 1956.

DYNAMICS OF PREFERENCE FOR VISUAL
ATTRIBUTES OF HOUSING ENVIRONMENTS

Peter G. Flachsbart, Ph.D.

University of Southern California

George L. Peterson, Ph.D.

Northwestern University

Abstract

    This paper reports results of recent empirical research which examines similarities and differences in the preferences of poor blacks and affluent whites for visual attributes of housing. Our interpretation of the results suggests a dynamic theory of preference which has implications for environmental design and research methodologies. Our results and assumptions suggest that an individual's preferences are influenced by his sensitivity to that of which he is deprived, thereby causing him to inflate its value, but only if (1) he perceives no other more pressing needs as he perceives his priority of needs, and (2) he has not adapted to that of which he is deprived to the extent that he no longer perceives a deprivation.

Problem Definition

    In recent years evidence has been brought forth to show that man's use of space is conditioned by his culture (1). As a consequence of this evidence a goodly number of environmental design researchers have undertaken empirical research to determine the differing environmental perceptions and preferences of various sub-cultural groups of society. Given the multiplicity of environments and the pluralistic nature of society, this area of environmental design research has been quite fertile. Researchers, in their hast to exploit this virgin field have often neglected the development of both theory and research design. The results of this neglect have been seemingly contradictory findings.

    The intent of this paper is to renew the development of theory and research design in the field of environmental design research. Ostensibly, the study reported in this paper is similar to other studies which attempt to determine the differing environmental preferences of various societal groups. Specifically, this paper reports results of recent empirical research which examined similarities and differences in the preferences of poor blacks and affluent whites for attributes of the external appearance of housing. Our interpretation of the results suggests a dynamic theory of preference which has implications for the design of research instruments which seek to measure human preferences.

The intent of this paper to renew the development of theory and research design cannot be overemphasized, as the danger exists that the findings reported herein may be misapplied to the population groups under consideration. The study should not be interpreted as an attempt to predict the consumer behavior of poor blacks and affluent whites in the housing market, because the sample holds little external validity for these two population groups. Our avoidance of this question, however, should not reflect on its importance.

Likewise, we have avoided the question of whether consumers are motivated more by the internal or external appearance of the housing environment. Again, our avoidance of this question should not reflect on its importance. As our intent is to propose a theory of preference, such theory if valid should transcend specific environments. Therefore, the selection of a particular environment for study, in this case the external appearance of housing, is arbitrary.

Hypotheses

Space does not permit a full discussion of the review of the literature concerned with human perception and preferences for housing environments. Studies which were reviewed to generate the working hypotheses listed below included those done by Wilson (2), Lamanna (3), Michelson (4), Rainwater (5), Hall (1,6), Peterson (7), Lansing and Marans (8), and Sanoff (9). To some extent the chosen wording of the hypotheses is arbitrary, reflecting some apparent contradictions in the literature. Each of the three working hypotheses has one or more sub-hypotheses (attributes). Each hypothesis is stated in terms of the preferences of poor blacks and affluent whites:

1. Both poor blacks and affluent whites are attracted to housing environments that possess:
    a. spaciousness;
    b. newness;
    c. naturalness;
    d. vegetation;
    e. cleanliness; and
    f. physical stability.
2. More poor blacks than affluent whites are attracted to housing environments that possess small scale.
3. More affluent whites than poor blacks are attracted to housing environments that possess:
    a. visual variety; and
    b. uniqueness.

Method of Inquiry

The selection of subjects proceeded as follows. First, names of undergraduate students with home residences within the United States were selected randomly from the 1969-70 student directories of two midwestern schools of higher learning (10). Second, selected students were contacted and asked to participate in an interview designed to determine their housing preferences. For this interview 151 students participated representing an 85% response rate.

Two interviews were later disqualified because the ethnic background of the subjects was Oriental and an additional three were not counted in the analysis because the subjects' ages were above the age range (17 through 22) which was arbitrarily established to hold stage in the life cycle relatively constant. The remaining 146 subjects were then classified according to racial background and self-reported annual family income. As the hypotheses were expressed in terms of the preferences of poor blacks and affluent whites, an annual family income of $10,000 was arbitrarily established as the dividing line between "poor" and "affluent". Consequently, 28 relatively affluent black subjects (annual family incomes $>$ $10,000) and 11 relatively poor white subjects (annual family incomes $\leq$ $10,000) were discarded.

Of the remaining 107 subjects whose interview responses were analyzed, 47 were relatively poor blacks whose families fell into the median income category of $3,000 to $6,000 and 60 were relatively affluent whites whose families fell into the median income category of $20,001 to $25,000. The then current home residences of these subjects existed in 26 mostly midwestern and eastern states. The large majority of poor black subjects (72.3%) lived within large cities; the affluent white subjects were almost equally distributed among residences in suburbs, large cities, and small towns. For this sample little external validity is claimed for larger societal populations of poor blacks and affluent whites.

Each subject was privately shown twelve pairs of projected, colored slides which pictured different types of housing environments. These photographs had been taken in various places in the eastern half of the U.S. and Canada. Each pair depicted two housing alternatives for which the subject indicated his preference as to which was more attractive. In an open-ended format the subject also stated his reasons for preference. These reasons were then content analyzed so that each sub-hypothesis could be tested quantitatively. To facilitate communication, to eliminate potential cues to the subject, and to randomize interviewer bias, black subjects were interviewed by one of six black interviewers and white subjects were interviewed by one of seven white interviewers who themselves were students. One subject was interviewed per session to enhance the uniqueness of his responses.

To some extent the method of photographic pairing as employed in this study was unique and unorthodox. Although the rationale for this method is beyond the scope of this paper, a full explanation does appear elsewhere (11). Nevertheless, some aspects of the method are important to stress. First, several pairs of photographs were used to test each sub-hypothesis to check the internal validity of the results. Second, in pairing photographs, each of the two photographs was selected to maximize the difference between them for one or two attributes and minimize the difference between them for all other attributes. This effect was created by pairing photographs in which a given attribute existed in its polar opposites in each photograph. Third, although attributes had to be imputed to photographs to enhance the probability that each sub-hypothesis could be adequately tested, final analysis of photographic preference and reasons for preference was based on subject-reported responses, i.e., attributes perceived in the photographs by the subjects. Fourth, the analytical statistic used to test each sub-hypothesis was the Chi Square (two-by-two contingency) statistic, with $p \geq 0.10$ as the criterion for non-confirmation of a sub-hypothesis.

Results

Race and Income as Predictors

Some justification needs to be given to the combination of race and income as joint predictors of preference for housing environments. To make this justification an analysis of photographic preference was undertaken in the following manner. First, the hypothesis was tested that for race held constant, photographic preference was independent of income. Second, the hypothesis was tested that for income held constant, photographic preference was independent of race. Finally, the hypothesis was tested that photographic preference was independent of a special combination of race and income, namely, whether a subject was poor and black or affluent and white. Each photographic pair, of which there were twelve, provided an occasion to test these three hypotheses, which should not be confused with the working hypotheses listed earlier.

Space does not permit a full discussion of the data used to test these hypotheses, although one should note that the sample of 146 subjects was used in this particular analysis, as that sample included both poor and affluent blacks and whites. Briefly, the results indicated that for the twelve photographic pairs the first hypothesis could not be rejected at all and the second hypothesis could be rejected only once. In contrast the third hypothesis had to be rejected for 50% or six photographic pairs and just barely missed being rejected at the 0.10 level of significance for two additional pairs. In our opinion these results lent justification to the combination of race and income as joint predictors of preference for housing environments.

On Confirmation of Working Hypotheses

Regarding the first hypothesis the assumption was made that a rural environment approximates a natural environment. Given this assumption most sub-hypotheses of all three general hypotheses tended towards confirmation. Since each sub-hypothesis was tested several times to check the internal validity of the results, the percent of the number of photographs for which the hypothesis was confirmed can be reported as shown in Table 1.

On Non-Confirmation of Working Hypotheses

Those photographs for which sub-hypotheses of the three working hypotheses were not confirmed are discussed subsequently.

Spaciousness and Visual Variety

Regarding spaciousness and visual variety the hypotheses were that both poor blacks and affluent whites are attracted to spaciousness and that more affluent

TABLE 1

Attribute	No. of Photographs Hypothesis Confirmed	No. of Photographs Hypothesis Tested	% Photographs Hypothesis Confirmed
1st Hypothesis:			
spaciousness	12	16	75.0
newness	5	6	83.3
ruralism	5	5	100.0
vegetation	16	23	69.5
cleanliness	1	3	33.3
physical stability	2	2	100.0
2nd Hypothesis:			
small scale	3	4	75.0
3rd Hypothesis:			
visual variety	5	7	71.4
uniqueness	2	3	66.7

whites are attracted to visual variety. These two attributes were contrasted in photographic pair #8. For this pair the results indicated that a significantly greater proportion of poor blacks (72.3%) preferred the right-hand photograph and a significantly greater proportion of affluent whites (70.0%) preferred the left-hand photograph. In giving reasons for these choices, significantly more poor blacks mentioned that the greater spaciousness of the right photograph, despite its greater uniformity, influenced their choice. For signigicantly more affluent whites the issue of visual variety v. visual uniformity was more important. They mentioned that the greater visual variety of the left photograph, despite its greater crowdedness, influenced their choice. These results are summarized in Table 2. The (L) and the (R) indicate the attribute applies to the left or right photograph, respectively. The (+) and (-) indicate that the attribute was valued either positively or negatively, respectively. "PB" refers to poor blacks; "AW" refers to affluent whites.

TABLE 2

Reasons	%PB	%AW	X^2	Level of Significance
crowdedness (L-) v. spaciousness (R+)	61.7	36.7	5.656	0.05
variety (L+) v. uniformity (R-)	49.0	78.3	8.810	0.005

Vegetation

For the attribute of vegetation the hypothesis was that both groups are attracted to vegetation. However, there were several photographs for which significantly greater proportions of affluent whites liked the vegetation. These photographs and the results for vegetation are given in Table 3.

2. VISUAL ATTRIBUTES OF ENVIRONMENTS

TABLE 3

Photograph	Reason	%PB	%AW	x^2	Level of Significance
1-Right	vegetation (+)	29.8	56.7	6.650	0.01
2-Left	vegetation (+)	12.8	35.0	5.778	0.05
5-Left	vegetation (+)	12.8	31.7	4.256	0.05
8-Left	vegetation (+)	12.8	48.3	13.575	0.001
9-Left	vegetation (+)	42.6	68.3	6.134	0.05
10-Left	vegetation (+)	34.1	55.0	3.857	0.05

Two observations can be made upon examination of these photographs. First, all of them have no front lawns as such. In place of lawns the vegetation consists of vines, shrubs, trees, and flowers. Second, the vegetation appears to be more natural or less cultivated. This latter observation may be more apparent than real, however, as poor blacks and affluent whites who were sensitive to vegetation merely stated that they liked the vegetation in the form of trees, lawns, shrubs, etc. They seldom commented on the degree to which the vegetation was cultivated. Hence, if they were reacting to the degree to which the vegetation was cultivated, then most of them were unable to verbalize their reaction.

Newness and Uniqueness

For the attributes of newness and uniqueness the hypotheses were that both poor blacks and affluent whites are attracted to newness and that more affluent whites are attracted to uniqueness. These two attributes were contrasted in photographic pair #6. The results for this pair indicated that a significantly greater proportion of poor blacks preferred the right-hand photograph whereas affluent whites were split as a group in their preference for the two photographs. Some affluent whites preferred the housing in the right photograph for its newness and other affluent whites preferred the housing in the left photograph for its uniqueness. Most poor blacks did not perceive the house of the left photograph to be unique. They said it was old and not desirable. These results are summarized in Table 4.

TABLE 4

Reasons	%PB	%AW	x^2	Level of Significance
oldness (L-) v. newness (R+)	53.2	30.0	4.972	0.05
oldness (L+) v. newness (R-)	12.8	33.3	4.994	0.05
uniqueness (L+) v. commonness (R-)	19.1	51.7	10.557	0.005

Scale

The hypothesis for scale was that more poor blacks than affluent whites are attracted to small scale structures. This hypothesis was not confirmed for photographic pair #7. For this pair similar proportions of each group stated positively that they preferred the small scale structure (2 stories) to the large scale structure (20 stories). This result, which is summarized in Table 5, suggests that both groups are attracted to small scale structures.

TABLE 5

Reason	%PB	%AW	x^2	Level of Significance
smallness (L+) v. largeness (R-)	49.0	40.0	0.530	---

Yet, close examination of the other photographic pairs for which scale was an issue reveals that each group may be attracted to a different small scale. The results for photographic pairs #6 and #11 are especially illustrative. For these photographs the houses which significantly more poor blacks preferred were single-story structures. The houses which significantly more affluent whites preferred were two- and three-story structures. These results, which are summarized in Table 6, suggest that some poor blacks are attracted to a scale which is smaller by a story or two than the scale preferred by some affluent whites.

TABLE 6

Photo Pair	Reasons	%PB	% AW	x^2	Level of Significance
6	3 story (L-) v. 1 story (R+)	34.1	11.7	6.550	0.05
6	3 story (L+) v. 1 story (R-)	19.1	30.0	1.121	-----
11	1 story (L+) v. 2 story (R-)	40.4	23.4	2.853	0.10
11	1 story (L-) v. 2 story (R+)	8.5	25.0	3.843	0.05

Cleanliness

The attribute of cleanliness arose for several photographic pairs. The results (Table 7) suggest at least a tendency for greater proportions of affluent whites to prefer housing environments which they perceived to be cleaner.

TABLE 7

Photo Pair	Reasons	%PB	%AW	x^2	Level of Significance
1	unclean (L-) v. clean (R+)	14.9	30.0	2.568	-----
2	clean (L+) v. unclean (R-)	6.4	21.7	3.714	0.10
4	unclean (L-) v. clean (R+)	27.7	48.3	3.897	0.05

Interpretation of Results

Our interpretation of the results is summarized in the following proposition, which we advance towards a theory of preference. An individual's preferences are influenced by his sensitivity to that of which he is deprived, thereby causing him to inflate its value, but only if (1) he perceives no other more pressing needs as he perceives his priority of needs, and (2) he has not adapted to that of which he is deprived to the extent that he no longer perceives a deprivation.

This proposition is advanced to explain the following results. First, more poor blacks tended to value positively housing environments which possessed greater spaciousness, more cultivated vegetation, newness, and smaller scale structures, usually single-story houses. We speculate that these attributes are uncommon to inner-city ghettoes where most of our poor black subjects had their home residence. Second, more affluent whites tended to value positively housing environments which possessed greater variety, uniqueness, less cultivated vegetation, two- and three-story houses, and cleanliness. Except for cleanliness we speculate that these attributes are uncommon to tract housing in suburbia where some of our affluent white subjects had their home residence. Regarding cleanliness, our poor black subjects may have adapted to a lower level of cleanliness in their environment to the extent that they no longer perceived a deprivation. Our affluent white subjects may have become accustomed to a higher level of cleanliness which they perceive as their current standard. Third, while both subject groups valued positively housing environments which possessed spaciousness and newness, attributes common to suburbia, only the affluent white subjects preferred greater variety to spaciousness and uniqueness to newness. We speculate that these results indicate a priority of needs such that the need for spaciousness and newness may have been satisfied in suburbia for affluent whites, but that the need for variety and uniqueness apparently has not been fulfilled.

If valid this proposition suggests that the variable of environmental experience may be more than just an explanatory predictor of environmental preferences. Indeed, the measurement of environmental experience may serve to accomplish at least two additional purposes. One, the measurement of environmental experience may serve as a baseline from which to determine the direction and magnitude of preferred environmental changes. Too often researchers measure preferences only. Two, measurement of environmental experience appears to be necessary to account for the effects of "environmental deprivation" and "environmental adaptation." Environmental deprivation appears to inflate preferences whereas environmental adaptation appears to deflate preferences.

Notes

1. Hall, Edward T., The Hidden Dimension (Garden City: Doubleday, 1966).

2. Wilson, Robert L., "Livability of the City: Attitudes and Urban Development," Urban Growth Dynamics, Eds. F. Stuart Chapin, Jr., and Shirley F. Weiss (1962), pp. 359-399.

3. Lamanna, Richard A., "Value Consensus Among Urban Residents," Journal of the American Institute of Planners, XXX, iv (November, 1964), pp. 317-321.

4. Michelson, William, "An Empirical Analysis of Urban Environmental Preferences," Journal of the American Institute of Planners, XXXII, vi (November, 1966), pp. 355-360.

5. Rainwater, Lee, "Fear and the House-as-Haven in the Lower Class," Journal of the American Institute of Planners, XXXII, i (January, 1966), pp. 23-31.

6. Hall, Edward T., "Environmental Communication," paper presented at the 135th meeting of the American Association for the Advancement of Science (1968).

7. Peterson, George L., "A Model of Preference: Quantitative Analysis of the Perception of the Visual Appearance of Residential Neighborhoods," Journal of Regional Science (1967), pp. 19-31.

8. Lansing, John B. and Marans, Robert W., "Evaluation of Neighborhood Quality." Journal of the American Institute of Planners, XXXV, iii (May, 1969), pp. 195-199.

9. Sanoff, Henry, "House Form and Preference," EDRA TWO, Proceedings of the 2nd Annual Environmental Design Research Association Conference, (Pittsburgh: Carnegie-Mellon University, 1970), pp. 334-339.

10. The two schools were Northwestern University and the National College of Education, both located in Evanston, Illinois.

11. Flachsbart, Peter G., "Race and Income as Bases for Preference for Attributes of Alternative Housing Environments," Department of Civil Engineering, Northwestern University, Evanston, Illinois. June, 1971 (unpublished Ph.D. dissertation).

THREE HUMAN RESPONSES TO THE NATURAL AND MAN-MADE PHYSICAL ENVIRONMENT

Chairman: Harvey M. Choldin, Dept. of Sociology, Univ. of Illinois, Urbana

Panelists: Edith E. Flynn, Dept. of Sociology, Univ. of Illinois, Urbana
Lars Lerup, Dept. of Architecture, Univ. of California, Berkeley
Nancy Marshall, Michigan State Univ., East Lansing
Erik Svenson, National Bureau of Standards, Washington, D. C.

Authors: Arn Henderson and Richard D. Bauman, "A Housing Analysis of the Cheyenne and Arapaho of Oklahoma"
Omer Akin, "Contextual Fittingness of Everyday Activity Encounters"
Robert R. Hahn, "Behavioral Evaluation of a Juvenile Treatment Center: Case Study of a Planning Methodology"
Robert B. Bechtel, "Types of Cities: The Subnational Urban Environment and Some Design Implications"
George L. Peterson, "Psychology and Environmental Management for Outdoor Recreation"

HUMAN RESPONSES TO THE NATURAL AND MAN-MADE PHYSICAL ENVIRONMENT: INTRODUCTION 3.0

Harvey M. Choldin, Session Chairman

Department of Sociology
University of Illinois, Urbana

The papers in this section indicate the varieties of social science research on design and policy problems. Of the five papers, four report original research and one is an essay based upon available literature. Three of the research papers are directly addressed to designers and two are addressed to policymakers concerned with housing and the built environment. The variety of topics and research methods offers an opportunity to consider the ways in which particular social science research methods may be applied to particular kinds of design problems.

Designers should look at selected points in the research papers, lest they be discouraged by technical points on research methods and theory-building at some distance from design processes. While this introduction is addressed primarily to designers, social scientists are directed to the same points of interest and they may focus upon methodological points in the papers.

One paper of particular note for design and general social science is about the housing of Cheyenne and Arapaho Indians in Oklahoma. The researchers, Henderson and Bauman, have demonstrated very special needs of these people, which are not met in the design of standard U.S. dwellings. These Indians have preferences regarding living near kinfolk which differ from those of persons in other U.S. social systems. They also prefer to live in the country rather than in towns of any type, unlike most other U.S. groups. Their housing needs relate differently to the course of the family life cycle because grandparents sometimes raise grandchildren. Such points are spelled out in the paper. The researchers also instruct us not to be deceived by the type of housing the Indians currently occupy; their preferences are radically different from their current housing types.

Henderson and Bauman have accomplished what most of us advocate but do not or cannot do. They have demonstrated _specific_ differences in social systems and cultures and the precise differences these make for housing design. Incidentally, the designer will profit most from the descriptive parts and conclusions in the report, while other researchers will want also to examine the extended discussion of research method.

A second paper which may be of use to designers is by Akin and includes a report of research on activities, preferences and movements of university students. It is entitled, "Contextual Fittingness of Everyday Activity Encounters". This research helps refine the idea of privacy to a list of which categories of roles are accepted and rejected by individuals in selected situations. The students reported to Akin on their activities in various locations. They also indicated what categories of persons were acceptable and not acceptable at those locations when they were performing those activities. For example, most reported that while they were studying, they would accept the presence of their boyfriends, their girlfriends, their roommate, or their teacher, and they would reject the presence of another friend, a stranger, or a dormitory staff member. Akin further examines particular settings, such as dorm rooms and libraries to see whether the acceptances and rejections vary by setting. Finally, he can combine setting, activity, and "appropriate" role-pairs. Akin then discusses various kinds of facilities in these terms and suggests reasons why some are heavily used and others are under-utilized.

This mode of analysis, matching place, activity, and role-relationship, will lend itself to systematic application to a variety of other settings. The concrete parts of this paper are the most valuable; study the research itself, beginning with the part on "Sample Population."

The third paper in the set which is directed at design is by Hahn, "Behavioral Evaluation of a Juvenile Treatment Center..." Hahn provides some relief for the often-attacked designer by attacking building users instead, the users being staff members in a juvenile treatment center. He examines the relationship between the stated treatment goals of staff members in an agency and the settings in which the treatment takes place. He also studies the extent to which the staff members know the locations of activities which support and detract from treatment goals. He reports the extent of their knowledge and ignorance in this regard.

He observed activities of the juveniles and the staff members in detail in selected small settings of the residential facility. He reported the frequencies of positive and detrimental activities in the settings. Finally, Hahn gives an interpretation of the data, relating design features of the settings to the behaviors which occurred in them.

The detailed observational method employed by Hahn has been used in other settings and is applicable to additional ones. His particular question, "Given a set of organizational goals, where do the positive and negative behaviors occur?", deserves additional trails. The additional question he asks, "What is it about this setting which promotes this kind of behavior?", is also important.

Bechtel, in his paper entitled, "Types of Cities...," takes data about a large number of U.S. cities and examines the ways in which they produce categories of cities with similar characteristics. He emphasizes one major substantive point arising from the statistical analysis, that there is one set of cities which have in common a set of characteristics related to old housing. He then considers the policy implication of this finding and he advocates a higher priority on dealing with the complex of problems arising out of the aging of housing inventory, including problems of financing, replacement, maintenance, and related processes. This paper is mistitled in that it does not present a typology of cities.

Designers should note that the studies of the Oklahoma Indians, the university students, and the boys in the treatment centers all employed different social science methods. The first was ethnography plus survey, the second survey alone, and the third observation. Each research product could have a different kind of use in the design process. Confronted with a client or user population with an unfamiliar culture and social structure, the designer could try to learn what Henderson and Bauman learned about the Cheyenne and the Arapaho. What is it about their way of life that pertains to the buildings they need and want? Hahn's observational method with its goal emphasis is applicable to a different kind of question. Recalling the client's and the designer's goals after a building is in use, they could ask, "Is the facility working the way we wanted it to?". Finally, Akin's methods are applicable to a third kind of question. Considering a category of buildings to be done in the long run, one might want to discover some combination of activities, role-combinations, and settings. Who does what where in an X (hospital, school, etc.)? Who do they prefer to be with and without when they do it? Akin's method is appropriate to this set of questions.

In each case there is no reliable short-cut to the answers. The three studies are detailed and represent many weeks and hours (at least) of data-collection and analysis. The designer may lament the size of the product relative to its cost in time but such is the nature of the scientific process.

A HOUSING ANALYSIS OF THE CHEYENNE AND ARAPAHO OF OKLAHOMA   3.1

Arn Henderson   Richard D. Bauman

College of Environmental Design   Civil Engineering Department
University of Oklahoma   University of Hawaii

Abstract

A major objective of this study was to generate a quantity of empirical data on family, housing, and environmental characteristics among members of the Cheyenne and Arapaho Tribes in western Oklahoma. A second objective was concerned with the determination of any other modes of physical community organization and house complexes that might be more satisfactory to Indians and to test a specific hypothesis: that a "micro-community" composed of several houses belonging to one extended family with certain shared facilities may be a desirable alternative to either present living patterns or existing federally sponsored housing programs. Depth interviews were conducted with 51 of 177 Indian families living in Blaine County, Oklahoma. Computerized statistical techniques, factor analysis and hierarchical clustering schemes were used to analyze the interview data. The focus of this paper is on the analysis procedures and results, particularly housing preferences.

Introduction

The American Indian today is the fastest growing minority group in the United States. On a national basis, the Indian birth rate is double the national average. Oklahoma, a former reservation area, has the highest Indian population of all the states. The 1970 United States Census reported the Indian population of Oklahoma to have been a little over 100,000. Low educational attainment seriously handicaps many Indians. In 1970, some 6.0 per cent of Oklahoma's Indians, 25 years and older, were reported as having had no schooling. At the same time, 58.9 per cent of Oklahoma's Indians had not gone beyond the eighth grade. Indians belong to the least affluent and least employed minority in the United States. A study conducted in Blaine County (western Oklahoma) in April 1966, by the Oklahoma Employment Security Commission revealed that 71.2 per cent of the 269 Indians who were interviewed were unemployed.(1)

Living and Housing Conditions

At present there is a serious nationwide lack of adequate housing in Indian communities. Although the federal government has pledged to solve housing shortages which now exist, Indian needs will continue to be great due to rapid population growth, formation of new families, population shifts, and deterioration of existing housing.

In rural western Oklahoma areas, most Indians live in single family dwellings which are old and in poor repair. The occupants are often unable to adequately heat, cool, lock, maintain or repair their homes. Also, there are no practicable means for thorough cleaning of these houses. Their structural parts are often in a weakened condition. Roofing and siding materials generally suffer from extreme weathering, dry rot, and termites. In addition to creating health hazards, these problems cause the houses to leak, and to be hot in summer and cold in winter.

Furnishings generally are minimal, and in poor condition. There is a chronic shortage of storage areas. Many older houses have no closets, and closet space is usually inadequate in newer structures, most of which are typical low-cost development houses or shells. As a result, family possessions and personal belongings often are stacked along a wall, in the yard, or in non-functional automobiles that sometimes dot the premises.

Lack of storage space also aggravates the effects of overcrowding, which is one of the most pervasive aspects of Indian housing, and is caused in part by high unemployment and low family incomes. Kitchens are often small and poorly equipped, making food preparation a time-consuming task for a large family. Dining facilities generally are too small to allow an entire family to eat together.

The common solution is to serve meals in shifts, or for some members of the family to eat in another room. Under these conditions, activity conflicts are inevitable. The living room is the only large family space, yet is usually far too small to accommodate simultaneously the varied activities of persons of different ages. Privacy for individuals is severely restricted. Children have no study or play space. Sometimes beds and pallets must be put in the living room. Sick persons and crying children may be kept in the same room with the rest of the family. All of these problems contribute to periodic high, intra-family tensions.

Sleeping arrangements are even more precarious. Beds and pallets usually are so concentrated in bedrooms that there is little room to move about. A living room couch may be used for a bed at night, or beds may be set up in the living room and used during the day as couches.

Disposal of garbage and other trash is often complicated and difficult. Most rural families have private dumps in a nearby ravine or creek bed. In small western Oklahoma towns, trash collection is sporadic or may not be provided by the community. Residents are usually expected to pay for trash collection or else haul their own trash to the city dump. As many low-income Indian families can neither pay for trash collection nor haul their trash on a regular schedule, it often is allowed to accumulate in unhealthy quantities in yards and other nearby areas.

In summary, our studies show that existing rural Indian housing in former reservation areas is not only inadequate to offer shelter from the environment, but also is instrumental in promoting conditions condusive to low health standards. Further, they lack esthetic amenities; while these are intangible factors, they contribute to the debilitating and demoralizing effects of housing conditions on Indians.[2]

Federal Housing Programs

The federal housing program most commonly utilized in western Oklahoma for low-income Indians is a modified Turnkey III - Mutual Help Home Ownership program. This is a program developed particularly for areas such as Oklahoma where Indians do not live on reservations. Under this program, each tribe must create a housing authority or be under the jurisdiction of another tribal housing authority in order to participate. The developer chosen by the housing authority provides all services from construction drawings to final finishing work, while the Indian participant earns equity credit by contributing one and one fourth acres of land upon which to build the house. In addition, the participant contributes approximately 400 hours of his own labor related to the construction of the house and receives a "sweat equity" credit.

One of the major disadvantages of the Turnkey III program for Indians appears to be the lack of cultural relevance in the design of housing. Oklahoma Indians, like many other low-income housing users, have had little opportunity to express their desires and needs in housing and must therefore rely on the architect's, designer's or bureaucrat's knowledge of the behavior of the occupants and uses of space. Since the decision-makers most often apply white middle-class values and ideas, the houses do not provide for the unique characteristics of a minority group, such as Indians living in rural areas of western Oklahoma.

The distinction between manifest needs and latent needs in low-income housing design has recently been expressed by John Zeisel[3]. The manifest needs include the basic parameters of a housing environment which will satisfy the fundamental physiological and psychological needs and provide protection against contagion and accidents. The manifest needs are obvious: a kitchen is a place to cook, a bedroom is a place to sleep. The manifest needs or functions of most spaces and their behaviors are superficially similar for different societal and cultural groups in the United States.

The latent need or function of a space can be both an expression of a user's life style and a reflex of his culture. It is through identification of the latent function of a space that cultural differences between groups can be taken into account in designed physical environments. Among Indian groups in western Oklahoma, the housing environment may function to accommodate a number of activities that are unique. Indian residences may function to fulfill religious needs. Examples of this would be peyote meetings of the Native American Church or prayer meetings associated with funeral services for an individual. Handgames, an important social activity, are frequently held in Indian homes, in sheds, and outbuildings on the property, or under arbors in warm weather. Some rural Indian homes also have dance grounds for summer pow-wows. The dance grounds often have an open shed or pavilion-like structure covering the dance area. Other activities frequently occuring in Indian homes are social functions such as practice sings, bingo games, box suppers, fund raising activities, and informal business meetings. Many Indian homes frequently function as a work space for beadwork or making shawls.

Use of outdoor space is a factor that must be acknowledged in any analysis of Indian housing. In warm weather, Indians who have facilities available spend much of their time out-of-doors, engaged in work, recreation, sleeping, and some cooking. Families may construct and maintain arbors in and around which these outdoor activities take place.

Overcrowding, one of the most conspicuous aspects of Indian housing, in Oklahoma or elsewhere, can at the simplest level, be called a result of insufficient amount of housing. A unique contributing factor to overcrowding among Indians is the involutional characteristic of extended family organization. As there is much loyalty and interaction between members of extended families, many Indians prefer to live near relatives. The kinship structure of Indians then has the effect of magnifying the degree of attachment of blood relatives and persons related only by marriage. As a result, persons whom most white Americans would consider first, second, or third cousins are called brothers and sisters, and great-aunts and uncles are called grandmothers and grandfathers. Such relatives are never refused food and almost never denied a place to stay, even when such services involve severe hardship for the host household. As long as present economic and social conditions exist, extended families apparently will continue to be important in Indian social structure.

Objectives

The objectives of this study were twofold. The first objective was to generate a quantity of empirical data on family, housing, and environmental characteristics among members of the Cheyenne and Arapaho tribes in western Oklahoma. A second objective was concerned with the determination of any other modes of physical community organization and house complexes that might be more satisfactory to Indians and to test a specific hypothesis: that a "micro-community" composed of several houses belonging to one extended family with certain shared facilities might be a desirable alternative to either present living patterns or existing federally sponsored housing programs.

Methodology

In 1867 the treaty of Medicine Lodge established a reservation area for Cheyenne and Arapaho Indians in what is now a nine county area of western Oklahoma. In 1891, the reservation was dissolved and all Cheyenne and Arapaho Indian received individual allotments of 160 acres each. In this study, Blaine County, Oklahoma was selected as the study area since it is representative of the population group as a whole, i.e., both Cheyenne and Arapaho Indians reside in the county.

Because of the complexities involved in family structures of many Oklahoma Indians, i.e., extended families and concommittant overcrowding, it was thought that a structure of the Indian population by households would provide an accurate and descriptive data base. Basic data required was the name and address, sex, age, tribe and blood quantum of the head of household. In addition, the same information was required for all other persons residing in the household as well as their relationship to the head of the household. An initial objective thus became one of locating Indian households and accurately delineating the family composition.

The 1970 Census does not provide the detailed information regarding identification, specific residence location, tribe, or blood quantum required for our analysis. Indian Health Service, which is responsible for the health and medical care for Indians, maintains clinics in two of the communities in our study area. Personnel in the clinics provided information regarding household location and family composition. The data was then carefully cross-checked and verified by several Indian people who were acquainted with Indian families in the study area. Finally, tribal rolls of the Cheyenne and Arapaho Tribes of Oklahoma dated October 31, 1967, were also checked as needed to verify data on age and blood quantum.

The sample design presented a major problem because attitudes toward a subject such as housing are not readily amenable to quantification in the statistical terms necessary for sample size determination. Since little research on Indian housing has been conducted, the quantifiable factors that Indian families related to housing were unknown. It was determined that, of the data available, differences in age and degree of Indian blood would be most likely to result in differences in attitudes toward housing. It was also our opinion that difference in tribal affiliation might reflect differences in housing attitudes. For this reason, a geographical cluster scheme was developed to reflect tribal affiliation. Finally, it was our opinion that families living in urban areas would have different responses toward housing than would families living in rural areas.

The study area, Blaine County, was divided into three zones each containing both rural and urban areas. We selected six simple random samples such that we estimated the mean age or degree of Indian blood in the relevant population with a 95 per cent level of confidence that the estimate would not vary from the proxy variables of age or degree of Indian blood by more than twenty per cent.

An interview schedule was developed that would permit analysis of basic parameters of family, housing and environmental characteristics and provide data on the acceptability of the "micro-community" hypothesis. Variables were grouped in the following categories: socio-economic characteristics, residence history, housing and environmental characteristics, exterior space characteristics, social networks, and micro-community participation. A total of 51 of 177 Indian families in the county were interviewed.

Analysis of Housing and Family Characteristics

The results of the questionnaire provided us with answers to 129 independent questions from each family suitable for statistical analysis. The problem of categorizing and analyzing approximately 6500 independent responses is best approached by means of computerized statistical techniques. But unfortunately many of the conventional statistical techniques applied to the results of a large questionnaire leave one with a bewildering array of data and many unanswered questions. Thus the question of how to approach the analysis was a major problem.

After considerable investigation, it was determined that the best approach for the analysis would be to use multivariate statistical techniques to isolate a limited number of "indicator variables", which we defined as those variables whose distribution patterns are representative of groups of the original variables. Theoretically, if we could find a few variables that were representative of all the variables, our analysis procedure would be greatly simplified because the statistical behavior of the indicator variables would be typical of the behavior of the variables.

Two separate procedures were used to select the indicator variables: Factor

Analysis and Hierarchical Clustering. Factor analysis uses the matrix of correlations among the original variables to define variable groupings. Each group of variables is associated with a new hypothetical variable called a factor. Each factor generally has strong enough linear correlations with the original variables that the factor represents much of the information contained in all the variables of the group. The loading of a variable upon a factor also serves as an indicator of the relationship of the variable to the factor. Our calculations were made with a factor analytic model known as principal component analysis and orthogonal varimax rotation. Computations were performed with an IBM 360 computer using Fortran IV factor analysis routines.

Hierarchical clustering is a relatively new technique that has been developed to present an investigator with a view of his data from a perspective different from that afforded by the factor analysis.[4] Whereas the factor analysis is based upon manipulations of the matrix of correlations among the variables, analysis by hierarchical clustering is based upon a process dependent on the rank order of the data.

To perform this analysis the data is first arrayed in a similarity matrix. Then the matrix elements are considered twice: first to find the smallest nonzero matrix element, and second to find the minimum diameter of two matrix elements. The routine uses this technique to search for compact clusters of variables in such a manner that when the search is completed, each cluster of variables has the smallest possible over-all diameter. The calculations were made with a model known as a hierarchical clustering scheme programmed for an IBM 360 computer.

After determining the indicator variables, hierarchical clustering was again used in the analysis in a slightly different manner to examine similarities and differences in the responses of the families to the indicator variables. In this manner, families were grouped into clusters based upon the way that each family had responded to the questionnaire.

Identifying the Indicator Variables

The first step in the identification process was to search the data for redundant or unusable information and for mutually exclusive information. Although the questionnaire was carefully designed and pre-tested, the statistical search when the fieldwork was completed indicated that some useless data existed in the file. Furthermore, Indians viewed some of the questions in a different light than we did. For example, data concerned with access to transportation such as travel time, trip frequency, and even annual mileage were very rarely answered even though we knew prior to the study that some families drive as much as 50,000 miles per year. Data concerning regularity of employment were mutually exclusive.

When the questions containing redundant or unusable data, mutually exclusive data, and data which only a few families answered were eliminated, 50 variables remained. These significant variables are shown in Table 1.

With the reduction of the data to 50 significant variables completed, the search for the Indicator Variables began in earnest. The response of each family to each of the 50 significant variables was considered as an observation. These response

3. HUMAN RESPONSES TO THE ENVIRONMENT / 117

Significant Variables Table 1

Data Concerned with the Socio-Economic Condition of the Family:

Tribe of the Male Head of Household	Age of Male Head of Household	Tribe of wife or Female head of Household	Age of wife or female head of Household
Total family size	Number in family 18 and older	Number of close relatives	Education level of Male Head of Household
Education level of wife or female head of household	Family Employment Characteristics	Regularity of Employment	Total family income
Percent of Indian blood of Male Head of Household	Percent Indian blood of Female Household head	Which relatives does the family have contact with	

Data Concerned with Potential Design Characteristics:

Where would you prefer to live	Does the family stay outside much	Does family want to live closer to relatives	Would family prefer to live closer to: Brother or sister Mother or father Children, cousins Nephews, nieces, Grandchildren
Would family share: Outdoor space; Workroom; well; Septic tank; Laundry; storage; Storm cellar; Hot water			

Data Concerned with Physical Condition of House:

Age of Dwelling Unit	Length of time family has lived in dwelling unit	Home ownership	House located on trust land
House have running water	Water available on site	Is toilet indoors or outdoors	Toilet waste disposal
Type housing	Square feet of space in the house	Type construction	Approximate age of the house

Data Concerned with Locational Environment & Access to Transportation:

Dwelling unit location	What type street is closest to the property	Car Ownership	Frequency of Hospital and clinic visits
Frequency of participation in handgames and powwows	Are the roads near the home ever impassable		

data were studied by both the principal component factor analysis model and the hierarchical clustering model. The purpose of using both models was to learn as much as possible about the data. As we learned more about the data, the intent was to identify that select group of variables that were descriptive of all the different responses of all the questions -- the Indicator Variables. All extra variables had to be eliminated because, if left in the family clustering analysis, they would have biased the analysis.

Selecting the minimum number of indicator variables began with application of the factor analysis model. A rotated matrix utilizing ten factors resulted in identification of 16 variables that had high factor loadings. The hierarchical model is not as suitable for identifying indicator variables as the factor model but it was used to confirm the suitability of the indicator variable selection. The 16 indicator variables selected by the factor analysis model and confirmed by the hierarchical clustering model are shown in Table 2. These variables were identified as the minimum number of variables whose distribution patterns were representative of the groups of original variables.

Table 2-The Indicator Variables

Tribe of Male Head of Household	Age of Male Head of Household	Tribe of Wife or Female Head of Household
Age of Wife or Female Head of Household	Regularity of Employment	Total Family Income
Home Ownership	Car Ownership	Family Size
Number of Close Relatives	House have Running Water	Dwelling Unit Location
Education Level of Male Head of Household	Education Level of Wife or Female Head of Household	Blood Quantum of Male & Female Household Heads
Square Feet of Space in House		

Clustering the Families

The process of clustering the families into groups with similar characteristics necessitated a look at the data from a different perspective than was used to select the indicator variables. During the search for the indicator variables, the responses of each family was considered to be an observation and each question was considered to be a variable. But, for the clustering analysis, each of the indicator variables was considered to be an observation and each of the families was a variable.

The hierarchical clustering model was used. This model is a particularly powerful tool for dividing individuals or groups of individuals into clusters depending upon the manner in which the individuals respond to a series of questions. It is powerful because the model starts with the total sample, on the first step splits the sample into two groups, on the second step splits the sample into three groups and continues the splitting process until the sample is reduced to singular data. Following this routine, the 51 families were grouped into 3 clusters: Cluster A was composed of 7 families; Cluster B was composed of 19 families; and Cluster C was composed of 25 families.

Descriptive Information

Some interesting phenomena began to appear when the average socio-economic characteristics of all 51 families. The most significant differences are shown in Table 3.

We observed distinct differences in the families of each cluster. The husband and wife in the families in Cluster B tend to be younger, better educated, have fewer close relatives, earn more and own more cars. However, they tend to live in houses that are both smaller and in poorer condition than those in Cluster A or C; and the houses tend to be rented whereas those in the other clusters are usually owned. The families in Cluster A are all full bloods, they have the least education, earn the least, have the smallest families, have lived in one home the longest, and live in the largest houses -- but their houses are the oldest and most are located on roads that are occasionally impassable. The families in Cluster C tend to have the oldest members and the largest number of family members 18 or older, their education level, income and car ownership are similar to Cluster A; but the number of households without a male household head or without a female head is the greatest of any cluster. Nearly all houses of all families in all clusters had running water and indoor toilets.

In spite of the significant differences in the socio-economic characteristics of the clusters of families, there are great similarities in the families' responses to the questions concerned with potential design characteristics. Nearly all families from all clusters indicated that they spent a great deal of time outside. They nearly all exhibited a preference for living near their brother or sister or children instead of their mother or father, cousins, nieces, or nephews. Nearly all families stated that they would be willing to share outdoor space, workrooms, wells, septic tanks, laundry facilities, and storm cellars but not hot water heaters. So it appears that regardless of the significant socio-economic differences of the families in the clusters, there is no significant difference in their preference for the potential design aspects investigated here.

The majority of all families would prefer to live in the country instead of a small town or city; of the families in Clusters A and B, the ratio is 2 to 1 in favor of the country. But in Cluster C, the ratio was approximately 50% in favor of the country and 50% in favor of a small town. The greater number of families in Cluster C with no male household head is probably the reason for the lower preference for living in the country.

The families in Clusters A and C have from 10-15 close relatives while the families in Cluster B have fewer, from 5-10. In all cases, sisters, brothers, or cousins are the most frequent visitors. The families in Cluster B are younger and have higher incomes. Correspondingly, the unemployment rate of either heads of households or spouses is considerably lower for Cluster B than Clusters A and C. Cluster B has 37 per cent unemployed while Clusters A and C have 86 per cent and 80 per cent unemployed, respectively.

Approximately 82 per cent of all individuals interviewed lived as a child in Blaine County. Another 10 per cent lived in adjacent counties during their childhood. The other 8 per cent grew up in other parts of Oklahoma or elsewhere. There is no significant difference for childhood residence location for any of the three clusters.

Nearly two-thirds of all homes in Clusters B and C are located on trust land and 80 per cent of Cluster A homes are on trust land. Approximately 75 per cent of the families in Clusters A and C own their homes while only 40 % of the Cluster B families own homes. Since the families in Cluster B have a higher employment rate and are more likely to rent, this would infer greater mobility for this group. The poor condition of housing for Cluster B, as indicated by the rating of structural points in Table 10, is in all probability, a function of the inadequate base of rental housing available to the Indian population throughout the area.

The frequency of participation in hand games or pow-wows is low for all families - averaging about once a month. The families in Cluster C participate even less, many of the younger female household heads never go to this type of social event.

Most families live in single family homes. In Clusters A and B, the proportion is about 75% single family - 25% multi-family. In Cluster C, more than 95% of the families live in single family homes. The typical single family home is constructed with asbestos shingles and was built between 1929 and 1940.

Conclusions

Validity of Sampling Procedure

As indicated earlier in the report, sample selection was a major problem because it was not known what family characteristics were significantly related to housing characteristics. It was hypothesized that the prime determinants in sample selection were mean age of the husband and wife in each family and their degree of Indian blood. It was further hypothesized that differences in tribal affiliation and urban or rural housing location were secondary determinants.

Table 3-Comparable Statistics of Clustered Families

Variable	51 Family Average	Cluster A Average	Cluster B Average	Cluster C Average
Age of Male Head of Household	51.5	53.6	44.1	58.3
Age of Wife or Female Head of Household	44.5	52.7	37.8	47.4
Total Family Size	6.0	5.1	5.7	6.5
Number in Family 18 and Older	2.7	2.4	2.7	2.9
Age of Dwelling Unit	34.3	46.3	31.6	31.7
Length of Time Family has Lived in Dwelling Unit	7.4	16.7	5.8	5.9
Education Level of Male Head of Household	9.6	8.4	10.1	9.4
Education Level of Wife or Female Head of Household	9.9	8.7	10.5	9.8
Total Family Income Exclusive of Commodities	4339	1827	6804	2890
Number of Cars	1.0	0.7	1.7	0.6
Square Feet of Space in House	699	936	555	741
Number of Rooms in House	4.3	4.1	3.8	4.6
Structural Points	33.4	30.3	35.5	32.6
Families with No Male Household Head %	26	29	10	36
Families with no Wife or Female Head of Household	12	14	10	12

The analysis does indicate that the original sampling hypothesis, i.e., mean age and degree of Indian blood as primary determinants of housing characteristics, is valid. However dividing the county into zones proved to be unnecessary. Our analysis indicates that there is no significant difference in the housing characteristics of families from either Cheyenne or Arapaho tribes. The analysis also indicates that the degree of significance of the location of the dwelling unit depends on whether you are interested in learning about how Indian people actually live or in learning about how Indian people would like to live. There are significant differences in the housing environment of Indian families living in urban areas when compared to those living in rural areas. But the differences are only great enough to influence family clustering in the case of families which have no male household head. The factor which is important to planners and designers is that the Indian families throughout the study area appear to have the same ideas about how and where they would prefer to live regardless of where they are living now. So, if one wants to learn about existing conditions, dwelling unit location should be considered. If one wants to learn about housing improvements Indian families would like, present dwelling location is not important.

Micro-communities

A major objective of this study was to conduct investigations into the feasibility of devising new community alternatives. There are a number of possible ways in which aggregations of Indian families might be organized in space. Our data suggests that one such form of organization might be a "micro-community". Such micro-communities could provide new and alternative patterns of proximal relationships that would offer more amenities at less cost and take advantage of Indian preferences for living close to kinsmen and co-operating with them in certain activities. In its simplest form, a "micro-community" might consist of a group or cluster of houses in close proximity to one another; certain facilities would be shared, all belonging to one extended family.

One form of micro-community could consist of several adjacent privately owned properties in a geometric configuration that would provide a common point of congruency. Each individual property would accommodate a single family residence. At the point of congruency or locus, would be a community area. This community area could be designed to accommodate the following: 1) outdoor activities, 2) common space for group activities, 3) functional and environmental amenities necessary for the maintenance of the micro-community.

References

[1] Oklahoma Employment Security Commission, Indians in Oklahoma: Social and Economic Statistical Data, 1966.

[2] Henderson, Arn and Richard Bauman, Indian Environmental Studies - the Cheyenne and Arapaho Tribes of Oklahoma, Norman, Okla., December, 1972.

[3] Zeisel, John, "Fundamental Values in Planning with the Non-Paying Client", Architecture for Human Behavior, Philadelphia AIA, 1971.

[4] Johnson, S.C., "Hierarchical Clustering Schemes", Psychometrika, Sept. 1967.

CONTEXTUAL FITTINGNESS OF EVERYDAY ACTIVITY ENCOUNTERS   3.2

Omer Akin

College of Architecture
Virginia Polytechnic Institute and State University
Blacksburg, Virginia

Abstract

This is a pilot study for the implementation of a method of analysing activity patterns of users with respect to the designed physical environment. Empirical data in the form of daily-diaries and preferences of college students is used in identifying the functional properties of a university setting. Based on the interaction requirements of the participants of the activities accommodated, individual built forms are classified into two categories: (a) single-modal, and (b) multi-modal. In congruence with this, the results of the diaries of users indicated dysfunctions in the utilization of physical and human resources. The pragmatic use of the conventional built form concept is identified as the cause of the dysfunctions observed.

Introduction

Change is the cause as well as the consequence of human motivations for advancement. Implementation of change is a conscious action based on human deliberation as opposed to spontaneous evolution. To promote desirable man-environment relations man develops strategies and methods towards the understanding of the nature of human settings. Consequently, he implements change altering the relationships of the components of these settings. Environmental design emerges as one of the means for promoting these transformations. In order to accomplish this with success, design methods are developed to categorize the properties of man-environment systems. Operationally three basic organizational categories exist: design parameters, independent variables, and dependent variables. (1)

Design parameters include all measurable properties of the designed setting. Activities, conglomeration of activities, and their physical containers which are used as the basic component elements of the design process in any large scale design problem are in fact design parameters. Independent variables are those properties of the system which are not within the direct and indirect control of the designer. For most design problems the human behavioral properties are the typical independent variables. Dependent variables take values, by definition entirely dependent on the values of the independent variables and the design parameters. After the values of the behavioral and physical resources are determined the functional performance obtained from the setting yields the values for the dependent variables.

Physical Contexture of the Environment

The physical resources of the setting and the functions accommodated in them make the basic parameters of the campus of Virginia Polytechnic Institute and State University. The functions and corresponding physical resources of the campus can be conceptualized at two different levels: (a) the micro-scale "activity" unit and its physical counterpart the "place," e.g., for studying-study hall, for eating-kitchen, for sleeping-bedroom, etc., (b) the macro-scale "functional structure" and its physical counterpart the "built form," e.g., for organized education-academic building, for retail-shopping mall, etc. These can be generically defined in the following fashion:

1. Activities: These are the specific behavioral manifestations of man-environment relations constituting a form of purposeful transformation of matter or information (e.g., sleeping, eating, reading, etc.).

2. Places: These are physical entities meant to contain and accommodate "activities" and their environmental requirements (e.g., classroom, bedroom, laboratory, etc.).

3. Functional Structures: These are specific conglomerations of "activities" and their environmental requirements serving societal function, (e.g., educational, retail, residential, etc.).

4. Built forms: These are physical entities consisting of various "places" and meant to accommodate specific "functional structures" and their spatial requirements (e.g., academic buildings, libraries, apartments, dormitories, shopping mall, etc.).

Human Behavior as a Determinant of the Environment

The cultural development of human societies is constantly moving towards "...greater diversity of function, increasing capacity to cope with complexity and change, heightened ability to profit from experience..."(2) Environmental design emerges as a means for promoting these transformations. Consequently a fuller understanding of man's motivations and goals within the functional, spatial context must be achieved. Man's encounters in his everyday surroundings are determined intrinsically more or less in the same fashion as most other animals.(3) Through the evolutionary process man has inherited his biological and emotional motivations as well as expanded his ability to control the environment by means of his intellectual motivations.(4) "The study of the environmental process from the point of view of the particular participant..." the surveyed individual "...creates a situation dichotomized into participant, on the one hand and all other environmental components on the other."(5) Within this framework the individuals' interactions with the environment are based on three basic orders: physical, social and cultural. When the limits set by the physical environment at a particular time and place are considered the organization and conduct of human activities manifest the greatest variance and freedom of choice among all other components of the setting. Consequently, one of the most significant aspects of such interaction is the interrelation of individuals. The individual is motivated by his vital, emotional and intellectual capacities almost simultaneously. In reality, these three different levels of motivation are very difficult to separate from one another. In fact, most circumstantial conflicts between individuals' interactions are generated by the combination of these motivations.

The respective concepts of spatial organization: territoriality, status hierarchy and role,(6) form the contexture of the individual's attitudes towards other persons. The descriptive phraseology used in everyday language representing personal attributes(7) are loaded with connotations pertinent to these three levels of organization. The symbol "girlfriend" or "teacher" or "custodian" implies specific meanings to an individual due to his/her previous experiences and mental structures. These are charged with subjective dispositions depending on the intellectual, emotional and spatial contexture of the individual. For instance, a college student may refuse to interact with an university official regardless of the occasion due to emotional reasons, while another student may find his presence as a transgression only in his own "turf," i.e., in his dorm room with his girlfriend, and still another student may find his presence inappropriate only while in need of personal privacy, i.e., studying. Hence, the determinants of such behavior may be intrinsic or contextual depending on the interplay of the three organizational levels: territoriality, status and role. In this work the opinion of individuals about the appropriateness of others as described above, shall be referred to as "contextual fittingness" regardless of their intrinsic or contextual nature. Attributes of typical persons evaluated by the subjects for their contextual fittingness shall be referred to as "personal attributes".

Performance of the Environmental Contexture

The basic functions of the built environment are fourfold. These define the different aspects of the performance required from the utility of any designed environment intended for the accommodation of human related activities. The four functions can be defined as follows:(8)

1. Climate Modification: This is the means for the displacement of external climate and ecology inputs and modification of sensory inputs into the human organism.

2. Activity Modification: This is the means to achieve a containing effect locating, inhibiting, and/or determining human activities.

3. Cultural Modification: This is the means to achieve a "cultural disposition" effect on the symbolic, cognitive value sets of individuals.

4. Resource Modification: This is the means for an addition of value to raw materials, a maximization of scarce resources of material and man power.

Sample Population

Two questionnaires were administered among the undergraduate students at Virginia Polytechnic Institute and State University, Blacksburg, Virginia. The sample population consisted only of undergraduates due to the fact that graduate students are not admitted to the dormitories as a policy of the university administration. The first questionnaire was completed during the Spring of 1972. To insure the return of an adequate number of completed forms the questionnaire was handed and collected during regular class hours.(9) The instructors of the courses assisted by verbal reinforcement of the significance of the survey and by collecting the returned forms. This cooperation was obtained by letting the subjects fill the forms outside class hours.

The 18 different courses in which the questionnaire was administered were randomly selected from the time table of classes. The sample represented 28 different departments of the university and class sizes varying from 7 to 92 students. The forms were distributed to 546 students out of which 227 or 41.5% of the original number responded. These forms were examined and the incomplete ones eliminated, which brought down the total number of forms evaluated to 95.

The second questionnaire was administered to complement the results of the first questionnaire. In this respect it was given to a sample population of 115. In all three classes in which it was administered the procedures used in the first questionnaire were utilized without change. Out of the 110 forms handed out 46 were returned within the set time limits. Out of these 44 forms, or 40% of the initial sample population, were completed and used for evaluation.

The sample was representative in terms of students' years of study and collegiate divisons. However, the female-male ratio of the sample was two times as great, i.e., two-thirds instead of one-third. The greater percentage of females compared to males returning completed forms accounted for this, i.e., 25% and 14% respectively.(10) Consequently, the percentage of males was 58 as opposed to the 70% of the university's total population.

The sample population of the second questionnaire consisted of 56% males and 44% females, which was consistent with the first questionnaire sample, yet, by virtue of this, significantly different from those of the total university population. Due to the instruments used in the second questionnaire the degree of familiarity of the population with the particular environment and its contents was important. Therefore the classes included in the survey were selected from sophomore, junior and senior level courses only. The sample consisted of 85% juniors and seniors for the females and 80% sophomores and juniors for the males. At least 83% of the total population had spent more than two years at the setting and 39% of all males and 53% of all females were currently living on campus. During the previous school year these percentages were 82 and 64 respectively for the same population.

Instruments

The first questionnaire was designed to deal with the patterns of the use of the campus and its facilities. The diary-form has been found to be the most appropriate means to accomplish this with relative ease and accuracy for large sample sizes. Response categories of the subjects, i.e., age, sex, year of study, department of study, and location of residence; and the diary of activities, locations, and corresponding times obtained from each form were coded for computerized analysis. A computer program in the PL/1 programming language was written to translate this information into tables of: (a) cumulative duration of use of physical resources, (b) cumulative duration of participation in daily activities, (c) numbers of trips generated by various physical resources and (d) interface matrix of daily activities and corresponding physical resources. These provided ample quantifiable information pertaining to the patterns of usage manifested by the subjects, in the particular setting.

In the second questionnaire the contextual fittingness of different "personal attributes" was measured to provide a means of classifying different activities. The criterion of measurement used was the subjects' tendency to reject different "personal attributes" from the context of a particular activity. In order to insure the use of those activities which the subjects could relate to intrinsically, rather than impartially, as the specific context of the rejection, the students were asked to identify the activities which they found personally fulfilling. Then they were asked to identify the most desirable setting(s) available to conduct these activities. The list of activities and locations provided in the questionnaire forms for these purposes were extracted from the results of the first questionnaire and they implied no restrictions on the choice of either the activities or the locations. Finally, the subjects were asked to simulate in their minds the selected activity settings and evaluate "personal attributes" listed in the questionnaire forms in terms of their "contextual fittingness" to these settings. Accordingly, the subjects "accepted" those "personal attributes" they found "fitting" by marking a "check" and "rejected" the others by marking an "X" in the spaces provided in the questionnaire blanks across the phrases representing each "personal attribute" listed.

A second PL/1 program was written to translate these results into: (a) activity-location matrices, and (b) histograms of reject-frequencies for each "personal attribute." Chi-square tests were carried out to measure the factor of correlation of "rejections" for each activity. This information was utilized to classify activities into three modes: solitary activities, social involuntary activities and social voluntary activities. Based on the distribution of these modes of activities to the corresponding locations, physical resources of the campus were classified into two: single-modal and multi-modal.

Results and Discussions

The results of the first questionnaire which dealt with the measurable properties of the physical setting and its organization were used in evaluating the design parameters and the dependent variables. The results of the second questionnaire, on the other hand, were utilized towards the qualification of the independent variables by means of the behavorial dispositions observed. The evaluation of the values of the dependent variables in the light of these dispositions yielded the criteria for appraising the design parameters.

<u>Activity and Place Structures</u>

The results of the first questionnaire verify that the built forms accommodated a varied number of activities at one time and were equipped with the appropriate place types, e.g., dorm room, classroom, reading hall, etc. However, no evidence was found for establishing a one to one correspondence between either activities and places or functional structures and built forms. It was also determined that the various built forms were physically separated from one another with an average, walking, time-distance of 10 minutes.

Apparently, the basic order used in the spatial allocation of these built forms was one of separation, instead of integration. For one thing, there existed a functional separation of various built forms, e.g., library, dining halls, infirmary, etc., and clusters of built forms, e.g., "agricultural quadrangle", "upper dormitory quadrangle", etc. For another thing, academic divisions existed in the form of physical distance between and/or within built forms.

<u>Fittingness to Context as a Measure of Activity Classification</u>

The activities elected by the subjects in the pilot questionnaire covered only four of the major classifications obtained from the results of the first questionnaire: study, eating, leisure and sports. "Leisure" in particular was mentioned in a variety of forms. The other four, which were personal hygiene, sleep, work and travel, were left out all together.(11) The selected built forms were inclusive of all types except for administrative and academic.(12)

The frequency of rejection of "personal attributes" from the given list of "attributes" indicated the following trends: (a) All frequency distribution revealed the existance of clear "reject" preferences as a function of the "personal attributes," (b) Equally clear trends of "rejects" existed as a function of the activities. For instance, chi-square tests indicated that rejections were consistent, for the activities: study, reading, entertaining opposite sex, watching TV, and outdoor leasure regardless of the "personal attributes." The rest of the activities showed considerable variance in the frequency of rejections as a function of the activities.(13)

The cumulative frequency of rejects obtained clearly reveals positions of the subject towards others. "Personal attributes" relevant to the context of the college life styles, which are girl/boy friend, best friend, and roommate, were ranked at the higher levels of the acceptance scale. Others with definite role and status implications who are placed at those positions by decisions external to the subject: teacher, same floor student, family member, same dorm student, ROTC-cadet, the president, dorm custodian ranked at the lower levels of the acceptance scale. Among the remaining those who ranked relatively higher: work associate, classmate, fellow student, young female, carried more favorable social and cultural connotations for the subjects than those which ranked lower: elderly person and stranger. These trends help in explaining the motivations behind "rejection" as a function of "personal attributes". In analysing the frequency of rejections in the context of each activity the inputs generated from the properties of encountered activities have been found to bear more significantly on the results.

TABLE 1

CHI-SQUARE TEST RESULTS USED FOR
CORRELATION OF REJECT FREQUENCIES OF
EACH ACTIVITY SETTING

Activities	Modes	X^2	D.F = 16 X = .01
Study		(11.69)*	23.54
Entertaining Opposite Sex	Mode	(21.12)	23.54
Reading	One	(16.98)	23.54
Watching TV		(21.00)	23.54
Outdoor Leisure		(11.20)	23.54
Dancing	Mode	25.01	23.54
Drinking	Two	26.32	23.54
Eating		29.72	23.54
Casual Social		59.84	23.54
Entertainment		43.65	23.54
Partying	Mode	34.30	23.54
Goofing Off	Three	24.91	23.54
Outdoor Sports		26.17	23.54
Indoor Sports		31.44	23.54
Movies		26.50	23.54
Bicycling		40.57	23.54

*Paranthesis indicates activities with significant correlation of rejects.

Three generic descriptions were used in classifying all activities into three modes: solitude, social voluntary and social involuntary.

Mode one: Solitude: Those activities which have significantly (14) consistent reject frequencies for all "personal attributes". These activities were studying, entertaining opposite sex, reading, watching TV and outdoor leisure.

Mode two: Social-voluntary: Those activities which are not significantly consistent in generating reject responses for all "personal attributes" and interaction with other persons was not vitally important for the execution of the activity. These activities were dancing, eating, drinking.

Mode three: Social-involuntary: Those activities which are not significantly consistent in generating reject responses and which require social interaction at the same time. These activities were casual social, entertainment, partying, outdoor sports, indoor sports, movies, goofing off and bicycling.

TABLE 2

ACTIVITY-BUILT FORM INTERFACE MATRIX: CLASSIFICATION OF ACTIVITIES AND BUILT FORMS

ACTIVITIES: Numbers Indicate Different Activity Modes. (See top of this page for the definitions of the "Modes")

Modes	Built-Forms	Studying	Entertain'g Opposite Sex	Reading	Watching TV	Outdoor Leisure	Dancing	Drinking	Eating	Casual Social	Entertaining	Partying	Goofing Off	Outdoor Sports	Indoor Sports	Movies	Bicycling
Multi-modal	Dormitory	1	1	1	1					3		3					3
	Apartment	1	1	1	1		2	2	2	3		3	3	3			3
	Fraternity		1				2	2		3		3					
	Student Center				1		2	2		3	3				3		
	Open Areas & Parks		1			1				3				3	3		
Single-modal	Academic Bldg.	1															
	Library	1		1													
	Dining Hall								2								
	Pub and Club						2	2									
	Movie Theatre										3					3	
	Sports Facilities										3			3	3		

Matching these activity modes with corresponding built forms yields two built form types, (a) those which house more than one mode: multi-modal, (b) those which house only one mode: single-modal. Indeed the single-modal built forms: library, academic buildings, movie theatres, clubs and pubs, dining halls, and sports facilities, generally are representative of specialized functions while the multi-modal built forms: apartment, fraternity house, student center, open areas and parks, dormitory and the campus are equipped to house a greater variety of activities.(15)

Functional Utility

Based on the four functions defined by Hillier(16) the results of the first questionnaire were interpreted. They revealed that the single-modal and multi-modal built forms manifested certain dysfunctions in terms of the utilization of human as well as physical resources of the campus. The two observations leading to these conclusions are:

1. The ratio of time spent at multi-modal built forms for study-eat-leisure-sports activities are significantly greater than those of single-modal built forms.(17)

2. Ratio of generated-trips-ratio to use-ratio are significantly greater for multi-modal built forms versus the single-modal built forms.(18)

TABLE 3

RELATIVE, CUMULATIVE AMOUNT OF TIME SPENT AT EACH ACTIVITY AND BUILT FORM

Modes	Built-Forms	Ratio of Time Spent for Each Activity Total Day-Hours									Coefficient of Conv. to Total Avail. Time Basis	Corrected Totals
		Study	Eat	Leisure	Sports	Personal Hygiene	Sleep	Work	Travel	Total		
Multi-modal	Dormitory*	13.2	0.0	5.0	0.0	2.3	18.5	0.0	2.7	41.7	1.00	41.7
	Apartment*	5.1	2.6	4.1	0.1	2.3	14.2	0.3	2.0	30.7	1.00	30.7
	Fraternity	0.1	0.0	0.0	0.0	0.0	0.0	0.0	0.0	0.2	1.20	0.2
	Student Center	0.0	0.1	6.0	0.0	0.0	0.0	0.0	0.4	6.5	1.40	9.2
	Open Areas & Parks	1.0	0.0	0.3	0.5	0.0	0.0	0.0	0.1	1.9	1.00	1.9
Single-modal	Academic Bldg.	10.2	0.8	0.3	0.0	0.0	0.0	0.0	0.8	12.1	1.71	20.6
	Library	1.1	0.0	0.1	0.0	0.0	0.0	0.5	0.1	1.8	1.45	2.6
	Dining Hall	0.3	2.4	0.2	0.0	0.3	0.0	0.4	0.3	3.9	3.84	15.0
	Pub and Club	0.0	0.1	0.1	0.0	0.0	0.0	0.0	0.0	0.2	3.00	0.6
	Movie Theatres	0.0	0.0	0.4	0.0	0.0	0.0	0.0	0.0	0.4	4.00	1.6
	Sports Facilities	0.2	0.0	0.1	0.3	0.0	0.0	0.0	0.0	0.6	1.50	0.9
	TOTALS	31.2	6.0	16.6	0.9	4.9	32.7	1.2	6.5	100.0		115.0

*SEE NOTE - TOP OF FOLLOWING PAGE

*TABLE 3 Note: Values are corrected to compensate for the bias created by dormitory and apartment based students who use only one of these built forms for domestic purposes while they share all other built forms.

TABLE 4

RELATIVE NUMBER OF TRIPS GENERATED BY BUILT FORMS
AND THEIR RATIO TO RELATIVE TIME SPENT FROM TABLE 3

Study-Eat-Leisure-Sports Activities Common with the Results of the First Questionnaire, Only

Built-Forms	"R" Ratio of Time Spent to Total Day-Hours	"X" Coef. of Conversion to Total Available Time Basis	XR Corrected Ratios	"T" Ratio of # of Trips Generated by Built Forms to Total # of Trips Gen'd.	XR ÷ T
Dormitory*	18.2	1.00	18.2	15.8	0.82
Apartment*	11.9	1.00	11.9	12.7	1.07
Fraternity	0.1	1.20	0.1	0.1	1.00
Student Center	6.1	1.40	8.6	4.7	0.55
Open Areas & Parks	1.8	1.00	1.8	1.0	0.56
Academic Bldg's.	11.3	1.71	19.3	6.8	0.28
Library	1.2	1.45	1.7	0.4	0.42
Dining Hall	2.9	3.84	11.1	4.8	0.23
Pub and Club	0.2	3.00	0.6	0.3	0.50
Movie Theatres	0.4	4.00	1.6	0.8	0.50
Sports Facilities	0.6	1.50	0.9	0.3	0.33
TOTALS	54.7		75.8		

*Values are corrected to compensate for the bias created by dormitory and apartment based students who use only one of these built forms for domestic purposes while they share all other built forms.

Following from these and the values of the independent variables and the design parameters, values for the dependent variables can be determined. These values reveal undesirable utilization of the campus and its contents. First of all physical separation of built forms reduces feasibility of regulating climatic and ecological effects of the physical environment of the total campus setting. This is significant not so much for voluntary activities but for activities and utilities generated from necessities, i.e., travel, services, etc.

Secondly, the concentration of activities at multi-modal built forms versus the specialization of activities at single-modal built forms create a dichotomy between built forms and related uses. This dichotomy hinders unbiased selection of activities for user participation. Consequently, it inhibits the user's free choice of information and expression of individual values. This is the basic heritage of today's educational system as much as it is the basis of design of cultures proposed for the future.(19)

Thirdly, **functionally detached** zones, i.e., dormitories versus academic buildings, physical separation of academic departments, decentralization and polarization of functions are the properties of the existing physical environment. The symbolic meanings of these properties are not representative of the contemporary developments either in the philosophy of education or in the trends of social life. Today, constantly growing emphasis is placed on interdiciplinary efforts and attitudes, free universities, integration of human interaction patterns and more equalized and homogenized social structures. The environments created by man must be responsive to meanings represented by these. Specialized built forms and detached structures each standing for their own integrity, leaves no room for the symbols sought for, above.

Finally, due to the functional specialization of built forms, generation of a greater number of trips by, and underuse of the single-modal built forms have been observed. As a consequence, not only has the efficiency of the use of physical resources been considerably low, but also, travel time has been in uncalled-for quantities. These signify waste of valuable work time of users: a vitally important factor for a campus of 15,000 student population.

Conclusion

The design parameters - the built form and its counterpart, the functional structure - as they are conceived through the practice of design professions today, are realized in the form of separate built entities. This is a consequence of the applicability gap between research and practice.(20) The use of any symbol standing for a built form i.e., the word "library," or "dormitory," or "dining hall," is symbolic of a constant set of activity and place types. The composition of these activity-place types as built forms is pragmatically accepted yet never sceptically questioned. Likewise the composition of the functional structures are untested but conventionally accepted.

It is clear that the relation between the individual activity-place units dictated by the values of the dependent and independent variables does not necessarily cluster them into "libraries," "dormitories" and "dining halls" separated by extensive walking distances. The release of arbitrary constraints governing the spatial allocation of activity-place units is the only way to remedy the problem.

To provide applicable and concrete recommendations to the design of meaningful physical settings the conventional built form and the functional structure concepts must be redefined. The relationship and requirements of activity-place structures as dictated by the values of the dependent and independent variables should be used as the criteria of definition.

This paves the way for the development of alternate research methods approaching the same problem with different frames of reference. Such an endeavour should initially deal with the identification and classification of the rich stock of various built forms and functional structures available in existing built environment. Special reference to the analysis of activity and space classifications in the university setting is already well underway.(21) The development of the psycho-social and cultural modification aspects of the various activity place structures, however, is not emphasized in the same manner as the physio-psychological and economical aspects are. In the macro design scale specific efforts in defining all performance criteria for activity-place structures must precede the strategies and tactics of three dimensional allocation of resources. Theoretical as well as empirical research is essential to evolve a new system of definitions for the current design parameters.

Further articulation of the methods described in this work is necessary to provide more concrete guidelines as well as developing new strategies. The instruments used in the first questionnaire which were initially developed by the Land Use and Built Form Studies Group were found to be extremely successful. The results obtained from more tightly structured diary forms as opposed to the free form diaries were nowhere near as accurate and complete.(22) Hence the considerable amount of time required for the evaluation of the results still remains as the only major handicap.

The first questionnaire presented a greater number of problems yet with less difficulty to remedy them:

1. Obtrusiveness: The results of the questionnaire rely heavily on the conscious generalizations made by the subjects while rejecting "personal attributes" which in reality may not, necessarily, comply with the unobtruded behavior of the subjects.

2. Variety and scale of instruments: The degree of "rejection" as well as the variations of "contextual fittingness" in the activity settings were not accommodated. Behaviors like "avoidance", "ignoring", etc. could be defined as overtones of "rejection" for the purposes of this survey.

3. Sample size and biases: The sample size especially in the second questionnaire was not adequate to justify valid generalizations. Despite various efforts, the elimination of self-selection was not possible all together. Especially, self-selection expressed through the difference between the ratios of returns for the two sexes was significant.

The following measures are recommended to improve both questionnaires:

1. The development of techniques for more efficient evaluation of the free-form diaries is necessary.

2. Semantic differential scales can be used in measuring "contextual fittingness" to yield more detailed results.

3. To expand further the scope of the "contextual fittingness" concept the physical and organizational properties of the setting and their acceptability by the user must also be measured.

4. The sample size needs to be increased and the survey to be reiterated at different intervals of time and at different settings. The many-sided evaluation of the results must be carried out classifying subjects of each survey by sex, year of study, field of study, and other appropriate criteria. This is expected to identify sampling biases originating from the use of academic courses as a basis for the definition of the sample population.

Firsthand information obtained from the users of a built environment supplies valuable feedbacks to the design process. The patterns of use and user preferences observed at Virginia Polytechnic Institute and State University clearly provide insights for research and design. However, it must be realized that such insights will not lead to pratical end results without further research at other similar settings. The particulars of the specific environment dealt with tend to lead to limited and biased results. In order to reach valid generalizations and feasible solutions all possibilities of the method proposed in this study must be exploited and the properties of potential research settings elaborated.

On the other hand, environmental management and behavioral modification aspects of implementing change must be recognized as well. Change is as much an outcome of the physical alterations as it is of manegerial and behavioral manipulations in the context of a designed environment. Consequently, designers are confronted today, more than any time in the past, with the task of expanding the scope of the definition of environmental problems and also the coordination of interdisciplinary efforts.

Notes

1. Levin, P.H., "The Design Process in Planning." The Town Planning Review. Vol. XXVII, No. 1 (April, 1966), 5-20.

2. Calhoun, John B., "Design for Mammalian Living." Architectural Association Quarterly, Reprinted from, Vol. I, No. 3. New York: Pergamon Press.

3. McBride, Glen, "Theories of Animal Spacing: The Role of Flight, Fight and Social Distance." Behavior and Environment. Edited by Aristied H. Esser. New York: Plenum Press, 1971.

4. Esser, Aristide H., "Ecological Contributions to Understanding the Human Use of Space." Man-Environment Systems. II, 2 (March, 1972), 106.

5. Proshansky, Harold M.; Ittelson, William H.; and Rivlin, Leane G., "The Influence of the Physical Environment on Behavior: Some Basic Assumptions." Environmental Psychology. Edited by Harold M. Proshansky, William H. Ittelson, and Leane G. Rivlin. New York: Holt, Rinehart and Winston, Inc., 1970.

6. Esser, Aristide H., "Ecological Contributions to Understanding the Human Use of Space." Man-Environment Systems. II, 2 (March, 1972), 106.

7. The list of "personal attributes" included in the second questionnaire was: girl/boy friend, best friend, roommate, work associate, classmate, fellow student, young female, young male, teacher, student from same floor of dorm, elderly person, cadet, stranger, the president, custodian.

8. Hillier, Bill; Musgrove, John; and O'Sullivan, Pat, "Knowledge and Design." Vol. II of Environmental Design: Research and Practice. Edited by William J. Mitchell. 2 Vols. Los Angeles: University of California, 1972. Sec. 29-3.

9. The pilot of the first questionnaire indicated an average return of 25%, which was for certain classes as low as 3%, when the collection of the forms was not formally structured.

10. This was checked by keeping a record of the number of males and females surveyed.

11. Compare activities listed in Table 1 and 2.

12. See Table 2.

13. See Table 1.

14. The results of the Chi-Square tests for correlation of reject frequencies are indicated in Table 1.

15. See Table 2.

16. Hillier, Bill; Musgrove, John; and O'Sullivan, Pat, "Knowledge and Design." Vol. II of Environmental Design: Research and Practice. Edited by William J. Mitchell. 2 Vols. Los Angeles: University of California, 1972. Sec. 29-3.

17. See Table 3.

18. See Table 4.

19. Skinner, B. F., Beyond Freedom and Dignity, New York: Knopf, 1971.

20. RIBA Research Committee, Members., "Stategies for Architectural Research." Architectural Research and Teaching. Vol. I (May, 1970), 3-5.

21. Land Use and Built Form Studies Group at University of Cambridge, Cambridge, England and Unit for Architectural Studies, University College Environmental Research Group at University College, London, England, have been working on activity and space classification studies for close to a decade. Their findings and guidelines have been used in this work as noted in the text.

22. Bullock, Nicholas; Dickens, Peter; Steadman, Philip; Taylor, Edwards; and Willoughby, Tom. Surveys of Space and Activities: Reading University. Land Use and Built Form Studies in the University of Cambridge, Vol. XL, Cambridge: University of Cambridge, 1970.

BEHAVIORAL EVALUATION OF A JUVENILE TREATMENT
CENTER: CASE STUDY OF A PLANNING METHODOLOGY (1)

Robert R. Hahn, Environmental Planner

Harold Lewis Malt Associates, Washington, D.C.

Abstract

A methodological perspective is suggested for understanding the relationship between self-regulating and non-infringing behavior and the physical settings in which these behaviors occur in a juvenile treatment center. The proposed method of study relies upon the convergent application of "behavior mapping," behavioral rating, and interview techniques. Individual settings within the center were found to have a marked predominance and scarcity of these behavior dimensions. Similarly, within some settings, there was a lack of "fit" between the staff's perceptions of the frequency of these behaviors and the observed results. The influence of the physical characteristics of each setting upon the two behaviors is discussed. The approach is stressed as having future meaning for the study not only of a juvenile treatment center but for the planning, design, and improvement of other programs as well.

Introduction

Planners and designers are slowly coming to realize the need for integrating into the evaluation phase of their work a more expressive set of behavioral requirements. This research study suggests a method for attempting to overcome some of the inherent problems associated with uncovering those hidden behavioral patterns, often ignored when evaluating the context of certain physical settings: defining meaningful goals; conceptualizing the physical setting; and deciding upon appropriate measurement devices.

While there are admittedly restrictions on the theoretical generalizations derived from this effort, we can contribute to our knowledge of planning as it relates to the individual -- the need, as F. Stuart Chapin indicates, for analyzing those very behaviors that should be the basis for the kinds of resources planned for. Defining individual treatment in a juvenile correctional center, is an example of such an attempt to understand in greater detail the "person moving within his daily round" and is an attempt to heed Chapin's advice to refine our planning predictors at the microlevel. (2)

Description of the Research Setting: Hypothesis for Study

Buckeye Boys Ranch, Incorporated, is a privately operated nonsectarian, interracial, residential treatment and rehabilitation center for delinquent boys, between the ages of 12 to 16 years, with serious social and emotional problems. Located on 80 acres of farm land near Grove City, Ohio in southwestern Franklin County, the property today reaches a paid value of over $500,000. Founded in 1950, by the Women's Juvenile Service Board, the Ranch receives its leadership from a board of trustees composed of business, professional, and civic leaders. This board, along with the Executive Director, and professionals in psychiatry, psychology, sociology, social work, and education, formulate policy and evaluate the program.

The program's emphasis on offering the boy a high degree of individual and personalized treatment is reflected in the two self-contained rustically-designed ranch houses. Perhaps the most influential factor in the physical environment of the Buckeye Boy's Ranch is the plan of the individual ranch houses themselves. At present, the two ranch houses, Argo House and Hislop House, with their emphasis on small group living units, handle 26 boys. Both houses are dominated by a large living-dining area. The glass windows in this setting offer a tranquil view on to the rustic setting of the ranch. Administrative and secretarial space is located in Argo House. This space is divided amongst the director's office, psychologist's office, social worker's office, assistant director's office, and a staff lounge. Argo House contains indoor recreation facilities (ie., ping pong table, pool tables) in the basement. Kitchen facilities along with sleeping arrangements are designed to accommodate 12 youth. These living arrangements include two double bedrooms and two 4-bed dorm rooms along with a common bathroom area in the hall.

Hislop House, on the other hand, is less "rustic" in appearance and approximates the typical "ranch burger" found in a middle-income residential subdivision. This house, in addition to containing its own kitchen and dining facilities, accommodates a "live-in" house couple. This apartment unit, off to the left of the entrance, has its own living room, kitchen, dining room and personal bathroom facilities. Hislop House also contains living and sleeping arrangements to accommodate 12 youth. This space is divided among two 4-bed dorms and four single bedrooms. In addition, this residence contains the campus school, off a wing in the basement, which consists of 3 classrooms, a library, teacher's office and lounge. A music room and arts and crafts area is located off the main basement which serves also as an area for indoor recreation.

The personnel at Buckeye Boys Ranch number between 24-28 staff members, some of whom are on a part-time basis. In addition to the Executive Director, whose position is primarily administrative, the professional staff is made up of the following group of clinicians: director of treatment services; a psychiatrist responsible for reviewing the quality of the treatment and conducting psychiatric therapy; a psychologist who provides psychological evaluations and conducts individual and group treatment; a chaplain; a social worker who works with the youth and families and assists in the treatment of youth; and 3 special education teachers in charge of developing the educational program.

The professional staff is supplemented by a body of child care personnel who include house parent couples, who supervise and conduct activities with the boys; a vocational counselor, who assists the boys in developing occupational and work skills; and night attendants, who are usually undergraduate college students with a background and interest in psychology and in the treatment of delinquent youth. Lastly, additional staff needed to provide the optimal level of services at the ranch include: secretarial staff and maintenance personnel.

Residential care at Buckeye Boys Ranch is coordinated with other services to the boy and his family. Accordingly, the therapeutically directed living experience is combined with individual and group therapy, family therapy, remedial and specialized education, medical care, and occupational work experience. The thrust of such treatment efforts is to return each youth to his own home, foster home, or perhaps a group home so each boy can learn more effective ways to resolve conflict.

To learn more about the treatment philosophy and the goals of the program, interviews were conducted with members of the professional staff. The results of these interviews were put in summary form. The range of responses varied slightly, as there was a high degree of consensus on the goals of the system.

While the long range treatment goal is one that encourages the youth to have the capacity to function, upon leaving the ranch, with as wide a degree of flexibility as possible, the following goals were cited as key facets of the treatment philosophy operating at Buckeye Boys Ranch:

1. Provide a milieu that encourages and develops the youth's feeling of worth.
2. Allow the youth the opportunity to improve his ability to test his decision-making powers commensurate with the norm for his age and level of maturity.
3. Encourage and teach the youth to learn to deal with stress so that the manner in which he resolves conflict doesn't infringe on others.
4. Encourage the youth to develop an increased self-awareness of himself and understanding of others so he can develop into a non-infringing person who respects and effectively relates to others.
5. Encourage the staff to develop, grow, and be sensitive to the youth's development.
6. Encourage the boy to assume responsibility for himself and accept the consequences of his behavior, so he can be self-sufficient and in charge of his life and not dependent on others for making decisions.

While I was interested in learning about the general treatment philosophy at the ranch, I was primarily interested in focusing on a series of treatment goals in reference to "ideal" behavior. That is, I was interested in having the staff define a series of goals that illustrated "successful" or "good" behavior on the part of the youth, as these would provide us with a set of indices that would reflect progress toward the previously defined treatment goals. Recognizing that the identified goals could be modified by the organization (ie., the staff learning to live with the constraints upon their treatment philosophy), goals were secured with the intention of being asked, "under the best condition" or the "most ideal situation," so as to resolve this organizational modification.

The following two behaviors were overwhelmingly cited by the members of the professional staff as being the two key behaviors underlying the treatment philosophy at the ranch: self-regulating behavior and non-infringing behavior.

Self-regulating behavior refers to giving the youth the opportunity to learn to make his own decisions; evaluate them; and anticipate the consequences of his actions. Providing the youth with both the autonomy to make his decisions and the presence of a pool of varied staff members who could give the immediate and appropriate feedback about the effect of the youth's decisions was seen as the major means for encouraging self-regulating behavior. Similarly, non-infringing behavior, with its hidden assumption that the values of the youth don't have to be the same as those of the staff or other boys, does suggest that his behavior should not infringe on the values of others within the treatment setting. The means for encouraging behavior, in this case, was more highly internalized than in self-regulating behavior. The basic individual relationship between the staff member and the youth, with the youth having the opportunity to "model" himself after different people in the setting, was defined as the major means for developing sensitivity to others and an understanding of the limits of one's behavior.

Having a clear statement of the behaviors to be examined and an explanation of why they are important for our consideration, we are in a position to set up the hypothesis. In developing the hypothesis, I relied on the staff and their formulation of the treatment goals to furnish me with the means for learning the content of the behaviors to be examined.

To demonstrate this research approach this study is aimed at examining the following hypothesis:

> Self-regulating and non-infringing behavior is found more frequently in certain kinds of settings within a juvenile rehabilitation facility and the staff of that facility are able to predict the settings where these behaviors are most likely to occur.

In testing this hypothesis, we will determine if literature dealing with the influence of the physical environment on behavior can give us any answers or suggestions as to the characteristics of each setting which is eliciting the observed behavior. That is, is the tendency for a particular behavior to occur in a particular setting, due in any way to the physical characteristics of that setting? Secondly, we will determine whether a set of specific conclusions for further study can be formulated about how the physical setting influences the described behaviors (ie., contributes or thwarts the accomplishment of the treatment goals). To the best of this author's knowledge, research, of a micro-ecological behavioral nature, has yet to be undertaken, which identifies a series of settings within a juvenile rehabilitation facility to determine whether these physical settings have a discernible effect upon the occurrence of self-regulating and non-infringing behavior.

Description of the Method

The method of analysis is based on the following two assumptions:

1. After securing a set of treatment goals, a particular set of behavioral categories can be developed and examined.

2. These defined behaviors will in fact reflect progress toward the treatment goals.

In comparing the staff's prediction of where these behaviors were most likely to occur with the volume count of each behavior in each setting, the following steps were given:

<u>Step I</u>: Infer a Set of Behavioral Categories Indicating the Dimension of Self-Regulating and Non-Infringing Behavior

In developing the actual behavior categories, the focus was on providing common-sense, easily observable, overt behaviors that could be easily recorded so as to explain the dimension of self-regulating and non-infringing behavior.
These behaviors were then generalized into short hand categories for observation. The table below provides a classification breakdown of the two behaviors in terms of its analytic and observational categories.

Analytic Category	Behavior	Observational Category
SELF-REGULATING	Staff makes a decision for boy	Staff decides
	Staff offers boy an opportunity to make his own decision	Youth decides
	Staff evaluates boy's decision	Staff evaluates
	Staff allows youth to evaluate the consequences of his decision	Youth evaluates
INFRINGING	Youth insults, swears at peer	Verbal infringement (peer to peer)
	Youth insults, swears at staff	Verbal infringement (peer to staff)
	Youth physically attacks peer	Physical infringement (peer to peer)
	Youth physically attacks staff	Physical infringement (peer to staff)
	Youth directs aggression to destruction of property	Property infringement

Rather than index every observed self-regulated act the youth demonstrated (ie., it would be difficult to decipher whether the act was youth initiated or staff instructed) it seemed more appropriate to focus on the means for encouraging this type of behavior -- immediate staff feedback. That is, indexing the staff's behavior and determine the degree to which they allow the youth the opportunity to make decisions; provide the appropriate feedback; and evaluate the boy's decision.

For our test purposes, we were in fact measuring infringing behavior -- behaviors that tend to infringe upon the values of others -- to determine those areas more likely to encourage non-infringing behavior. Non-infringing behavior will be defined as that behavior where the least amount of observed infringing behavior occurs. Observing infringing behavior required that we record the interpersonal relations of both youth and staff.

Step II: Define a Series of Settings Within the System

A list of all possible settings within the ranch were made. This included settings within the two ranch houses and outside on the grounds. This list was generated with the help of staff members who were more fully acquainted with the ranch. Prime focus was on those areas which were most likely to elicit the behaviors of interest. The preliminary list to be investigated included the following:

Argo House: Entrance, Dining area, Living room area, 4-Bed dorms, Hallway leading to dorm area, Basement, Kitchen.

Hislop House: Dining area, Classrooms, Library, Basement, Laundry room, Music room, Wrestling room, Kitchen.

Step III: Interview Staff to Acquire Their Ranking of What Kind of Behavior (Self-Regulating or Infringing) Is Most Likely to Occur in Each Setting

The Likert summated format was used in order to provide comparisons with the results obtained with the observation techniques. Of a possible 28 staff members, 24 were interviewed. Of the latter number, 20 interviews were completed in the correct manner. Staff members were asked to estimate how often the dimension of self-regulating or infringing behavior could be expected to occur in each of the defined study settings. There were 4 choices for each setting: Almost always/ Usually/ Sometimes/ Very rarely. Point values of 4, 3, 2, 1 were assigned to these responses.

Step IV: Measure by Observation the Defined Study Settings. Index the Frequency with which Self-Regulating and Infringing Behavior is Elicited in Each Setting

Of the original list of 16 settings, 12 were retained to focus attention on during observation. The laundry room, music room, wrestling room, and kitchen in Hislop House were discarded due to their sporadic and limited use by both staff and youth.

The location and timing of observations involved coverage of the 12 settings on a time-sampling basis. The 20 one-hour periods were randomly selected from a pool of 60 one-hour slots that could occur between Monday May 15 to Friday May 19, 1972. The 20 one-hour periods were evenly distributed between the morning, afternoon, and evening hours. Observations began as early as 8:00 am and lasted no later than 8:00 pm. An average of 4 one-hour observations were conducted each day so as not to alter the behavior of either the youth or the staff.

Neither the staff or the youth knew precisely the purpose of the observations. Staff members did not suspect that their behavior was being evaluated along with the kids. In fact, the staff thought the observations were of the youth only. Observations were recorded on data sheets for quick and easy use. Observers were asked to record in the boxes provided, the number of times a youth or staff member engaged in each of the behavior categories. In addition, observers were asked to record on the data sheets, the total number of youth and staff present in each setting during the 5 minute period.

<u>Step V</u>: Validate the Results

Kendall's Rank Correlation r (tau) was used to test the strength of the association of the staff's ranking of the settings with the observed ranking (frequency count) of the settings for each behavior dimension. In addition, the significance of r was tested.

<u>Testing of the Hypothesis</u>

In correlating the staff's ranking with the observed ranking of the frequency with which each behavior occurred in each setting, the following correlations were calculated:

<u>Correlation I</u>:   The Staff's Ranking with Observed Ranking of the Frequency with which Infringing Behavior Occurs (r=.11)

<u>Correlation II</u>: The Staff's Ranking with Observed Ranking of the Frequency with which Self-Regulating Behavior Occurs (r=.52)

<u>Correlation III</u>: The Staff's Ranking of Self-Regulating Behavior with the Observed Frequency of "Youth Decides" and "Youth Evaluates" (r=.43)

In testing the significance of r, it was found that in the population of settings from which the sample was drawn: (a) the staff's predicted ranking and observed ranking of the occurrence of infringing behavior are not associated in the population of settings which this sample was drawn; (b) the staff's ranking and observed ranking of the frequency with which self-regulating behavior occurs is associated; (c) the staff's ranking of the predicted frequency of self-regulating behavior and the observed ranking of the actual frequency of "youth decides" and "youth evaluates" are associated.

3. HUMAN RESPONSES TO THE ENVIRONMENT / 145

Analysis of the Findings: Influence of the Physical Environment

There is little, if any literature, either psychological or in the man-environment field, which has systematically researched the relationship between general design features and its effect upon either the dimension of self-regulating or infringing behavior. The situation is even dimmer when trying to isolate this relationship in the setting of a juvenile treatment facility. Rather, the majority of literature dealing with the linkage of behavior to specific environmental features or patterns of physical elements has tended to focus on such behavioral dimensions as "personal space" (3); privacy (4); crowding (5); isolation (6); propinquity, friendship, and neighbor relations (7); and passivity (8). Since the literature offers little in explaining the influence of the physical setting of a juvenile rehabilitation facility upon the behaviors under study, we turn to our observations and early interviews with members of the professional staff for drawing some generalizations relating to the influence of the physical settings.

Some interesting and significant differences were previously revealed between the staff's perception of the predicted frequency of the two behaviors and their actual observed occurrence (especially infringing behavior). It is relevant to consider whether these differences are in any way attributed to the influence of physical-environmental features. The analysis is organized around the following distinctive settings:

Entrance to Argo House

The entrance to Argo House is the visitor's first encounter with the Buckeye Boys Ranch. A low frequency of infringing behavior was discovered in this setting. In fact, the entrance had a total frequency of one recorded infringing behavior during the entire observation period. This corresponded closely to a low ranking by the staff of 1.75. The close agreement between the staff's ranking and observed rankings as to the non-infringing nature of this setting can be attributed, in part, to the influence of the physical characteristics of this setting (i.e., dominance of administrative and staff office space, trophy showcases; and plaques honoring ranch founders and benefactors).

The administrative influence of the entrance doesn't provide for an easy and most natural setting for encouraging this behavior. While the atmosphere is comfortable and a mood of friendly welcome, the immediate accessibility of the staff's offices lends a formal mood which connotes to the visitor and possibly to the youth alike, that of all the settings in the ranch, a base line of expected behavior and "decorum" is required here.

The findings go on to indicate a high degree of self-regulating behavior. Interestingly enough, there was less agreement between the staff's rankings and the observed rankings as to the expected frequency of self-regulating behavior. The close location of the staff's offices, the staff lounge, and the general administrative nature of this setting could be a factor in contributing to the greater frequency of self-regulating behavior. In fact, as the findings indicate, this behavior tends to be mainly in the realm of the staff deciding and staff evaluating, with little opportunity for the youth to initiate more control in his decision-making (i.e., the entrance was ranked first in terms of the observed frequency with which the staff makes and evaluates decisions for the youth). There is the added possibility that the office and administrative character of the setting suggests to the individual staff member the need to exert greater control over the decision-making of the youth, if in fact, the desired and appropriate "showcase" behavior of the setting is to be maintained. Robert Bechtel's findings on the public housing environment are relevant here. (9) He found in the public housing environment that the heavy hand of management in all decisions fostered a passive and dependent attitude in its residents.

Living Room in Argo House

This setting, located off to the left of the entrance, is the major living area for the youths residing in Argo House. This setting clearly establishes, in the minds of those who enter, the residential character of the ranch (i.e., T.V.; bookshevles; comfortable and informal couches; brick fireplace; hanging "moose-head" etc.). While the staff's ranking agreed with our observations corresponding to the highest frequency of infringing behavior, the staff ranked this setting fifth in its expected degree of self-regulating behavior. The high frequency of infringing behavior could be attributed to the small scale of the setting and the many competing uses it serves. This high frequency is partly due to the number of youth present in the setting. As Figure 1 discloses, the greatest percentage of infringing behavior occurred when 4-7 youth were present. Interestingly enough, the presence of staff (which tends to remain constant over different settings with different frequencies of infringing behavior) is less of an influencing factor than the number of youth present.

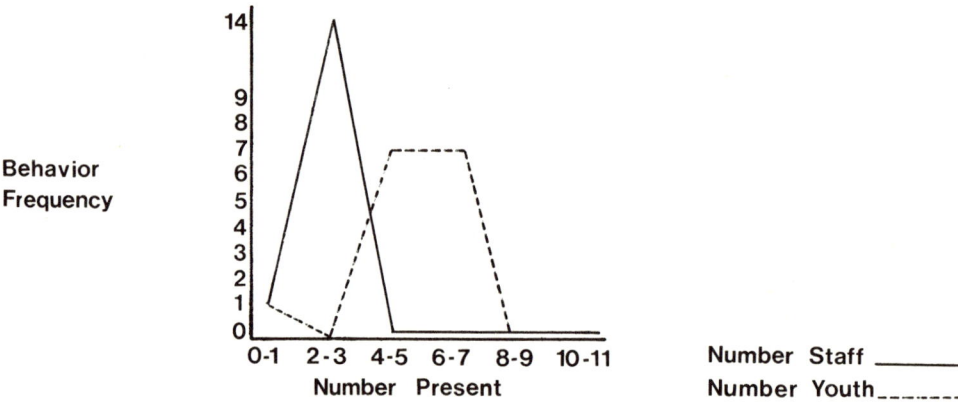

Figure 1: The Distribution of Infringing Behavior (Living Room, Argo).

A study entitled, "Toward a Psychological Theory of Crowding," would suggest that crowding could be related to the high occurrence of infringing behavior within this setting. Its author, J.A. Desor, contends that "being crowded" is "receiving excessive social stimulation and not merely a lack of space." (10) These findings would suggest that decreasing the density of people is by no means the only method of alleviating the high occurrence of infringing behavior within this setting.

Lastly, the living room showed, in contrast to the entrance, a greater opportunity for the youth to play a stronger role in making and evaluating his decisions (i.e., 55% of the total amount of observed self-regulating behavior was associated with the staff giving the youth a greater freedom in decision-making).

Classrooms

Of all the twelve settings, the classrooms received the most consistent agreement (observed and predicted) in terms of the frequency with which self-regulating behavior could be expected to occur. No doubt, the image of a classroom and its associated physical accouterments of desks, blackboards, etc. coins an easier image in the minds of the majority of the staff as being a most appropriate setting for encouraging self-regulating behavior. Also, to be considered, are Barker's and Gump's findings, on the relationship between the size of a school, size of its settings, and the associated behavior patterns that are produced. (11) Buckeye Boys Ranch, a small institution, creates small settings that possibly produce an undermanned classroom setting that, in contrast to a larger public school arrangement, allows the youth to have a greater responsibility in his decision-making, commensurate with his ability.

Stanley Milgram, discusses the concept of "overload" which could provide an additional link to explaining the high occurrence of self-regulating behavior in this setting. (12) Milgram uses the term to depict the experience of the urbanite, as characterized by a wide variety of adaptations to an overloaded social environment which results in the allocation of less time to each input (i.e., disregard of the needs, interests, and demands of others). This setting consisted of a reduced range of situations determined very little by adaptations to overload. As a result, this could explain the increased time, on the part of the staff, for allowing and encouraging the youth to develop his cognitive and decision-making functions.

Four-Bed Dorms in Argo House

The staff, when interviewed, cited the 4-bed dorms in Argo as the one setting in the ranch most likely to contribute to infringements between the youth. This was borne out by the sum of the staff's rankings (3.00) indicating this setting as the second most likely setting of the 12 where infringing behavior could be expected to occur. During the entire 20 one-hour observation periods, 0 infringing behaviors were observed. While our observations were on a time-sampling basis and admittedly selective, too much emphasis might be placed on the associated advantages of the single and 2-bed dorm rooms versus the 4-bed dorms. While single units might tend to increase a feeling of ownership by the youth, its associated advantage for encouraging non-infringing behavior might be over emphasized.

While the findings of the low frequency of infringing behavior in the 4-bed dorms contradict the prevailing opinion among those who design single rooms as being preferable to larger dormitories for permitting privacy and reducing the competition of inmate influences, (13) they are in agreement with the work of Ittelson and Proshansky. (14) These researchers found that the proportion of isolated-passive behavior rises regularly with the size of the bedroom in a psychiatric ward. They discovered more interaction (and might we add, more possibility of peer to peer infringement) in a 3-bed room of a psychiatric ward than in a 6-bed room. This finding suggests that those involved in the treatment of the juvenile delinquent might reconsider the neglected treatment benefits of a 4-bed dorms.

Conclusion and Limitations of the Study

The lack of an adequate enough data base in the remainder of the settings makes it difficult to assess and draw any appreciable conclusions as to the effects of the physical setting upon the two behaviors. While it may be possible to suggest the application of the method used in this investigation to study other settings, it is not possible to assert much certitude as to the validity and reliability of our findings to other social systems, even other juvenile treatment centers. I suspect it is highly possible that both the ecological structure and the normative patterns among the youth and the staff are in many ways unique and unlike those found in other juvenile treatment settings. Secondly, the practical application of our results is somewhat clouded by the degree to which infringing ("bad behavior") and self-regulating ("good behavior") are correlated. Similarly, as is typical of a case study, we found it necessary to structure our research methods and analysis, sometimes at the expense of blurring our results. For instance, an adequate statistical test was not administered to show that the observed behavior departed from a random pattern or one that was simply a function of the number of the persons present in the room.

While the differences in the frequency of observed behavior are due in part to the influence of physical environmental features, no doubt, other factors (sociological and administrative) are to be considered. However, in a limited study of this nature, it was difficult to isolate the normative influence in our results. In summary, the proposed method of investigation can be a valuable approach to the planning profession as long as the methodological restrictions are made explicit.

Notes and References

1. This paper is based on a thesis submitted by the author in partial fulfillment of the requirements for the degree of Master of City Planning from The Ohio State University. The research was supported, in part, by an OLEPA grant (No: 380-00-J-70) provided by the Program for Study of Crime and Delinquency, Columbus, Ohio. The author is indebted to its Director, Dr. Harry E. Allen along with Professor Raymond Mills, Professor Jerrold Voss, and Dr. Richard

Klimoski, The Ohio State University.

2. Chapin, F.S., and Brail, R.K., "Human Activity Systems in the Metropolitan United States," Environment and Behavior, 1, 2, December, 1969, pages 107-130.

3. Sommer, R., Personal Space: The Behavioral Basis of Design, Englewood Cliffs, Prentice-Hall, 1969.

4. Kira, A., "Privacy and the Bathroom," in Ittelson, W.H., Proshansky, H.M. and Rivlin, L.G., Environmental Psychology: Man and His Physical Setting, New York City, Holt, Rinehart and Winston, 1970, pages 269-275.

5. Desor, J.A., "Toward a Psychological Theory of Crowding,"Journal of Personality and Social Psychology, 21, 1, 1972, pages 79-83.

6. Glaser, D., "Architectural Factors in Isolation Promotion in Prisons," The Effectiveness of a Prison and Parole System, Bobbs-Merril, 1964.

7. Gans, H., "Planning and Social Life: Friendship and Neighbor Relations in Suburban Communities," Journal of the Americal Institute of Planners, 28, 7, 1961.

8. Bechtel, R., "A Behavioral Comparison of Urban and Small Town Environments," in Eastman, C., and Archea, J., EDRA 2, Pittsburgh, 1970, pages 347-353.

9. Bechtel, R., "The Public Housing Environment: A Few Surprises," Environmental Research and Development Foundation, 1971, page 4.

10. Desor, J.A., "Toward A Psychological Theory of Crowding," Journal of Personality and Social Psychology, 21, 1, 1972, page 79.

11. Barker, R., Gump, P.V., Big School, Small School, Stanford University Press, 1968.

12. Milgram, S., "The Experience of Living in Cities," Science, 167, March, 1970, pages 1461-1468.

13. Glaser, D., "Architectural Factors in Isolation Promotion in Prisons," The Effectiveness of a Prison and Parole System, Bobbs-Merril, 1964.

14. Ittelson, W.H., Proshansky, H.M. and Rivlin, L.G., "The Environmental Psychology of the Psychiatric Ward," in Ittelson, W.H., Proshansky, H.M. and Rivlin, L.G., Enviornmental Psychology: Man and His Physical Setting, New York City, Holt, Rinehart, and Winston, 1970, page 432.

TYPES OF CITIES 3.4

THE SUBNATIONAL URBAN ENVIRONMENT AND SOME DESIGN IMPLICATIONS

Robert B. Bechtel

Environmental Research and Development Foundation
Kansas City, Missouri

Abstract

Eighty-five United States Standard Metropolitan Statistical Areas (SMSAS) were classified by the B C TRY cluster analysis system using 68 1960 census variables. Fifteen city types were derived that had greater discriminatory power than 10 other city classifications.

There were two "super-types" of cities: Those within a relatively small geographical area and those that covered a large part of the United States. This finding suggests that United States cities can be arranged along a bipolar continuum with purely local geographical factors at one end and economic-functional factors at the other. The 15 city types define empirical sub-units of this continuum.

Results are only tentative and must be verified by re-analysis using 1960 census data. Yet speculation on the significance of certain findings leads to important questions.

The chief discriminating set of variables define older households. Age of households as measure of environment quality has significant bearing on design policy. Building without regard for maintenance costs may be a principal cause of urban problems, contributing to decreasing revenues, racial impaction, and housing abandonment. The design solution is to create the maintenance-free, low cost house. But will economic and social considerations permit this, or will designers have the courage to create designs that change national policy?

Introduction

Up until recently, the chief practical excuse for attempting to classify cities has been to use the economic differences among cities as part of the justification for city programs and budgets (1). The main burden of classification has been carried by THE MUNICIPAL YEAR BOOK, a publication of the City Manager's Association. Economic classifications of United States cities have been made since the 1945 edition of this publication and the most recent was in 1967.

In a more theoretical sense, cities were first classified by Harris (2) and more recently by Hadden and Borgatta (3) and Berry (4). These attempts have always been few.

With the rise of urban problems, the need for new and better classifications became urgent. Many government agencies began to demand that more attention be given to social data. Bauer (5) accused the government of "economic philistinism," and various reports of commissions and agencies such as the report of the special commission on the social sciences of the National Science Board (6) and the Secretary of HEW (7) the use of social indicators in planning the nation's needs. The National Academy of Sciences (8) even urged the Department of Housing and Urban Development to use a typology of United States cities in planning its programs.

Despite the urgency of problems, the classification of cities has lagged considerably. There are several reasons why this is so. One reason is the problem of methodology. To handle such enormous amounts of data, the researcher is forced to use complex computer systems that are difficult to understand. One methodology has dominated the field -- factor analysis.

It is not necessary to know all the complexities of factor analysis to understand why some of its results are less than satisfactory. Hadden and Borgatta (9) refused to classify United States cities because the factors derived were simply too closely related to each other. In other words, by the requirements of factor analysis, cities do not fall into separate categories like apples and oranges -- they seem to be hybrids of only one or a very few related factors. But factors that are related are not really "true" factors at all so Hadden and Borgatta (10) gave up on the task.

In addition to the problem of related (technically call "oblique") factors even when unrelated (or "orthogonal") factors are obtained, one is left with the problem of how to interpret what the factor means. Since factors are constructed by correlating several variables together, there is often the problem of deciding how the variables are related. For example, Maloney (11) correlated percent voting (inverted (12)), percent Negro, murder rate, low education, low income, percent party change (inverted), percent voting Republican. This factor was named "southern syndrome." With a certain reasonableness, one can easily see how few people voting, high proportion of Negroes, high murder rate, low education, low income, and few party changes could be typical of southern cities. However, a high percentage voting Republican fits this pattern for only the 1960-1964 elections, and on close examination one could ask why this would not just as well fit many northern ghetto cities such as Chicago. The only reason, in fact, that it was called southern syndrome was that certain cities such as Shreveport, Mobile, Birmingham, New Orleans, and Memphis did load highly on the factor. The assumption in factor analysis is that one dimension underlies a factor and it is still very unclear what single dimension underlies the seven variables mentioned above.

Despite these difficulties, however, the use of factor analysis continues in classifying cities (13), with the results that some useful data are obtained but it is usually in the form of listing cities according to how they rank on some variable such as racial ghettoes or manufacturing. The problem of how to classify cities using large numbers of variables has not been much improved since Hadden and Borgatta's 1965 study.

The New Method

Robert C. Tryon (14) developed a method of classification that circumvented many of the problems of factor analysis. Tryon termed it cluster analysis. Cluster analysis, as a general technique, is not new and it has been used frequently in the classification of animal species (15). Tryon's technique went further than any previous method by combining the advantages of both factor analysis and cluster analysis. He eliminated the assumption of an underlying dimension to a factor by organizing related variables around a single variable. This means that the name of the cluster is the same as the first, or key, variable. In addition to this, Tryon's system eliminated from analysis variables that were not highly correlated and eliminated variables that appeared in more than one cluster. The result was a much cleaner scheme for classification.

Finally, the B C TRY system which was developed by Tryon and Bailey (16) combined over 30 overlapping computer programs to produce an exhaustive classification of objects or persons. The researcher feeds in his data and gets out a complete classification scheme with options on how he wants to improve it. Virtually no factor analysis program provides this feature. The method for classifying objects, in this case cities, is called 0-typing and is based on scores that each city gets on the clusters. The city classification described below will make the use of the system clearer.

Cluster Analysis

Sixty-eight variables were chosen as the ones most likely to provide social indicators of urban problems and urban structure. These variables were also chosen because they had been used by Tryon (17) earlier and by other researchers such as Shevky and Bell (18) and Maloney (19). Because data on crime, unemployment and other variables were not available, they were left out. Most of the data were from the 1960 census.

The 68 variables were grouped into clusters according to Table 1. Cluster one is called older households because the key variable, moved in '58-'60 is reflected, meaning there are very few moved in '58-'60. The other variables are generally consistent with the key variable; moved in '50-60 is reflected, built in '39 or earlier and moved in '39 or earlier, and also moved in '40-'53 are positively related. These variables are clearly consistent with older households. Also connected with this cluster as a definer is the ownership of a different house in the same city. The remaining variables are in the cluster but are not considered important enough to be called definers.

Cluster two is a middle income cluster because the key variable for low income is reflected. Cluster three is segregated white, cluster four is students, cluster five persons alone, and cluster six the use of auto transportation.

It should be noted here that cluster one, older households, has not been found by other studies as a key indicator of city differences. More will be said about this later.

City Types

Only 85 cities were selected for classification. The chief reason for this was to compare the classification with Maloney's (20) work which used the same 85 cities. Using Tryon and Bailey's (21) O-Type analysis format, 15 types of cities were derived according to the six clusters mentioned above (see Table 11). Six cities were omitted from the classification because they did not fit into any of the types homogeneously. This included New York Parts 1 and 11 because no data was available for the separate parts in the later comparison with other systems of classification (see Table 11).

One thing should be noted from this classification. There are two "super" types of cities. One super type are the cities of close geographical proximity such as type 15, the Pennsylvania cities, or type 7, the northwest cities. The other super type is the "far-flung" or those cities not in geographical proximity such as type 2, El Paso, Phoenix, Sacramento, Miami, and San Jose. It appears then that some cities do have a geographical proximity as an element of their typology while others do not.

Comparison with Other Typologies

Before going further to draw conclusions from this typology, it is necessary to compare it with other typologies. Even though attempts to classify cities have been few, it may be possible that other attempts have been more successful. The chief problem in making such a comparison is to get the same variables and the same cities. Since many other studies did not use the same cities as this one, the number of similar cities cannot be constant across all comparisons. It is possible to have the same variables by selecting an entirely new set of variables for all the cities from the County and City Data Book (22) published by the U.S. Government Printing Office. Forty-nine variables were selected from that document.

In selecting city classifications to compare with the Tryon methodology, an attempt was made to use the classification schemes that best represent the different points of view in classifying cities.

Duncan (23) represents the functional or economic viewpoint from which to classify cities. Cox and Zannaras (24) represent an attempt to classify cities by popular views of the geographical region to which they belong. Maloney (25) represents a purely factorial approach and has two classifications if one counts those cities loading both high and low on his factors. Forstall (26) represents the City Manager's Yearbook method of classification and actually has three methods of classification. Nelson (27) classifies cities according to a service criterion. While Hadden and Borgatta (28) did not intend to classify cities, their 13 factors provide a way to do so.

Finally, the BC TRY system is by no means exhausted with the 15 O-Types derived. If one wants to consider a purely descriptive classification that is closely tied to the unique characteristics of a particular city, he can classify cities by the key variable in the first cluster. This means analyzing each city by the data in its own census tracts and then lumping cities together that have similar key variables in their first clusters. This produces 11 types. It should be noted that cities with the second and third variables similar form subgroups of the key variable classification. For example, Baltimore, Chicago, Washington, Dallas, and Philadelphia form one subgroup under Negro concentration cities while St. Louis, Fort Worth, Detroit, Los Angeles, Houston, Mobile, Richmond, Charlotte, and Flint form another (29).

The 11 different typologies of cities are compared in Table III. Each typology was tested for its ability to show a significant difference on each of the 49 variables from the County and City Data Book (30). The number of differences below the .01 level of significance (by F Test) are listed beside each method. As can be seen, the BC TRY typology with 15 types shows the greater differentiating power. The Hadden and Borgatta typology is next best with Forstall II and Duncan and Maloney following. Thus it is clear that the BC TRY method edges the others out in its ability to show differences among its types.

Conclusions

A note of caution must be injected here. The results of this study are only tentative in nature. The work was done on 1960 census data and needs to be repeated on 1970 census data to test its reliability. All further remarks are largely speculative until this research is done. Nevertheless, reflection on the results is instructive.

This study shows, among other things, that the classification of cities involves two main elements. One element comprises all those variables which focus on a particular geographical area. These may be related to such factors as location of coal in the Pennsylvania cities type or the proximity of the ocean as in the northwest cities type. The second element undoubtedly involves what Duncan and Forstall call the functional or economic -- that is what the city principally does by way of manufacturing or retailing or providing some service. These types are obviously interwoven in the location and structure of each city. Another way of viewing this is to say there are two kinds of subnational urban environments, the regional-geographical and the functional.

What this means in theoretical terms is that both the functionalists and the sociologist geographers are right. Cities are formed by their geography and their function except that some are grouped more by geography and others more by function. The geographical to the functional forms a continuum of types.

The results show that for purposes of maximizing differences, the variable of older households is probably the best single indicator of city differences. Not much attention has been given in the past to this set of variables and some reflection is needed to consider what it may mean.

First, it means that older households tend to need repair (as opposed to cluster two which had households not needing repair), have fewer automobiles (walk to work), and seem to be comprised of working class people (negative on professionals, high on unskilled workers), and has few vacancies. This suggests a number of related problems that need to be investigated. Does the cluster really refer to what are commonly called gray areas? If so, then these may comprise the lowest common denominator across all the cities studied. They may possibly point to a problem of even larger future magnitude than the so-called ghettoes. Further investigation needs to be done in regard to whether this is also the group most in rebellion against property taxes, welfare, and bussing (31).

Whatever these relationships may be, the age of the household is an important variable to consider for design and planning purposes. For example, does the designer or planner calculate the service life of buildings as an integral part of his scheme? Airplanes and trucks have maintenance schedules based on how long it takes for the basic equipment to wear out in use. These schedules are presented to the customer at time of delivery. Houses are seldom programmed beyond initial cost of construction. Few lower cost houses today are expected to last the life of the mortgage. In fact, it is more profitable to have quick refinancing of mortgages than to wait out the life of the mortgage. Based on this principle, Hood and Kushner (32) have shown how certain low cost housing plans of HUD permit doubling of investments in two years. Newer multiple dwellings have a legal requirement for a maintenance reserve fund to keep the units in repair but few low income heads of single dwelling households can afford even the minimal repairs. The result is that the more numerous low cost aging households can quickly become a burden on a city's resources and perhaps may be the central source of a city's problems when one also considers they are the widest basis for property tax as well. Central cities have experienced a decline in relative returns from property tax (33).

Perhaps the solution to many of these problems lies in having the city or some governmental agency take responsibility for aging houses before they become a public liability through abandonment and code violation. One way to alleviate the problem is to insist on higher standards of construction. A house of better materials will not deteriorate as quickly. Still another solution is the code enforcement technique, but this has always penalized the poorer home owner, especially the elderly who have difficulty getting loans for repairs (34).

Another suggestion is the compulsory maintenance reserve or a similar plan (35) which would require each property owner to contribute to a maintenance fund or to have maintenance insurance. The theoretical result of this would be that no building would be without funds to keep it in repair. But many social and economic difficulties need to be solved before this solution can be tried. For example, how would poor people and those on fixed incomes be able to afford it? Even more important, would the American public be willing to accept another compulsory slice out of their incomes? Still another solution is the design of a maintenance-free house. But will our society, let alone the design professions, let such a product be created?

Whatever the possible solution, the problem is ever present in design and planning and designers must some day decide whether they want to take a lead on creating policy for more durable housing on whether they will continue to be led by the purely financial considerations of quick turnover.

The critical test for future research in this area is whether the age of households increases as a discriminator of city types for the 1970 census data. Should that be the case, then it would be imperative to examine the whole question of age of housing as a causal element across many urban problems and there would be a strong case for maintenance requirements as one of the most critical of design and planning elements. The age of the housing stock may be an indicator of the social problems of any city and accurate knowledge of the location and amount of the older stock may be the prime basis for calculating need for remedial programs. The designer who failed to consider maintenance as the most critical of elements for his programming would then be considered a prime contributor to the ills of urban life.

TABLE I
Six Oblique Clusters (Preset) Of 68 Variables
(D) indicates defining variable*

Cluster 1 (.9798)**	Cluster 2 (.9703)**	Cluster 3 (.9751)**
Older Households	Middle Income	Segregated White
-65 (D) moved in '58-'60	-41 (D) $5,999 or less	-56 (D) nonwhite, owner occup.
-60 (D) built '50-'60	43 (D) $10-25,000 fam. inc.	55 (D) white, owner, occup.
62 (D) built '39, earlier	44 (D) median income	1 (D) native white
68 (D) moved in '39, earlier	22 (D) $100 or more rent	26 (D) occuped d.u.
67 (D) moved in '40-'53	42 $6-9,999 family inc.	45 white, under 5 years
40 (D) different house, same	23 house value $15,800+	+34 couples in household
-61 built '40-'49	25 d.u. not needing repair	-64 1.01 or more persons/room
-58 vacant, any cond. d.u.	17 female clerical-sales	-28 no heating
59 walk to work	66 moved in '54-'57	-31 Negro
3 male, 25 older	15 male craftsmen, foremen	-18 female craftsmen
48 male, unskilled worker	-49 female unskilled	-46 divorced, separated, white
4 female, 25 older	- 8 male not in labor force	
21 .5 or less pers. rm.		
-13 male professionals		
-29 Puerto-Rican born, parentage		
2 Female, all classes		
-16 Female, professionals		

3. HUMAN RESPONSES TO THE ENVIRONMENT

TABLE I (Continued)

Cluster 4 (.9537)**

Students

12 (D) students
5 (D) high school edu.
-39 (D) same house as in '59
37 enrolled in high school
38 enrolled in college
36 (D) enrolled in kindergarten, elementary

Cluster 5 (.7617)**

Persons Alone

27 (D) person alone
-35 (D) unrelated indiv.
20 3-6 or more persons/ household
33 pop. per household
9 female not in labor force
-32 pop. in group qtrs.
10 male employed
-14 male clerical

Cluster 6 (.9645)**

Auto Transportation

50 (D) auto transportation
-51 (D) rail, subway, bus
24 (D) family detached d.u.
19 (D) owner occupied d.u.
-47 single, whites

*Only defining variables are used in computing cluster scores

**Reliability of definers only, reliability of other variables available from author

TABLE II
Types of Cities Among 78 U.S. SMSAs, 1960 Census*

Type 1

34 San Diego
43 Denver
69 Los Angeles

Type 2

3 El Paso
15 Phoenix
42 Sacramento
53 Miami
85 San Jose

Type 3

7 Knoxville
33 New Orleans
47 Norfolk-Portsmouth

Type 4

1 Birmingham
25 Mobile
27 Houston
28 Chatanooga
39 Shreveport
48 Richmond
50 Beaumont-Port Arthur
51 Memphis
62 Nashville
72 Atlanta

Type 5

5 Baltimore
17 Grand Rapids
20 Albany
22 Trenton
31 Philadelphia
45 Syracuse
54 Cincinnati
56 Buffalo
59 St. Louis

Type 6

13 Akron
77 Gary-Hamond-East Chicago

Type 7

23 Tacoma
38 Spokane
64 San Francisco
67 Seattle

Type 8

8 Davenport-Rock Island-Mobile
11 Kansas City
49 Dayton
55 Columbus
65 Minneapolis-St. Paul
70 Salt Lake City
71 Omaha
73 Des Moines

Type 9	Type 11	Type 14
30 San Bernadino-Riverside-Ontario	4 Portland	36 Milwaukee
52 Tampa-St. Petersburg	10 Duluth	37 Rochester
60 Oklahoma City	18 Lansing	83 Cleveland
66 Dallas		
78 Fort Worth	Type 12	Type 15
80 Fresno		
	12 Tulsa	9 Johnstown
Type 10	57 Wilmington	19 Erie
	68 Wichita	21 Utica-Rome
2 Youngstown-Warren	74 Charlotte	24 Wilkes-Barre-Hazelton
14 Louisville	79 Indianapolis	41 Pittsburgh
16 Flint		46 Reading
26 Toledo	Type 13	75 Harrisburg
29 Centon		76 Allentown-Bethlehem-Easton
58 Peoria	61 Chicago	
82 Detroit	84 Newark	

*6 cities omitted from type because of lack of homogeneity. New York Parts I & II omitted because no data was available as separate parts on the 49 variables.

TABLE III
Comparison of Different Typologies*
Differentiating Power on the 49 Variables

Typology	No. of Types	No. of Cities*	No. of Variables Significant (p .01)
1. Maloney's high factors	8	29	31
2. Maloney's low factors	8	27	1
3. Cox and Zannaras	8	64	13
4. Duncan	7	49	31
5. Forstall I	4	66	11
6. Forstall II	8	63	32
7. Forstall III	6	63	3
8. Nelson	9	52	5
9. Key Variables	11	82	3
10. BC TRY (six clusters, 15 types)	15	78	38
11. Hadden & Borgatta	11	50	36

*Since each comparison must involve cities from the original list of 85 SMSAs, and most studies did not include all 85, the number of cities is less than the total. Also, clustering and O-type analysis eliminated cities that were not homogeneous with types. Generally, eliminating cities increases mathematicl chances of significantly different types. Further, in all comparisons, except the BC TRY, when cities did not fall into categories specified by the author, or scores were tied, the cities were eliminated. This greatly increased differentiating power for Hadden and Borgatta especially.

Notes and References

1. Arnold, D., Classification as Part of Urban Management in Berry, B. (ed.) City Classification Handbook: Methods and Application, Wiley, 1972, pages 361-378.
2. Harris, C., A Functional Classification of Cities in the United States, Geographical Review, Vol. 33, No. 1 (January) 1943, pages 86-99.
3. Hadden, J., and Borgatta, E., American Cities, Rand McNally, 1965.
4. Berry, B., (ed.), City Classification Handbook: Methods and Applications, Wiley, 1972.
5. Bauer, R., (ed.), Social Indicators, MIT Press, 1966.
6. National Science Board, Knowledge Into Action: Improving the Nation's Use of the Social Sciences, National Science Foundation, 1969.
7. Department of Health Education and Welfare, Toward A Social Report, U.S. Government Printing Office, 1969.
8. National Academy of Sciences, A Strategic Approach to Urban Research and Development, Printing and Publishing Office, 1967.
9. Hadden, J., and Borgatta, E., American Cities, Rand McNally, 1965.
10. Hadden, J., and Borgatta, E., American Cities, Rand McNally, 1965.
11. Maloney, J., Metropolitan Area Characteristics and Problems, paper delivered at the Convention of Associated Press Managing Editors, October, 1967.
12. Inverted variables, or negative loadings on a factor, mean that the variable is related in a negative fashion. For example, in this factor a negative loading on percent voting means few voted rather than many.
13. Berry, B., (ed.), City Classification Handbook: Methods and Applications, Wiley, 1972.
14. Tryon, R., and Bailey, D., Cluster Analysis, McGraw Hill, 1970.
15. Sokol, R. and Sneath, P., Principles of Numerical Taxonomy, 1963.
16. Tryon, R., and Bailey, D., Cluster Analysis, McGraw Hill, 1970.
17. Tryon, R., Identification of Social Areas by Cluster Analysis, University of California Press, 1955.
18. Shevky, E., and Bill, W., Social Area Analysis: Theory, Illustrative Application and Computational Procedures, Stanford University Press, 1955.
19. Maloney, J., Metropolitan Area Characteristics and Problems, paper delivered at the Convention of Associated Press Managing Editors, October, 1967.
20. Maloney, J., Metropolitan Area Characteristics and Problems, paper delivered at the Convention of Associated Press Managing Editors, October, 1967.
21. Tryon, R., and Bailey, D., Cluster Analysis, McGraw Hill, 1970.
22. County and City Data Book, U.S. Government Printing Office, 1967.
23. Duncan, O., Metropolis and Region, John Hopkins Press, 1960.
24. Cox, K., and Zannaras, G., Designative Perceptions of Microspace: Concepts, A Methodology and Applications, Proceedings of EDRA II, 1970.
25. Maloney, J., Metropolitan Area Characteristics and Problems, paper delivered at the Convention of Associated Press Managing Editors, October, 1967.

26. Forstall, R., A New Social and Economic Grouping of Cities in Municipal Year Book, 1967, The International City Manager's Association, pages 102-159, (1970).
27. Nelson, II., A Service Classification of American Cities, Economic Geography, Vol. 31, 1955, pages 189-210.
28. Hadden, J., and Borgatta, E., American Cities, Rand McNally, 1965.
29. Readers wishing a list of cities classified by key variables into 11 types should write the author at Environmental Research and Development Foundation, 4948 Cherry Street, Kansas City, Missouri 64110.
30. County and City Data Book, U.S. Government Printing Office, 1967.
31. American Jewish Committee, The Reacting Americans: An Interim Look at the White Ethnic Lower Middle Class, American Jewish Committee Institute of Human Relations, 1969.
32. Hood, E., and Kushner, J., Real Estate Finance: The Discount Point System and Its Effects on Federally Insured Home Loans, UMKC Law Review, Vol. 40, No. 1, 1971, pages 1-23.
33. Douglas, P. (ed.), Building the American City, U.S. Government Printing Office, 1968.
34. Douglas, P. (ed.), Building the American City, U.S. Government Printing Office, 1968.
35. Grigsby, W., Suggestions for a Housing Maintenance Program for Cities given at the NAHRO Codes and Rehabilitation Conference, Washington, D.C., March, 1972.

PSYCHOLOGY AND ENVIRONMENTAL MANAGEMENT FOR OUTDOOR RECREATION[*]

George L. Peterson

Associate Professor
Department of Civil Engineering
The Technological Institute
Northwestern University
Evanston, Illinois 60201

Abstract

Recent research in outdoor recreation is reviewed, and it is demonstrated that many of the questions being raised involve psychological relationships between man and his physical environment. Psychologists are needed for joint ventures in the environmental psychology of outdoor recreation. Unfortunately, the demand for environmental psychology has not been as articulate as the military, industrial and aero-space demand for engineering psychology, and the response has been slow. Most of the research in outdoor recreation has been done by non-psychologists. An outline of important problems is presented as part of an appeal for more help from psychology in developing a theoretical and methodological framework for future research.

Introduction

There are many problems in the planning, designing, delivering and management of public service systems and resources that are in need of more help from psychology. This paper focuses on outdoor recreation. The purpose is to review some practical problems that are not being solved effectively, and to show that there is need and opportunity for psychologists to help remedy the situation. The subject might as well be housing, transportation, residential environment, or any of numerous public aspects of the physical environment, because a similar list of needs and opportunities can be cited for each. Recreation is peculiarly interesting, however, due to the evasive nature of many of its variables which are distinctly psychological in character.

---

[*]Preparation of this paper was supported in part by Grant No. EC00301 from the U.S. Public Health Service. Cooperative Research Project No. 13-253 with the North Central Forest Experiment Station, U.S.D.A., Forest Service, also contributed. The paper is based on a verbal presentation prepared for the Symposium on Integration of Behavioral Science into Public Works Engineering, American Psychological Association Meetings, Honolulu, Hawaii, September, 1972.

Lest anyone underrate the importance of recreation as an object for research and as a matter for great public concern, the facts speak for themselves. Driver (1) has convincingly demonstrated the economic and social importance of recreation. Jensen (2) has summarized numerous arguments that lead to the conclusion that outdoor recreation has established itself economically and socially as an activity of major importance.

But whatever the tendentious justifications, millions of people consume vast quantities of time, spend billions of dollars, and exhibit great energy in pursuit of enjoyment through sport and play. If only to make those expenditures more efficiently rewarding, recreation is more than justified as an object for research. Much of this research must be concerned with environmental psychology.

The Man-Environment Problem

From the point of view of one who plans, designs, delivers or manages man's physical environment, there are several important facets to the problem. Whether the environment in question is a tool, constraint, or final demand, it bears a functional relationship to human activity. If this human activity and its purposes are to be enhanced, the functional relationship between man and environment must be understood. One the one hand, this might result in criteria for modifying, designing or controlling the environment so that it better serves human purposes. On the other hand, it might require services such as information, education or assistance which improve man's ability to use the environment or to adjust himself to it (Cypra & Peterson (3)). There may also be a need to control the behavior of the people who interact with a particular environment. The choice of manner of adjustment is, of course, a matter of policy; but whatever the choice, the human being is a primary variable in the problem, and he should be understood. The object is to enhance human activity by improving the cooperation between man and his physical environment. This is as true in recreation as it is in housing or transportation.

The Contribution of Engineering Psychology

Psychologists have been working together with engineers for a long time, and have made remarkable contributions to the enhancement of man-maching interaction. As explained by Grether (4), "Engineering psychology is concerned with designing equipment and systems to match the capabilities and characteristics of people. Less commonly, engineering psychology is concerned with procedures for operation of equipment, with testing of equipment to determine its suitability for human operation, or with prediction of personnel and training requirements created by newly designed equipment."

Notable contributions have occurred during the past two or three decades in defense, aerospace and industry (Grether (5)). It is unfortunate that as much concerted effort has not been devoted to the domestic front, where the people who are being protected in war, who watch the moon walk on television, or who pay for the goods produced by industry live out their daily lives. This is not to belittle what has been learned about "human factors" in these "military-industrial" endeavors, because psychology has contributed remarkably to the goals of these programs, and much of what has been learned is applicable to domestic problems —when someone gets around to applying it. But much remains to be done, both basic and applied, and the engineers are as slow to respond as the psychologists in applying what has been learned already. In McCormick's excellent survey of human factors engineering (6), the chapter on "Living Environment" is short and superficial. In his words, "to contemplate the human factors aspects of this living environment requires a significant jump or two from the conventional human factors context of airplanes, industrial machines, and automobiles." Let's spend a little more time making that jump or two!

Why Has the "Living Environment" Been Neglected?

Psychologists and engineers are a lot like everybody else in at least one way: they tend not to do things that nobody wants to pay for, and this is probably the answer to the question about why the scientific study of "living environments" has been neglected. In discussing the reasons why psychology has not contributed more to the improvement of consumer products, Grether (7) cites an inability of the public to articulate its demand for or to judge the benefits of safety, ease and efficiency of operation, and ease of maintenance. Recently, we are seeing movement in the other direction.

A list of the obstacles facing "environmental psychology" on the domestic front might include the following:

    1. <u>Diffuse and inarticulate (unorganized) demand</u>: There is no federal agency or wealthy industry offering large-scale contracts and grants for research. This is a matter of public and industrial policy responding to the absence of organized public pressure. Emergency and/or unified national priority and/or profitable markets largely motivated the advances in engineering psychology. People are simply not expressing their domestic needs in a market situation, or in terms of political pressure. Does this mean that the need is not there, or that the nature of the problem prevents its articulation? Probably the latter.

    2. <u>Complexity, lack of structure and uncertainty of response</u>: When the aspiring researcher casts about for a problem, money is not likely to be his only consideration, especially if he has academic intentions. He is likely to be more interested in the currency of prestige, gained in academia by publishing large numbers of definitive papers in a short period of time. Naturally, the advantage will go to the researcher who picks manageable problems in conventional disciplines

and grows up under the tutelage of an established professor who has accumulated an inventory of solvable problems. Consequently, the "system" is biased against the kind of innovative research needed.

The Recent Growth of Environmental Psychology

In spite of these obstacles, or perhaps as a symptom of shifting circumstances, psychologists have begun to take an interest in domestic man-environment problems. Craik (8) has recently written a review of environmental psychology in which he brings together related work from a wide variety of disciplines. Wohlwill (9)(10) also has contributed significantly from the "inside," and the book by Prohansky, Ittleson and Rivlin (11) is a useful, though incomplete, collection of related papers. The effort by Gary Winkel and others to establish the journal Environment and Behavior also indicates the growing interest in the field.

However, most of the applied work to date has been done by a variety of researchers in a variety of disciplines, as demonstrated by Kates and Wohlwill (12), Craik (13), and Prohansky, Ittleson and Rivlin (14). What little theory that exists is either incomprehensible to "outsiders" or non-operational (verbal-conceptual). The applied work consists mainly of a collection of tentative, partial, or non-generalizable answers to a diverse spectrum of largely unrelated questions. Many of these studies have been conducted in a theoretical vacuum, behind well-guarded provincial walls. As Los Angeles has been called a collection of suburbs looking for a central city, so the current state of environmental psychology can be called a collection of answers looking for a theory, as though spawned by frivolous curiosity.

A Brief Review of Psychology in Outdoor Recreation Research

Potentially, one of the most fertile prospects for environmental psychology is outdoor recreation. Here the "recreator" is frequently involved in a direct transaction with the physical environment, and is profoundly influenced in his behavior and satisfaction by the condition of the environment and his perception of it. A multitude of problems requiring the help of psychology exist, as is well-illustrated by a brief look at some of the recent studies by recreation researchers.

In attempting to answer problems associated with the management of wilderness and wildlands, researchers inevitably find themselves dealing with psychological questions. Stone and Taves (15) studied wilderness camping groups in the Quetico-Superior for the purpose of presenting "...a view of man in the wilderness and an agenda of research problems that concern sociology." In spite of their avowed sociological interest, a significant portion of their work was concerned with "imagery of wilderness" and "motivation of wilderness travelers." Subsequently, Bultena and Taves (16) focused more specifically on the "motives inducing vacationers to visit a forested recreation area, how they interpreted their visit, and the implications of this for forest policy and management." Their work included a "social-psychological explanation" of an apparent conflict between wilderness fascination and a desire for conveniences or improvements.

Later studies by Lucas (17)(18)(19)(20)(21)(22)(23)(24) included concern with (a) perception of wilderness, (b) visitor reaction to specific activities and attributes (e.g., fishing and timber harvesting), (c) social carrying capacity (i.e., crowding), (d) conflicting uses, (e) conflicts between values and perceptions of managers and users, (f) the effect of perception on behavior, and other questions with obvious psychological relevance.

Shafer has been involved in several studies that sought answers to psychological questions and employed psychological methods. (Shafer and Burke (25); Shafer and Mietz (26); Shafer (27); Shafer, Hamilton and Schmidt (28); Rutherford and Shafer (29)). These studies focused directly on preference, perception and motivation in the context of forest recreation. In their work on "Wilderness Users in the Pacific Northwest," Hendee, Catton, Marlow and Brockman (30) studied tastes, preferences and attitudes. They used simple attitude measurement techniques to develop a Wildernism-Urbanism scale, and were able to differentiate between "wilderness-purist" and "urbanist" users. They also analyzed user attitudes toward alternative types of management policy. Later work by McKechnie (31) and Craik (32) has pursued the question of "environmental disposition" in more depth, and has resulted in The Environmental Response Inventory. Catton's work on wilderness motivation (33) raises many psychological questions that call for more attention by psychologists.

Recent work by Gilbert (34), Peterson and Gilbert (35), and Gilbert, Peterson and Lime (36) aims at developing a mathematical model for predicting canoe travel behavior in the Boundary Waters Canoe Area. The need to understand the canoeist's travel decision process leads to many interesting questions of motivation, perception, preference, attitude and behavior. Stanky's doctoral dissertation (37) includes a concern with the psychology of crowding in the wilderness setting. Specifically, he studied the sensitivity of user satisfaction to encounters with other users and with other modes of travel. A study by Peterson (38)(39)(40) explores the usefulness of psychological inventories for describing and analyzing wilderness motivations, perceptions and satisfactions. Among other things, motivation and perception inventories are used together to measure the degree of congruency between user "aspiration and the perceived reality of experiences" as called for by Bultena and Klessig (41). The inventories are also used to compare managers with users, and to assess the significance of situation and sampling bias in wilderness surveys.

Outside the realm of wilderness and forest recreation, the psychological questions in outdoor recreation research have been equally abundant. Only a few examples will be noted here. Jensen (42) lists (1) developing appreciation for nature, (2) enhancing individual satisfaction and enjoyment, (3) providing opportunity for diversion and relaxation, (4) developing physical fitness, and (5) developing desirable social patterns as objectives of outdoor recreation. This shows clearly that the management of recreation facilities and resources requires a great deal of information about the psychology of recreation. As further illustration of this point, Dattner

(43) discusses "The Psychology of Play." He unfolds the process by which children develop intellectually, and briefly explains the roles of various forms of play in the process. Mercer (44) has written a brief review of "The Role of Perception in the Recreation Experience," and provides a list of over one hundred references. Wholwill and Carson (45) devote approximately one-sixth of their book on Environment and the Social Sciences to psychological aspects of recreation.

Recent work in outdoor recreation focusing on urban or generalized issues and having strong psychological components has been published by Brown (46), Martens (47), Neulinger and Breit (48)(49), Neumann (50)(51), Peterson and Neumann (52), West and Merriam (53), and Bishop, Peterson and Michaels (54). Brown (55) views participation in recreation as discretionary behavior that is sensitive to changes in economic sentiments —that is, changes in the recreation consumer's perception of the state of the economy. Martens (56) examines "...the individual's motives for participation in such group activities as football, softball, volleyball, soccer and basketball." He identifies two distinguishable types of motives, task motives and affiliation motives, and finds that individuals generally are more task-motivated than affiliation-motivated. He also examines the effect of team success on the strength of these two kinds of motives.

Neulinger and Breit (57)(58) worked toward the development of an instrument for measuring leisure attitudes. They state that "there is little doubt that leisure attitudes are closely linked to the core of personality." Peterson and Neumann (59) propose an explanation of recreation preference in terms of personal characteristics on the one hand and attributes of the recreation opportunities on the other. From a pilot study of the visual characteristics of beaches, they report that preference is sensitive to such visual attributes as vegetation, sand quality, crowding, etc. Two types of users are identified: those seeking scenic natural experiences and those seeking urban social experiences. The two types respond differently to the visual environment and have conflicting aspirations, but use the same facilities simultaneously. West and Merriam (60) note a weak relationship between family cohesiveness and participation in outdoor recreation, which suggests that the recreation activity somehow modifies the way family members relate to themselves and to each other. Finally, Bishop, Peterson and Michaels (61) studied the preferences of children for the play environment. Their study focused on four problems: (1) testing the reliability of measurement methods; (2) testing the behavioral validity of measured preferences; (3) description of preferences, including sex and ethnic differences; and (4) evaluation of the congruity between child preferences and the beliefs of adult designers about child preferences. (Bishop, 1971 (62); Peterson et. al., 1972 (63)).

Obviously, a great many more examples could be cited; but the above list serves to demonstrate that recreation research is beginning to intersect with psychology, and would be greatly enhanced by more involvement from psychologists.

An Outline of Significant Problems

An obvious problem with much of this psychology-related research in outdoor recreation is that as a whole it is disjointed, piecemeal and lacking in unifying structure. Each piece of work seems to attack a specific problem, often in a specific place, without reference to generalized objectives or unifying theory. This should not be taken as a blanket criticism of individual projects. Rather, it is indicative of the growing need for a joint venture in "recreational psychology," perhaps as a subdivision of what has come to be called "environmental psychology." Enough independent problems have appeared to indicate a need to bring them together as parts of a unified whole. When this whole has been defined, together with its parts and the interrelationships among those parts, a second generation of research can proceed more maturely, with order and a higher degree of effectiveness. In other words, the first generation of research has consisted of a lot of independent answers to unstructured questions. The second generation ought to proceed from a logical overall structure from which important questions can be identified and ordered in priority.

This paper does not pretend to fill this gap by providing the matrix for future research. It does propose, however, that one way to progress toward order is to examine the unanswered and poorly answered questions confronting the public effort to plan, design, deliver and manage outdoor recreation opportunities. Admittedly this is an unorthodox approach (although we don't like to admit it). It means applying available methods to important problems, rather than applying selected problems to important methods.

In a perceptive but brief discussion of the potential contributions of psychology to recreation resource management, Driver (64) identifies eight responsibilities of the manager (here paraphrased for brevity): (1) to assess the recreation resource; (2) to protect the resource; (3) to develop and manage the resource; (4) to provide opportunities that satisfy user needs and preferences; (5) to enhance the experience of the user; (6) to publicize and market the program; (7) to function effectively in the decision process and organizational hierarchy; and (8) to assist in adapting social institutions to changing needs. In each of these categories, he demonstrates needs and opportunities for psychological study.

The problems associated with management process and institutional arrangements (7 and 8, above) are beyond the scope of this paper. Focusing more specifically on the psychological aspects of the opportunity itself and its users, the management problems can be simplistically categorized as (1) public policy and priority, (2) management of resource, (3) management of user, and (4) manager-user interaction.

Public Policy and Priority: This is a matter of deciding whether, what kinds and how much recreation opportunity to provide as public service, as well as how to stimulate or otherwise control the production of private service in the recreational domain. The problem certainly includes articulation of public values, needs, desires and aspirations. Ideally, the matters of value should be resolved through the political process, but the measurement of values and attitudes can be a useful lubricant. It is of great help in identifying conflicting groups and in conflict resolution, as well as in describing the values, attitudes and objects themselves, and the extent and nature of demand. As Wingo (65) puts it, "The social responsibility of recreation policy, then, is to assure that the production and distribution of opportunities for play are consistent with the constantly evolving meaning of the good life." In more concrete terms and with reference to wilderness, Lucas (66) says that an important policy question is "...whether any wilderness areas should be officially established." If it is decided to provide officially designated wilderness areas, then one has the problem of how much, what kind, where and for whom.

Assessment of demand thus becomes an important problem. It is not sufficient to describe and project "user days." Providing more of what people now use can lead to the unhappy situation in which people motivated by the desire for solitude in a scenic natural setting are provided with more "urban-social" beaches that are crowded and "plastic." (Peterson and Neumann (67)). What is needed is an understanding of the reasons why people make their recreational choices, including their motivations, the constraints under which motivations influence behavior, and the processes by which motivations are formed (Peterson, Hanssen and Bishop, 1971 (68)). A projection of "who wants what" is better, but not much better, than a projection of "who does what." What we need is a theory of recreational choice from which demand can be _explained_ and _predicted_, particularly _latent_ _demand_ and response to heretofore unexperienced alternatives.

Another policy matter closely tied to the problem of whether, how much, and what kind of recreation opportunity should be provided with public funds is the problem of evaluating the social and personal benefits to be generated by the investment. By "benefits" I do not mean "user days" or the relatively meaningless contribution to the gross national product. What I do mean are the therapeutic, "satisfactional" and socially constructive products of the investment. For example, the Kerner Commission reports that deficiencies in urban recreation opportunity may be a cause of riots and disorders (Driver (69)). Another example is the family cohesiveness impact of outdoor recreation reported by West and Merriam (70). What are the social, physical and psychological processes by which recreation produces personal satisfaction and public welfare? Related to this need to allocate public investments in recreation wisely, is the need to justify the investment so as to obtain public "permission." An investment or preservation that cannot be defended usually will fall prey to more concrete alternative demands. It would help if behavioral scientists could develop ways of articulating the specific benefits of specific forms of recreation.

Management of Facilities and Resources: Even when public policy has provided a recreation agency with funds for a specific program —say a wilderness area —and guidelines or goals for its management, there remain numerous decisions and acts of management that would benefit from psychological input. For example, there is the decision of what kind and how much management service to provide in the control, development and/or design of facilities and resources. In the Boundary Waters Canoe Area (BWCA), Congress has specified that the Area shall be managed as wilderness, but has not been very specific about defining what wilderness is. The psychologists

can help again by identifying the spectrum of specific motivations and perceptions
that bring people to an area or type of recreation —or better still, the alternative
motivations that might be fulfilled by an area or type of recreation under various
management options. If these motivations can be expressed in terms of specific environmental and social conditions, the manager is in a good position to adopt the
options that maximize benefits. He is also in a better position to provide a spectrum or variety of opportunities and facilities, rather than catering to aggregate
demand. The need for variety that matches the spectrum of demand has been argued by
Wagar (71). It is also necessary to know how much of the demand is generalized and
how much is specific to a place, activity or type of environment (Shafer & Burke (72)).

In addition to deciding what opportunities to provide and how to manage the attributes of those opportunities, the manager often has to make capacity decisions. In
the case of playgrounds, parks, or auto campgrounds, for example, it is a question
of deciding how many or how much of each kind of facility (Lucas (73)). With a
wilderness area like the BWCA, it is a matter of determining carrying or use capacity (Lucas (74), Lime (75)). One aspect of this problem is the capacity of an area
to withstand use without significant ecological damage. The other side of the problem is of more interest to psychologists: how does user satisfaction vary with level
of use, i.e., crowding and evidence of crowding, and how does this vary from user to
user? In heavily used areas where capacity cannot be increased without damaging the
resource, it may be necessary to limit use.

The psychology of user reaction to crowding in wilderness areas is complicated,
whereas most research attempts to deal with the problem have treated it simplistically. Several different types of motivation are conceivable. One is a desire
simply to be alone. Another is a desire to be unique, to feel that one has done
something that other people don't or can't do. A third is to avoid the consequences
that follow from crowding: competition for campsites, delay on portages or trails,
noise, litter, depletion of fish, etc. Fourth, to some extent a trek into the wilderness might be an escape from anxieties and tensions that result from interpersonal relationships, and the presence of other people may constitute a reminder or
threat. A fifth possibility is a desire to avoid certain "kinds" of people whom one
perceives to have unsympathetic values or incompatible behaviors; for example, the
conflict between paddle canoeists and motor canoeists in the BWCA (Lucas (76)).
Several doctoral dissertations could be written on the psychology of the reaction
to crowding in wilderness areas, and the result would be much more meaningful definitions of "social carrying capacity."

<u>Management of Users</u>: The notion of ecological and social carrying capacity in
wilderness recreation suggests that where demand exceeds capacity, users should be
controlled. The need to manage demand or control behavior occurs in many recreational situations. Protection of the resource by restricting use levels or modifying specific behaviors is one problem. A second is the need to preserve the quality
of the experience vis-a-vis user conflict and interference with each other. A third
is to attract users that may not be aware of potential benefits, or, alternatively,
to discourage people who are there for the "wrong" reasons and might be better satisfied elsewhere. Fourth, it may be desirable to inform, train or educate users
so as to enhance their ability to use and enjoy a facility.

All of these problems concern the control of behavior, perceptions, or belief, and are thus of direct concern to the psychologist. In the BWCA, for example, over-use on certain heavily travelled routes already has prompted restriction on these lakes (Rupp (77)). Zoning to separate motors from paddle canoes has been in effect for several years. In order to control behavior effectively, the manager needs to predict the effects of alternative options. In the BWCA one requirement is for an explanation of travel behavior that is sensitive to the alternative control methods available (Gilbert, Peterson and Lime (78)). To describe behavior is not sufficient, because several options operate on the reasons for travel, rather than on the travel itself. An understanding of the recreationists' decision process is clearly needed, which is another problem in psychology.

Manager-User Interaction: Rupp (79) calls for "...increased 'involvement' of all concerned parties, in the decision making process" in the management of the BWCA. There are many reasons why recreationists as well as other users or potential users of recreation resources ought to participate in the management process. Only one of these is communication as discussed earlier in connection with articulation of values. However, even though management may understand users' recreation needs, preferences and perceptions, they may still find themselves with political problems or simply disgruntled publics, if these publics are not understood and treated properly. What is "proper treatment" is a complicated question, as every public official knows, but "community participation" is a hot issue in all aspects of public planning and management these days. We need more help from psychology in understanding the dynamics of user-manager interaction.

Part of the difficulty occurs because the manager and user simply don't understand each other's values or points of view. Several authors have recently described value and perception conflicts between managers and users in recreational settings (Lucas (80); Clark, Hendee and Campbell (81); Peterson (82); Peterson (83)). In the case of the author's experience, what began as an uninformed user's mistrust of BWCA management ended, after objective study, as a greater appreciation of the manager's broader perspective, his greater technical knowledge and his unique values and motivations. Conflicts remain, but now they are in the open and can be discussed, negotiated or accepted intelligently. The manager may have different views, but many of his reasons now have become clear.

Finally, there is the matter of user reaction to the results of management, specifically to the controls imposed. In wilderness, an overt control of behavior may be as inappropriate as an educational Visitor Center on a remote lake. Controls and other manifestations of management should be unobtrusive, at least in wilderness areas, and this calls for two more inputs from psychology: (1) to what extent do alternative programs and controls intrude, psychologically, into the recreational experience; and (2) how does one control behavior unobtrusively in various settings? In the BWCA, for example, advance registration for campsites would be an obtrusive control. It is direct regimentation of behavior, which may be contrary to the user's reasons for visiting the area. Unobtrusive control might consist of subtle alterations of "travel friction" within the area, or information dissemination at the access points (vis-a-vis crowding, fishing, etc.) (Gilbert, Peterson and Lime (84)).

Closure

Thus, it is clear that psychological attention to the "living environment" has been sparse, often superficial, usually conducted by non-psychologists, and in the disjointed context of specific applied problems. A unified attack by psychologists on domestic environments is needed badly. Outdoor recreation is a particularly interesting candidate for such activity, because its psychological dimensions are legion. The principal contributions that psychologists can make are in the areas of (1) unifying theoretical structure, (2) application of measurement and experimental methods, and (3) application of the considerably specialized knowledge already available about human behavior. Perhaps a more pressing need, however, is for institutional revisions that allow the demand to be more focused, including appropriate academic programs in environmental psychology and more funding of research in these areas.

Notes

1. Driver, B.L., "Potential Contributions of Psychology to Recreation Resources Management," in Environment and the Social Sciences: Perspectives and Applications, edited by J.F. Wohlwill and D.H. Carson. Washington, D.C.: American Psychological Association (1972), pp. 233-248.
2. Jensen, C.R., Outdoor Recreation in America. Minneapolis: Burgess Publishing Company (1970).
3. Cypra, K., and G.L. Peterson, "Technical Services for the Urban Floodplain Property Manager: Organization of the Design Problem," Natural Hazard Research Working Paper No. 12. Toronto: Department of Geography, University of Toronto (1969).
4. Grether, W.F., "Engineering Psychology in the United States," American Psychologist, Vol. 23, No. 10 (1968), pp. 131-157.
5. Grether, W.F. See Reference (4).
6. McCormick, E.J., Human Factors Engineering. New York: McGraw-Hill (1970).
7. Grether, W.F. See Reference (4).
8. Craik, K.H., "Environmental Psychology," in New Directions in Psychology 4. New York: Holt, Rinehart & Winston (1970), pp. 1-121.
9. Wohlwill, J.F., "The Physical Environment: A Problem for a Psychology of Stimulation," Journal of Social Issues, Vol. XXII, No. 4 (1966), pp. 29-38.
10. Wohlwill, J.F., "The Emerging Discipline of Environmental Psychology," American Psychologist, Vol. 25, No. 4 (April, 1970), pp. 303-312.
11. Prohansky, H.M., et. al., Environmental Psychology. New York: Holt, Rinehart & Winston (1970).
12. Kates, R.W., and J.F. Wohlwill, "Man's Response to the Physical Environment," Journal of Social Issues, Vol. XXII, No. 4 (October, 1966).
13. Craik, K.H. See Reference 8.
14. Prohansky, H.M., et. al. See Reference 11.
15. Stone, G.D., and M.J. Taves, "Research into the Human Element in Wilderness Use," Proceedings of the Society of American Foresters (1956), pp. 26-32.
16. Bultena, G.L., and M.J. Taves, "Changing Wilderness Images and Forestry Policy," Journal of Forestry (March, 1961), pp. 167-171.
17. Lucas, R.C., "Visitor Reaction to Timber Harvesting in the B.W.C.A.," Research

Note LS-2, U.S.D.A. Forest Service, Lake States Forest Experiment Station. St. Paul, Minnesota (1963).

18. Lucas, R.C., "The Status of Recreation Research Related to Users," Proceedings of the Society of American Foresters. Boston (1963), pp. 127-130.

19. Lucas, R.C., "Wilderness Perception and Use: The Example of the Boundary Waters Canoe Area," Natural Resources Journal (January, 1964), pp. 394-411.

20. Lucas, R.C., "Recreational Use of the Quetico-Superior Area," U.S.D.A. Forest Service Research Paper LS-8, Lake States Forest Experiment Station. St. Paul (1964).

21. Lucas, R.C., "The Recreational Capacity of the Quetico-Superior Area," U.S.D.A. Forest Service Research Paper LS-15, Lake States Forest Experiment Station, St. Paul Minnesota (1964).

22. Lucas, R.C., "The Importance of Fishing as an Attraction and Activity in the Quetico-Superior Area," U.S.D.A. Forest Service Research Note LS-61, Lake States Forest Service Experiment Station. St. Paul, Minnesota (1965).

23. Lucas, R.C., "The Contributions of Environmental Research to Wilderness Policy Decisions," Journal of Social Issues, Vol. XXII, No. 4 (October, 1966), pp. 117-126.

24. Lucas, R.C., "User Evaluation of Campgrounds on Two Michigan National Forests," U.S.D.A. Forest Service Research Paper NC-44, North Central Forest Experiment Station. St. Paul, Minnesota (1970).

25. Shafer, E.L., Jr., and H.D. Burke, "Preferences for Outdoor Recreation Facilities in Four State Parks," Journal of Forestry, (July, 1965), pp. 512-518.

26. Shafer, E.L., Jr., and J. Mietz, "Aesthetic and Emotional Experiences Rate With Northeast Wilderness Hikers," Environment and Behavior, Vol.1, No. 2 (1969), pp. 187-197.

27. Shafer, E.J., Jr., "Perception of Natural Environments," Environment and Behavior, Vol. 1, No. 1 (June, 1969), pp. 71-82.

28. Shafer, E.L., Jr., J.E. Hamilton and E.A. Schmidt, "Natural Landscape Preferences: A Predictive Model," Journal of Leisure Research, Vol. 1, No. 1 (Winter, 1969), pp. 1-20.

29. Rutherford, W., Jr., and E.L. Shafer, Jr., "Selection Cuts Increased Natural Beauty in Two Adirondack Forest Stands," Journal of Forestry (June, 1969), pp. 415-419.

30. Hendee, J.C., et. al., "Wilderness Users in the Pacific Northwest - Their Characteristics, Values, and Management Preferences," U.S.D.A. Forest Service Research Paper PNW-61. Northwest Forest and Range Experiment Station, Portland, Oregon (1968).

31. McKechnie, G.E., "Measuring Environmental Dispositions with the Environmental Response Inventory," EDRA TWO, Proceedings of the 2nd Annual Environmental Design Research Association Conference, edited by J. Archea and C. Eastman. Pittsburgh (1970).

32. Craik, K.H., "The Environmental Dispositions of Environmental Decision-Makers," Annals of the American Academy of Political and Social Science, Vol. 389 (1970), pp.87-94.

33. Catton, W.R., "Motivations of Wilderness Users," Pulp & Paper Magazine of Canada (December 19, 1969), pp. 121-126.

34. Gilbert, C.G., "The Use of Markov Renewal Theory in Planning Analysis: An Application to the Boundary Waters Canoe Area." Ph.D. Dissertation, Department of Civil Engineering, Northwestern University, Evanston, Illinois (June, 1972).

35. Peterson, G.L., and C.G. Gilbert, "Application of Markov Renewal Theory to Travel Behavior in the Boundary Waters Canoe Area," Proceedings of the Fall IEEE Electronics Conference (October, 1971).

36. Gilbert, C.G., G.L. Peterson, and D.W. Lime, "Toward a Model of Travel Behavior in the Boundary Waters Canoe Area," Environment and Behavior, Vol. 4, No. 2 (June, 1972), pp. 131-157.

37. Stanky, G.H., "The Perception of Wilderness Carrying Capacity: A Geographic Study in Natural Resources Management," Ph.D. Dissertation, Department of Geography,

Michigan State University (1971).

38. Peterson, G.L., "Motivations, Perceptions, Satisfactions and Environmental Dispositions of BWCA Users," Research Report to the North Central Forest Experiment Station, St. Paul, Minnesota. Northwestern University, Evanston, Illinois (1971).

39. Peterson, G.L., "Attitudes, Perceptions and Environmental Dispositions: Comparison of Managers and Canoeists in the Boundary Waters Canoe Area," (Forthcoming, 1972).

40. Peterson, G.L., "Evaluating the Quality of the Wilderness Experience Congruity among Preferences and Perceptions," (Forthcoming, 1972).

41. Bultena, G.L., and L.L. Klessig, "Satisfaction in Camping: A Conceptualization and Guide to Social Research," Journal of Leisure Research, Vol. 1, No. 4 (1969), pp.348-354.

42. Jensen, C.R. See Reference (2).

43. Dattner, R., "The Psychology of Play," in Design for Play. New York: Van Nostrand 1969), pp. 23-30.

44. Mercer, D., "The Role of Perception in the Recreation Experience: A Review and Discussion," Journal of Leisure Research, Vol. 3, No. 4 (Fall, 1971), pp. 261-276.

45. Wohlwill, J.F., and D.H. Carson, editor, Environment and the Social Sciences: Perspectives and Adaptations, American Psychological Association (1972).

46. Brown, P.J., "Sentiment Changes and Recreation Participation," Journal of Leisure Research, Vol. 2, No. 4 (Fall, 1970), pp. 264-268.

47. Martens, R., "The Influence of Success and Residential Affiliation on Participation Motivation," Journal of Leisure Research, Vol. 3, No. 1 (Winter, 1971), pp. 53-58.

48. Neulinger, J., and M. Breit, "Attitude Dimensions of Leisure," Journal of Leisure Research, Vol. 1, No. 3 (Summer, 1969), pp. 255-261.

49. Neulinger, J., and M. Breit, "Attitude Dimensions of Leisure: A Replication Study," Journal of Leisure Research, Vol. 3, No. 2 (Spring, 1971), pp. 108-115.

50. Neumann, E.S., "Evaluating Subjective Response to the Recreation Environment," Ph.D. Dissertation, Department of Civil Engineering, Northwestern University. Evanston, Illinois (June, 1969).

51. Neumann, E.S., and G.L. Peterson, "Perception and Use of Urban Beaches," EDRA 2 (1970), pp. 327-333.

52. Peterson, G.L., and E.S. Neumann, "Modeling and Predicting Human Response to the Visual Recreation Environment," Journal of Leisure Research, Vol. 1, No. 3 (Summer, 1969), pp. 219-237.

53. West, D.C., and L.C. Merriam, "Outdoor Recreation and Family Cohesiveness: A Research Approach," Journal of Leisure Research, Vol. 2, No. 4 (Fall, 1970), pp. 251-259.

54. Bishop, R.L., G.L. Peterson, and R.M. Michaels, "Measurement of Children's Preferences for the Play Environment," Proceedings, 3rd Annual Environmental Design Research Association Conference, edited by W.J. Mitchell (1972), pp. 6-2-1 — 6-2-9.

55. Brown, P.J. See Reference (46).

56. Martens, R. See Reference (47).

57. Neulinger, J., and M. Breit. See Reference (48).

58. Neulinger, J., and M. Breit. See Reference (49).

59. Peterson, G.L., and E.S. Neumann. See Reference (52).

60. West, D.C., and L.C. Merriam. See Reference (53).

61. Bishop, R.L., G.L. Peterson, and R.M. Michaels. See Reference (54).

62. Bishop, R.L., "Towards the Synthesis of Environmental Design Criteria from Children's Preferences for the Visual Appearance of Urban Play Environments," Ph.D. Dissertation, Department of Civil Engineering, Northwestern University, (1971).

63. Peterson, G.L., R.L. Bishop, R.M. Michaels, and G.J. Rath, "Children's Choice of Playground Equipment: Development of Methodology for Integrating User Preferences

into Environmental Engineering," Journal of Applied Psychology (1972, in press).
64. Driver, B.L. See Reference (1).
65. Wingo, L., Jr., "Recreation and Urban Development: A Policy Perspective," Annals of the American Academy of Political and Social Science, Vol. 352 (1964), pp. 129-140.
66. Lucas, R.C. See Reference (23).
67. Peterson, G.L., and E.S. Neumann. See Reference (52).
68. Peterson, G.L., J.U. Hanssen, and R.L. Bishop, "Toward an Explanatory Model of Outdoor Recreation Preference," paper prepared for the Symposium on Consumer Behavior and Environmental Design, American Psychological Association Meeting, Washington, D.C. (September, 1971).
69. Driver, B.L. See Reference (1).
70. West, D.C., and L.C. Merriam. See Reference (53).
71. Wagar, J.A., "Quality in Outdoor Recreation," Trends in Parks and Recreation 3 (July, 1966).
72. Shafer, E.L., Jr., and H.D. Burke, "Preferences for Outdoor Recreation Facilities in Four State Parks," Journal of Forestry (July, 1965), pp. 512-518.
73. Lucas, R.C. See Reference (24).
74. Lucas, R.C. See Reference (18).
75. Lime, D.W., "Research for Determining Use Capacities of the Boundary Waters Canoe Area," Naturalist, Vol. 22, No. 4 (Winter, 1970), pp. 9-13.
76. Lucas, R.C. See Reference (21).
77. Rupp, C.W., "Boundary Waters Canoe Area Management," Naturalist, Vol. 21, No. 4 (Winter, 1970), pp. 2-7.
78. Gilbert, C.G., G.L. Peterson, and D.W. Lime. See Reference (36).
79. Rupp, C.W. See Reference (77).
80. Lucas, R.C. See Reference (24).
81. Clark, R.N., J.C. Hendee, and F.L. Campbell, "Values, Behavior and Conflict in Modern Camping Culture," Journal of Leisure Research, Vol. 3, No. 3 (Summer, 1971), pp. 143-159.
82. Peterson, G.L. See Reference (38).
83. Peterson, G.L. See Reference (39).
84. Gilbert, C.G., G.L. Peterson, and D.W. Lime. See Reference (36).

FOUR ENVIRONMENTAL RESEARCH AND DESIGN FOR DIFFERENT AGE GROUPS

Chairman: Jerome T. Durlak, Dept. of Sociology, York Univ., Canada

Panelists: Thomas Byerts, Gerontological Society, Washington, D. C.
Robin Moore, Dept. of Arch., Univ. of California, Berkeley, California
Harold Proshansky, Environmental Psychology, C.U.N.Y.
Fred W. Vondracek, Human Development, Penn State Univ.

Authors: Joseph Muntanola Thornberg, "Child's Conception of Places to Live In"
Leanne G. Rivlin, Maxine Wolfe and Marian Beyda, "Age Related Differences in the Use of Space"
Edward H. Steinfeld, "Physical Planning for Increased Cross-Generation Contact"
Paul G. Windley, "Measuring Environmental Dispositions of Elderly Females"
Sandra C. Howell, "Researching the Behavioral Implications of Residential Design: The Case of Elderly Housing"

ENVIRONMENTAL RESEARCH AND DESIGN FOR DIFFERENT AGE GROUPS 4.0

Jerome T. Durlak

Assistant Professor
Urban Studies Program
Division of Social Science
York University

In Toronto there is a new Science Centre which has been designed with many facilities that permit the users to "learn by doing". If you go there on a weekend <u>sedate</u> children use the facilities in the way the designer intended. The main reason the children use these facilities in the "correct" way is because their parents are present. During the week when parents are usually not present the Science Centre is filled with <u>excited</u> children playing hide and seek, sliding down elevator railings and generally using the physical environments in ways for which the designer had never planned.

Environmental researchers and designers are interested in manipulating built environments or having the users manipulate environments in ways which will allow people to do the things they want to do and also to create new possibilities of activity. The researcher-designer and the design team must concern himself, herself or themselves with how the physical setting, together with the formal or informal organizational rules and the concepts and attitudes of the users, organize the activities of the users and allow the users to modify their environments. The built form may attempt to instruct, facilitate or support the user and the user may adapt, change or be unaware of the physical environment. Finally, as Kevin Lynch points out elsewhere, the behaviour of the users and the physical setting "may be antagonistic, stable or fluid, demanding or permissive, repetitive or open ended".

The researchers and designers at this session realize that understanding how people use and value the spatial environment is the key to planning built environments that fit human purposes. At the same time researchers and designers get frustrated very quickly because of the time involved in conducting research, the paucity of available data, the costs and efforts involved, and the difficulty of conducting research in a way that comes up with design requirements that are useful, feasible, generalizable and readily understood by everyone involved.

The following research presentations demonstrate that research in environmental behavior is beginning to have an important role to play in the design of environments for different age groups. It is also important to note the cross fertilization that is taking place. The architect-researchers in these presentations are using psychological theory and techniques to answer their questions and the psychologists are sitting down and talking with architects, planners and public officials in the process of conducting their research.

4. ENVIRONMENTS FOR DIFFERENT AGE GROUPS

Jose Muntanola is concerned with the ways in which children develop "conceptions of places to live in." He takes the theoretical works of Piaget on the development of space-time conceptions in children and has different age groups of children build a place for a family to live in and then has the same children draw a graphic representation of the best place for people to live in.

For two and a half years, Rivelin, Wolf and Beyda, who are environmental psychologists, have been doing a comprehensive study of the utilization of space in a new children's psychiatric hospital. Their present research report, which is done by naturalistic observation, is concerned with the use of identical house areas occupied by two different age groups. Their findings imply that if the hospital desires to encourage the development of specific behaviors as part of the treatment goals, spaces for younger children may require stronger definitions than those occupied by older children.

Steinfield, an architect, uses interviews to assess the interest that older adults have in making contact with younger children and the incentives needed by older adults to facilitate participation. He then looks from the senior adults point of view at physical features of proximity, privacy and location that would encourage contact between senior adults and young children.

Paul Windley, an architect, is concerned with whether older people have disengaging dispositions toward the physical environment. For the study he interviewed a sample of older people and a sample of young students with a series of verbal and pictorial measures to develop scales of preference for privacy, environmental complexity, environmental stability and manipulation of the environment.

Sandra Howell is concerned with the design research needs and behavioral implications of housing for the elderly. By reviewing the literature and consulting with architects planners and public officials, Howell determines architectural problems in housing elderly people such as building density and location of community space and then presents possible solutions to the problem along with possible social and behavioral implications of the proposed solutions.

CHILD'S CONCEPTION OF PLACES TO LIVE IN   4.1

Joseph Muntanola Thornberg

Research Associate, University of California, Berkeley
Doctoral Degree in Architecture--University of Barcelona, Spain

Abstract
The paper analyzes, on experimental basis, the genesis of architectural forms in children and adolescents, and such a genesis is considered as "Conception of Places to Live In". Using the structural developmental models of Jean Piaget, the experiment detects four main stages in the child's conception of places to live in. Then, some general assumptions about the interpenetration of perceptual and mental abilities in relation to architectural creativity are stated. Finally, some consequences are established, on one hand, in relation to design education for everybody, and, on the other hand, in relation to design languages considered as tools to build human places.

Introduction
This paper contains the core of a more extended work dealing with the relations between architecture and design theories on the one hand, and on the other hand, dealing with some psychological and educational theories. Its aim is to analyze the significance of architectural forms by "seeing" the evolution of conceptions of places in children. We have only a few valuable works dealing with the genesis of architectural forms (1) and the purpose of this study is also to prove that this kind of research needs more attention and interest. Diagram-1 shows the main stages in space-time conceptions in children, including the stages of mental development of conceptions of places to live in. All this classification into stages is made according to the works of Piaget on space-time conceptions in children.(2)

Method
Each subject (see diagram-2 for distribution by ages) was presented with a small family of wooden dolls representing a man, a woman, two girls, two boys, a dog and a car. Immediately, he was asked to build a place for the family (for these people) to live in with some concrete material for construction situated on the same table where the small family of dolls was standing. Some transformations in the physical material were made by the experimentor, mainly by constructing a round foundation (basis) and saying to the subject: "Can you finish this place for these people to live in?"

All the subjects made their places successively of two different kinds of material: A) Wooden blocks and B) With clay (several balls of soft clay). Then each subject was asked to draw a graphic representation of the best place for the people to live in, either from inside or from outside, without any indication of correlation between the places which they have built and the drawings. Of course, the buildings were never standing while the subject was drawing.

Analysis of the Stages

Diagram-2 shows the distribution of subjects by ages and stages. Figures 1 to 11 show some schematic representations of the products of each stage in conceptions of places.

Stage-I: Ritual Conception of Places

Places made by children in this first mental stage, between 3 and 5 years of age, have the shapes indicated in Figure 1. They are "solid" and "massive" places: Lines (children call them trains), columns (statues), or heaps of clay (mountains), always with people following the physical form of the place (see Figure-1). Children have no reaction with the round or square form presented by the examiner, except putting the small dolls inside the built foundation (see Figure-5), or repeating the same place that they built before. At this age, they begin to draw round or square forms (see drawing 30-1, in Figure-6) and they can recognize the topological differences in objects by manipulation (3), but, in spite of this, they cannot build real "hollow" places with blocks or with clay. Clear examples are the performances of Alice (30) and Ammy (36).

Alice (30) (3 years and 3 months) ...When she sees the small family she says: "I have a dog like this" and after: "This is me; how can I have a different dress?" (She is looking at the small girl). She begins drawing (see drawing 30-1). Afterwards, she builds the place silently, and says: "Will mommy be here, now?" She says that the column is "A statue". (see the place in Figure-1). With the square foundation she begins to put some pieces above (see Figure-5), but immediately she builds again the same place as before, near it. She puts the people on her place and says: "They are on the roof." With clay she sticks the dolls in the clay and says: "They are looking outside..."

Ammy (34) (3 years and 10 months) ...She says: "I cannot draw a place." Asked for an inside drawing she says: "I can't." With blocks she builds place 34 in Figure-1. With the round basis she puts the people inside, and builds this place again and says (pointing out her place): "My house is this. It is a train." She does not allow the experimentor to destroy her place.

The ritual conceptions of places have the following characteristics:
A) Places are massive places (see Figure-1)
B) The process of coordination of actions upon the material depends on the physical properties of the material itself. The blocks, therefore, are "trains" or straight lines, and the clay is used as "mountains" or piles of material
C) Representation has a symbolic space-time neither logical nor physical. Piaget has called this situation "animistic", in that, objects are symbols of living things. He has already pointed out that the topological space-time in this stage-I, cannot distinguish between the "content" and the "container."(2) In order to differentiate between content and container, you must use some kind of operational conservation, for instance, angles, parallelism, perpendicularity, etc. None of these operational spatial properties will be developed until stage-II. Without these operational conservations the topological images of children remain static and they cannot be used in the construction of hollow places. The results are the "massive" places built by putting one piece of material after another, and by creating simultaneously some physical form and some symbolic representation of it (statue, mountain, train, ship, etc.). This kind of behavior is what I have called

"ritual." The ritual of Alice (36) can be detected in her first reaction with the square foundation: She intends to repeat her coordination of actions upon the material (to put blocks one above the other) but she notices the difference in relation to the previous place she built, and she repeats exactly the same performance as before using the same kind of blocks. Ammy (34) puts blocks "in line" and this is "her place."

Sociologically, the differentiation between me-you and he follows the same mental pattern. Alice is looking for her mother in the place (her place) that she has built. The place where she lives and the place she has built IS THE SAME ONE.

Stage-II: The Functional Identity of Places
Places made by children throughout stage-II look like those of Figures-2 and 5. They are made on the basis of an "architectural unit" (AU) with a real inside-outside structure. I have room here only to present two main points: Why a cubic AU? and How this AU works from the beginning (substage a), to the end (substage c) of stage-II? The first question must be answered by looking at the products in substage a (see Figure-2). Here are some extracts of protocols.

Helen (43)
(4 years and
3 months)
...Helen begins the construction with blocks making the ground of the house (basis) (see Figure-2). She builds a small garage (with roof and without ground-basis), afterwards the room for people with high walls without windows, because (she says), "This wall is at the back of the house." She has a lot of trouble in the manipulation of the material, and the blocks fall down several times. She does not care about the roof. With clay she sticks the dolls totally inside the clay and says, "They are in jail...."

Ethan (32)
(3 years and
10 months)
"...He builds with clay sticking people in the material and he says: "They are looking outside." With the drawings he is a real artist. First of all he says: "I cannot draw a place. I can only draw a window." Drawings 32-1-6, and 32-1-8 (Figure 7) are examples of his work. As can be seen in drawing number 32-1-6 he can finally draw a house. With blocks, he builds a square basis and begins the walls as Figure-5 indicates. The experimentor builds a ground basis and Ethan builds the same square form inside. This time he puts people on the walls and says: "They are looking outside...."

Emily (48)
(4 years and
4 months)
"...Emily is a good draftswoman, but she has a lot of trouble with construction, and she notices that. She builds with blocks twice. The first time, she builds a house starting with a ground basis and two walls, but she cannot put on a roof, and the walls fall down several times (see Figure-2). The second time she builds a shouse with a roof (a very rough one) but she forgets totally to build the wall which stands on the other side of the place, in relation to her position at the table (she builds only what she can easily see)..."

All the children in this stage II-a have the same difficulties in the manipulation of the material, and all of them begin their place with the "ground basis" and use this basis (always square with blocks, and sometimes round with clay) as basis for the construction. Everybody builds a cubic AU (architectural unit) except Kelly (seven years and two months) who asks immediately: "Can I build a round house with clay?" (And she begins with a round ground-basis.) Drawings in this stage II-a are already very creative and complex, with animals, people, etc. Following the work of Piaget(4) I call this situation, "Functional Identity of Places." First of all, I want to say that all of stage-II is still pre-operational: nobody builds the round basis presented by the examiner. Figure-5 shows the reactions of the people which are: to repeat their place, to divide the round basis into four parts, etc., but they don't build as children in stage III will do. Then, how do they build? We must analyze changes in relation to stage I. The AU is built thanks to some "Functional Identifications" between some physical properties in the material, detected through manipulation, and some formal characteristics of simple figures which they can easily detect in our real places and represent by imitation. A closer description of the performances of Emily, Helen and Ethan will discover these "Functional Identities." Places are not only "Rites" or sequences of actions with a symbolic value, as in stage I, but they are, in stage-II, a continuous effort to identify, through manipulation and imitation at the same time, some basic similarities in the use of the material. The similarities that children identify can be seen in the places made by Emilly, Helen or Ethan, and in the rest of drawings in Figures 2 and 5, and they are: the square basis, the two parallel walls on two opposite sides of this square basis (see Emily and Ethan), two walls "in angle," making a corner of the place, (this is the first step in Helen's house) etc. All these "similarities" are previous steps for building a complete AU. With this complete AU, children construct the more complex places in sub-stages b and c by using the same "functional identity" between some physical properties of the material and some formal characteristics of representative figures. In the drawing 74-2 (Figure 8) we can notice this more complex "functional identity." The cubic AU is used here at two different and homological levels. The first one is the whole cubic house, and the second one is the sub-division of the whole house into four cubic rooms (four AU inside one AU).

Places in stage-II develop in the same way as the euclidean conservations do: by enclosures, or by "boxing" a sequence of homological forms one inside the other(2). Piaget has pointed out that this procedure is actually the unique path to mental operational thought. Sociologically, functions begin to diversify in stage II. At the beginning everybody is sleeping. Later in sub-stage b and c, places are strongly "family-centered". Each person has a constant "social role" inside his concrete place: mommy in the kitchen or with the children and daddy always watching the TV set.

Stage-III: Concrete Conception of Places
From 8-9 years of age, conceptions of place suffer a deep change in structure. Figures-3 and 5 show the products of this stage, and they say a lot of things by themselves. The round form built by the examiner is always constructed at stage-III proving a great advance in the logic space-time reality. Floor plans and "above-plans" (roof plans) emerge, showing the effects of the mental operational thought (see plan 117-1, Figure-10). The main new feature is the complex "Architectural Unit" as we can see in Figure-3. The nature of this complex AU can be seen in Erik's performance in his second day session (see drawing 91-2, Figure 9). He begins by designing the square form of the place, and afterwards draws the door of the place, by indicating a hole in the close square form. Then, he "enters mentally" the place: <u>designing the inside topology as he walks in the place</u>. So the place is full of corridors and "white areas," and he must invent

strange functions in order to fill the space. We can notice that the coordination of actions has become much more mentally anticipated thanks to the growth of abstract thought. Judith (11 years and 10 months), on the other hand, does the opposite: she designs a complex "itinerary" of functions (see drawing 110-1, Figure-9), but she builds nothing afterwards, because the form is too complex and she becomes lost. The problems of construction are too much for her abilities. The preceding descriptions reflect the advances and the limitations of the concrete-operational conception of places to live in. The coordination of actions upon the material become anticipatory thanks to the existence of a logic-operational space-time. This anticipation has striking effects both in the use of images and in the coordination of functions through "mental itineraries." Places built by children at the end of stage III (see Figure-3, schemes 116 and 118) are constructed by using <u>all</u> the material and <u>only</u> all the material, and without "trial and error" by constructing the place on the first attempt. The construction of the round form develops the same mental pattern. Paolo hesitates at first, and decides to build a beautiful form (similar to number 116, Figure-3) saying, "<u>I saw this roof in Tilden Park.</u>" Here imitation can become anticipatory thanks to the operational ability in the coordination of actions upon the material.

The weakness of the concrete-operational stage in relation to stage-IV is the lack of flexibility in the use of the logic-operative space-time models. Geometric concepts (euclidean or projective) are still incomplete at a concrete-operational mental level(2). Children use rectangular frames of reference (see places in Figure-3), and they can coordinate mentally different views upon the same object, even if this object is not a simple cubic AU which was the most elaborated product in stage-II. However, when the material is not "at hand" they have a lot of difficulties in linking sequences of functions (itineraries) and sequences of forms (complex AU), as I have pointed out in the performances of Erik and Judith. They lack a final and important mental product which will be obtained in stage IV: the experimental use of formal-operational conceptions of places. Sociologically, the change is also complete in relation to stage-II. Children create new functions following their own hobbies and ideological differences begin to create struggles between them.

Stage-IV: Formal Conception of Places
It is not strange that very few people can be included in stage-IV as Figure-4 shows. A formal conception of places demands a lot of perceptual and mental abilities as well as an emotional and aesthetic equilibrium. The present analysis has focused its attention on children and this stage IV has been seen only as an upper level of development. Therefore, I will describe merely the new directions we can see in adolescents at the beginning of stage-IV. Difficulties of creating an architectural form in stage-IV are the same as in stage-III plus a new mental perspective which I have described before as "an experimental attitude."(5) Following this attitude, adolescents use some formal conceptions of places to create a place simultaneously at a social and at a physical level. Figure-4 shows the place built by Jim (114) (14 years). He begins explaining to a fellow what he is doing by saying (looking at the experimentor), "<u>I am doing this for some psychological reasons and some architectural reasons at the same time.</u>" He says "<u>I must be really creative.</u>" Drawing 141-3 (Figure-11) is made by Jim too. The whole place is here a continuous genesis of forms and functions. The Architectural Unit

is conceived independently of the material in itself (the material in itself is created in some way) and independently from none previous form (square or round in some clear fashion). The result is some kind of "maze" or "labyrinth."

However, the AU, clusters or round forms, are used here in a very different way than in stage-II or -III; they serve as "working-tools" and not as the unique buildable form. People in stage-IV chose forms by creating them and they know that at the same time they are "enclosing" social functions. Design becomes language, but language which is created in each and through each socio-physical situation.

Basic Assumptions About the Development of Perceptual and Mental Abilities for Architectural Creativity
This brief and condensed overview of the material allows us to evaluate the consequences of the child's conception of places in design education and in design languages. Data have proved to be consistent with the stages and with the theories that J. Piaget has developed in space-time logical conceptions of children, either in relation to perception or to the mental levels of abstraction. In order to apply the preceding analysis of conception of places, I must uncover the physico-logic model which has underlied it, using some of the Piaget's own explanations.(6) The forms or "paradigms"(7) of our space-time conceptions are carved out of a progressive differentiation between what Piaget calls structures of causality and the logic-operative structures growing in our mind. The structures of causality are articulations of our own actions upon the objects inside a physical space-time reality; from them arise some physical properties such as velocity, elasticity, etc. The logic-operative structures are coordinations among the interactions of the objects in themselves, which are constructed by abstracting some physical properties inside a logical space-time reality. The pure model of these structures is mathematical space-time.

In order to obtain the progressive differentiation between causality and logic-operative structures children use simultaneously two kinds of processes: A) An effort of representation, through which different perceptions made in different space-time situations are linked together in a representative space-time "schema." B) A process of coordination, by which different actions or interactions subect-environment are interrelated through the action itself. The result is the "internalization" of these interactions in reversible operations, always with the help of the representative processes. Consequently, "paradigms" are neither a simple application in the physical space-time reality of some logic-operative structures built "a priori" in the mind of the children, nor a consequence of imitation of the physical space-time reality abstracted afterwards in a logic-operative model. The real process observed by Piaget is more complex: each stage of mental development contains a double equilibrium between causality and logic-operative structures on the one hand, and, on the other hand, between representation and coordination. Place-conceptions have been analyzed following this model. Each stage of conceptions of places has revealed its own mental level and nature, in which representation, coordination, causality and logic-operativity are conceived.

It is obvious that our conceptions of place are physico-logical and socio-logical at the same time and this is exactly what children express to us. E. Cassirer(9) points out the fact by saying, "...the differentiation of places serves as a basis

for the differentiation of contents; of the I, you, and he on the one hand, and of
the physical objects on the other hand...." I must stop my argument here. I hope
that I have made clear the complexity of the structures which underlie architectural
design processes considered as conceptions of places to live in. Architecture,
then, is the expression of the whole process of symbolization in man-environment
relationships. And the architectural differentiation of places to live in is the
point from which this differentiation is created and perpetuated in architectural
types and languages.

Physico-Logical Consequences in Design Education
Design education must animate the four basic processes which bring about the architectural creativity described in the last chapter. Here are some conditions of
this four-fold animation: A) Logic-operative training (geometry, mathematics)
should accompany the representative and the coordinative processes in design
education. The visual-manual activities become very important if the learning
process involves, not only the logical abstract side of our conceptions of places,
but the causal side of them which develops inside the concrete and physical space-
time of our experience. The optimum of animation is reached when children (and
adults) can link their experience in space-time with new possibilities of logic
anticipation or architectural forms and structures. B) Throughout my experiences
with children I have noticed the disappointment caused by the lack of "causal
animation" which still fills up our schools. There are two complimentary fields
of work, the first one is the knowledge of the materials most used in our places
for living. The second one is the special and individual animation in the use of
some materials which each person can select. C) Representative abilities in the
use of "schemas" can be increased through training: drawing, songs, painting,
etc. However, the analysis of the conceptions of places shows a basic issue which
is the kernel of design education: the interrelation between coordination and
representation, or the synthesis of "schemas" and "schemes." <u>This constant flow
from representation to coordination, and from coordination to representation, is
the best condition for increasing design abilities</u>.

Consequences for the Implementation of Design Languages
Design languages should be seen as "working-tools" for conceiving places to live
in. The main condition in any language is the distinction between the "signs" as
media and the "signs" as message. However, in place conceptions, this media and
this message are simultaneously created in the place and through the place itself.
As a result of this very specific linguistic nature of design, the child's conceptions of places explain, in my opinion, why we have now as many <u>design languages</u>
as researchers who are trying to avoid or to reject the "old situation," in which
each architect-artist had his own <u>language of design</u>.

Later stages-III and -IV show that places (and design languages) are no longer a
total consequence of cultural heritage or of a direct response to previous environmental facts. Design is created through a socio-physical dialogue between users
and place, and a physical setting and a social setting are simultaneously conceived. Between the "languages of design" of small experiences of participation
and construction of places, on the one hand, and, on the other hand, the "design
languages" of our topological, ecological, anthropological, sociological, mathematical, economic, etc. devices, children use both at the same time. Following

child's conceptions of places to live in, one can say that the good design languages should not be catalogues of pre-determined "socio-physical events," but "schemas" and "schemes" of a socio-physical environment which should not emerge from the total pre-determination of behavior or of physical forms. On the other hand, "languages of design" must be places, real and human places.(10)

References

1. Abercrombie, J.J., <u>Perception and Construction</u>, (In Design Methods, Ward and Broadbent, 1969 Lund Humphries)
   Stern, A., <u>Encyclopedie Practique De L'Education Artistique</u>, Delauchaus Niestle-1963.
2. Piaget, J., <u>Child's Conception of Geometry</u>, Routledge and Paul-1960.
3. Piaget, J., <u>Child's Conception of Space</u>. Routledge and Paul-1962.
4. Piaget, J., <u>Epistemologie at Psychologie De L'Identite</u>, PUF-1968.
5. See this experimental attitude in: Bachelard, G., <u>L'Experience De L'Espace Dans La Physique</u>, Alcan-1937.
6. Piaget, J., <u>Theories De La Causalite</u>, PUF-1971.
7. Kuhn, T., <u>The Structure of Scientific Revolutions</u>, Univ. Chicago Press-1962.
8. Diagram-2 is made of the analyses of Piaget and of the clear paper of Jonas Langer, <u>Stages of Temporal Schemetizing</u>, Berkeley-1972.
9. Cassirer, E., <u>The Philosophy of Symbolic Forms</u>, Yale Univ. Press-1955.
10. See: Hillier, <u>Musgrove and O'Sullivan</u>, <u>Knowledge and Design</u>, EDRA III-1972.

Diagram-1 (See note number 8)

Stages in Conceptions of Places in Relation to Space-Time Intellectual Development

Ages	General Intellectual Development	Levels of Spatial Organization	Levels of Time Organization	Levels in Conceptions of Places to Live In
Adolescence and beyond	Stage-IV Formal-Operational	Formal-Operational Space	Formal-Operational Time	Stage-IV Formal Places
Middle Childhood	Stage-III Concrete-Operational	Concrete-Operational Space	Concrete-Operational Time	Stage-III Concrete Places
4-7	Stage-II Intuitive-Pre-Operational	Pre-Operational Functional Space	Functional Time	Stage-II Functional Identity of Places
2-5	Stage-I Transductive Pre-Operational	Symbolical Space	Representational Time	Stage-I Ritual Places
Infancy	Stage-0 Sensori-Motor	Sensori-Motor Space	Presentational Time Practical-Flow of Time.	Stage-0 Sensori-Motor Places

Diagram-2 Distribution of Subjects Tested in the Stages of Conceptions of Places

Stages \ Ages	I	IIa	IIb	IIc	II Total	III	IV	Total by Ages
3	6	(1)			1			7
4	4	(5)	(1)		6			10
7		(2)	(5)	(4)	11			11
9		(1)	(3)	(3)	7	2		9
11		(1)	(0)	(1)	2	7		9
14-15		(1)	(0)	(0)	1	2	1	4
More than 15						3	1	4
Total by Stages	10	(11)	(9)	(8)	28	14	2	54

4. ENVIRONMENTS FOR DIFFERENT AGE GROUPS / 187

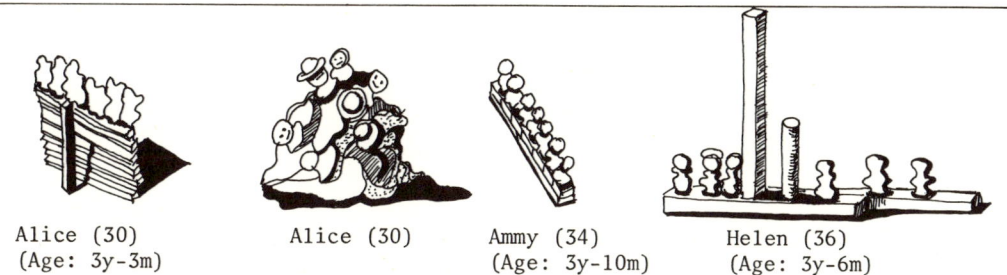

| Alice (30) | Alice (30) | Ammy (34) | Helen (36) |
| (Age: 3y-3m) | | (Age: 3y-10m) | (Age: 3y-6m) |

FIGURE 1   STAGE-I: The Ritual Conception of Places

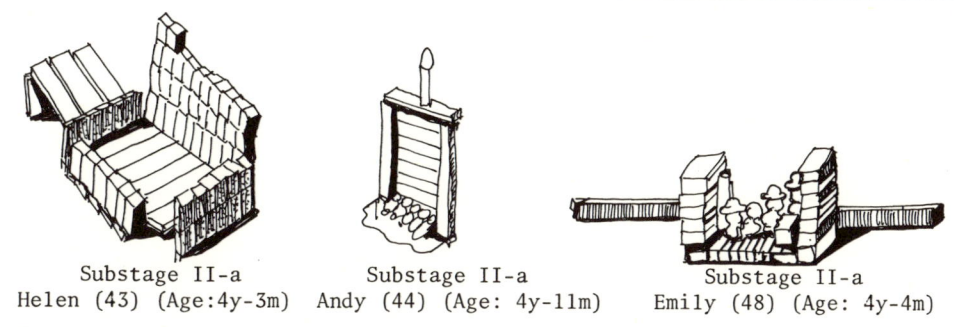

Substage II-a
Helen (43) (Age: 4y-3m)

Substage II-a
Andy (44) (Age: 4y-11m)

Substage II-a
Emily (48) (Age: 4y-4m)

FIGURE 2   STAGE II: The Functional Identity of Places

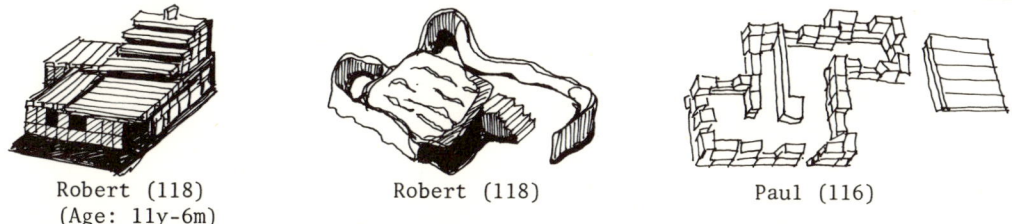

Robert (118)
(Age: 11y-6m)

Robert (118)

Paul (116)

FIGURE 3   STAGE III: The Concrete-Operational Conception of Places

STAGE IV: The Formal-Operational Conception of Places

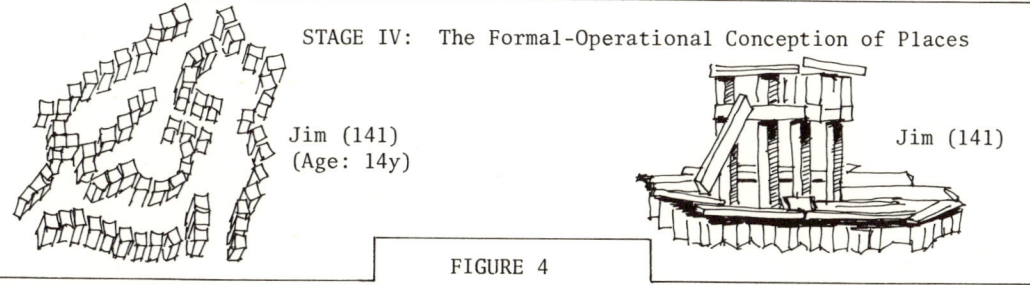

Jim (141)
(Age: 14y)

Jim (141)

FIGURE 4

188 / ENVIRONMENTAL DESIGN RESEARCH, VOL. 1

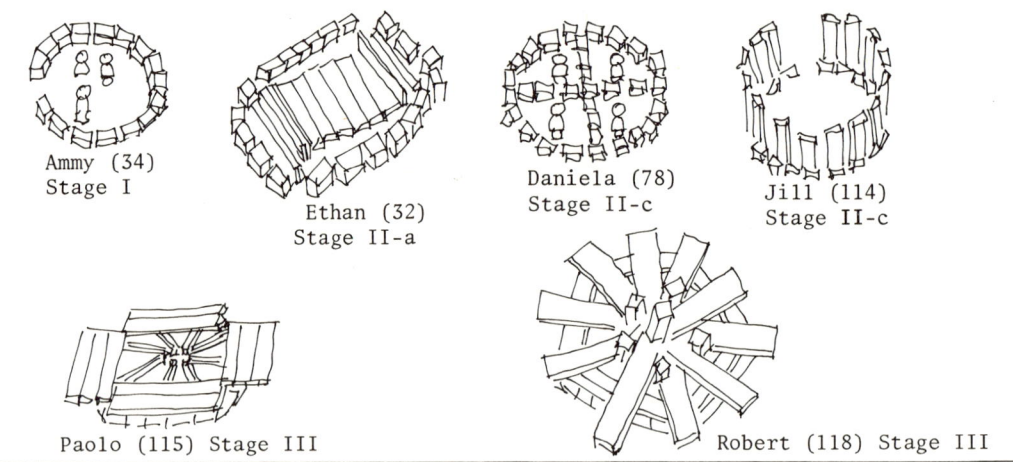

FIGURE 5   The Round Form in All the Stages

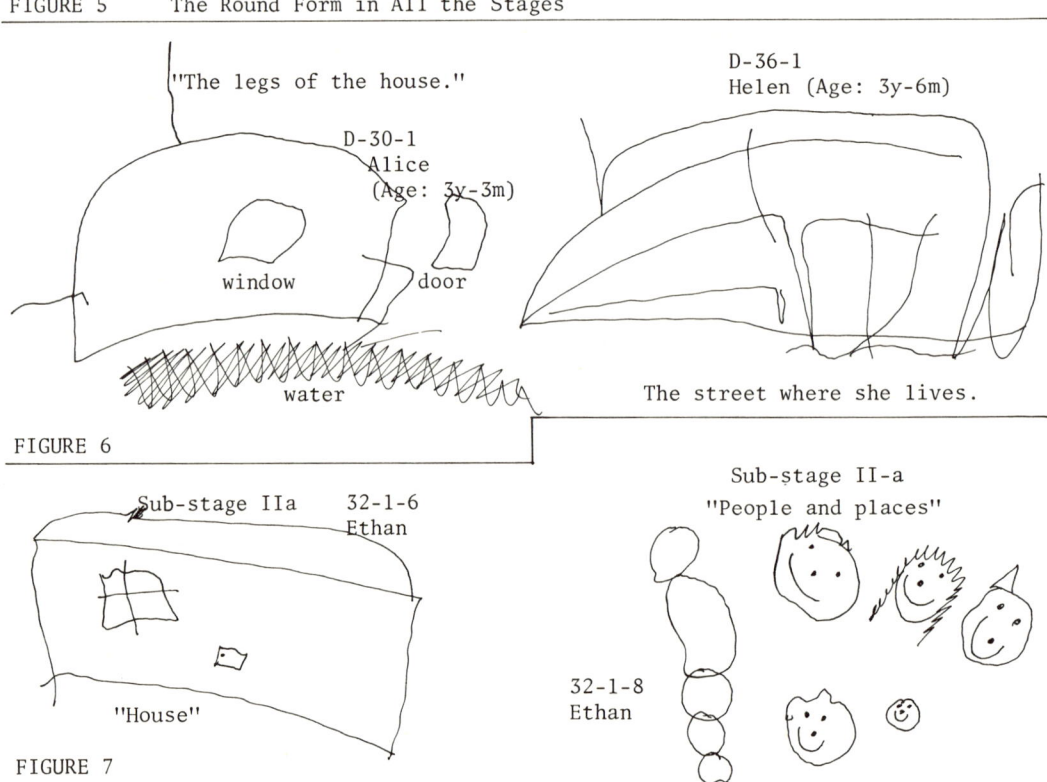

FIGURE 6

FIGURE 7

4. ENVIRONMENTS FOR DIFFERENT AGE GROUPS / 189

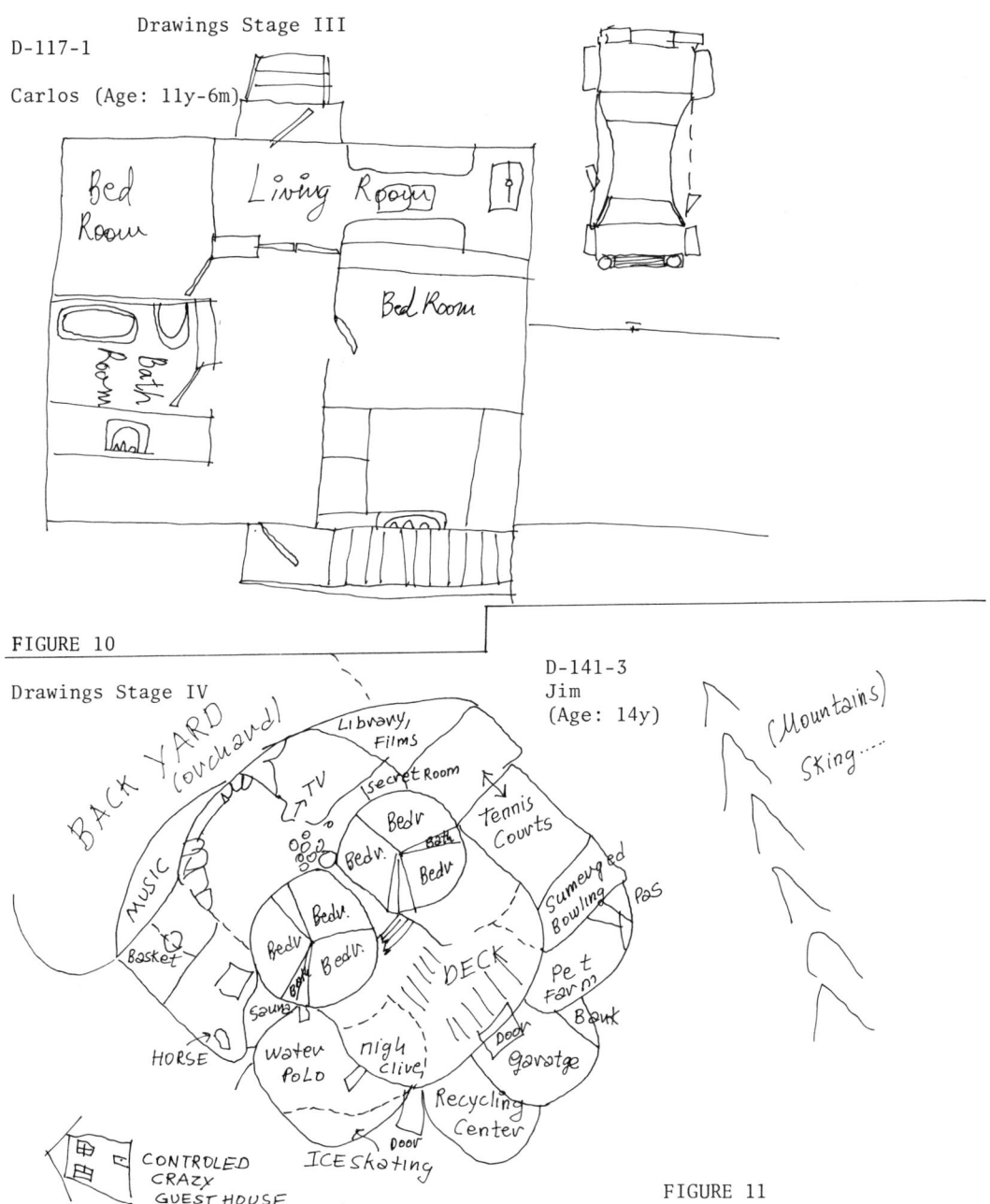

FIGURE 10

FIGURE 11

Sub-stage II-c

D-74-2

FIGURE 8

Drawings Stage-III   D-91-2
Erik

D-110-1
Judith
(Age:11y-10m)

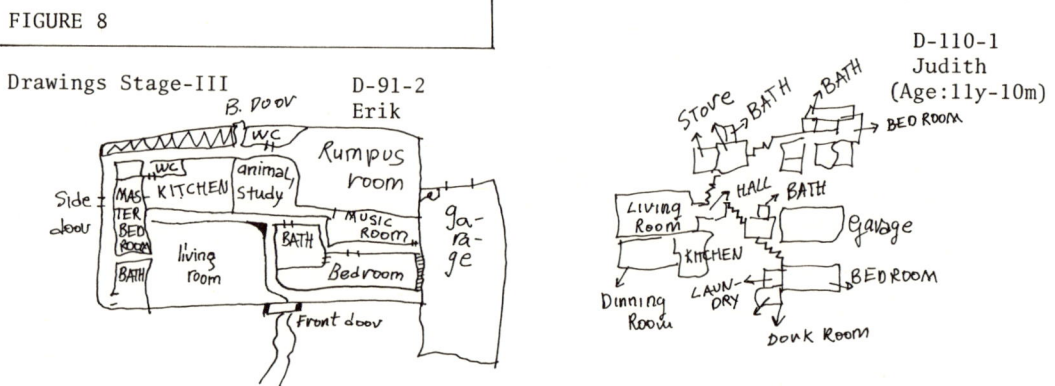

FIGURE 9

AGE RELATED DIFFERENCES IN THE USE OF SPACE   4.2

Leanne G. Rivlin, Maxine Wolfe and Marian Beyda

Environmental Psychology Program
City University of New York

Abstract

The present research, part of a comprehensive, longitudinal study of a new children's psychiatric hospital, is concerned with age-related differences in the use of space. Observations were made in identical house areas used by two different groups, one, ages 8 through 12, the other, 13 through 16. While the general types of activities in the house as a whole did not differ for the two age groups, the location of specific activities and the distribution of groups did differ. These differences were considered in terms of developmental differences in space use and their design implications.

For the past two and one-half years we have been studying the relationship between the physical design of a new children's psychiatric hospital and the patterns of behavior and use of space that evolved over time (12, 13). A critical aspect of this study involved an analysis of various personal characteristics including age, sex and whether the child was a day or full-care patient and their relation to space use. The present paper focuses on age and space utilization patterns. There is a tendency, when considering the design of psychiatric institutions, including those for children, to focus on pathology as the major component dictating design decisions (2). On the other hand, in the design of facilities for normal persons, the age of occupants is frequently a major consideration. A question in the present paper is whether there are age related differences in the use of space in a children's psychiatric facility, despite the idiosyncratic effects of pathological behavior and the conformity resulting from long-term institutionalization.

Age differences, as they relate to the way children function, have a long history of interest in developmental psychology. Many of the early developmental studies were directed to establishing age-normative patterns for a variety of different abilities (see 3, 7, 8), ranging from motor capacity to conceptual abilities. Although these studies have provided considerable documentation of specific developmental stages, they generally have not been concerned with details of the physical setting in which behaviors appear. Conversely, except on a very general level, the knowledge of stages has not been incorporated into the design of facilities for children. A review of child development literature has pointed up some areas in which age differences would seem to have direct implication for the design of children's environments. The early studies of Midwest (1) showed substantial differences in the activity patterns of younger children when compared with adolescents. There was a greater tendency for younger children to shift from one activity to another, to do things in sequence, one at a time, and to leave an activity before completion. Older children, on the other hand, tended to remain in an activity for a longer time, to carry on multiple actions, and to remain in an activity until completion. In addition, as the age of children increased, so too did the complexity of their friendship networks as well as the number of children doing things together. Coupled with this was a decrease in the number of adults present during activities.

There is also ample developmental data on age-related differences in environmental adaptation. The work in cognitive development, especially that growing out of Piagetian theories (10), has pointed up the changing conception of the world and capacity to handle aspects of the world over the developmental continuum. An excellent analysis can be found in Hart and Moore's review (5), which indicates that even without strict adherence to Piaget's theories, there is evidence to support the view that children change in their perception of the world and their capacity to deal with spatial relationships as they move from the more concrete earlier years to the skills in generalizability that come with environmental experience.

There has been some research on specific environmental design components that appear to be related to age. Bayes (2), for example, reviewed color preference research and cited several differences, some supported by empirical data, others speculative in nature. In particular, he states that research has indicated that up to age six, warm colors are preferred over cool colors. However, this is an area where differences in opinion exist with other researchers citing the adolescent years as the change point. At the very least this literature suggests that children do have color preferences, and that the range of preferences increases with age, especially colors that are disliked. On the more speculative side is the question of differing shaped rooms as related to the needs of different age groups in view of the lack of concrete evidence.

There is increasing awareness of the need for research directed to the design implication of developmental differences (11, 4). There are many ways these age differences can be studied, and the approach of the present research involved intensive analysis of an existing facility used by children ranging in age from 6 to 16. By focusing on a study of living areas that were identical in design, occupied by two different age groups, it was possible to compare patterns of use with age the only differentiating variable.

In order to understand differences in the use of space, it is necessary to provide a brief picture of the design and functioning of the children's hospital, specifically in relation to age-related aspects.

Description of the Hospital

The children's psychiatric hospital is a state facility designed to provide inpatient care for 192 children ranging in age from 6 to 16. Combining residence facilities, a school, areas for occupational therapy, recreation, diagnosis and treatment, it shares an extensive tract of land with a large state psychiatric hos-

4. ENVIRONMENTS FOR DIFFERENT AGE GROUPS / 193

pital for adults. The children's building is divided into four areas: a living area, a recreational area, administrative area, and the school. Eight houses, or ward units, arranged in four pairs, one above the other, provide facilities for 24 children each. The house consists of a central corridor leading off to a dayroom with a kitchenette, a nurses' station with an adjacent bench area, a laundry room, a seclusion room, and three apartments called "living units." (See Figure 1 for a plan of the houses.) Each living unit has an entry foyer leading to its own living room and four bedrooms, two single-bedded rooms, one two-bedded room and a four-bedded room. The bathroom in each house was not observed. Each set of houses shares a common dining room on the main floor. Behind each house is an outdoor play area. The houses were furnished identically, with a single exception - the television was located in the dayroom of the younger children's house while the living rooms of the older children's living units contained the TV sets

Age Related Characteristics of the Design

The hospital as constructed provided for houses that were identical in design. The original space program requirements specified the age range that would be accommodated in the facility, although no indication was made as to how they would be distributed. In an attempt to understand the basis of the details of the hospital's design, we interviewed the architect responsible for the plan. He reiterated the lack of specificity of the program with reference to age groups. In order to provide for a wide range of ages, he designed the houses in three sets with one play area appropriate for young children, another for children of a middle age group and the third an unstructured outdoor garden for adolescents. A fourth area was set aside for kindergarten or preschool children. Aside from child-sized facilities in the kindergarten section and the playgrounds behind each of the houses, the interiors of the houses bear no indication of appropriateness for any particular age, although the architect assumed that each set of houses would, in fact, have children assigned according to age. He suggested that the canteen, a public room near the houses, would be a good "hang-out" for an adolescent age group.

Age Related Details of the Hospital's Functioning

A series of previous observational and interview studies has pointed up a number of trends: 1) the actual uses of the facility did not always reflect the intentions of the architect and planners, due in part to the medical model design of the facility which ran counter to the community-oriented treatment philosophy of the administration; 2) space utilization patterns evolved over time and were influenced by the design, the program, the treatment philosophy, the nature of the occupants and a variety of unexpected events; 3) specific characteristics of the occupants, e.g. sex and resident status (full or day care) influenced activity patterns and use of space. Age differences were also viewed as influences on use of the facility.

Due to insufficient funds, the hospital opened with 16 children (only four of whom were full-care) and an equally limited staff. Obviously, the original intentions to separate children by age could not be carried out, and all the children were placed within one living unit of one house. The house chosen for occupancy was close to other functioning areas of the hospital (recreational and school). The play area behind that house happened to be the one designed for the youngest age group, and consequently was infrequently used. The kindergarten suite, consisting of a large room with child-sized equipment and furnishings, bathroom and an adjoining play area was used by two day-care children. The canteen was not in use at all.

Over the two and one-half years of occupancy, several age-related changes have occurred. A steady increase in the number of children (to 57 at the time of the present study) enabled age separation in the living areas. Children aged 8 to 12 were placed in one house while the older children, 13 to 16, were placed in a second house. Both houses are adjacent to the play area designed for a group aged 8 to 12. The canteen is used mainly as an adjunct arts and crafts area and only occasionally provides the physical setting for casual group encounters. The kindergarten suite is no longer used for children. A number of factors, including its physical separation from nearby living and recreational areas, the small number of children accommodated in a rather large room, and the shortage of space for staff needs led to its conversion into a staff library.

Method and Procedure

In order to study the use of space in the children's hospital, a series of four mapping studies were conducted, covering two and one-half years of functioning. Data for the present paper were obtained from the fourth mapping study, undertaken immediately after the age separation of children into two separate houses. The older group, ranging in age from 13 to 16 consisted of 8 day-care and 19 full-care children. The younger children, occupying the other house, ranged in age from 8 to 12, and consisted of 17 day-care and 13 full-care patients.

The basic approach in quantifying and describing behavior patterns and the use of the house areas was the behavioral mapping technique (6). This standardized observational method, developed in previous studies, provides a profile of all activities observed, the participants, the time of the activity and specific location. Trained observers record activities into 69 coded behaviors which, for the purposes of data analysis, were combined into 19 analytic categories. These categories are: 1) aggression, 2) high-energy physical release, 3) high-energy organized, 4) high-energy unorganized, 5) low-energy organized, 6) isolated active, 7) isolated passive, 8) domestic, 9) media, 10) music, 11) orientation, 12) meetings, 13) exploration, 14) cuddling - holding a child, 15) idiosyncratic behavior, 16) talking, 17) telephoning, 18) traffic, and 19) miscellaneous. (See Table 1 for a description of the categories. For the present study reliabilities were sufficiently high to justify use of the observational procedure.

Data for the present study were obtained by having an observer make a predetermined tour every fifteen minutes, covering all areas of each house. For each room or area, the observer recorded ongoing activity, the number, residence status (full or day-care) and sex of patients, the number of staff present, and if a room, whether it was locked or unlocked. Observation periods extended from 10 A.M. to 9 P.M. Data were gathered for a full week.

TABLE 1
DESCRIPTION OF CATEGORIES OF BEHAVIOR

Category	Description of Behaviors Included
1. Aggression	Aggression toward objects, people, self and undirected; arguing; disturbing an activity; restraining a child
2. High-energy physical release	Chasing; climbing; jumping off things; roughhousing; running
3. High energy organized	Dancing, marching, rhythmicizing; exercising; active games, sports
4. High energy unorganized	Play; fantasy play; roller skating; bike riding
5. Low-energy organized	Arts and crafts; sitting games; teaching; playing with or caring for animals
6. Isolated active	Looking out window at an activity; personal hygiene; reading; wandering; watching an activity; writing
7. Isolated passive	Crying alone; hiding; lying awake; sitting alone; sleeping; standing alone; looking out window
8. Domestic	Eating and drinking; housekeeping; food preparation
9. Media	Phonograph, radio, television
10. Music	Play instrument; singing
11. Orientation	Looking for a person; looking for a place
12. Meeting	Group meetings; patient-staff conference
13. Exploration	Explore objects; explore places
14. Cuddling/holding a child	Self-explanatory
15. Idiosyncratic behavior	Pathological behavior which is specific to the person. Including head banging, hallucinations, etc.
16. Talking	Self-explanatory
17. Telephoning	Self-explanatory
18. Traffic	Walking through, to or from an area with no other concomitant behavior
19. Miscellaneous	Arriving; leaving; hospital routine; preparing for activity; transporting objects; waiting

Results

The observational data were analyzed to obtain both the frequency with which specific activities were observed and the total numbers participating in each activity. For the present paper, frequency of activity will be reported since these data better reflect the details of what is actually occurring. In addition, since our interest is the relationship between activity and age of patient, behaviors involving staff alone will not be included.

Overall Distribution of Activities

Table 2 represents the frequency and percent of all activities for the children within each house analyzed according to location. It should be noted that the day-care children were rarely in the house so that these data essentially represent the activities of full-care children, 19 in House 4 and 13 in House 3. A chi-square analysis indicated that the distribution of activities across rooms was significantly different when the two houses were compared ($p < .01$). For the younger children, the highest percent of activity occurred in the dayroom (38.66%), while for the older children, the highest percent of activities occurred in one of the living units (23.85%). Excluding the rooms with infrequent use by children in both houses (nurses' station, seclusion room and laundry room), of the remaining rooms (benches, corridor, dayroom), it is apparent that the older children's activities were distributed much more evenly across the house than those of the younger children. In fact, there was a "peaking" of activity in the younger children's dayroom with more than twice as much activity there as the next most active room or area. If we compare the percent of activities in specific rooms across houses, we find that the younger children engaged in twice as many activities in the dayroom as did the older children. The other obvious difference was in the use of the bench area around the nurses' station. The older children engaged in twice as many activities in this area than did the younger children. This area seemed to serve as a kind of "hang-out" for the older children. The older children engaged in almost as many activities in the bench areas as in the dayroom, although the latter was a much larger area with table, a kitchenette, windows, etc. The younger children, on the other hand, engaged in approximately six times as many activities in the dayroom as in the bench area.

TABLE 2

FREQUENCY AND PERCENT OF CHILDREN'S ACTIVITIES WITHIN THE YOUNGER AND OLDER HOUSE BY ROOM

	Younger House f	% [1]	Older House f	% [1]
Laundry	0	0	1	.11
Seclusion Room	16	2.39	8	.88
Nurses' Station	16	2.39	14	1.54
Benches	42	6.27	117	12.86
Corridor	94	14.03	117	12.86
Dayroom	259	38.66	138	15.16
Total Public Rooms	427	63.73	395	43.41
Living Unit 1	90	13.43	147	16.15
Living Unit 2	47	7.01	151	16.59
Living Unit 3	106	15.82	217	23.85
Total Living Units	243	36.27	515	56.59
Total House	670	100.00	910	100.00
Within Living Units				
Public Areas	110	45.27	245	47.57
Bedrooms	133	54.73	270	52.43
Total LU	243	100.00	515	100.00
Number of full-care children	13		19	

[1]Percentages are based on the total number of activities within each house; in the case of the living units, on the total number of activities in the living units within each house.

If we combine all the data for the rooms outside the living units, i.e. the public rooms, and compare these to the frequency of activity within the living units, another interesting distinction is found. The older children engaged in significantly more activity in the living units than outside the units, while for the younger children, the reverse was true. Clearly, one important factor in the location of activities within the two houses was the placement of television sets. In the younger house it was in the dayroom, while the older children had a set in the living room of each living unit. Despite the fact that the older children had a TV in their living rooms, the drawing power of the benches combined with the dayroom was at least equivalent to the drawing power of the living rooms. Of the 910 activities in the older children's house, 117 took place in the bench area, 138 in the dayroom (both non-TV areas), and 212 in living rooms of the living units. That is, a total of 255 activities took place in non-TV areas versus 212 in the TV areas. In contrast was the younger children's overwhelming use of the dayroom where the TV was located. Of 670 activities in the younger children's house, 259 took place in the dayroom, 42 in the bench area, and 82 in the living rooms. Comparing TV to non-TV areas, we see that for the younger children the television set was a much stronger environmental "magnet" with 259 activities in the TV area as compared with less than half that number (124) in the non-TV areas.

Another view of how children distribute themselves within house areas is through the size of groups involved in activities within these areas. A comparison was made of the number and percent of activities involving 1, 2, 3, 4 and 5 or more children in each room within each house. If we consider the house as a whole unit, there was little difference between the two houses. Both younger and older children engaged in a preponderant number of one-person activities (82.40% for the younger, 83.51% for the older children). In order of frequency, this was followed by two, three, four and larger groups. Two-person groups accounted for 11.80% of the younger children's activities and 13.51% for the older children. The high percentage of one-person activities must be viewed in the context of the way the houses functioned within the total institutional setting. Our previous research (12, 13) indicated that early in the two and one-half years of occupancy and con-

4. ENVIRONMENTS FOR DIFFERENT AGE GROUPS / 197

tinuing through to this study, the house areas became crystallized as a place for isolated behavior, especially during the daytime hours. Thus, the preponderance of one-person activity in both houses would be expected. In light of this, the data providing the most information would be those involving a comparison between houses of the use of specific areas for activities involving differing numbers of children.

An interesting difference is found in comparing the sizes of groups in the public areas and in the living units. Although one-person activity was most commonly observed in both places, the younger children one-person activity accounted for 82.35% of the total activity in the public rooms, while for the older children it accounted for 73.01%. The older children had almost twice as much two-person activity in the public rooms as did the younger children (21.80% vs. 12.38%). In the living unit, on the other hand, this finding reversed itself. There was 90.45% of activity involving only one person for the older house, 80.86% in the younger house; 11.96% of the activity involved two people for the younger children and 7.95% for the older children. Within the living units, foyers and living rooms showed the same pattern of activities of varying group size. One-person activity again was common, but the older children showed more of it than did the younger children (92.80% in contrast with 83.90%). In addition, the younger children had almost twice as many two-person activities (13.56% in contrast with 6.67%).

One further point should be noted about these data. The younger children revealed a similar distribution of numbers involved in activity groups regardless of the room they were in. For the older children, the number in an activity was much more a function of the specific room in which they were located, i.e., the public rooms showing the greatest amount of two-person activities, followed by the foyer and the living room within the unit, and the lowest in the bedrooms.

It is clear from even these preliminary data that there are significant differences in the use of identical physical areas when adolescents are compared with younger children. The adolescents were more likely to use the living unit or apartment than were the younger children, but also seemed to use the bench area in a very special way. The bench area is smaller in size than any other area or room observed, including the single-bed room. Aside from seating space, it is not structured in terms of function. Yet it normally held more children than could possibly be seated. Although adjacent to the nurses' station, we have never observed the staff summonsing or directly encouraging adolescents to use that particular spot.

For the younger children, the dayroom seemed to be the center of activity. A major factor here seems to be the presence of the television set in that room, although the television was not the only activity taking place.

There is an indication that the younger children were more general in their use of space for different kinds of interactions. Whether they were in a bedroom or the dayroom, the distribution by group size was similar. The older children tended to define rooms as appropriate for specific kinds of interactions, the dayroom being most appropriate for two-person activities while the bedrooms functioned most frequently for one-person activities.

Types of Activities

The results cited indicate that the distribution of activities within identical spaces was different for the younger and older groups. The question remains whether the types of activities in which the children engaged were also different for the two age groups. If we look at the house as a whole, the general distribution of predominant activities is the same for the two age groups (see Table 3). It is clear that for both the younger and older children the five most frequent activities are identical in order with isolated passive ranking highest, followed by talking, media, isolated active and traffic. These results reiterate the single-person nature of much of the house activity. With the exception of talking, four of the five most frequent activities were engaged in by one person, including media (mainly watching TV). The small differences that did appear occurred in the frequency of low-energy organized behavior (sitting games and arts and crafts) and high-energy unorganized behavior (mainly play). For the younger children, low-energy organized shared fifth rank with traffic accounting for 7.01% of activity while for the older children it was far below fifth position at 2.64% of activity. For the younger children high-energy unorganized followed closely at 5.37% while for the older children only 1.21% of this activity occurred.

TABLE 3

FREQUENCY AND PERCENT OF SPECIFIC TYPES OF ACTIVITIES[1] OBSERVED IN THE ENTIRE HOUSE[2]

	Younger House			Older House		
	f	%	Rank	f	%	Rank
Isolated Passive	149	22.24	1	243	26.70	1
Talking	118	17.61	2	188	20.66	2
Media	112	16.72	3	134	14.72	3
Isolated Active	82	12.24	4	120	13.19	4
Traffic	47	7.01	5	84	9.23	5
Low-Energy Organized	47	7.01	6	24	2.64	7
High-Energy Unorganized	36	5.37	7	11	1.21	11
Total number of activities observed in each house	670	100%		910	100%	

[1]This and subsequent tables contain only categories in which a significant amount of activity was observed.

[2]Living units and public areas combined.

The house areas have been divided for purposes of analysis into two separate areas or groups of rooms: the living units containing the bedrooms, living room and foyer, and the public areas containing the corridor, dayroom, bench area, nurses' station, seclusion room and laundry. Children infrequently used the laundry, the seclusion room or the nurses' station. Therefore, the public areas to be reported will include the dayroom, bench area and corridor. Further, rather than present data from the living units as a whole, we will focus on the bedrooms in comparison with the living rooms in order to pinpoint more adequately the location of specific activities.

As mentioned earlier, the older children's televisions were located in the living rooms of the living units. An examination of the distribution of activities in those living rooms (Table 4) reveals that media ranked highest at 45.28% of activi-

TABLE 4

FREQUENCY AND PERCENT OF SPECIFIC TYPES OF ACTIVITIES OBSERVED IN THE LIVING UNITS

	Younger House			Older House		
	f	%	Rank	f	%	Rank
Living Rooms						
High-Energy Unorganized	16	19.51	1	-	-	-
Low-Energy Organized	11	13.41	2	11	5.19	6
Traffic	10	12.20	3	13	6.13	5
Isolated Active	9	10.98	4	18	8.49	4
Isolated Passive	9	10.98	4	31	14.62	3
Talking	9	10.98	4	33	15.57	2
High-Energy Organized	6	7.32	5	-	-	-
Media	6	7.32	5	96	45.28	1
Total number of activities observed in living rooms in each house	82	100%		212	100%	
Bedrooms						
Isolated Passive	62	46.61	1	145	53.70	1
Talking	24	18.04	2	40	14.81	3
Isolated Active	18	13.53	3	46	17.04	2
High-Energy Unorganized	8	6.02	4	6	2.22	5
Low-Energy Organized	6	4.51	5	-	-	-
Traffic	-	-	-	10	3.70	4
Total number of activities observed in bedrooms in each house	133	100%		270	100%	

ty. Watching TV was always a one-person activity. When TV is combined with the other one-person activities occurring in the living rooms (isolated passive and isolated active), we find that this accounts for 68.39% of overall living room behavior. The second most frequent activity was talking, but it accounted for only about 15% of activity (one-third that of media). For the younger children the living rooms without the TV apparently had a very different function. There was no peaking of behavior, and the most frequent activity, high-energy unorganized, accounted for less than 20% of all behavior. The other frequent activities accounted for between 11 and 13% of all activity. Examining the distribution of frequent activities in the living rooms, one finds that the younger children spread themselves over a broader range of possible behaviors than did the older children. There was much more diversity in the activities accounting for a substantial amount of the younger group's behavior. In addition, the younger children's range of behaviors in the living room was more diverse than their behavior in the dayroom (see Table 5) and more diverse than the older children's behavior in either the living rooms (Table 4) or dayroom (Table 5). For the younger children there was a rather high use of the living room as a traffic area, and talking ranked low despite the absence of TV. The two highest ranking activities, high-energy unorganized (play) and low-energy organized (sitting games) were always two-person activities, although a more structured means of social interaction than talking. Excluding media, the amount of isolated behavior in both living rooms was similar at about 23% but with a difference in <u>type</u>. The older children had more isolated passive behavior while the younger children had an equal frequency of active and passive behavior. However, adding the frequent "one-person media" in the older children's living rooms, the picture emerges of much more one-person activity.

Since the younger children's TV was located in the dayroom, it seems reasonable to compare this room with the older children's living room. The presence of the TV as expected resulted in a high degree of viewing, which accounted for almost 40% of all behavior (Table 5), although 5% below the media of the older children in their living rooms (Table 4). Interestingly, talking ranked second in the younger children's dayroom, a pattern similar to that in the older children's living rooms. The presence of the TV for the younger children, then, did not eliminate talking as a possible additional or concurrent activity. Yet, in the younger children's living rooms, without the TV, talking ranked **fourth** while it was first in the older dayroom. Two factors seem to contribute to the different distribution of activities for the two age groups compared. One aspect of this appears to be a developmental factor, that is, what each group apparently does with-

TABLE 5

FREQUENCY AND PERCENT OF SPECIFIC TYPES OF ACTIVITIES OBSERVED IN THE PUBLIC AREAS

	Younger House			Older House		
	f	%	Rank	f	%	Rank
Dayroom						
Media	103	39.77	1	22	15.94	3
Talking	43	16.60	2	31	22.46	1
Low-Energy Organized	27	10.42	3	-	-	-
Isolated Active	24	9.27	4	14	10.14	4
Isolated Passive	23	8.88	5	13	9.42	6
Domestic	16	6.18	6	26	18.84	2
High-Energy Organized	-	-	-	13	9.42	5
Total number of activities observed in dayroom of each house	259	100%		138	100%	
Bench Area						
Isolated Passive	18	42.86	1	37	31.62	1
Isolated Active	13	30.95	2	18	15.38	3
Talking	7	16.67	3	32	27.35	2
Low-Energy Organized	2	4.76	4	6	5.13	5
Traffic	1	2.38	5	-	-	-
Miscellaneous	1	2.38	6	-	-	-
Media	-	-	-	7	5.98	4
Total number of activities observed in bench area of each house	42	100%		117	100%	
Corridor						
Isolated Passive	28	29.79	1	9	7.69	4
Talking	24	25.53	2	34	29.06	1
Traffic	16	17.02	3	30	25.64	2
High-Energy Organized	4	4.26	4	7	5.98	5
High-Energy Unorganized	4	4.26	4	-	-	-
Miscellaneous	4	4.26	4	-	-	-
Isolated Active	-	-	-	20	17.09	3
Total number of activities observed in corridor of each house	94	100%		117	100%	

out the structure of a strong environmental prop -- the television set. The younger children did not engage in much unstructured social interaction (talk), but instead engaged in structured social interactions, essentially substituting other props, i.e. games and toys. The older children engaged in more unstructured social behavior, talking and two-person food preparation. Furthermore, the older children utilized another media (listening to the phonograph) in the absence of the TV.

The bedroom functioned mainly as a place for isolated passive behavior for both groups, accounting for about one-half of all behavior (Table 4). Talking and isolated active behavior were also in evidence, although at a much lower level. Yet, the total amount of isolated behavior for the younger children was lower than for the older children (60.14% as opposed to 70.74%) and, although representing a considerable drop from the first three activities, high-energy unorganized and low-energy organized also occurred with some frequency.

In examining the remaining public areas (Table 5) it is necessary to recall that the bench area was used much more frequently by the older children than by younger children. There was also a clear difference in the type of activity occurring within that area when it was used. While isolated passive behavior was the most frequently observed activity for both groups, the younger children clearly used the area more for this behavior than did the older children. In fact, if we combine the two major types of isolated behavior, the younger children revealed 73.81% while the older children showed 47.00%. In addition, the older children engaged in almost twice as much talking in this area. It is also clear that the two age groups used the corridor differently. Talking was at about the same level for both groups, yet the younger children showed substantially more isolated passive behavior in this area (29.79% as compared to 7.69% for the older children). Total isolated behavior was similar for both groups, but again the type differed with the older children using the corridor for more active single-person behavior (e.g. reading and watching an activity) while the younger children were using it for passive behaviors (sitting or standing with no concurrent activity).

Discussion

One might be tempted to consider the results of this study as relevant only to a very special group of children observed in a special setting. Although these results are preliminary, it is our feeling that many of the age-related differences found in this milieu are likely to be relevant to settings that ordinary children occupy. One would expect that adaptation to a total institutional setting implies an absolute uniformity of behavior despite age or other personal variables. To some extent this is true, since there are predominant behaviors associated with institutionalization. For example, the high degree of isolated behaviors found in the children's hospital is characteristic of adult psychiatric institutions as well (6). However, in spite of a programmed day in the children's hospital (similar for both age groups), an explicit set of rules governing behavior, the use of drugs for treatment of both age groups and an identical physical design for both houses, there were patterns of differences between the two groups. These patterns, which were not evident in the overall distribution of types of activities when the

4. ENVIRONMENTS FOR DIFFERENT AGE GROUPS / 201

houses were viewed as a whole, were apparent in the size of the groups involved in activities, in the dispersion of activities across rooms and in the extent to which different rooms were the locus of different types of activities. One might predict that in a less structured situation even more age-related differences in the use of space might appear.

One factor which could be operating to produce different patterns of use might be a tendency on the part of the staff to encourage the use of different spaces for the two age groups. Our observation does not support this possibility. There seemed to be no overt pressures on either age group to conform to staff expectations of where they should be. By looking at the specific differences between the age groups, it is possible to assess their design implications.

The activities of the older children tend to be more evenly distributed across all available spaces. While the location of the TV set clearly defined the major activity for one type room (the living rooms), it did not limit the locus of other activities or the use of other spaces. That is, the older children used the living room for its implicit purpose, but also used the other spaces for their purposes (i.e., bench area and dayroom). This may reflect the development of a capacity to match more skillfully what they wanted to do with the total range of spaces within the house and to select a space on the basis of this match. For younger children, the "magnetic force" of the TV set seemed powerful, leaving some areas of the house with relatively low use. Thus, we can look at the TV set as a very strong environmental component which drew the younger children to a specific location and defined a large measure of their behavior there (as it did for the older children). Yet, when the younger children used the living rooms (without the TV), a rather wide diversity of activities was observed. In fact, the younger children showed more variety of activity in the living rooms than they did in the dayroom and more than the older children showed in any house area. The younger children seemed to use unstructured space more spontaneously and more diversely. Their activities may be more a reflection of immediate reaction than a deliberate review of the total range of environmental possibilities. When there is some external definition, this may act as an immediate influence on choice of activity. These observations are parallel to others we have made in open school settings. Older children seem much more capable of finding suitable spaces for activities they want to pursue -- a capacity less developed in younger children. In unstructured classrooms the younger children seem to have more difficulty in appropriately locating themselves. These findings seem to be related to Piaget's descriptions of the cognitive capacities of children of different ages (9,10).

The location of isolated passive behaviors (sleeping, lying awake and sitting doing nothing) reveals further interesting differences for the two age groups. The bedroom was the major place for isolated passive behavior for both groups, although it was used for this somewhat more by the older children. The bedroom certainly was an appropriate and expected place for withdrawal and private activity. While the bench area ranked second in isolated passive behavior for both groups, this space was more frequently used for this purpose by the younger children. The bench seat itself, again, would make the behavior not totally unexpected in this area. The younger children, however, also used the corridor and the dayroom much more than the older children for this isolated passive activity. In fact, in terms of the more public areas, it was only in the living rooms that the older children showed somewhat more isolated passive behavior. The more frequent use of publicly defined spaces for more personal and private behaviors (sleeping, etc.) by the younger children -- especially the corridor and dayroom -- underline the less functionally apparent use of space by that age group. These areas contained larger numbers of people than other areas, and there was generally more traffic in or adjacent to them -- factors which might make them less appropriate for private activity. Older children seem to be more aware of these factors, confining their isolated behavior much more to the bedroom and somewhat less to the bench area.

Although there is a clear need for further investigation of age and space use patterns, the data presented lead one to speculate on the planning implications. There is evidence that age affects use of available facilities, but only future studies can define specifics of these age differences. At the very least, it would seem that facilities to be used by different ages should consider the need of younger children for clear definitions of spaces. If planners, teachers, group leaders or administrators desire to have spaces used in specific ways, younger children may require explicit environmental cues. A seating area in an exposed sector of a ward may have been designed as a social area, but it could be and was used for other functions as well. This does not imply that space alone accounts for age-related behavior. In fact, subtle differences in supervision of children may be operating. However, in areas where supervision was minimal, our data still uncovered age-group differences. This raises the issue of whether the environments designed for younger children should be matched closely to their age needs. Our own data, preliminary as they are, do not argue for overmatching. Instead, they indicate that when specific behaviors are the goals and where the physical environment is regarded as an important factor in reaching these goals, the design components must communicate the message at the child's level of understanding.

Notes

This research was supported in part by U. S. Public Health Service grant MH 18010. Data collection and analysis were assisted by Marian Beyda, Joan Burress, Karen Franck, D. Geoffrey Hayward, Alan Sommerman, Linda Wallach, Sheree West, Geoffrey Weiland and Walworth Wentworth.

1. Barker, R. & Wright, H. Midwest and its children. N. Y.: Harper & Row, 1955.
2. Bayes, K. The therapeutic effect of environment on emotionally disturbed children. Surrey, Eng.: Gresham, 1967.
3. Gesell, A. & Ilg, F. Infant and child in the culture of today. N.Y.: Harpers, 1943.
4. Hart, R. Environments for the developing child. Los Angeles: EDRA III Proceedings, 1972.
5. Hart, R. & Moore, G. The development of spatial cognition: A review. Place Perception Report No. 7. Chicago: Environmental Research Group, 1971.
6. Ittelson, W., Proshansky, H. & Rivlin, L. The use of behavioral maps in environmental psychology. In H. Proshansky, W. Ittelson & L. Rivlin (Eds.) Environmental psychology: Man and his physical setting. N.Y.: Holt, Rinehart & Winston, 1970, 658-668.
7. Kuhlen, R. The psychology of adolescent development. N.Y.: Harpers, 1952.
8. Piaget, J. The language and thought of the child. N.Y.: Humanities Press,1952.

9. Piaget, J. & Inhelder, B. The child's conception of space. N.Y.: W. W. Norton, 1967.
10. Piaget, J. & Inhelder, B. The psychology of the child. N.Y.: Basic Books, 1969.
11. Pollowy, A. & Bezman, M. Design-oriented approach to developmentsl needs. Los Angeles: EDRA III Proceedings, 1972.
12. Rivlin, L. & Wolfe, M. The early history of a psychiatric hospital for children: Expectations and reality. Environment and Behavior, 1972, 4, 33-72.
13. Wolfe, M. & Rivlin, L. The evolution of space utilization patterns in a children's psychiatric hospital. Los Angeles: EDRA III Proceedings, 1972.

PHYSICAL PLANNING FOR INCREASED CROSS-GENERATION CONTACT(1)

Edward H. Steinfeld

National Bureau of Standards

Abstract
Research sought to determine the feasibility of increasing beneficial social contact between older people and young children and to identify physical features that will support such contact. An interview survey of 96 elderly residents of two age-segregated housing projects assessed their interest in contact with young children and identified features of proximity, conditions of privacy and location features that would encourage it. A set of scale models were used as a survey instrument. It appears feasible to plan for increased cross-generation contact in a wide variety of community settings. Three community planning strategies and several design implications are suggested.

Introduction and Background
The purpose of this study was to identify and describe physical environments that will help increase mutually beneficial social contact between the very young (0-5 years old) and older adults (about 60 years old and older).

There are three reasons for increasing such contact: 1) a need for more service-giving roles for the old, 2) the pervasiveness of a negative stereotype of older people in our society, 3) a growing lack of opportunities for cross-generation social contact.

Although one pattern of successful adjustment to aging is a gradual disengagement from obligations and activities, a pattern of sustained activity also can be successful (2). The social context of a society, however, can enforce disengagement on an individual, even when it is not desired (3). Contrasting the role of the aged in modern and primitive society, L. W. Simmons observed that, in the former, the individual makes adjustments with age, while, in the latter, society makes the adjustments for him (4). Providing enough service-giving roles for elders in a community can allow them the greatest potential for expansion of interests and acceptance of their life as appropriate and meaningful (5). It could also help overcome the continued persistence of poverty among the older population (6).

Presently, the image of the aged is far from positive. Studies show a "stereotype of old age as a period of physical, social and emotional decline" (7). A change of these existing attitudes could influence the younger generation's behavior toward its elders (including political behavior related to approving expenditures for services and facilities). It might also create a better psychological climate for the elderly to develop healthy strategies of coping with the aging process.

Attitude change is not likely to come about through reading books or public relations campaigns. It is fostered through direct experience, i.e., social contact with people having a positive self image.

In primitive societies, the very old and very young often have intimate associations. Since both groups have low mobility, they are left together at home where elders, both men and women, instruct children in everything from myths to sex (8). About the only institution in a modern industrial society than can offer an opportunity for such meaningful cross-generation contact is the family, through the roles of parent and grandparent. Studies of inter-generational relations, however, indicate that older parents have an essentially dependent relationship to their grown up children (9). Grandparenthood, in the family context, moreover, is not an important substitute for the parent's role--it is merely an extension of it (10). Now, and probably more so in the future, community child care services and high residential mobility of families from city-to-city (11) reduce the potential for face-to-face contact. Thus, the modern family today evidently does not provide much of a chance for older people to give services to others and have frequent and intimate associations with the young.

The Foster Grandparent Program provides an example of how planning for increased cross-generation contact may benefit society. The Foster Grandparents, whose incomes fall below the poverty line, work a 20-hour week and are assigned to one or two children in orphanages, homes for the mentally retarded and other institutions. Evaluations of the program have shown that it has measurable benefits to physical and mental health for both children and older people. Moreover, it provides an excellent resource for augmenting institutional services (12).

Opportunities for developing contact between these two groups should be expanded into many more settings within the community and to many more types of children and senior adults. The Foster Grandparent Program documents much about social planning to increase cross-generation contact. The purpose of this study was to identify physical planning strategies that can be interfaced with social planning strategies to develop more comprehensive approaches.

Bringing disparate age groups together in physical proximity is not likely to be enough. In such situations, contact will not develop spontaneously. Rosow found that older people living in age-integrated settings had few friends among their neighbors of different ages (13). Lipman found that people in three "old people's homes" that were located in close proximity to social amenity facilities such as shops, bus stops, parks, etc., had no informal contact with neighborhood residents (14). Barker, in his ecological analysis of a midwest town found that older people were often present in the same settings as other age groups but in a more passive, vicarious state of social involvement (15).

None of these studies focused on preschool children and senior adults or on systems that were explicitly planned to increase contact between the two age groups. Moreover, none included an attempt to measure desire on the part of the elderly for more contact with younger age groups. However, findings of these researchers suggest that increased cross-generation contact will not be successful unless the encounters are planned--that is, unless an environmental (social and physical) context is designed to occasion and reinforce such behavior. Such an approach has been implemented in Syracuse, N. Y. Here an apartment building for the elderly was built in conjunction with a university dormitory. Reports indicate that it has been successful in inducing beneficial social interaction between the elderly residents and students (16).

These physical factors are likely to be particularly important in increasing cross-generation social contact: 1) proximity of the groups to social contact settings (17) and 2) domains of privacy as control of involvement with others (18). Several related factors are crucial considerations in planning for such contact: 1) extent of interest among senior adults, 2) incentives needed to encourage interest, 3) variability of interest among the older population and 4) popular activities through which social contact may take place.

Research Activities and Findings
An interview survey of elderly residents of age-segregated apartment buildings was undertaken. The survey had two phases: 1) the interest phase--an assessment of interest on the part of residents in various kinds of planned contact with young children and incentives toward participation, and 2) the in-depth phase--an assessment of physical features of proximity and privacy that would encourage contact between elderly people and young children and an identification of differences in background characteristics between people interested and disinterested in formal social contact with young children.

Interest Phase
Two high-rise apartment facilities in Flint, Michigan were selected as sites for the survey. The only important difference between the two buildings is their rent. Building A is a public housing facility and building B is a non-profit building. The residents of building B have to be more affluent to pay the rent. Those in building A have to be poor to qualify for residence. The sample was limited to white residents over age 60. Moreover, severely handicapped residents were eliminated from consideration. A random age-stratified sample of 90 respondents (building A, N = 49, building B, N = 41) was selected at a sampling rate of 1:2. The response rates were 84% in building A and 80% in building B. The reasons for no response included being on vacation, being in the hospital, or, in very few cases, not answering the door. The "interest" interview schedule of questions was as brief as possible, requiring from 15-30 minutes to complete. No one was paid to answer questions.

The interest survey demonstrated that 43% of those interviewed desired increased social contact with young children, through formal child-care programs. More people were initially interested for pay than as volunteers (38 percent versus 23 percent). Feeling "not capable" for health or other reasons was the predominant reason for disinterest. There was a strong indication that easy access to the activity setting and relatively few obligations on performance would be incentives for encouraging and maintaining interest. When such qualifications were mentioned, general interest jumped to 51%. The majority of the interested and undecided respondents would prefer to participate in child-care activities in their own neighborhood. However, most would be willing to go to other neighborhoods if transportation were provided. The most popular activities for participation were welcoming children in the morning to a preschool program, entertaining (e.g., storytelling or playing music), helping children play games, and being a "substitute grandparent."

There was a significant difference between income groups (buildings) in interest: the low income group was more interested both in general ($p < .04$) (19), as well as for pay ($p < .03$). Moreover, there was a significant difference in interest with age ($p < .04$)--older people were less interested.

In-Depth Phase

The in-depth sample consisted of 56 people chosen from the "interest" sample placed within four age-matched analysis groups of 14 people each; two groups were chosen from each building--one interested in formal social contact, one disinterested. The matching provided a total sample tending toward older late adulthood in age.

These background characteristics of interest groups were investigated: marital status, size of family, sex, existing contact with young children, occupational status, mobility, activity engagement, length of residence, life goals, subjective health. Overall variety of activity engagement was determined by summing affirmative responses with respect to participation in 15 different daily activities. Variety of social engagement was determined by a sub-scale using the 8 of the 15 activities which involved social interaction.

Preferences for physical features were assessed by using three-dimensional scale models of activity settings (Fig. 1-3). Features of the models were abstracted to emphasize the relevant variables. Each scale model represented a setting for one type of activity: 1) a park (recreation), 2) a community center (social meeting), 3) a neighborhood (housing). The type of setting and each element in the models were explained to the respondents as it was presented. Wherever they were needed, color cues and simulated features were used, but care was taken to be uniform in the use of such cues to avoid confounding results by choices based on attractiveness.

The park was used to measure desirable proximity in terms of the physical distance to children. (Fig. 1). Three benches were positioned with identical accoutrements (tree, walk) at three different distances from a playground for children. It was explained that each of the benches was exactly the same walking distance from the point at which the walk diverged. Respondents were then asked to indicate at which bench they would want to sit if they were walking into the park alone or with a friend.

The community center was used to measure desirable proximity in terms of accessible distance to children (Fig. 2). It was explained that the three models represented three different ways in which a community center could be built to house a nursery

208 / ENVIRONMENTAL DESIGN RESEARCH, VOL. 1

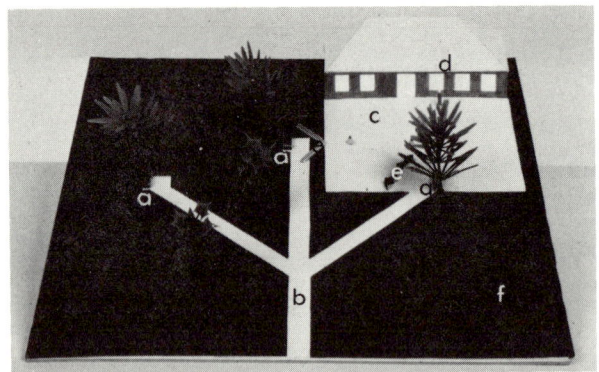

Figure 1. Model of the Park (Recreation Setting).
   a. benches
   b. walk
   c. playground
   d. nursery school
   e. barriers (place into position only after preference was recorded--data not reported in this paper)
   f. grass (simulated)

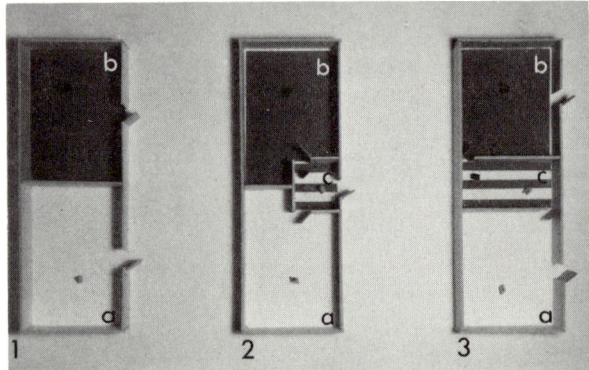

Figure 2. Models of the Community Center (Social Meeting Setting)(20)
   1. plan with complete separation
   2. plan with shared entrance
   3. plan offering a choice of contact
   a. activity area for young children
   b. activity area for senior adults
   c. shared space

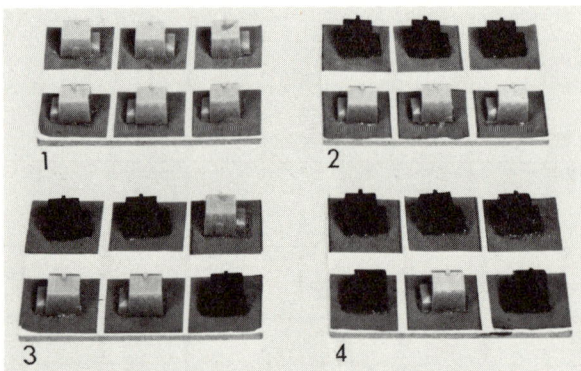

Figure 3. Models of Neighborhood (Settlement Pattern)
   1. age segregated neighborhood
   2. high density of senior adults, separation
   3. high density of senior adults, mixed
   4. low density of senior adults
light houses-older people
dark houses-families with young children

school and a senior citizens' activity center. The differences in accessibility were also explained, i.e., in the top model one group would have to go outside to visit the other, in the middle model they would both share a reception room and could visit without leaving the building, in the bottom model they would both have separate entrances but could choose to be together in the middle room if they so desired. Respondents made two choices and the pattern of their preferences was recorded. The pattern of choice as a whole was used as an indication of preference.

The neighborhood models were used to measure desirable proximity in terms of the degree of age-segregation of older people in the settlement pattern (Fig. 3). Four cards were prepared with six plastic model houses on each, arranged on both sides of a street on individual lots. Each neighborhood or block had a different mix of senior adults and families with young children. Respondents were asked to choose the block they would like most to live on and the block they would like least.

Concerns about privacy with respect to children were determined by asking open-ended questions regarding the reasons for respondents' choices in each model. Responses were then content-analyzed for attitudes toward control of involvement with others and coded in a standard way.

The entire in-depth interview lasted about one hour, probably the limit of tolerance for many senior adults.

As a whole, the sample had these background characteristics: 1) very few of the respondents had never been married (11 percent), 2) most were widowed (71 percent), 3) eighty percent were women, 4) sixty-two percent were from blue collar families, 5) families were fairly large, 6) 48 percent never come in personal contact with young children.

Mobility was relatively low. Data on mode of travel, trips from home, and number of active drivers demonstrated that 67% of the respondents had only 3 trips per week and only one-quarter of the respondents had the independence engendered by driving themselves or with their spouse. Low mobility is also demonstrated by the fact that almost 60 percent of the respondents spent three hours or less outside the apartment building when they went out.

There were only three background characteristics in which interest groups differed. In life satisfaction goals, the interested group chose social interaction goals most often and the uninterested group chose good health most often ($p < .02$). In rating their present health, sixty-eight percent of the uninterested respondents felt that their health was not good or fair, while only 39 percent of the interested group felt this way ($p < .04$). Social engagement scores were higher for the interested group (one way analysis of variance by ranks, $p < .04$); thus, the interested people had a greater scope of social interests (although not necessarily of total interests nor a greater degree of involvement in the social activities they pursue).

The pattern of preferences in the recreation setting indicates that almost two-thirds of the sample were willing to position themselves in locations where informal contact was likely to occur (Table 1). The rank correlation between scores on an involvement-reserve (privacy concerns) continuum, and a close-distant continuum (proximity based on physical distance) was .82 ($p < .01$). This indicated an association between preferences for physical distance and the concern for control of social interaction with children in recreation settings.

In the social meeting setting, the interested group desired more accessibility to the children's side of the building (Table 3). There were significant differences between interest groups with both proximity (Kolgomorov-Smirnov Test, $p = .01$) and privacy concerns ($p < .02$). Another important finding was the higher overall degree of concern for more privacy in this setting than in the recreation setting (Table 4). Seventy-six percent were concerned with "reserve" and "autonomy" versus only 47 percent in the recreation setting. The rank correlation between accessibility preference and privacy concerns was .50 in this setting.

Additional attitudes toward accessibility-based proximity were culled from responses to questions posed about the possibility of facilities for young children being built within the respondents' buildings, or next door. Most people approved of the next door site (79 percent). The within-building site, however, evoked significant differences ($p < .02$) between interest groups.

Interest groups had similar preferences for most liked settlement pattern (Table 5). More than half the respondents preferred some degree of age integration (59 percent), although almost all of these people desired a high density of older people. Of those who preferred total age segregation, almost two-thirds did not mind families with younger children living down the street, or on the next street over from them. A few more did not mind such arrangements if there was a park barrier between them. Respondents split decisively in their least-liked preference, either selecting the low density (of older people) neighborhood (59 percent) or the totally age-segregated one (39 percent). The rank correlation between most-liked preference scores on a continuum from total age segregation to low density of older people (settlement pattern) and scores on a continuum from reserve to variety of contact (privacy concerns) was .83 ($p < .01$). It should be noted that, in the privacy concerns, respondents emphasized "variety" of neighbors rather than actual social interaction (see Table 6).

Although, in some cases, there were no significant differences for preferences between interest groups, each group by itself, as well as the total sample, did have significant preferences for proximity features. These findings (see Table 7) demonstrate that, on the whole, the sample preferred: 1) close physical distance to children in the recreation setting, 2) neighborhoods with a high density of older people. They also pointed out that interest groups held opposite preferences for accessibility to children in the social meeting setting.

There was a degree of consistency in respondents' preferences for proximity in the three settings (coefficient of concordance, $W = .50$, $p = .02$); the same holds true for the privacy concerns (coefficient of concordance, $W = .50$, $p = .02$). Of more

TABLE 1. Proximity (Physical Distance) Preferences In the Recreation Setting

Proximity Desired	Interest In Child-Care Programs					
	Positive		Negative		Total	
	%	(n)	%	(n)	%	(n)
Distant	29	(8)	43	(12)	36	(20)
Intermediate	11	(3)	14	(4)	12	(7)
Close	61	(17)	43	(12)	52	(29)
Total	100	(28)	100	(28)	100	(56)

No significant differences.

TABLE 2. Privacy Concerns in the Recreation Setting

Privacy Theme	Interest In Child-Care Programs					
	Positive		Negative		Total	
	%	(n)	%	(n)	%	(n)
Reserve	8	(2)	22	(6)	15	(8)
Autonomy	23	(6)	41	(11)	32	(17)
Involvement	69	(18)	37	(10)	53	(28)
Total	100	(26)	100	(27)	100	(53)

No significant differences.

TABLE 3. Proximity (Accessibility) Preferences In the Social Meeting Setting

Proximity Desired	Interest In Child-Care Programs					
	Positive		Negative		Total	
	%	(n)	%	(n)	%	(n)
As much as poss.	29	(8)	18	(5)	23	(13)
Some	36	(10)	21	(6)	29	(16)
Choice	14	(4)	7	(2)	11	(6)
Limited Only	7	(2)	4	(1)	5	(3)
Limited, if at all	11	(3)	4	(1)	7	(4)
As little as poss.	4	(1)	46	(13)	25	(14)
Total	100	(28)	100	(23)	100	(56)

* $p = .01$ (Kolgomorov-Smirnov Two Sample Test)

TABLE 4. Privacy Concerns In the Social Meeting Setting

Privacy Theme	Interest In Child-Care Programs					
	Positive		Negative		Total	
	%	(n)	%	(n)	%	(n)
Reserve	27	(7)	50	(13)	38	(20)
Autonomy	35	(9)	42	(11)	38	(20)
Involvement	38	(10)	8	(2)	23	(12)
Total	100	(26)	100	(26)	100	(52)

* $p < .02$ (Chi-Square Test)

TABLE 5. Proximity (Settlement Pattern-Age Segregation) Preferences in the Housing Setting

Neighborhood Type Desired	Interest In Child-Care Programs					
	Positive		Negative		Total	
	%	(n)	%	(n)	%	(n)
Segregated	32	(9)	54	(15)	43	(24)
High Density/ Separation	36	(10)	25	(7)	30	(17)
High Density/ Mixed	32	(9)	14	(4)	23	(13)
Low Density	--	(0)	7	(2)	4	(2)
Total	100	(28)	100	(28)	100	(56)

No significant differences.

TABLE 6. Privacy Concerns in the Housing Setting

Privacy Theme	Interest In Child-Care Programs					
	Positive		Negative		Total	
	%	(n)	%	(n)	%	(n)
Reserve	22	(6)	32	(8)	27	(14)
Autonomy	11	(3)	20	(5)	15	(8)
Limited Variety of Contact	48	(13)	24	(6)	37	(19)
Unspecified Variety	18	(5)	24	(6)	21	(11)
Total	100	(27)	100	(25)	100	(52)

Test for differences not possible.

TABLE 7. Summary of Preferences for Proximity Conditions: Attained Significance Levels for Kolgomorov-Smirnov One-Sample Tests

Item	Interest In Child-Care Programs		
	Positive	Negative	Total
Proximity (Physical Distance)	.05	n.s.	.05
Proximity (Accessibility)	.01	.05	n.s.
Settlement Pattern (Preferred)	.05	.01	.01

TABLE 8. Rank Correlation Matrix

Privacy Variables	Proximity Variables		
	1	2	3
1	.82	.24	.28
2	.27	.50	.43
3	.19	.34	.83

importance, privacy variables correlated most strongly with proximity variables that referred to the same settings (Table 8). Since each setting was constructed to represent a certain kind of proximity, it was expected that there would be differences in privacy concerns in each. This proved to be the case. Evidently, certain privacy concerns are associated with different kinds of proximity even though preferences for proximity and privacy concerns in different settings may be consistent.

Conclusions and Implications

Many older people living in age-segregated housing appear to be interested in participating in child care programs. The foster-grandparent type of relationship seems to be a good match with their interests. It appears that informal social contact might also be successful. The background characteristics of older people associated with interest in formal contact (planned programs) with young children are low-income, younger age, good subjective health, desires for increased social interaction, and a varied pattern of engagement in social activities. People with these characteristics should be the target population for planning programs.

The respondents, as a group, have relatively low mobility, especially when compared to the mobility reported by single respondents in a study by Lansing, et al, of 10 suburban and urban communities. In the Lansing study, single people (including widowed and divorced) averaged 5.4 trips per day, almost all by automobile (21). This and other findings imply that location of settings in the immediate housing environment or provision of transportation would be an incentive for older people to participate in child care programs.

Accessibility to children in a social meeting setting was the only physical feature for which preferences of interest groups differed significantly. A much greater proportion of the sample was concerned with reserve and autonomy in that setting than in the others. Thus, proximity to children out-of-doors or in site planning of buildings may encourage informal contact even among people disinterested in formal contact. Since accessibility appears to be a more potent factor than physical distance and settlement pattern in insuring control of social involvement, accessibility is more likely to discourage contact if privacy concerns are not met. For those adults who have a positive disposition toward involvement with children in social meeting settings, the findings (see Tables 3 and 4) suggest that there should be a continuum of available privacy conditions to encourage the maximum amount of cross-generation contact. This continuum could have three domains: 1) "territories" for senior citizens within children's settings (22), 2) a "common ground" for compatible activities, 3) a "refuge" for the older people with controlled access. Each should be separated from the other in ways that maximize control of involvement (see Fig. 4). In the recreation setting, there was a clear preference in the total sample for close physical distance. Thus, providing settings for older people very close to children's recreation activities would probably be successful in encouraging informal social contact between the two groups. The still important privacy concern for autonomy (32%), however, suggests that some of the senior adults' settings should be planned carefully to provide freedom of choice over involvement (see Fig. 5). In the housing setting, a majority of the sample preferred some degree of age integration (57%), yet the

4. ENVIRONMENTS FOR DIFFERENT AGE GROUPS / 213

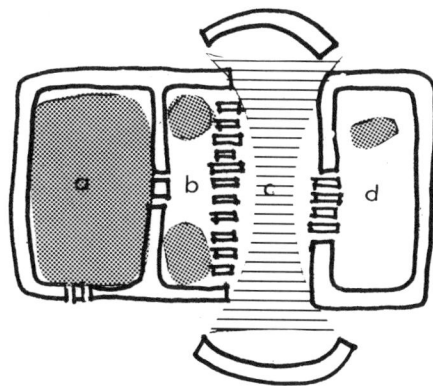

Figure 4.

Key to graphic symbols:

  Locus of activity designed specially for older people.

  Boundary of setting.

  Controlled "gate" in boundary.

  Partial boundary or "filter," such as gates, planting, changes in level.

Public space.

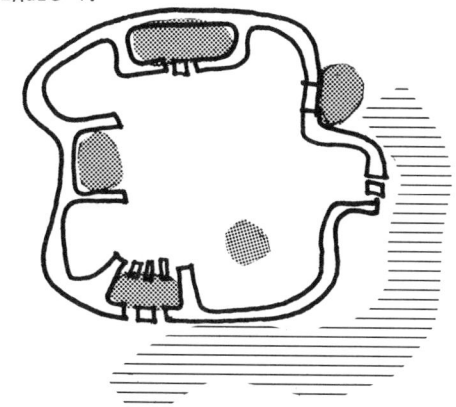

Figure 5.

Figure 4. A social meeting setting for both groups.
a. area exclusively for old people - refuge
b. autonomous zone - common ground
c. public zone
d. area for children with territories

Figure 5. A recreation setting for children with several different types of settings for older people within it and contiguous to it.

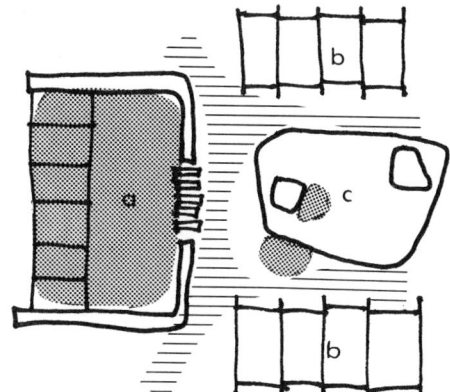

Figure 6.

Figure 6. A housing setting with an "enclave" of units for older people.
a. "enclave"
b. houses for families with young children
c. playground

entire sample showed a preference for a high density of older people. There was also a clear preference for total age integration as the least-liked settlement pattern. These findings suggest that a large number of people who live in age-segregated housing for the elderly would rather live in neighborhoods where there would be some opportunity for informal contact with other age groups. They do not, however, like a large degree of age integration. A neighborhood with a high density of older people yet a substantial and accessible younger population may suit their desires (see Fig. 6).

Three strategies for community planning can be generated from the implications discussed above. First, a <u>network of "host" settings</u> in the community can be linked together with housing settings for older people by a transportation and communication system that would place interested elders in child-care settings needing their services. Each of these "host" settings could have specially designed areas for use by older people to support the kind of formal role in which they would like to engage. For example, a day-care center might have a "grandma's kitchen" which would be an activity focus under the personal control of older women who liked to cook. Second, a <u>continuum of age-integration</u> could be built into the community housing system providing a range of informal contact opportunities in the general neighborhood and in housing-oriented recreation and social meeting sites. The continuum could extend from age segregated apartment buildings through smaller, insulated enclaves of 20-50 people to blocks where 50% of the dwelling units might be for older people. The architecture and site design of each type of housing could help insure personal control over interaction by use of gates, planting, changes in level, and other kinds of "filters." Third, <u>multi-purpose neighborhood facilities for all ages</u> could provide all types of settings, as shown in Figures 4-6 and discussed above.

In summary, this study demonstrates that: 1) many older people wish to increase their social contact with the very young, 2) the physical environment can be an important factor in providing opportunities for that to happen, and 3) providing environmental opportunities for such contact must be tempered by design of settings to provide adequate control over involvement.

Several limitations of the research should be noted. First, the small sample used may restrict generalizability of the results to similar kinds of people, in similar housing settings. Second, the findings regarding interest in formal contact should be interpreted from the point of view of a disposition rather than a commitment. Third, the expertise of social planning agents would be very important in the success of formal or informal cross-generation contact. Simulation of behavior choices with scale models promises to be a highly useful research tool. Of course, observations of actual behavior are preferable for predicting behavior. However, when innovative planning and design is a goal, suitable environments for observation are not usually available. Simulation methods can be substituted to obtain insight for planning and design decisions. The predictability of the methods can then be assessed through later observations in the innovative environment.

Notes

1. This study was done while the author was a doctoral student in Architecture at the University of Michigan.
2. Reichard, Susan F., et al., "Adjustment to Retirement," in Middle Age and Aging, ed. by Bernice L. Neugarten (Chicago: University of Chicago Press, 1968), pp. 178-180.
3. See E. Cumming and W. E. Henry, "A Formal Statement of the Disengagement Theory" in Growing Old (New York: Basic Books, 1961), p. 213.
4. Simmons, Leo W., "The Role of the Aged in Primitive and Civilized Society," in Psychological Studies of Human Development, ed. by R. G. Kuhlen and G. G. Thompson (New York: Appleton-Century-Crofts, 1952), pp. 260-265.
5. These are the two developmental goals of the last two stages of life as seen by Erik H. Erikson, and described in his book, Childhood and Society (New York: Norton, 1950), pp. 267-268.
6. Brotman, Herman B., "A Profile of the Older American," (paper presented at the Conference on Consumer Problems of Older People, Hudson Guild-Fulton Center, New York, New York, 1967), p. 5.
7. Hickey, T., Hickey, L., and Kalish, R., "Children's Perceptions of the Elderly" in Journal of Social Psychology, XXXVII, (1953), pp. 249-260; see also Margery Calhoun and Leonard Gottesman, "Stereotypes of Old Age in Two Samples," (paper presented at the Midwest Psychological Association meetings, 1963). p. 5.
8. Simmons, Leo W., The Role of the Aged in Primitive Society, (New Haven: Yale University Press, 1945), p. 199.
9. See, for example, Marvin Sussman, "The Family: Social Change in Family Life," in Growing Old in Tomorrow's Cities (unpublished manuscript) ed. by Leon A. Pastalan and Wilma Donahue; see also F. A. Davis and W. P. Hawkinson, "Wish, Expectancy and Practice in the Interaction of Generations," in Older People and Their Social World, ed. by Arnold M. Rose and William A. Peterson.
10. Boyd, R.R., "The Valued Grandparent: A Changing Social Role," in Living In the Multi-Generational Family (Ann Arbor: Institute of Gerontology, University of Michigan-Wayne State University, 1969).
11. See, e.g., Survey Research Center, "Residential Mobility, 1949-50, in the Ten Largest Standard Metropolitan Areas in the United States," Detroit Area Study, Project 843, no. 1211, University of Michigan, 1967.
12. Saltz, Rosalyn, "Foster Grandparents and Institutionalized Young Children: Two Years of a Foster-Grandparent Program,"Detroit: Merrill Palmer Institute, 1969 (Mimeo), p. 1; Robert Lefferts, An Evaluation of the Foster-Grandparent Program, New York: Greenleigh Associates, 1966, p. 48.
13. Rosow, Irving, Social Integration of the Aged (New York: Free Press of Glencoe, 1967).
14. Lipman, Alan, "Old People's Homes: Siting and Neighborhood Integration," in The Sociological Review, XV, No. 3, (November 1967), p. 232.
15. Barker, Roger G. and Wright, Herbert F., Midwest and Its Children. Reprint of 1951 Edition, (Hamden, Connecticut: Shoestring Press, 1971).
16. Verbal presentation by Walter Beattie, Environments for the Aged Conference, Gerontological Society, San Juan, Puerto Rico, December 1971.

17. See, for example, William Michelson, "The Case for Spatial Determinism," in <u>Man and His Urban Environment</u>: (Reading, Massachusetts: Addison-Wesley, 1970); see also Powell Lawton and M.A. Simon, <u>Gerontologist</u>, Summer, 1968.
18. See, for example, William H. Ittelson, Harold M. Proshansky, and Leanne G. Rivlin, "The Environmental Psychology of the Psychiatric Ward," in <u>Environmental Psychology</u>, p. 439; see also, Alan Lipman, "Chairs as Territory," in <u>New Society</u> (April 29, 1967).
19. All tests are Chi-Square tests by the likelihood-ratio method, unless noted otherwise.
20. Six "patterns" of preferences were possible based on first, second, and third preferences.
21. Lansing, John, Marans, Robert, and Zehner, Robert, <u>Planned Residential Environments</u> (Ann Arbor: Survey Research Center, University of Michigan, 1970), p. 16.
22. In paper for a practicum in the psychology of aging on observation at a Foster-Grandparent Project, Susan Sachs reported that foster-grandparents took stationary positions and their "charges" circulate around them.

MEASURING ENVIRONMENTAL DISPOSITIONS OF ELDERLY FEMALES

Paul G. Windley

Department of Architecture
Kansas State University

Abstract

This study develops the theoretical concept of environmental disengagement thought to be reflected in the environmental dispositions of elderly people, and also develops a methodology for disposition measurement. To explore the usefulness of the theoretical concept, an investigation involving a sample of elderly females and a sample of female students is outlined, and score differences between the samples on four environmental disposition scales are compared. Three of the four scales showed adequate reliability and validity, and significant differences were found between the two samples on two of the four disposition scales. The methodology was shown to be adequate in assessing environmental dispositions, while the theoretical concept was only partially adequate in explaining environmental dispositions. Some research implications of the findings are discussed.

Introduction

The overall purpose of this paper is to demonstrate the importance of the simultaneous development of theory and method. It seems imperative to this author that in a developing field such as environmental behavior, measurement methods should be closely linked to theoretical positions to prevent an overproliferation of both. Although theories serve a valuable ordering and predicting function and greatly reduce trial and error effort necessary for the progress of science, the constructs of a theory which cannot be measured or tested in the real world render it useless. Likewise, to develop elaborate measurement methods independent of theory results in a random effort to relate measured concepts to each other, often with little consequence.

The specific objectives of this study are: 1) to state a theoretical position and appropriate assumptions concerning the environmental dispositions of elderly people; 2) to initiate the development of a methodology capable of assessing such dispositions; and 3) to undertake an investigation to test the reliability and validity of the methodology, and corroborate the theoretical assumptions and assess their usefulness for environmental design and research.

The elderly were chosen as a target population in this study because their dispositions reflect years of experience in adjusting to the physical environment, and therefore, should be well defined and easily identified across a variety of physical settings. In addition, the environmental problems of this special population have not been a major concern of society in the past.

Environmental Disengagement

The underlying rational in this study is drawn from the psychology of personality: that people relate to the physical environment in unique and stable ways, much as they relate to themselves and others. It is believed that environmental dispositions reflect such individual relationships with the real world. Environmental dispositions (or traits) are defined as relatively enduring and related attitudes toward objects and settings in the natural and man-built environment. The complex link between attitudes and overt behavior is best summarized by Nunnally (1): "attitudes are the cutting edge of future action." Clearly, a person saying he hates a building is often a precursor to his leaving it or defacing it.

The concept of environmental dispositions is supported by two well known scholars in the field of personality: Lewin (2) (3) and Cattell (4) (5). The value of Lewin's work is his recognition that a person's attitudes and behavior are a function of the total situation (life space): the <u>person</u> surrounded by his <u>psychological environment</u> (including relevant physical objects and settings) and the <u>non-psychological environment</u> (irrelevant objects and settings). Lewin suggests that because the boundaries between the three parts of the life space are permeable in both directions, the man-environment relationship is an interactive, reciprocal system.

The value of Cattell's work is his recognition that attitudes toward physical or social objects are subsumed under a number of more general traits and lie at the very foundation of personality. Such traits account for the broad complexity of behavior and are the strongest predictors of future action.

Central to this study are the developmental aspects of dispositions or the changes which they undergo with increasing age. For Lewin, dispositions are integral parts of the person and develop through his interaction with the psychological environment. With increasing maturity there is greater differentiation (an increase in the number of parts of a whole) both of the person and of the psychological environment. According to Lewin, the number of dispositions and attitudes toward social and physical objects increases with age, but at the same time they become more rigid in the attempt to satisfy more complex needs.

For Cattell, changes in dispositions with age result from an attempt to reconcile biological and cultural changes. Metabolic rates, sensory powers, and the general homeostatic capacity of the body decline with age. These factors combine with the loss of occupation and social devaluation to increase emotional insecurity and instability which lead to changes in personality and disposition.

Other scholars report that the aging process engenders a movement toward egocentricity, and shifts from extraversion to introversion, from activity to passivity, and from dominance to submission (6) (7). Neugarten (8) reports a change from the ability to manipulate the world for need satisfaction to being manipulated by the world; while Kuhlen (9) argues that the older person perceives the world as more ambiguous, more complex, and more rigid.

The common theme in these varied theoretical orientations and empirical findings is a phenomenon of "disengagement" with increasing age, that is, a slowing down process, a change in style and way of looking at the world. The concept of disengagement is discussed by Cumming and Henry (10) who state that aging in the average person is a mutual withdrawal between the person and the social system to which he belongs. When the aging process is complete, the equilibrium which existed in middle life between the individual and society has given way to a new equilibrium. Lawton's (11) "docility hypothesis" is useful in further understanding the disengaging behavior of the elderly. This hypothesis states that the greater the competency of the human organism, the less will be the proportion of variance of behavior due to elements in the physical environment. Almost without exception, the older person has experienced some reduction in various areas of competence and is, therefore, more affected by the physical environment and may refrain from interacting with difficult settings when possible. This concept or phenomenon of disengagement is highly relevant to environmental problems of older people because of its implications for how older people look and feel about their physical environment. Is there an environmental disengagement as well, and can it be identified through environmental dispositions? The assumption in this study is that disengaging dispositions toward the physical environment will accompany disengaging personality shifts with age. The following four environmental dispositions thought to reflect disengagement are under investigation in this study.

Environmental privacy. This disposition taps preferences to control access to others by using features of the physical environment. Territorial behavior is thought to be positively related to preferences for privacy and together comprise the definition of this disposition (12) (13). The hypothesis is that as individuals age, the need for more privacy and greater manifestations of territoriality will be expressed in their attitudinal responses.

Environmental complexity. This disposition taps preferences for informational variability in the physical environment. It is suggested that an environmental setting or artifact can be located on a continuum from simple to complex. Its position on this continuum is determined by a combination of the setting's complicatedness and its ambiguity (14) (15). The assumption is that every individual has an optimum preference for environments somewhere along this simple-complex continuum. The hypothesis is that older persons' attitudes will more frequently be expressed for settings closer to the simple end of the continuum than will the preferences of younger populations.

Environmental change or stability. This disposition taps preferences for physical spaces and artifacts in a time dimension, i.e., preferences for modern vs. traditional settings or objects. The assumption is that modern cities, buildings, and objects are images of change from what has been, and that preferences for past over present environments show a form of rigidity toward change in the physical environment. The hypothesis is that older people's attitude preferences will be toward the traditional settings more than the preferences of younger populations.

Environmental manipulation. This disposition taps preferences consistent with one of two themes: man over environment, or environment over man tendencies (16). In the former tendency, the physical environment is viewed as challenging and as something to manipulate and use for one's own purposes. The latter expresses the tendency to see the environment as sacrosanct, such that the individual must harmonize with or subjugate himself to it. The hypothesis is that as persons age, their attitudinal responses reflect the environment over man tendency more often than the man over environment tendency (17) (18).

The Measurement Methodology

The aim of the measurement method is to tap the four dispositions discussed above. Two basic standards used in evaluating the usefulness of any measure are its reliability and its validity. The reliability of a measure depends on the amount of measurement error present in the instrument and can be assessed by calculating a coefficient of stability, a coefficient of equivalence, and an internal consistency coefficient (19). The reliability coefficient "Alpha" is in most cases adequate for estimating all three types of reliability (20).

The validity of a measure concerns what the test measures and how well it does so. Validation always requires empirical investigations relating the measure to other outside variables and is thus an unending process. The purpose for which a test is used determines which one (or combination of) the three validity categories is to be considered: content validity, predictive validity, or construct validity (21). To achieve construct validity is to substantiate the existence of constructs (dispositions) thought to be theoretically related. Since this paper is interested in substantiating the existence of environmental disengagement, construct validity is most significant.

Although several indirect methods such as the galvanic skin response and the TAT have been used to measure attitudes, the most valid approach is to ask people directly what their attitudes are (self-report). Once individual attitudes toward environmental objects are assessed, common attitudes toward a variety of objects are grouped together to form a disposition scale. However, scale construction first involves the development of individual attitude statements or items, and then one of several strategies can be employed to group the items to form scales.

Two types of attitude items were developed for use in this study: verbal items and visual items (22). The verbal items consisted of statements such as, "I often need a place to go where I am completely alone," to which the subjects would agree, feel neutral about, or disagree. The visual items were paired graphic illustrations of various environmental settings and objects to which the respondent was asked to indicate his preference (Figure 1). A numerical score for both items was determined for each respondent depending on his response or preference. Approximately forty verbal and twenty visual items relevant to each disposition were developed, with the exception of Environmental Manipulation which contained only verbal items (23). The

WHICH HOUSE DESIGN DO YOU MOST PREFER?

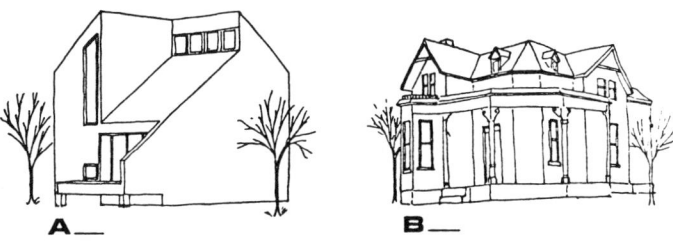

A___   B___

FIG. 1. An example of visual items for Stability Scale.

definition of each disposition was then explained to five judges who were asked individually to select from the pool of items for each disposition those which they felt best fit the definition. Only those items on which the judges were unanimous or nearly so were retained to represent each disposition.

After items are developed, scales are usually constructed according to one of three strategies (24): the internal method (factor analysis of large item pools), the external method (aligning items with some external criterion), or the intuitive method. The intuitive method was used in this study. With this approach, the investigator has some attribute or disposition in mind and he selects items which he believes will relate to this disposition, depending on his intuitive understanding of the attribute to be assessed. Many times a formal psychological theory is used to guide one's understanding. The items are then administered to a group of subjects and then correlated with the total score on the proposed scale. Those items correlating most highly are retained for the final scale. A total of twenty items composed of fifteen verbal and five visual items was selected for each scale. The sample size used in developing each scale ranged from forty to eighty people which included both university students and elderly people.

The Investigation

The main purpose of this study is to demonstrate the usefulness of developing both method and theory concurrently. Since the method and theory under investigation involve developmental differences in people, age is the independent variable. Therefore, attempts were made to control the effects of extraneous variables such as sex, SES, and race. Only white females in the middle to upper SES levels were studied at this point in time. This decision was based on the availability of a large sample of elderly women living in two very similar retirement homes in a midwestern city, the availability of female students on a university campus, and the similarity in SES levels of both groups.

An elderly sample (N = 80 with a mean age of 81) was selected from the two retirement homes. A young, student sample (N = 80 with a mean age of 21) was selected from sophomore and junior women majoring in a variety of subjects. The four disposition scales and a biographical questionnaire were then administered in a similar manner to both samples with the exception of a portion of the students who responded in a class situation. Since age related differences on the four dispositions were assessed by numerical scale scores, the anticipated scoring direction for each scale is given as follows: elderly females will receive lower scores than student females on the Complexity, Stability and Manipulation Scales, and higher scores on the Privacy Scale.

Findings

Reliability and Independence of the Scales

TABLE 1

MEAN SCALE SCORES AND STANDARD DEVIATIONS
FOR ELDERLY AND STUDENT SAMPLES

Sample	N	Privacy		Complexity		Stability		Manipulation	
		Mean	S.D.	Mean	S.D.	Mean	S.D.	Mean	S.D.
Elderly	80	60.3	6.0	47.7	6.4	53.6	8.6	60.6	6.6
Student	80	58.6	7.8	66.1	7.4	51.6	11.6	63.4	6.0

Table 1 displays the mean scores and standard deviations for all respondent groups on all scales. The reliability coefficients (Alpha) calculated for each of the four scales on the elderly and students samples (N = 160) are: Privacy, .40; Complexity, .78; Stability, .75; and Manipulation, .50. The reliability for the Privacy Scale on the combined samples is lower than desirable (25) suggesting that a great deal of measurement error exists in the scale. The reliability for the Manipulation Scale is moderately low and no doubt suffers, although to a lesser degree, from some of the same problems as does the Privacy Scale. The Complexity and Stability Scales are homogeneous and stable enough to invoke confidence in their potential validity.

Table 2 displays the product moment correlation coefficients between all scales. These correlations suggest that the scales are independent of each other and are most likely tapping separate dispositions, with the exception of the correlation (r = .35) between Complexity and Manipulation. Although small in magnitude, this correlation implies that these dispositions are related in a minor way.

TABLE 2

INTER-CORRELATION OF SCALES
N=160

	Privacy	Complexity	Stability	Manipulation
Privacy	1.00			
Complexity	-.15	1.00		
Stability	-.04	-.03	1.00	
Manipulation	-.03	.35*	.17	1.00

*p.<.001.

Comparison of Respondent Groups

The direction of the mean scores for all scales is in the hypothesized direction with the exception of the Stability Scale which is slightly reversed. The mean scores on all scales for the elderly and student samples were submitted to a one-way analysis of variance for independent means to determine if there was a statistically significant difference between the groups (Table 3).

TABLE 3

ONE-WAY ANALYSIS OF VARIANCE FOR ELDERLY VS. STUDENT
MEAN SCORES: PRIVACY, COMPLEXITY, STABILITY, MANIPULATION

	Source	df	MS	F
Privacy:	Between Samples	1	122.5	2.5
	Within Samples	158	48.3	
Complexity:	Between Samples	1	13524.0	280.2**
	Within Samples	158	48.3	
Stability:	Between Samples	1	172.2	1.6
	Within Samples	158	108.9	
Manipulation:	Between Samples	1	299.8	7.5*
	Within Samples	158	39.9	

Note.--N=80 for each sample. *p.<.05. **p.<.001.

The differences between the elderly and student samples for Privacy and Stability were not significant, while the differences for Complexity and Manipulation were significant. The female elderly's attitudinal preferences for environmental complexity, as defined in this study, is markedly less than the female student preferences. Likewise, the elderly tend to express the environment over man tendency more than the student sample, although the magnitude of this score difference is not great.

Validity of the Scales

To show validity for each scale, scale scores were correlated with biographical and personological data. Also, the verbal items were correlated with the visual items for each scale to determine if a verbal-visual transference was made on the scale items (Table 4). In addition, the two respondents who scored highest and

TABLE 4

CORRELATION OF VERBAL WITH VISUAL ITEM SCORES
N=160

Visual Scores	Verbal Scores		
	Privacy	Complexity	Stability
Privacy	.12	.14	-.01
Complexity	-.06	.52*	-.01
Stability	-.09	.02	.47*

Note.-- No visual items were developed for the Manipulation Scale.
*$p < .001$.

lowest on each scale were interviewed six weeks later about their past experiences related to the environment, and their feelings about their present living environments.

The low reliability of the Privacy Scale renders any correlation with biographical or interview data undependable. Those who scored high on Complexity tended to spend most of the day away from their room or apartment, have had a private room when growing up, describe themselves as more active, have lived most of their lives in large cities, and have lived in several different houses and cities. They also have low fear for selected environmental phenomena (e.g., floods, high density living, and heights), and they have had more formal education. No significant correlations with biographical data were found for the Stability Scale. Those receiving high scores on the Manipulation Scale describe themselves as energetic, are presently married, have received more formal education, and have low fear for selected environmental phenomena. Although the correlations were significant ($p < .05$, $r = .20$ and above), most were low in value (between .20 to .38) suggesting that only a minor relationship exists between the scale scores and the biographical data.

Table 4 presents the verbal item score correlations with the visual item scores for the three scales. The low correlation (r = .12) for Privacy suggests that little verbal-visual transference was made. The correlations for the Complexity and Stability Scales, however, show that an individual's verbal preferences are moderately related to his preferences for visual representations of environmental objects and settings.

All interviewees unknowingly corroborated their high or low scale scores with a self-report during the interview. For example, those respondents who scored high on Complexity thought they could tolerate high levels of complexity in the environment; while those scoring low preferred low levels.

Discussion

The Measurement Method

Data collected from the biographical questions, the interviews, and correlations of verbal with visual scores demonstrate a degree of validity for the Complexity, Stability, and Manipulation Scales. Thus the methodology employed in this study has been adequate in measuring the dispositions as defined. Additional stimulus media such as colored slides, photographs, together with sound, odor, and textures should be investigated. Alternate measurement techniques such as the interview, TAT, or physiological responses would be a valuable means of comparison.

Design Implications

It should be pointed out that because this study was necessarily exploratory, the design implications discussed below are more indicative of future areas to be investigated than of immediate and definitive design decisions. Environmental Complexity appears to be defined along two dimensions: 1) the density of buildings, people, and streets; and 2) the level of information manifested by the organization of a given physical setting. Unlike the female student sample, elderly females say they like less dense, simple, predictable, and well defined settings with low human activity. Symmetrical spaces might be provided to reduce the amount of information processing required. The use of subdued colors, simple materials, and spaces free from a too frequent interplay of void and solid, or of changes in level might be considered.

The design implications of Environmental Stability appear to be primarily symbolic communication involving emotion, organization of the environment, and its utility. Some females regard modern settings as clean, cheerful, spacious, and easy to maintain; while others associate traditional settings with feelings of security, respect, and consider them a link to the past. Clearly, many individuals may incur long-term adjustment problems when faced with new environmental images because of involuntary relocation.

The primary design implication for Environmental <u>Manipulation</u> appears to be the capacity of a setting to accommodate the manipulator on the one hand, and the non-manipulator on the other. In a home for the elderly or in a student dorm, this might mean providing the following options: movable walls and component furniture, kitchenettes, space for indoor or outdoor gardening, individual heating units, easily cleaned or maintained rooms, and space in which to park an automobile; or conversely, fixed walls and furniture, central heating, staff prepared meals, maid service, and public transportation

<u>Theoretical Considerations</u>

The adequacy of any theory depends on its degree of construct validity (26) or how explicitly the relationship between its constructs is stated, and how observable these relationships are in the real world. Before either of these requirements is met, or before individual differences on measured constructs can be identified, one must be confident that one has measured the hypothesized constructs. Such confidence is usually supported by evidence of convergent and discriminant validity as suggested by Campbell and Fisk (27). Convergent validity is demonstrated when two or more independent methods of measuring the same constructs or traits correlate sufficiently high to encourage further examination of validity. Discriminate validity is demonstrated when two or more independent methods of measuring the same trait or traits correlate higher than correlations obtained on different traits by the same independent methods. Of course, the confidence derived by demonstrating both types of validity stems from the knowledge that scores obtained on given traits are free from method variance. Campbell and Fiske are careful to point out that the prerequisites for fully demonstrating convergent and discriminant validity are: that the methods be developed independently of each other and that they be administered at different times. Given that the verbal and visual items used in this study might be considered two measurement methods, neither of the two criteria are met in this study. The first step, therefore, in further validating the environmental disengagement theory should be to administer the two methods together with other methods to as many different data sources and under as many different experimental contexts as possible. It would be well to begin a systematic sampling of environments ranging from the natural to man-made and from large to small-scale. Such individual differences as sex, SES, and race should be examined. Finally, each scale score's relationship to other variables, particularly different environmental settings, should be subjected to experimental treatment, and include age group differences and developmental changes over time.

Tentative findings based on reliabilities in this study suggest that the Stability and Privacy dispositions as defined should no longer be considered constructs of the environmental disengagement theory as far as elderly females are concerned. Also the Complexity and Manipulation dispositions were not shown to be independent.

Notes and References

1. Nunnally, J. Psychometric theory. New York: McGraw-Hill, 1967.
2. Lewin, K. A dynamic theory of personality. New York: McGraw, 1935.
3. Lewin K. Principles of topological psychology. New York: McGraw, 1936.

4. Cattell, R.B. Personality: A systematic theory and factual study. New York: McGraw-Hill, 1950.
5. Cattell, R.B. Personality and motivation structure and measurement. New York: World Book, 1957.
6. Chown, S.M. Personality and aging. A conference on Theory and Methods of Research in Aging, West Virginia University, May, 1967.
7. Gutman, G.M. A note on the MMPI: Age and sex differences in extroversion and neuroticism in a Canadian sample. British Journal of the Society of Clinical Psychologists, 1966, 5, 128-129.
8. Neugarten, B.L. Adult personality: Toward a psychology of the life cycle. In B.L. Neugarten (Ed.), Middle age and aging. Chicago: University of Chicago Press, 1968.
9. Kuhlen, R.G. Developmental changes in motivation during the adult years. In B.L. Neugarten (Ed.), Middle age and aging. Chicago: University of Chicago Press, 1968.
10. Cumming, E., & Henry, W.H. Growing old: The process of disengagement. New York: Basic Books, 1961.
11. Lawton, M.P. Ecology and aging. In L.A. Pastalan & D.H. Carson (Eds.), Spatial behavior of older people. Ann Arbor: The University of Michigan, 1970.
12. Westin, A. Privacy and freedom. New York: Atheneum, 1967.
13. Pastalan, L.A. Privacy as an expression of human territoriality. In L.A. Pastalan & D.H. Carson (Eds.), Spatial behavior of older people. Ann Arbor: The University of Michigan, 1970.
14. Kaplan, S., & Wendt, J.A. Preference and the visual environment: Complexity and some alternatives. Proceedings of the 3rd Annual Environmental design.
15. Rapoport, A., & Kantor, R.E. Complexity and ambiguity in environmental design. American Institute of Planners Journal, 1967, 33, 210-221.
16. Kluckhohn, C. Dominant and variant value orientations. In C. Kluckhohn & H.A. Murray (Eds.), Personality in nature, society and culture. New York: Alfred Knopf, 1962.
17. Meresko, R., Rubin, M., Shontz, F.C., & Morrow, W.R. Rigidity of attitudes regarding personal habits and its ideological correlates. Journal of Abnormal and Social Psychology, 1954, 49, 89-93.
18. Wesley, E. Perseveration behavior, manifest anxiety, and rigidity. Journal of Abnormal and Social Psychology, 1953, 48, 129-134.
19. Nunnally, J. Psychometric theory. New York: McGraw-Hill, 1967.
20. Cronbach, L.J. Coefficient alpha and the internal structure of tests. Psychometrika, 1951, 16, 297-334.
21. Nunnally, J. Psychometric theory. New York: McGraw-Hill, 1967.
22. The verbal items were written by this writer or obtained from other tests (cf. G. McKechnie, Measuring environmental dispositions with the environmental response inventory. Proceedings of the 2nd Annual Environmental Research Association Conference, Pittsburgh, Pennsylvania, 1970; G.H. Winkel, R. Malek, & P. Thiel, The role of personality differences in judgements of roadside quality. Environment and Behavior, 1968, 1, (2), 199-233; N. Marshal, Personality correlates of orientation toward privacy. Proceedings of the 2nd Annual Environmental Design Research Association Conference, Pittsburgh, Pennsylvania, 1970) in which case some items were altered to better fit the environmental dispositions in this study. All

visual items were sketched by the author.
23. The attempt to evoke attitudes toward dynamically manipulating the environment with static environmental representations proved unsuccessful.
24. Hase, J.D., & Goldberg, L.R. Comparative validity of different strategies of constructing personality inventory scales. Psychological Bulletin, 1967, 4, 321-348.
25. J. Nunnally, Psychometric theory. New York: McGraw-Hill, 1967, suggests that if in the initial stages of scale development the reliability coefficient does not attain a level of at least .50, it is not usually profitable to continue the development of the scale as presently defined.
26. Cronbach, L.J., & Meehl, P.E. Construct validity in psychological tests. Psychological Bulletin, 1955, 52, (4), 281-301.
27. Campbell, D.T., & Fiske, D.W. Convergent and discriminant validation by the multitrait-multimethod matrix. In D.N. Jackson & S. Messick (Eds.), Problems in human assessment. New York: McGraw-Hill, 1967.

RESEARCHING THE BEHAVIORAL IMPLICATIONS OF RESIDENTIAL DESIGN: THE CASE OF ELDERLY HOUSING   4.5

Sandra C. Howell, Ph.D.

Florence Heller Graduate School for Advanced Studies in Social Welfare
Brandeis University, Waltham, Massachusetts

Abstract

The practical utility of social and behavioral science research for architects, planners and housing policy makers depends upon the availability of research information in a familiar translation. Using the elderly and disabled as a population prototype, a format is proposed which quickly relates psycho-social outcomes to alternative design solutions. The format also has the potential of generating further research hypotheses.

The existing behavioral information pool, which may be of utility to planners of housing for aging and disabled populations, is now contained in publications and research reports principally directed to the academic community or to grant agencies. Its value for decision making at the applied level is, thus, limited. Behavioral information has not been systematically and appropriately translated for the designer in order that its validity in relation to specific design solutions may be tested.

As a consequence of this hidden repository, public policy and private design decisions are continuously made which will most certainly affect the social and personal alternatives available to aging Americans at least for the forty year mortgage lives of the structures that have been erected.

This paper is an attempt to specify areas of applied research in environment and behavior that appear particularly relevant to future decisions on housing and neighborhoods congenial to human aging. The content is based upon several different approaches to determining the behavior-information needs of designers and planners. They include:

- a. a review of the behavioral science literature relevant to environments and human aging and an assessment of its inclusion in trade magazines and professional publications[1];

- b. a series of consultations with architects and planners, at various phases in the development of their specific project proposals;

- c. interactions with public officials in relation to interpretations of federal, state and local policies;

- d. participation in and reports of meetings formally developed to elicit information needs from practitioners and/or usable research data from

d. social scientists.(2)

The complexity of the problem of <u>identifying</u> information needs is suggested by the following facts which emerged from the above experiences:

1. Architects, developers and planners do not, typically, think in terms of the specific <u>behavioral</u> <u>outcomes</u> of their work. This is not a revelation but is of sufficient concern that at least one major planning agency, the New York Urban Development Corporation, is attempting to formulate performance specifications in social and behavioral terms.

2. Assumptions about expected user behavior, when consciously made, are typically based upon stereotypes, partial knowledge or inappropriate applications of past experiences.

3. Attempts by the planning professions to survey needs, preferences and life-styles of intended users are often cursory, unsystematic and based upon inadequate understanding of the evaluative capacities of respondents.

4. Design decisions which may critically affect behavior are weighed the same as those decisions which are of probably secondary relevance to living habits. There are, in other words, few guidelines which indicate a performance hierarchy, or a set of social priorities available for use in construction trade-offs or in contract negotiations relative to cost.

In order to assist the design professions to develop an understanding of some of the possible outcomes of their decisions, I am developing a format based upon common alternative design solutions. Table I is a composite of some of the more recurrent architectural problems, solutions to which may have moderate to major behavioral impact on the residents of housing for the elderly or the disabled. If it is considered that such a resident population may be a most extreme model for behavioral impact research, it can then be suggested that less vulnerable populations, as well, would exhibit some characteristic environmental effects to the same or similar design decisions.

I have chosen particular examples of either a selected (A) or a rejected (B) solution illustrating real decisions and some key behavioral research issues. The vertical arrays of problems and solutions are independent of one another; (i.e., represent separate situations). The strategic column of the table is (4) which specifies some of the behavioral implications of the solutions (cols. 2 and 3) and indicates areas where further research effort needs to be made and dependent variables need to be more clearly defined operationally.

A review of the problems and implications listed in Table I immediately suggests that were total single design situations displayed, the individual solutions would certainly not be independent. Where instrumental facilities, (e.g.: laundries) or shared spaces are placed and how many are appropriate for a given population density are obviously interrelated and, in turn, associated with the height of a building, its configuration, and, as well, to the existing services and their use in the surrounding neighborhood.

The major issue justifying this classification attempt is, however, whether the behavioral implications of design decisions can be built into the official guidelines for housing? For example, I recently reviewed a design for 300 units of housing for the elderly. It is to be an L shaped, eleven-story structure in which a single laundry facility will be situated in the basement of one corner of the L. In another case, a laundry was placed off a ground floor elevator lobby requiring a person to cart dirty clothes in front of the entrance. As one elderly citizen commented, "Do they expect me to do that in my kimono and curlers?" Neither of these solutions are optimum for an elderly or disabled population but those which might have been proposed were not even entertained by the architects.

The behavioral implications for spatial decisions in the case of elderly and/or disabled is not at all simple to untangle. For example, in the case of a lobby related laundry, there may be a strong argument in favor of this situation as an incentive to isolated elderly to dress and socialize, if only for the act of doing laundry.

Beyond Implication

The design of an environment behavior research strategy which would result in "hard data" of use to planners requires a critical reappraisal of the ways in which we, as behavioral scientists, have been posing hypotheses.

Illustrative is the case of the behavioral impacts of age segregated versus age integrated residential enclaves. Two questions have been posed in the existing research. One relates to the social interaction and morale of elderly in relation to peer density of residential settings [3]; the other to preferences by elderly for contact with other age groups [4]. The research designs in these cases specified that (a) same age-different age preferences would refer to the presence of young children and adults of undetermined number, (b) the conclusions were based upon verbal responses to questions about behavior and preference and (c) age density alternatives were not systematically employed as independent variables. In fact, I have yet to see a study of interactions between adult age groups, in medium to high density residential settings, that explored the observed natural activity patterns of participants with the exception of Joan Shapiro's study of New York City's slum residential hotels [5]. What I am saying is that the conclusions about augmentation of social interaction through extreme age segregation has become a determinate public housing policy without a full exploration of either the range of site and age density alternatives or of the possible long run distortions of social relations that might result from such a policy. There is, for example, a controversy current among gero-psychologists as to the need for stimulation and problem solving challenge for the aging. To what extent does the service supported age-cloister diminish the capacities of the older individual to solve problems of daily living wherein neighbors are recruited into service? An equally important long term cultural question is that of the impact on youth of viewing the aged (the grandparent and great-grandparent generation) as isolated or isolable?

It would seem that we have confused the expediency of need for low-cost housing for the elderly with the issues of social roles and behavioral adaptations in both the formulation and translation of research in housing for this group.

Table 2 spells out for one frequent architectural problem a few specific independent and dependent variables which would be required in order to begin clarifying appropriate alternative solutions.

Again, it is important at this point in time that practicing architects and developers, as well as the HUD and housing authorities which review their projects, at least raise questions about behavioral impacts. This, in my consulting experience, has been the cardinal omission in the planning process.

It should further be stated that only through systematic observation of how, in fact, specified tenant groups use the spaces and hardware that are built will behavioral implications emerge and alternative solutions be derived.[6]

A final illustration of the unplanned design decision is that of the form and placement of emergency call systems thought to be so essential to the safety and security of elderly tenants. We have virtually no follow-up on the actual use of such monitors and alarms. Only by chance did I happen onto the fact that elderly tenants in one new building had, in a majority of cases, hidden the emergency call string in their living rooms behind pictures and mirrors where its usefulness in time of crisis would be zilch to crashing. Why is not a telephone system with a lit emergency button a much more culturally acceptable elicitor of appropriate crisis response?

TABLE 1

PSYCHOSOCIAL IMPLICATIONS OF ARCHITECTURAL SOLUTIONS

(1) Architectural Problem	(2) Solution A	(3) Solution B	(4) Possible Social and Behavioral Implications of Solution A
Age homogeniety	Age segregated adults	Age integrated adults	stimulate social interaction (e.g. new friendship formation); encourage on-site activity; support community withdrawal and tenant isolation; evolve into quasi-institution; require increasing external service inputs; increase youth prejudices against old; provide captive setting for victimization.

TABLE 1 (Cont.)

(1) Architectural Problem	(2) Solution A	(3) Solution B	(4) Possible Social and Behavioral Implications of Solution A
Building density	150 units	25 units	provide greater opportunity for friendship choices and casual encounters; allow better utilization of community space - economy of scale; allow for greater variety of activities; reduce participation (e.g.: number of activities/person) permit withdrawal and invisibility; promote impersonality; suggest to community, separatist and "deviant" population.
Per cent of varied sized living units	No 2-bedroom	10% 2-bedroom	exclude certain classes of adults (e.g.: disabled requiring live-in care) force unnecessary institutionalization; limit life style choice; prevent extended family arrangements.
Height of building	15 floors	4 floors	reduce internal mobility (i.e.: anxiety about elevator use); increase options and spatial use choices; encourage floor level friendships; increase utilization of on-site community space; decrease utilization of on-site community space; encourage self-government in shared use of facilities (e.g.: laundry); create antagonisms in use of shared facilities; promote depersonalization; result in diminished security and responsiveness to intruders and to problems of individual tenants (e.g.: late awareness of emergencies.)

TABLE 1 (Cont.)

Architectural Problem	Solution A	Solution B	Possible Social and Behavioral Implications of Solution A
Location of entrance lobby	On public thoroughfare	Off public thoroughfare	promote use of space for observation; promote use of space for development of friendship and casual encounters; produce avoidance of space.
Location of community space	On-site ground floor	Off-site within ½ mile	discourage use by off-site neighbors induce underutilization of common space; promote use by already socialized; promote use by withdrawn or formerly isolated because of proximity to living unit.
Floor area around mailboxes	10 square feet	20 square feet	reduce total social interaction spaces (non-utilized) limit activity at space; permit control of activity;
Visibility of mailboxes from street	Visible	Obscured	permit security of acts; allow visual penetration by outsiders; allow visual review of outside.
Location and number of laundry facilities	(1) ground floor proximate to lobby	(5) alternate floors	limit privacy for function; increase opportunities for social interaction; require formalization of pre-action behavior (dressing for social contact); reinforce anxiety of isolate; create conflict over use of machinery; control cost of maintenance; result in less reinforcement of casual encounters because of chance times of use;

TABLE 1 (Cont.)

Architectural Problem	Solution A	Solution B	Possible Social and Behavioral Implications of Solution A
			Allow for casual encounter; reduce opportunity for floor-friendships.
Common (congregate) dining room and kitchen on site	Inclusion	Exclusion	limit alternative spatial use; provide for choice behaviors; reduce economic stability of structure (because of under-utilization); allow development of resident clique which excludes neighborhood; provide secure in-house meal facility; augment social interaction opportunities; encourage opportunities for tenant management; allow natural development of needed support during periods of psychological and physical incapacity; provide nucleus for institutional environmental effects; counter undernutrition behaviors of isolates.

TABLE 2

Social and Behavioral Implications of Location of
Mailboxes in High Density Dwellings and
Hazardous Neighborhoods

Some Proposed Research Variables

Independent	Dependent
(time specified where appropriate)	
Street traffic	social interaction at mailboxes in lobby

Proximity of building entry:
    to mailboxes
    to elevators
    to lobby
	assault/robbery rate (resident-non resident)

Visibility of boxes
    from street
    from lobby
	non-resident pedestrian street traffic at peak mail hours;

Density of dwelling	occupancy of mail space

Mean no. of residents:
    in lobby
    at boxes	

4. ENVIRONMENTS FOR DIFFERENT AGE GROUPS

Notes

1. An annotated bibliography and key word index of environment-behavior research is being developed by Leon Pastalan, University of Michigan, Department of Architecture, but its applicability to the trade has yet to be evaluated.

2. The American Gerontological Society, One Dupont Circle, Washington, D.C., is supporting such meetings through a grant from HEW, Administration on Aging. NIMH funded a pilot design-evaluation workshop at Brandeis University, June 1972 and the American Institute of Architects has co-sponsored several conferences.

3. Carp, F.M., A Future for the Aged: Victorial Plaza and its Residents. Austin, Texas, University of Texas Press, 1966.
Rosow, I., Social Integration of the Aged, New York, The Free Press, 1967.

4. Hamovitch, M.B. and Peterson, J.E., "Housing Needs and Satisfactions of The Elderly". Gerontologist. Spring, 1969, 9.
Sherman, S., "The Choice of Retirement Housing Among the Well Elderly". Aging and Human Development. 1971, 2.

5. Shapiro, J., Communities of the Alone. New York: Association Press, 1971.

6. Lawton, M., "Public Behavior of Older People in Congregate Housing," in John Archea and Charles Eastman's (eds.) Proceedings of the 2nd Annual Environmental Design Research Association Conference, October, 1970, Pittsburgh, Pennsylvania, pp. 372-379.

FIVE   ENVIRONMENTAL COGNITION

Chairman: Roger M. Downs, Dept. of Geography, Penn
              State Univ.

Panelists: Reginald Golledge, Dept. of Geography,
              Ohio State Univ., Columbus
              Robert Hershberger, Dept. of Arch. Arizona
              State Univ., Tempe
              Donald Demko, Dept. of Geography, Queens
              Univ., Kingston, Canada
              James Blaut, Dept. of Geography, Univ. of
              Illinois, Chicago, Illinois

Authors: Basil Honikman, "Personal Construct Theory
              and Environmental Evaluation"
              Roger B. Howard, Sara D. Chase, Mark
              Rothman, "An Analysis of Four Measures
              of Cognitive Maps"
              Rachel Kaplan, "Predictors of Environmental
              Preference: Designers and 'Clients'"
              Stephen Kaplan, "Cognitive Maps, Human Needs
              and the Designed Environment"
              H. Stephen Leff and Paul S. Deutsch, "Con-
              struing the Physical Environment:
              Differences Between Environmental
              Professionals and Lay Persons"

ENVIRONMENTAL COGNITION: INTRODUCTION   5.0

Roger M. Downs, Session Chairman

Department of Geography
The Pennsylvania State University

All too often a session chairman's remarks are rightfully disregarded as an attempt to manufacture coherence among an amorphous collection of papers, and to praise progress where none is evident: in other words, he must pretend that the whole session is at least equal to the sum of its parts. In this instance, such pretence is unnecessary. These papers are as representative of the current state of the art in environmental cognition studies as one could expect from five unsolicited pieces of work, and, what is more important, indicative of some of the significant changes in methods, thinking, and objectives that the field is experiencing. In this third consecutive EDRA session devoted to environmental cognition, the obligatory hopes expressed by earlier participants are being answered, and we can find evidence of progress and achievements on three fronts, those of methodology, cumulative knowledge, and the link with environmental design.

The first (and overlengthy) enthusiasm for taking quick, exploratory looks at the shores of these terrae incognitae of the mind has been blunted by the formidable barriers of research design and methodology. A roomful of willing architecture students armed with pencils, sketch pads, and imagination is no longer sufficient. Valid and telling criticisms from both within and especially without have resulted in a necessary but less exciting focus on ways of structuring and posing research questions. This is reflected in the four empirical papers by Honikman, Howard and Chase, Rachel Kaplan, and Leff and Deutsch. The careful, thorough exploration of the scope and value of one fashionable methodology, that of George Kelly's Personal Construct Theory, is demonstrated by both Honikman, and Leff and Deutsch. Howard and Chase show the importance of employing complementary and partially overlapping methods to answer a single research question, an approach all too rare in environmental cognition studies. Another methodological impetus has come from the use of multidimensional scaling techniques to get a simple spatial plot of complex data sets, and Rachel Kaplan's study of environmental preference makes use of one such procedure, the Guttman-Lingoes smallest space analysis.

Methodological sophistication alone is insufficient: research findings only have a final meaning within a commonly agreed upon framework of concepts and theories. Although there are still no explicated, tested theories, in the sense accepted in psychology, some of the confusions in concepts and terminology which plagued earlier work in environmental cognition are beginning to disappear. The distinctions between perception and cognition, attitude and preference, image and cognitive map, are being clarified, and Stephen Kaplan's paper assists in this process. He shows how the cognitive map concept can be used to throw light on Kevin Lynch's original terms, legibility and imageability, and links together environmental needs and preferences. Although Kaplan's paper is purely conceptual,

Howard and Chase use the cognitive map concept to generate and test hypotheses about the nature of distance cognition. We are witnessing a gradual weaving together of ideas and data, giving at least a coarse net if not the fabled rich tapestry! The previous EDRA sessions themselves have assisted in this process, and these five papers show some of the benefits of and needs for more cumulative knowledge.

Given the aims of EDRA one would expect the papers presented to have a strong focus on problems of environmental design. This has not always been the case; instead the limp ending that 'it is hoped that designers will find. . .' has been all too prevalent. It is reassuring to find genuine attempts to tie more traditional academic research interests with such pragmatic questions as what is preferable or how does this affect. . .? Honikman, Rachel and Stephen Kaplan, and Leff and Deutsch have all phrased their efforts in a design context and simply applied superficial decoration after the event. Such praise should not be mistaken for that perennially optimistic euphoria which still pervades EDRA sessions. Research on environmental cognition cannot give immediately useful design prescriptions but these papers do offer some valuable additions to our sparse knowledge. For example, both Leff and Deutsch, and Rachel Kaplan differentiate responses by 'designers' and 'non-designers or client groups' to similar environments, and Honikman ties cognitive attributes to those of the cognized environment. Such steps have not been present in much of the previous work on environmental cognition.

As much as is possible in five unsolicited papers, this **session** is a good reflection of what is happening in research on environmental cognition. They are good illustrations of the ways in which research is being conducted and ideas developed. They are not representative of the range of content areas being investigated--thus, for example, the developmental studies of Blaut, Hart, Moore, and Stea, and the work on symbols and structuring by Appleyard and Kreimer are not covered. Nevertheless, this session is a record of some progress, some hope, and the frightening spectacle of what remains to be done.

PERSONAL CONSTRUCT THEORY AND ENVIRONMENTAL EVALUATION

Basil Honikman

School of Architecture and Urban Design
University of Kansas
Lawrence, Kansas 66044

Abstract

This paper focuses on two aspects of the results of applying a Personal Construct Theory approach to environmental evaluation. It concludes with discussion about possible applications of the approach and how it may be combined with other theories to expand our understanding of how people interpret and use environment.

The two aspects concern the complexity with which people construe environment and the relationships which form between abstract or evaluative constructs and specific physical features of the particular environment.

Introduction

At the EDRA III Conference at UCLA I presented a paper which dealt with the adaptation of Personal Construct Theory for use in environmental evaluation(1). The purpose was to describe this approach conceptually and to explain its processes in both theoretical and operational terms. An indication of the kind of information which could be obtained by examining how people construed an environment was also included.

In this paper my intention is to present two interesting aspects of the fully analyzed results. In particular attention is paid to 'plotting environmental elements in the informants construct space' and to the 'lattices of linear and implicated links' which are formed between constructs. The location of 'elements in the construct space' enables us to gain some idea of the complexity or simplicity with which an environmental event is construed. The 'lattices of linked constructs' show how relationships are formed between evaluative criteria (e.g. informality or happiness) and physical characteristics (e.g. rough brickwork).

The paper concludes with suggestions as to how construct theory and other approaches to environmental analysis, may be combined in even more fruitful inquiry.

Before dealing with the results, however a brief review of the salient points of personal construct theory and how it was used in my study will obviate the need of the reader to refer to the proceedings of EDRA 3.

Personal Construct Theory

'The Psychology of Personal Constructs' (Kelly) (2) describes the development of a theory and associated techniques for dealing with personality problems. Kelly's approach enables the psycho-therapist to observe the way his patient 'makes sense' of the complex patterns of people and events that influence his behavior. It seemed that this approach could be adapted to examine the way people made sense of the complexities of environment and that the results of this kind of examination would constitute a reasonable basis for theorizing about environmental evaluation. Kelly reasoned that man was capable of being his own 'scientist'. Each and every experience as it was 'absorbed' by the person, was interpreted in the light of previous experiences, expectancies and anticipations until it 'made sense'. (Bannister) (3). Kelly called this process 'construing' and the things or ideas which were construed, he called 'constructs'. He maintained that people construed events in the world using their previous experience as a basis. When they approached an event which was wholly new then it was construed using previous experiences as a 'guide' to understanding it. When this happens the new event is said to fall within the 'range of convenience' of the person's process of construing. Sometimes, however, previous experience proves inadequate in helping the person make sense of an event. He may find that his systems of belief and understanding are seriously confounded by the new experience. In these cases the event falls outside 'the range of convenience' and anxiety or even hostility may result. (2)

Stringer identifies 'man the architect' as being Kelly's 'man the scientist' in an environmental context. (Stringer) (4). In making sense of the continuing stream of events and experiences provided by the environments which surround them, human beings transform accommodation into 'homes' by designing, modifying, decorating and arranging their houses until they match their construing of 'home'. Sometimes the building falls outside the inhabitant's range of convenience for construing 'home' and then if he is forced to consider living in it he may become unhappy or anxious. If the building falls within his range of convenience he may still disapprove of it but its inadequacies should be evident in his system of construing it and he should be able to either adapt to them or be able to overcome them.

It seemed that if I were able to examine how people construed an environment then I would be able to see how its various characteristics were themselves construed. I should, therefore, be able to see which of the physical characteristics were significant to the person and how his assessment of them contributed to his overall environmental evaluation.

Using personal construct theory as a basis I conducted an initial study. Interviews with informants resulted in the compilation of a list of adjectives which they personally used to describe their homes. Semantic differential and factor analysis techniques were used to see if differences between scores given firstly for an environmental idea such as 'cottage' and secondly for a color picture of a 'cottage', could be explained in terms of the physical characteristics of the cottage. This proved impossible because the use of one set of rating scales by all the informants meant that their individual construct systems could not be identified. There was also no way in which the physical characteristics identified by the researcher could be linked into the evaluations that were established. I was only able to say which of the pictures most closely matched the informant's idea and not why. (Honikman) (5). A deeper experimental process which could identify the details of a construct system was clearly necessary.

Experimental Process

The development of the experimental process using Kelly's (2) construct eliciting and repertory grid techniques, together with Patrick Slater's (6) Ingrid '67 principal components analysis and Hinkle's (7) construct laddering, resistance to change and implication grid techniques is described in the proceedings of EDRA 3 (Honikman) (8).

The following is a brief outline of the experimental sequence.

Construct Eliciting and Laddering

This is the process whereby each informant's supper and subordinate constructs are identified. Eliciting was carried out using 15 color slides depicting a variety of living rooms, as the elements.

Repertory Grid

Each of ten selected elements were given scores on a 1 - 7 scale in terms of each of the informant's personal set of 10 superordinate bi-polar constructs.

Resistance to Change and Implication Grids

The first of these grids establishes the status of each informant's 20 elicited and laddered constructs in his hierarchical construct system.

The implication grid establishes whether each construct implies any other construct and whether the implication is reciprocated or merely one directional.

Forty informants contributed to the study and the object was to try to establish what it was about one of the living room photograph (element 2) that accounted for its unanimous approval by the informants in my initial study.

Selected Aspects of the Results

"Elements in the Construct Space"

The Ingrid principal components analysis program determines both the spread of variance between the principal components and the 'loadings' by which each element and construct relate to them. Figures 1 and 2 show the elements and constructs plotted in the construct space for informants 6 and 29. The X-X and Y-Y axis represent the two major principal components. (i.e. those including the largest

5. ENVIRONMENTAL COGNITION / 245

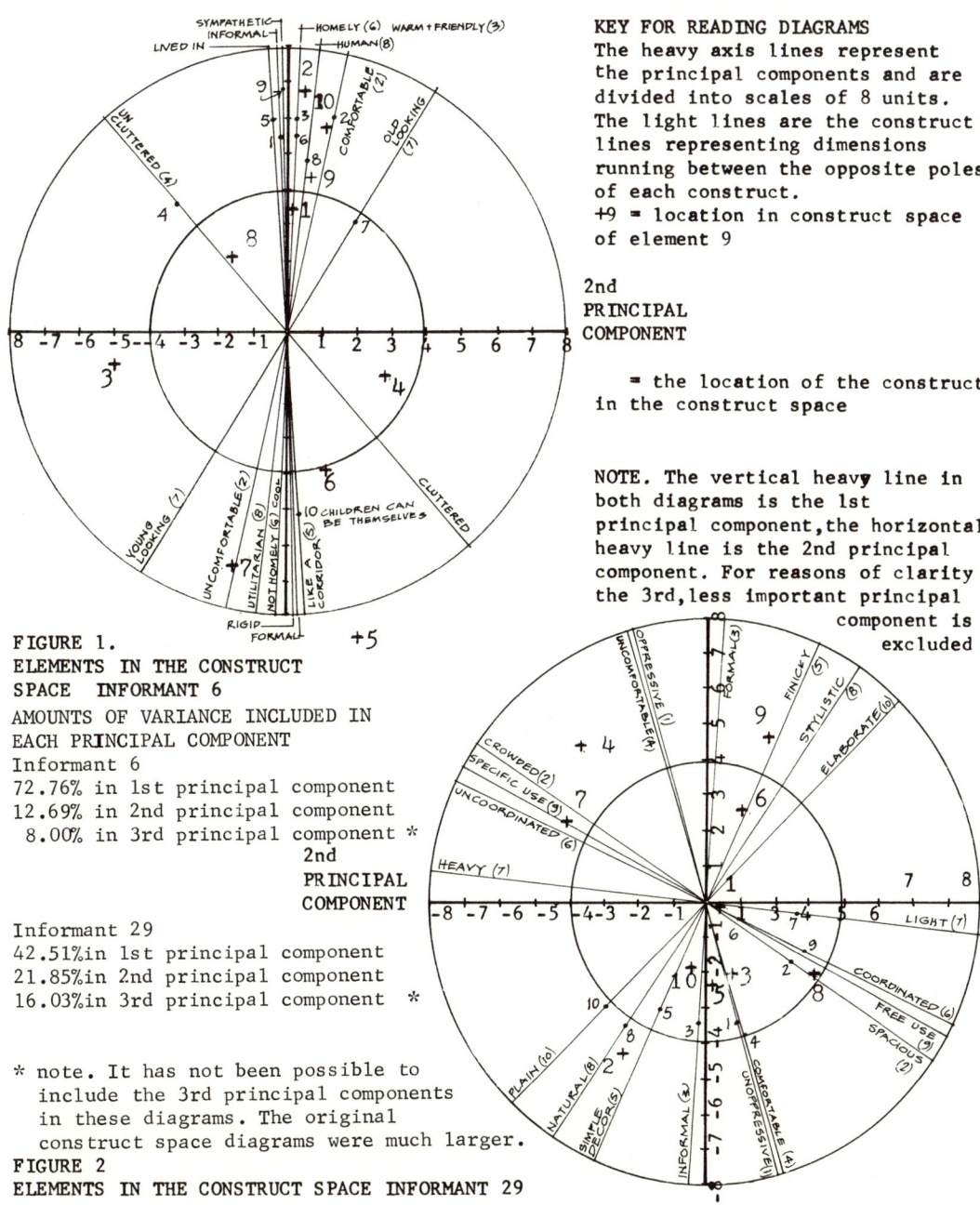

FIGURE 1.
ELEMENTS IN THE CONSTRUCT
SPACE   INFORMANT 6
AMOUNTS OF VARIANCE INCLUDED IN
EACH PRINCIPAL COMPONENT
Informant 6
72.76% in 1st principal component
12.69% in 2nd principal component
 8.00% in 3rd principal component *

Informant 29
42.51% in 1st principal component
21.85% in 2nd principal component
16.03% in 3rd principal component *

* note. It has not been possible to
  include the 3rd principal components
  in these diagrams. The original
  construct space diagrams were much larger.
FIGURE 2
ELEMENTS IN THE CONSTRUCT SPACE INFORMANT 29

KEY FOR READING DIAGRAMS
The heavy axis lines represent
the principal components and are
divided into scales of 8 units.
The light lines are the construct
lines representing dimensions
running between the opposite poles
of each construct.
+9 = location in construct space
of element 9

2nd
PRINCIPAL
COMPONENT

= the location of the construct
in the construct space

NOTE. The vertical heavy line in
both diagrams is the 1st
principal component, the horizontal
heavy line is the 2nd principal
component. For reasons of clarity
the 3rd, less important principal
component is
excluded

amounts of variance). The third principal component is represented by the line at the bottom of the diagram. Each construct and each element are plotted using their principal component loadings as co-ordinates to the XX and YY axes. If as in the case of informant 6 almost all the variance is included in the first principal component and most of the constructs also cluster close to it, we can suggest that he evaluated the elements in terms of one major context. The meanings of the ten constructs do not differ greatly. The informant may be said to have made up his mind on the basis of one parameter reflecting, to a large degree, the meanings of most of the constructs.

Informant 29 on the other hand (figure 2) had three principal components each including a significant amount of variance. His constructs were spread more evenly in the construct space and related to the three components.

Examination of these two diagrams reveals the extent of the complexity with which the informants evaluated the elements. Informant 29's evaluation involving three clear parameters was considerably more complex than that of informant 6.

In scoring the repertory grids informant 6 used much more of the scoring range than 29 did. This suggests that the complexity with which the latter construed resulted in a moderate evaluation. We can say that he is able to tolerate minor 'faults or inadequacies' because low scores on the parameters (or constructs) to which they relate, are balanced by higher scores in other parameters. His view of the elements was therefore balanced. The converse is true of informant 6. So powerful is the influence of his major principal component that there are no other significant parameters which can balance or moderate his evaluations. In other words his entire evaluation is based on only one major but simple factor while the judgements of informant 29 depend on at least three criteria. Generalizing, we can postulate that when the extremes of the scoring range are used, the informant construes simply and probably in terms of one major environmental parameter.

Two opinions lend to support this argument. Canter who has looked at the relationship between cognitive complexity and satisfaction with environment also concluded that people with few simple criteria for discrimination, tended to make more extreme judgements. (Canter) (9).

Bieri theorized:

"That a complex cognitive structure allows for a higher differentiation among persons than a simple cognitive structure" (Bonarius) (10).

Plotting elements on the construct space offers a convenient and revealing way to look at the complexity of environmental construing.

If we examine the location of the elements in the 'construct space' we can see how they were considered in terms of each construct. We can begin to suggest which constructs amount to reasons for high or low evaluation. In the case of informant 6 it is clear that elements 1, 2, 10 and 9 are 'liked' and 7 and 6 are disliked. On the other hand the more complex construing of informant 29 means that none of the elements occupy extreme positions in his construct space indicating that their evaluation is not established by one super important parameter.

The 'elements in the construct space' diagrams enable us to see both the complexity of each informants construing and the way in which each of the elements he evaluates relates to the hierarchical systems of constructs which comprise the principal components of his evaluation. These systems are illustrated by interrelated diagrams describing the various ways that the informant's constructs are linked to each other.

Linear Link Diagrams and Implication Grids

Linear link diagrams (figure 3) show the different link types by which element 2, principal components and constructs were connected in the first part of a hierarchical system of construing. The second part of the system is shown by the implication network (figure 4).

The construing of element 2 by informant 29 is described to show the kind of information afforded by the experimental process.

Informant 29 is chosen for this outline because his evaluation is spread more equally between the contexts (principal components or parts of evaluation) than most of the other informants.

The constructs listed as relating to each principal component in the linear link diagrams give a sense of meaning which reflects the context represented by the principal component. The constructs in principal component 1, (figure 3) each seem to identify areas of judgement which are relatively independent from each other. Perhaps the theme which could be argued as connecting them is one of general 'living' or even 'domestic performance'.

The constructs in components 2 and 3 seem to be connected by meanings less general and more specific in character. The constructs in the second principal component suggest a context of 'uninhibited space' and in the third principal component the construct 'co-ordinated' followed by 'informal' and 'free use' generates an impression of 'organized informality'.

Having indicated some idea of the scope of each principal component within the informant's construct system, the linear link diagram and implication network show how the subordinate constructs contribute to it.

The linear links of laddered constructs are easy to follow from the diagram. The suggestion that the first principal component represented a context of 'general living room character' is supported because only one physical characteristic construct contributes to it. Physical characteristic constructs are the most specific of all constructs so that it is reasonable to expect them to relate more frequently to more specific principal components rather than to general ones.

"Spacious" is obviously a major centre of informat 29's implication network (figure 4), followed by 'simple decor', 'flexible', 'comfortable' and 'unco-ordinated'. None of the physical characteristics constructs figure in this network.

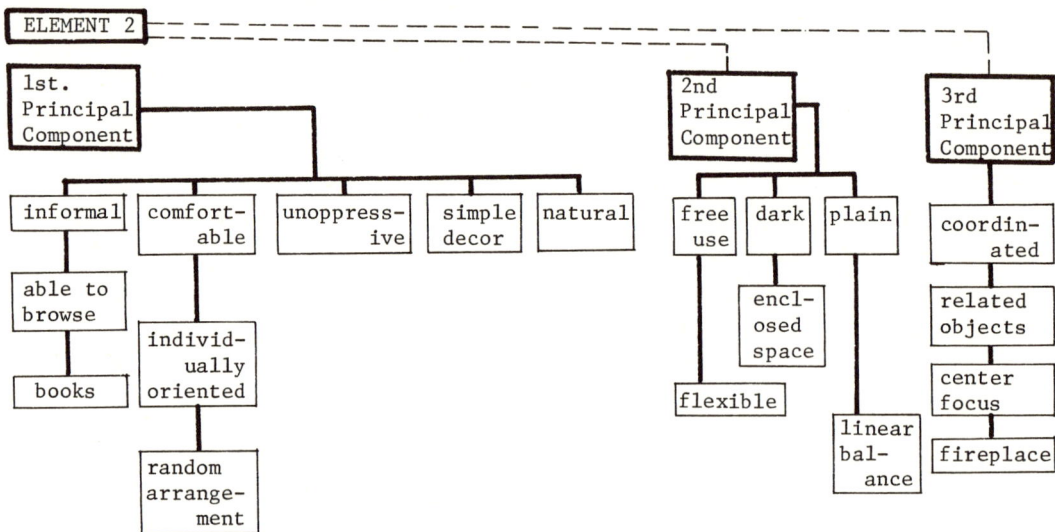

FIGURE 3
LINEAR LINK DIAGRAM FOR INFORMANT 29

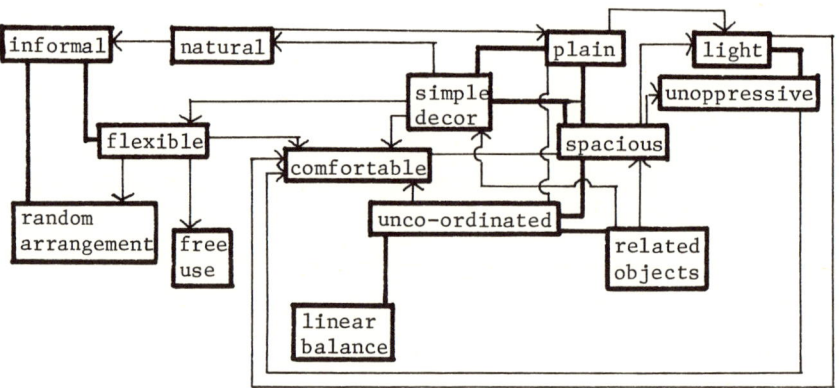

FIGURE 4
IMPLICATION NETWORK FOR INFORMANT 29

The informant does not consider that physical characteristic constructs, unqualified and on their own, imply very much for other, more descriptive or evaluative constructs. In other words, implication links are formed only when a physical characteristic has been linked to more superordinate evaluative constructs. Physical characteristics imply other constructs for the informant only after he has made an initial evaluation or judgement about them. In addition to 'books' and 'fireplace' constructs which are purely physical, informant 29 identified, 'random arrangement', 'simple decor', 'centre of focus', 'related objects' and 'enclosed space', as constructs which are partly evaluative (in that they reflect his personal interpretation) and partly physical. For example the construct 'related objects' consists of the 'objects' which exist as tangible physical items and 'related' which describes an evaluation the informant makes about the object.

The fact that these are not purely physical characteristic constructs does not preclude them from representing, together with books and fireplace, the significant ingredients the informant identified in the element.

Collectively the diagrams identify the principal components of the informant's system of construing and the superordinate constructs which are closely associated with them and which collectively represent their meaning. The physical characteristics which were significant in the way the informant construed the living room photographs are also established and the various links by which each of these physical characteristics are connected to more superordinate, evaluative non-physical constructs may be traced.

In this way a schematic diagram of the informants network of constructs and links may be compiled which graphically demonstrates how he sees and interprets the particular living room environment represented in the photograph. (element 2).

Discussion

Speculation about the application of a Personal Construct Theory approach in a realistic context is not difficult. Clearly further research is needed to deal with operational difficulties. Questions such as what is technically involved in replacing photographic elements with real environments should not be insoluable.

'Time' for instance, is an influential factor in the construing process. People are continually experiencing new events and new constructs are constantly being admitted to their construct systems. These may change the nature of the construct system and initial rejection or disapproval can easily mature after long acquaintance into warm approval. These and other issues are important in the development of personal construct theory as a tool for environmental analysis however this discussion is devoted to looking at the possibilities of combining my adaptation of construct theory technique firstly with another approach by Anna Bridge and secondly with a 'territoriality' approach by Duncan Joiner.

Anna Bridge (unpublished) a member of the Architectural Psychology Research unit at Kingston Polytechnic elicited room constructs without using photographs or exposing her informants to real environments. (11)

She selected 5 main areas of a home, such as 'kitchen', bathroom, living room etc., and the informants were asked to consider an example of each of these that they liked and one that they disliked. In this way 10 abstract elements were defined but each informant could rely on his personal experience of real environments for construct eliciting. The results of repertory grid testing and analysis enable both elements and constructs to be plotted in the construct space. In this way she is able to see whether her informant's consider rooms similarly because they accommodate the same function or because they have similar visual or spatial qualities.

If linear link and implication network diagrams were prepared in conjunction with Bridge's results, we would be able to see which physical characteristics contribuated to functional, formal and any other major environmental contexts significant in the informants systems of construing.

A study could be set up as follows: Informants would be asked to compile a list of room types. They would then be asked to write down two examples of each type, one which they liked and one which they disliked. Each example would be drawn from rooms with which they had had personal experience and would be specifically nominated. Constructs would be elicited by the triad method. A triad could consist of a 'liked' kitchen, a 'liked' bathroom and a 'liked' bedroom or a 'liked' study, a 'disliked' study and a 'liked' living room. The 'sorting' of various triads combining 'liked' and 'disliked' examples of the same room type with other room types would result in sets of constructs covering both the quality of the room and its ability to accommodate the activity. A 'liked' dining room may be preferred to a 'disliked' dining room because better dining could take place in it and it might also be preferred to a 'liked' living room because of spatial quality or informality. It is easy to imagine all kinds of combinations between activity and evaluative constructs. The plotting of elements and constructs in the construct space would reveal relationships between construct groupings and elements. One would be able to propose that an informant approved of a room not only because it was an appropriate place for its designated activity but also because of a number of other environmental qualities. A 'liked' study might be construed similarly to a 'liked' bathroom because of qualities not necessarily associated with bathing or studying. On the other hand a 'liked' kitchen might be preferred to a 'disliked' kitchen simply because it was a better 'cooking machine'.

At this stage the linear link diagrams and implication networks would be compiled and would show how the physical characteristic of the rooms fitted into the informants overall construct system. Some informants might link the cooker in a kitchen directly to the activity of 'preparing food' which might be a superordinate construct highly loaded to the 'liked kitchen' element. The implication grid, however, could show that the cooker had many reciprocal implications for evaluative constructs, such as 'homeliness' or 'practicality' and we would begin to be able to see how the performance of a room in accommodating an activity was construed on the basis of specific characteristics. We would also be able to see that although certain physical characteristics of the room impeded the specific activity associated with it, it was nonetheless a 'liked' room. It would be particularly interesting if a multi-activity room like a kitchen was construed with principal components relating to different activities. The first major principal component, accounting for most of the variance, might be approving while subsequent principal components reflecting more specific sets of kitchen criteria could be disapproving. In this way the experiment would begin to demonstrate how the interactive roles of 'form' and 'function' influenced environmental evaluation.

Clearly the other people using the rooms would also figure in an informant's construing process. The way they operate on and within the room affects the way it is seen to function. In my experiment the use of photographs meant that the informants could not include the living room users in their construct systems, and consequently their attitudes to personal space and aspects of territoriality could not be reflected in their construct systems.

Kelly's sociality corollary states. "To the extent that one person construes the construction process of another, he may play a role in a social process involving the other person." (2)

Bannister explains that in this corollary Kelly is concerned about interpersonal relations. (Bannister and Mair) (12). In the environmental sense it is clear that the interpersonal relations could well affect the way a pupil, for example, construes a headmaster's study. Most of the pupil's construing of the environment, which includes the headmaster as a 'physical characteristic', will be related to making sense about how the headmaster is construing him. In a study of personal space and social ritual in small office spaces, Joiner suggests that the use of the offices and organization of the furniture in them may be related to sustaining social relationships (Joiner) (13). The headmaster might well position his desk in relation to the window, the door and the lights so that a visitor will be conscious of intruding into the territory of a superior, more powerful being. In this case the headmaster would be anticipating how, by the arrangement of his environment, he can reinforce and enhance his status in the mind of the visitor.

Joiner's study uses a participant observation method. Evidence upon which opinions and conclusions are based is compiled from programmed observations. The way in which the person psychologically uses the particular environment is deduced from watching him behave in it. No attempt at understanding his cognitive interaction with the environment is made.

Question of territoriality, interpersonal relations in environment and personal space are important aspects of man-environment theories but researchers such as Joiner tell us little about how the informant understands his territory and the people in it. Consider the combination of a study such as Joiner's using observation techniques, with a personal construct approach for analyzing the process and components of evaluation. The results would tell us much more than that people used their accommodations as aids and props in their relationships with others. We would begin to understand which qualities a person looked for in an environment and how its ingredients contributed to them. We would be able to say that the status he felt he gained from, for example, the positioning of his desk was because the desk itself, in association with the color of the walls, the thickness of the carpet and the kind of people who came to see him were part of a network of constructs closely related to the network with which he construed his rank. We would further be able to tell what it was about the desk in the particular room which made it a better prop or aid to his purpose, than one of another type.

The results of the combined study would extend our practical understanding of the particular environment because we would be able to predict what would happen if for example, we moved the person to a smaller room and what measures we could take to make this move more acceptable to him.

More generally contention is that useful man-environment theory will develop from combinations of this kind. The examination of the details of a construct system can expose the way the physical environment and its cognition are associated.

Notes

1. Throughout this paper the word 'environment' is used to mean a particular piece of territory which may be natural or constructed, designed or accidental and internal, external or both.

2. Kelly, G.A.; The Psychology of Personal Constructs, W.W. Norton and Company, New York. 1955.

3. Bannister, D.; A Psychology of Persons, In New Society. April 12, 1969.

4. Stringer, P.; Architecture, Psychology the Game's the Same, Architectural Psychology - Proceedings of the Conference held at University of Strathclyde. 1969. Canter, D.V. Editor.

5. Honikman, B.; The Investigation of a Method of Relating the Personal Construing of the Built Environment to the Designer, AP 70. Proceedings of Architectural Psychology Conference. R.I.B.A. Publications and Kingston Polytechnic. 1970.

6. Slater, P.; Notes on Ingrid 67; Biometrics Unit, Institute of Psychiatry, University of London. 1967.

7. Hinkle, D.N.; The Change of Personal Constructs from a Viewpoint of Theory of Implications, Unpublished Ph.D. Thesis. Ohio State University. 1965.

8. Honikman, B.; An Investigation of the Relationship between Construing of the Environments and its Physical Form, Proceedings of E.D.R.A. 3 Conference at U.C.L.A. 1972.

9. Canter, D.V.; Should we Treat Building Users as Subjects or Objects, Architectural Psychology. Canter D.V. Editor. R.I.B.A. 1970.

10. Bonarius, J.; Research in the Personal Construct Theory of George A. Kelly, Progress in Experimental Personality Research, Volume 2. Maher B. Editor. Academic Press 1965.

11. Bridge, A. (Unpublished); A Study of Construct Differences between Rooms of Similar Intended Function, Unpublished Study Architectural Psychology Research Unit. Kingston Polytechnic, U.K. 1972.

12. Bannister, D.; Mair, J.M.M.; The Evaluation of Personal Constructs, Academic Press. 1968.

13. Joiner, D.; Social Ritual and Architectural Space, Architectural Research and Teaching. Volume 1, Number 3. 1971.

AN ANALYSIS OF FOUR MEASURES OF COGNITIVE MAPS   5.2

Roger B. Howard
Sara D. Chase
Mark Rothman

Department of Psychology
Colgate University

Abstract
If cognitive mapping is to become a useful tool in environmental design, we must have measures which are reliable and valid. The four measures studied here are highly reliable. The data also suggest that the subjective distances between points depend primarily on (1) the actual distances, and to a lesser extent on (2) the paths traveled by the subjects from one point to the other, and (3) the frequency of environmental features along the path. The four methods thus have equal construct validity to the extent that the cognitive-map construct allows these interactions.

Adequate measuring techniques are a prerequisite for the collection of meaningful data in any science. It is not surprising therefore that behavioral scientists have generated a considerable body of theory and methodology relating psychological constructs to observable behavior.

Many areas of environmental research are concerned with human behavior. The problems in these areas bear a strong resemblance to the problems in more traditional areas of the behavioral sciences. It thus seems likely that the maximum rate of progress in environmental research would be achieved if the problem of developing adequate measuring techniques were considered with respect to the already existing information. This paper is concerned with the application of some of the important concepts of measurement theory to cognitive maps. The implications of this analysis are illustrated in a series of experiments designed to test the adequacy of four methods for measuring psychological distances.

Criteria for Adequate Measures of Cognitive Maps
There is no universal agreement on the criteria which characterize good methods of measuring. Nevertheless most psychometricians seem to agree on two classes of criteria which should be considered: Validity and reliability. A valid method is one which measures what it is supposed to measure. An intelligence test should measure intelligence and a measure of cognitive maps should measure some aspect of cognitive maps. Reliability refers to the repeatability of the measurements made by the method. If two equivalent people are given the same intelligence test under the same conditions, a good intelligence test should give the same score for both. Likewise, an adequate measure of cognitive maps should allow two people with equivalent cognitive maps to produce equivalent data about their maps. In essence we are asking that the method reduce error variance to a minimum.

Although reliability and validity are related, they are not equivalent constructs. A proposed method for measuring cognitive maps may actually measure intelligence if the instructions are difficult to understand. The test may produce equivalent scores for two subjects with equivalent intelligence and therefore be highly reliable. At the same time, we would have good reason to question its validity as a measure of cognitive maps.

There are different types of reliability and validity, not all of which are applicable to any particular area of research. The different forms of reliability are primarily dependent on the form of the data and will not be considered here. Nunnally (2) describes three forms of validity. <u>Predictive validity</u> requires that we know precisely what a cognitive map is good for and <u>content validity</u> requires that we know exactly what a cognitive map is. Since none of us is capable and most of us are unwilling to meet these conditions, predictive and content validity will not be considered further in this paper, although they should be extremely important in future research. Construct validity, however, deserves careful consideration.

The above analysis points to one of the major problems of research on cognitive maps. We would like to develop measures of cognitive maps which would eventually lead to practical consequences such as the development of better environments. Yet to measure the content or predictive validity of these measures, we must be able to either define precisely what the construct is we are measuring (content validity) or to tell what it is good for (predictive validity). In the case of cognitive maps most researchers have the general idea that they include cognitive representations of the external environment, but we know very little about their form or content or what they may eventually be used for in environmental design. In short, we have two unknowns, the measure and the construct it is supposed to measure, but only one equation defined by the data.

The procedure which has been developed to handle the simultaneous validation of measures and the development of definitions for the constructs being measured is called <u>construct validation</u>. It is not a simple linear procedure leading from theory to measure to validation, but a feedback loop in which one reconsiders what is meant by a cognitive map in light of the data collected to develop the measure, and then collects more data which answer more precise questions, and so on until what is being measured and its relationship to the construct are defined.

The first step toward construct validation is some general statement about the nature of the construct to be studied. The term cognitive map has been used in many ways. At its narrowest it refers to the cognitive representation of some part of the external world (Cf. Hart and Moore [3], Ladd [4]). Many researchers have been primarily concerned with its map-like properties and have emphasized the spatial relationships among the elements of the map. At its broadest, the term can refer to much more. Kaplan (5) suggests that the entire cognitive structure of the subject may be considered a cognitive map composed of the representations of objects and their associations. The spatial relationships in the external world constitute one form of association between representations, but objects may be associated along many other dimensions such as their physical similarity or their functional value to the subject. Thus while part of the cognitive map may contain representations associated in part by their spatial relationships, this by no means reflects the true richness of the cognitive map.

Fortunately it is not necessary to choose among the definitions. Kaplan's definition subsumes those structures proposed by others. Since it is impossible to sample the entire range of cognitive structures in one set of experiments, it is sufficient to consider those elements of the map which are related spatially, as a first approximation to the complete structure. In other words, we will begin the process of construct validation by studying cognitive maps of some part of the environment. In the experiments reported below, our subjects provided information about their cognitive maps of their campus environment.

In order to complete the first step it is also necessary to consider what attributes make up the cognitive map and which of them we will begin to study. The cognitive maps of different people presumably represent different things in different patterns of association. We cannot study all with one set of measures in one series of experiments. Thus it is necessary to select some subset of the possible variables that characterize cognitive maps for initial investigation. Since the part of the map we are considering is spatial, the distances between points would be the most obvious (though not necessarily the most interesting) attribute of the map. Our subjects were asked to use their cognitive maps to judge the distances between points in the environment.

The second step in the construct validation procedure is to select some methods of measurement which may reflect the attributes of the map selected under the first step. In the present case the attribute is subjective distance between points. There are many scaling and psychophysical methods which have been developed which are applicable to distances. We have selected four of these which seem representative of the methods which are available. The first is the _method of reproduction_ which was developed to study psychophysical quantities. The subject is asked to draw a map of the environment and thus to reproduce his representation of it. The second is the _method of modeling_, in which the subject is asked to place scale models of the objects defining the distances on a surface. The third and fourth methods are the _method of absolute judgments_ in which the subject is asked to give estimates of the distances of interests, and the _method of ratio estimation_ in which the subject is asked to mark a line so that it is proportional to a line representing a fixed distance. Both are scaling methods which do not require the subject to produce a physical analog of his map and hence they do not force consistency onto his data as do the previous two methods.

In the third step of the validation process data are collected which show to what extent each method of measurement reflects the attributes of the cognitive map. The success of each method depends in part on the nature of the construct. Since the attribute being measured in this example is subjective distance, the better the relationship between the true distances and the subjective distances, the better we might suppose the method as it is very unlikely that the true distances should be approximated by the subject unless the true distances were present in his cognitive map.

It is also possible for the method to be sensitive to other characteristics of the cognitive map besides the main atrribute of interest. The type and frequency of environmental features encountered along the path between points might produce systematic distortions in the perceived distances. In this case the validity of the measure must depend on the construct being measured. If the theorist believes that subjective distances should be a function of the environmental features encountered, then the measure which reflects this relationship will be the more valid measure. On the other hand, if his construct of a cognitive map does not include the possibility of such interactions, then clearly the method in question is less valid.

The important concept here is that a method may be better than another for some purposes and not for others. It is the empiricist's task to determine what the instruments are measuring so that the theorist may decide which is the most valid measure of his construct. In the second and third experiments reported below, we attempt to relate some of the variance in the data collected by the four methods to the type and frequency of objects encountered, both in the environment and in the cognitive map.

The fourth step toward construct validation involves reconsideration of the entire process, including revision of the concept, further tests of the methods (or the development of new methods), and the gradual expansion of the domain of variables being studied to other populations of subjects, attributes, spaces, and other types of cognitive map.

In summary, of the three forms of validity considered, construct validity is the most relevant when both the construct being measured and the characteristics of the measure are not well defined. The first and second steps in construct validity are to determine the general nature of the construct, the attributes to be studied, and the domain of measurement methods which might be used to study them. In the research reported here, the distances between points in a spatial cognitive map are taken as one aspect of the construct to be studied by four representative measures. The third and fourth steps are to evaluate the measures in the light of what the construct would predict, and then to either revise the construct or select the best measures as defined by the construct for further testing and data collection.

Experiment I

    Method. The subjects were 130 undergraduates from two psychology courses who participated as part of their course requirement. The subject judged the distances between eight points: Five buildings on the Colgate University campus, the University Football field, a street corner, and an inn in the surrounding town of Hamilton, New York. The points were selected to define a wide range of distances (from 200 to 1700 yards) and there was no direct line-of-sight route either by foot or automobile between most pairs of points.

Thirty subjects were run in the ratio estimation condition. Each subject was asked to mark 54 lines so that the ratio of their lengths to the length of a fixed line was the same as the ratio of the 54 distances being judged to a reference distance (the distance between points 1 and 5, which was 300 yards.).

Thirty-two subjects participated in the <u>absolute judgment condition</u>. They were given the actual yardage between points 1 and 5 and were then asked to give absolute judgments of each of the 54 straight-line distances in any units that they wished to use.

The twenty-nine subjects in the <u>reproduction condition</u> were asked to draw a partial map of the environment which would indicate the location, shape, orientation, and main door (if relevant) of the eight points, which were listed for them on a sheet of paper. They were given a reference line representing the distance between points 1 and 5.

The remaining thirty-one subjects were run in the <u>modeling condition</u>. Each subject was given eight three-dimensional plexiglass models of the structures defining the points. The models were scaled to correspond to the scale established by the distance between points 1 and 5. All other materials and procedures corresponded to those in the mapping condition.

<u>Results</u>. There are many different sets of distance data collected in this and the subsequent experiments. The data in this experiment are subjective and actual line-of-sight distances between the eight points. Since it is cumbersome to use the full titles, these data sets will be referred to as <u>subjective</u> and <u>actual LOS distances</u>.

The data from the ratio estimation and absolute judgment conditions contain two estimates of the distance between each pair of points. These were divided into distances toward and distances away from the center of campus. Several different center points were used in a preliminary analysis. None of them showed any particularly insightful deviations from the others. Since most students live on the campus or spend a substantial part of their time there, it seems reasonable to use the center of the campus as the point from which to define the two sets of subjective distances. The two subsets of data from the ratio and absolute judgment conditions thus contained the same number of subjective LOS distances as the data from the reproduction and modeling conditions.

The reliability coefficients for the six sets of subjective LOS data were computed from the between and residual mean-squares obtained by a single-factor repeated-measure analysis of variance on the data adjusted for anchor points as suggested by Winer (6). They are shown in Table 1 and are quite impressive. The minimum value is .987 while the maximum is .995.

A problem arises when these coefficients are interpreted as predictive statistics. The distances are not independent of each other in the reproduction and modeling conditions. Once the subject has determined two of the three distances between three points the third is fixed by the two-dimensional surface upon which he is constructing his map. It is unlikely that this is a problem with the other two methods which do not require consistency. Nevertheless, seven independent distances were selected which has maximum variance from the regression lines for

Table 1
Reliability Coefficients

Subjective LOS Distances:	Rp*	M	R-T	R-A	Ab-T	Ab-A
Complete Data (k = 27)	.995	.995	.987	.988	.988	.991
Independent Data (k = 7)	.987	.990	.980	.979	.981	.987

Traveled Distances	Toward	Away
Complete Data (k = 27)	.996	.993
Independent Data (k = 7)	.996	.996

Subjective FEFs: Distance #	1	2	3	4	5	6	7
Independent Data (k = 7)	.825	.856	.847	.892	.883	.869	.905

*A key to the abbreviations may be found at the foot of Table 2.
k = number of treatment levels

the complete set of subjective LOS distances, and reliability coefficients were computed for them. The results are also shown in Table 1, and are essentially equivalent to those for the entire set of data, despite the fact that the reliability coefficients are strongly affected by the length of the measure (26 for each of the complete sets of data which contain 27 distance estimates per subject, seven for each of the data sets based on the seven independent distances). Thus the reliabilities of the four measures are all quite high and do not allow us to distinguish among them.

The validity of the data depends on the degree to which the subjective LOS distances can be explained by hypothesized characteristics of cognitive maps. For this experiment we assumed that the actual LOS distances would be our predictor and that the more valid a measure the higher the correlation between the actual and subjective distances. This means that we are interested in the correlation of the 27 actual LOS distances with each of the six sets of subjective LOS distances. The 27 mean subjective LOS distances were therefore computed for each of the six subjective data sets and each of the six sets of 27 mean subjective LOS distances was correlated with the set of 27 actual LOS distances. The data shown in Table 2 are thus the correlations between means and not between individual subjects.

The correlation coefficients among the complete sets of 27 distances are shown in the upper off-diagonal matrix of Table 2. The correlations between the actual and subjective LOS distances are shown in column seven and are all extremely high. The minimum correlation is .980.

Since the nonindependence of the distances is also a problem in correlation, the seven independent distances selected previously were intercorrelated. These coefficients are shown in the lower off-diagonal matrix of Table. The correlations between the actual and subjective LOS distances are shown in row seven. In general they are slightly lower.

As would be expected from the high correlations of the subjective LOS distances with the true LOS distances, the intercorrelations among the measures are also quite high (Table 2) for both the complete set of 27 distances and the seven independent distances. Thus it would appear that there are no real differences among the reliabilities and validities of the measures within the limits of these data.

Table 2
Correlation Coefficients

	Rp	M	R-T	R-A	Ab-T	Ab-A	Ac	Tr-T	Tr-A	AFEFs	SFEFs-T
Rp		.998	.987	.981	.991	.983	.980	.984	.984	.825	
M	.998		.982	.975	.985	.976	.987	.985	.985	.822	
R-T	.910	.883		.998	.993	.992	.986	.952	.954	.825	
R-A	.910	.900	.993		.995	.996	.980	.944	.943	.812	
Ab-T	.956	.941	.958	.961		.997	.980	.962	.965	.828	
Ab-A	.948	.929	.986	.984	.991		.987	.944	.944	.838	
Ac	.973	.962	.959	.959	.964	.977		.948	.948	.837	
Tr-T	.978	.978	.905	.928	.967	.949	.952		.999	.834	
Tr-A	.974	.972	.900	.909	.972	.946	.941	.991		.841	
AFEFs	.900	.880	.959	.967	.967	.975	.962	.915	.907		
SFEFs-T	.897	.863	.971	.964	.931	.965	.964	.889	.849	.975	
SFEFs-A	.891	.870	.973	.946	.905	.942	.948	.847	.848	.892	.943

Subjective LOS Distances: Rp = reproduction, M = modeling, R-T = ratio estimation--toward campus, R-A = ratio estimation--away from campus, Ab-T = absolute judgment--toward, Ab-A = absolute judgments--away; Ac = actual LOS distances; Tr-T = traveled distances--toward; Tr-A = traveled distances--away; Frequencies of Environmental Features (FEFs): AFEFs = actual FEFs, SFEFs-T = subjective FEFs--toward, SFEFs-A = subjective FEFs--away.

Experiment II

One of the more interesting parts of construct validation is the attempt to discover what variables explain the variance found in the data collected by each measure. In Experiment I it was shown that most of the variance in the subjective LOS distances obtained by all four methods appeared to be explainable by the actual LOS distances between points. This distance, however, is frequently confounded with other variables. One of these is the distance between points that would be traveled by a subject. Although traveled distance is sometimes equivalent to the actual LOS distance considered in Experiment I, it is at times quite different and therefore may be a better predictor of subjective LOS distance or may explain some of the variance in the data which was not explained by the actual LOS distances.

The paths used by the subjects in traveling from one location to another may also be used to define the features of the environment which he sees. Laboratory studies show that lengths which are broken rather than continuous appear longer (cf. Howard [7]). The same process may also operate on the experience of distance traveled. Since the frequency of environmental features experienced by the subject is partially a function of the traveled distance, frequencies might also explain additional variance in the data from Experiment I.

   Method. The traveled distances between the eight points were obtained from 110 volunteers from the same population described in Experiment I. Each was given a set of maps of the campus and town and asked to draw lines showing the route which they typically used in traveling from each of the eight locations to the others. The procedures were the same as for the ratio estimation condition of Experiment I.

The data were analyzed by measuring each route and noting the path the subjects took. Although there were slight variations the average paths were easy to define. Each of these was then retraced by the experimenters and the frequency of seven categories of features noted: Large buildings, small buildings, corners, curves, grades, fields, and trees. Because of the extremely high correlation between paths towards and away from the center of campus between any two points, only one path was usually distinguishable. Thus only one set of frequencies was tabulated for each distance.

   Results. Reliability coefficients were computed for the traveled distances toward and away from the campus for the complete set of 27 distances and for the seven independent distances described above. The results are shown in Table 1. As can be seen, they are at the same level as were those for the subjective LOS distances.

The intercorrelations of the traveled distances with the previously collected data are shown in columns and rows eight and nine of Table 2. The traveled distances correlate well with the actual LOS distances, but at a slightly lower level than the actual LOS distances.

Before complete data analysis was begun, various combinations of the seven frequency categories for environmental features were studied by multiple regression. The frequency of trees and buildings contributed most to the combined frequencies scores. Nevertheless, the combined frequencies appear to be the most stable and meaningful scores which predict the most variance. Thus they will be considered here. The scores will be called the actual frequencies of environmental features to distinguish them from the subjective frequencies studied in Experiment III, and will be abreviated as actual FEFs.

The intercorrelation of the actual FEFs with the subjective LOS distances are shown in row and column ten of Table 2. These correlations are lower than the correlations of the subjective with the actual LOS distances for both the complete and independent sets of distances.

In order to obtain further information, the data from experiments were submitted to multiple regression analysis. Various combinations of the different variables were studied. The most interesting are shown in Table 3.

Multiple Regression Coefficients

			Dependent Variables				
Independent Variables		Rp*	M	R-T	R-A	Ab-T	Ab-A
Complete Data: (N = 27)	Ac+Tr-T	.984	.987	.986	.982	.983	.987
	Ac+Tr-A	.984	.987	.986	.981	.983	.987
	Ac+AFEFs	.985	.995	.990	.992	.988	.995
	Ac+AFEFs+Tr-F	.990	.995	.994	.997	.995	.998
	Ac+AFEFs+Tr-A	.990	.995	.994	.997	.995	.998
Independent Data: (N = 7)	Ac+Tr-T	.980	.981	.970	.962	.970	.980
	Ac+Tr-A	.979	.980	.971	.962	.971	.979
	Ac+AFEFs	.990	.990	.975	.972	.974	.986
	Ac+SFEFs-T	.986	.976	.974	.964	.965	.979
	Ac+SFEFs-A	.986	.989	.970	.967	.964	.979
	Ac+AFEFs+Tr-T	.995	.994	.985	.988	.990	.990
	Ac+AFEFs+Tr-A	.994	.995	.980	.985	.989	.995
	Ac+AFEFs+SFEFs-T	.990	.990	.976	.975	.975	.989
	Ac+AFEFs+SFEFs-A	.990	.991	.975	.974	.974	.990

*A key to the abbreviations may be found at the foot of Table 2.
N = number of means compared.

In general, adding either the traveled-distance or the actual-FEFs variables to the actual LOS-distance variable increases the multiple regression coefficient and thus the amount of variance accounted for in the subjective LOS-distance data. The highest coefficients are for a combination of all three variables. This suggests that the subjects in all four measuring conditions were influenced by the paths they took and the frequencies of feature which they passed in addition to the actual LOS distances. Since the data from each measure are described about equally, the construct validity of each of the four measures for predicting subjective from actual LOS distances and actual FEFs is about the same.

Experiment III

In Experiment II we found that the actual FEFs along a route explained a small amount of the variance in the subjective LOS distances obtained by all four measures. The subject, however, does not experience the actual features while using his cognitive map. Thus it msut be either that the subjective LOS distances were distorted at the time they were experienced by the subject or some representation of the features (or at least the FEFs) existed at the time of the measuring which biased their responses.

In all likelihood both processes occur. However, it is possible to get some suggestive evidence on the relative strengths of each process by asking subjects to report the FEFs represented in their maps and then comparing the amount of variance in each set of subjective LOS distances accounted for by the subjective FEFs with the variance accounted for by the actual FEFs. If it is primarily the subjective FEFs which are involved, then their contributions should be high. If most of the distortion occurred as the distances were encoded, then we might expect the actual FEFs to be better predictors.

Method. Because of the quantity of information requested, two groups of 32 subjects were asked to imagine the route they would travel between each of the seven independent pairs of points studied above and to count the number of features in each of the seven categories used in Experiment II. The direction (toward or away from campus) was randomized for one group. This order was counterbalanced by the other group.

Results. Separate reliability coefficients were computed for each of the seven distances toward and seven distances away from the campus. Table 1 shows that they are not as high as in Experiment I, but they are high enough to assure us of the robustness of the data and to justify further consideration.

Rows 11 and 12 of Table 2 show the correlation coefficients for the subjective FEFs. As can be seen, the actual FEFs are more highly correlated with the subjective LOS distances than are the subjective FEFs.

The data were further analyzed by multiple regression as in the previous experiment. The major results are shown in Table 3. They suggest that the actual FEFs account for more variance than the subjective FEFs. This implies that most of the distortion occurred in the perceptual process rather than the memory or reconstruction process.

General Discussion
The results of these experiments suggest that all four measurement methods are equivalent for the set of variables studied. All four show very high reliability of about equal values. Likewise, their construct validity would appear to be about equal under these conditions. All four show a strong linear relationship between subjective and actual LOS distances, and the actual FEFs encountered along each route. Each shows lower correlations between the subjective LOS distances and the distances that would be traveled to get from one point to the other and each shows a high correlation with the others. To the extent that these results fit one's construct of how cognitive maps ought to behave, then the methods are valid. In any event, the data suggest that the four methods are equally valid or invalid according to the definition one accepts.

Assuming complete equivalence of methods from one series of experiments, however, would be like accepting the null hypothesis in statistics: It is a very dubious procedure indeed. As indicated in the introduction, the process of construct validation is a long and complicated one involving many cycles over different parts of the domain of empirically observable phenomena. We have studied one population of subjects on one set of distances in one environment with one set of measures on one major attribute of one type of spatial cognitive map which Kaplan (8) suggests is one segment of the complete human cognitive structure. The generality of these conclusions will depend on further research. In the meantime practitioners and scientists interested in exploring cognitive maps would do well to consider the processes of construct validation as part of their use of any measurement method.

References

1. Preparation of this paper was supported in part by grants to the first author from the Sloan Foundation and the Colgate Research Council.

2. Nunnally, J.C., Psychometric Theory. New York: McGraw-Hill, 1967.

3. Hart, R.A., and Moore, G.T., The Development of Spatial Cognition: A Review. Clark University, Worcester, Mass., Place perception research report #7.

4. Ladd, F.C., Black Youths View Their Environment: Neighborhood Maps. Environment and Behavior, 1970, 2, 74-79.

5. Kaplan, S., Cognitive Maps In Perception and Thought. In R.M. Downs and D. Stea (eds.), Cognitive Mapping: Images of Spatial Environments. Chicago, Illinois: Aldine, in press.

6. Winer, B.J., Statistical Principles In Experimental Design. 2nd ed. New York: McGraw-Hill, 1971.

7. Howard, R.B., Neurophysiological Models of Figural Aftereffects and Visual Illusions. Psychonomic Monographs, 1971, 4, #51.

8. Kaplan, S., Cognitive Maps In Perception and Thought. In R. M. Downs and D. Stea (eds.), Cognitive Mapping: Images of Spatial Environments. Chicago, Illinois: Aldine, in press.

PREDICTORS OF ENVIRONMENTAL PREFERENCE: DESIGNERS AND "CLIENTS"   5.3

Rachel Kaplan

Department of Psychology
University of Michigan

Abstract

It has previously been suggested that preference for slides of the outdoor environment is only partially accounted for on the basis of complexity ratings. The present study explored two other variables, coherence and mystery. Further, the prediction of preference was studied for people who differed in prior training in design-related professions.

Highly significant differences were obtained in the preference patterns of the three samples: students in Architecture, Landscape Architecture, and the College. Coherence and mystery were found to be relatively independent of each other; each was strongly effective as a predictor of environmental preference.

Of the various factors that help in the understanding of environmental preference, complexity has received the most attention. Craik (1) discusses several studies dealing with the simplicity-complexity dimension, Rapoport (2) has extended his previous statements on the preference for an optimal level of complexity, and Wohlwill (3) asserts that he has demonstrated such an inverted-U relationship between complexity and preference of physical environments. However, in a study reported at EDRA last year and more fully detailed by Kaplan, Kaplan and Wendt (4), it was shown that Wohlwill's assertion is perhaps overstated and that complexity was found to have limited utility in explaining preference for environmental displays. While complexity and preference were indeed linearly related within a nature and within an urban domain, the overwhelming preference for nature material could not be explained in terms of complexity.

In his EDRA paper, S. Kaplan (5) proposed a "tentative model" which dealt with complexity as one of four informational factors pertinent to the prediction of preference. The purpose of this paper is to present results of a study that investigated two of these, "coherence" and "mystery." Furthermore, the prediction of preference was studied for people who differed in prior training in design-related professions.

   The tentative model. Briefly, the part of the model pertinent to the present study consists of two independent "sources of information," each further divided according to the "degree of inference required." The two sources of information differ in how readily accessible the information is. When the transmission of information is rapid and present in the setting, one can more easily figure out what is going on. This dimension is quite similar to Lynch's (6) "legibility" concept.

Kaplan divides this category into two further components, coherence and identifiability. The latter deals with "making sense out of what is depicted," with recognizing it for the object or setting that is intended. "Coherence," which requires relatively less inference, depends on redundancy of the elements and textures that help make the display "hang together."

The second major category, "predicted information," consists of those situations where the information acquired would be increased by studying the scene either for a longer time because of the "complexity" of the material or from a different vantage point in the case of "mystery." A scene high in mystery would promise more information if one could step into the picture to "see around the corner" or behind the foliage.

Kaplan contends that "both legibility and predicted information are important in landscape preference. Further, the landscape represents sufficiently diverse patterns of information that both can be and often are present in the same setting." People desire settings where they can make immediate sense out of the general context but there is also attraction to needing more information to fully comprehend and appreciate the setting.

The study reported here is based on subject ratings of preference, mystery, and coherence. A series of 60 slides of the outdoor environment was used. These covered a wide range of contents and different organizational properties.

Method

Subjects. A total of 107 subjects participated in the study. They consisted of three sub-samples: the advanced Architecture students (n=38) were enrolled in a required course in landscape architecture; the Landscape Architecture students (n=30) were in the second year of a three-year graduate program; the College students (n=39) were enrolled in an upperclass psychology course. The last of these, lacking any specific training in design, might be considered to represent the public or the client.

Slides. The 60 slides were all monochromatic, with half consisting of graphic renditions and the other half photographic. Since graphics often depend more heavily on contour than do photographs, an effort was made to select graphics that had a stronger sense of texture as well as some sketchier line drawings.

Four major content areas were sampled: those with predominant paths or highways, those with a predominance of natural areas, those depicting a grouping of related buildings, and those where the focus was on a part of a building. That is not to say that there was no overlap in these categories. In particular, the "part building" scenes included a certain amount of landscaped natural area. This category could also be subdivided in terms of the kind of buildings depicted: a private residential dwelling or a public (including apartment unit) building. An effort was made to take photographs of content areas for which graphic material could be found, although it was not possible to match these on a one-to-one basis. This proved particularly difficult in the case of nature areas where few graphic renditions were available and for oblique views of building complexes where the artist is at an advantage.

The slides, or "environmental displays" using Craik's (7) terminology, can also be categorized in terms of organizational qualities. They were approximately evenly divided in terms of a deep/shallow designation. A display can be considered "deep" if one can see far into the picture, while it is "shallow" if it is mainly foreground or if the foreground object or mass prevents one from "entering" the scene. They were also approximately evenly divided in terms of an open/enclosed designation. Here the difference is the openness of the depicted space. The feeling of being enclosed can be created by the trees and foliage, or in some instances by the juxtaposition of buildings or parts of buildings. In a few instances a "shallow" display could not be categorized in terms of "open" or "enclosed," but generally speaking all combinations of these organizational dimensions were represented, with between nine and sixteen instances of each.

A panel of judges served to make the content and organizational judgments and the final 60 slides were selected to reflect a balance of these criteria. For purposes of presentation, the slides were in random order and were presented in a different order each time.

Response format. For all three ratings -- mystery, coherence, and preference -- a five-point scale was used ranging from "not at all" to "a great deal." The instructions indicated that the rating of mystery meant "to what degree do you think you would learn more if you could walk deeper into the scene." Coherence was defined as "to what degree does it hang together." And for preference, the subjects were asked to indicate "how pleasing you find the scene; how much do you like it."

The response sheet consisted of both sides of a single page with identical instructions on both sides. Each side had 35 numbered lines and three banks of numbers from 1-5 so that the subject circled the appropriate number in each of the three labeled columns. (The pages had extra lines to prevent the subject from knowing when the last slide was shown.)

Procedure. The slides were presented in a classroom setting. After general discussion of the meaning of the three ratings, three practice slides were shown and rated, and further discussion was permitted. The 60 slides were then shown for 20 seconds each, with a brief rest after the first 30.

Results: Preference Domains

Following the procedure outlined by R. Kaplan (8), dimensional analyses were performed using the preference ratings of all 107 subjects. Both the Guttman-Lingoes Smallest Space Analysis III (SSA-III), a non-metric factor analysis, and the hierarchical cluster analysis program ICLUST were used to determine the main preference domains. Lingoes (9, 10) explains that the SSA-III entails a rank-ordering of the original correlation matrix thus making the procedure non-metric, while the subsequent procedures using this transposed matrix are basically factor-analytic. This procedure has been found to yield highly stable results even when the variables are somewhat altered or when different samples are studied. The ICLUST procedure, developed by Kulik, Revelle, and Kulik (11), is fast and efficient and complements the results obtained from the SSA-III.

Based on these analyses, three non-overlapping domains were identified and these form the basis for subsequent analyses of the data and comparisons of the three subject groups. The same analyses made it possible to identify some inadequacies in the environmental display sampling process which necessitated the elimination of a large number of the slides from further analyses. Both the nature of the three domains of preference that were found and the nature of the displays that did not load on these dimensions will be discussed briefly.

The nature domain. This consisted of seven slides which have in common that they represent natural settings with few indications of man-influence. Where parts of houses or paved paths are visible they in no way detract from the basically "woodsy" feeling. In all cases, the displays give an enclosed (as opposed to open) feeling and with one exception, they were characterized as "deep" (as opposed to "shallow").

Only two other slides in the total set fit these descriptions of being strong on nature, deep and enclosed. One of these shows a private residence with a side yard that is densely wooded, but the nature does not obscure the building, and the other is a graphic rendition which seems to have communicated a too romanticized impression to appear on the same dimension with the photographs. Other nature scenes which did not load on this dimension were either basically open in their organization or basically shallow, or both.

Part-buildings with nature. This domain consisted of seven slides which have in common that they depict parts of public (including apartment) buildings in a distinctly natural setting; they are all "open" in organization, but both shallow and deep displays are included among them. The landscaped area, being open in character, does not obscure the built component but provides a setting.

The examples of part-buildings which did not load on this dimension can be characterized by being clearly residential, lacking a clear setting, lacking the feeling of openness, or combinations of these. In addition, several of the part-building displays were in a context of a group of buildings and these too did not join this domain. It is particularly striking that the various scenes depicting parts of residential dwellings in an enclosed natural setting did not form a dimension of their own. As with the Kaplan et al. (12) study, it would seem that subjects make rather fine discriminations when it comes to residential scenes.

Building complexes. Seven displays in this domain depict architecturally striking groupings of related buildings that are graphic renditions. The eighth display is a photograph of an architecturally striking masonry canopy extending from the side of an obviously major building. Half of the displays include definite landscaping features; the other half lack these. They can all be characterized as relatively "shallow" in organization. The scale of these graphic displays would be very difficult to accomplish photographically since the appropriate "view from a hill" is rarely available. The artist or modeler is not restricted by such difficulties.

Three displays of building complexes did not load on this dimension. All three differed in being "deep" in organization, in suggesting a stronger feeling of topography, and in two cases, in being more difficult to decipher.

The displays that were predominantly of paths or highways did not form any coherent domain probably because they represented a great variety of contents, strong differences in scale, and no uniformity of organization.

   Graphics vs. photographs. Few of the graphics are included in the three preference domains and in general the graphics and photographs did not mix in the dimensional results. While the explanation for this is necessarily after the fact, it can nonetheless be helpful in guiding future attempts with such material. Many of the graphic displays that showed no clear results were difficult to understand -- they were low in identifiability. They were too sketchy or too crowded with detail that could not be understood in the 20 seconds of presentation. The same problem of identifiability plagued some of the photographs. Four in particular were of extremely stylized, stark, and unfamiliar settings.

The display sampling problem is a difficult one. One has to have criteria for selection and while a panel of judges can reliably rate the slides in terms of content and organization, the subjects are basing their judgments in terms of many other attributes as well. Thus "part building," with or without a major path, with or without a landscaped setting, and with or without the feeling of openness, still ignores that the "part-building" itself can be of a great variety of structures and that this too makes a difference.

Results: Group differences

Before comparing the three subject samples in terms of their patterns of preference, it is interesting to note the relationship between the ratings of graphic and photographic material as a whole, and between the 22 slides which comprise the three preference domains and the remaining 38 slides that did not load on these dimensions. Based on the entire sample, the mean preference, mean coherence, and mean mystery ratings for the 30 graphic displays as opposed to the 30 photographic displays were virtually identical (3.06 vs. 2.98; 3.29 vs. 3.26; 2.92 vs. 2.90, respectively)! Even within each sub-sample there were no significant differences between graphic and photographic ratings. Comparably, the differences between ratings of the slides used for the subsequent analyses and those eliminated by the dimensional analyses were also nonsignificant.

Two basic modes of analysis were used to examine the differences in the subsamples with respect to their preference patterns. Comparisons were made based on the separate ratings (preference, mystery, and coherence) with respect to each preference domain (nature, part-building, and building complex). In addition, correlational analyses were used to determine the role of mystery and coherence in the prediction of preference.

Preference. With respect to each of the three preference domains the ratings of the groups were significantly different. The difference among the groups was smallest for the nature domain (F=3.56, df=2, 104, p<.05) where the Architects had the lowest mean rating and the other two groups were much higher and at roughly the same level. The groups differed most strongly in their preference of the building-complex domain (F=8.31, df=2, 104, p<.001) where the Architects were at the high end, the College students at the low end, and the Landscape Architecture students a close second. For the landscaped part-building domain, (F=6.81, df=2, 104, p<.005) the Landscape Architecture students had the greatest appreciation, the College students had the lowest mean rating, and the Architecture students fell right inbetween.

For all three groups, the part-building domain was the least preferred. For the Architects the building complexes were by far the most preferred while for the College students the nature domain took a strong lead. The Landscape Architecture students liked the building complex dimension nearly as much as did the Architects and the Nature domain nearly as much as did the College students.

Mystery. The three groups showed no difference in their ratings of the mystery component of the nature domain or the building complexes. For the part-building domain (F=3.24, df=2, 104, p<.05) the College students felt less mystery was evident than did the other two groups. All three groups rated the nature domain as by far highest in mystery, the part-buildings as lowest, and the building complexes right inbetween.

Coherence. The Architects found significantly less coherence in the nature domain than did the other two groups (F=3.43, df=2, 104, p<.05), while the groups did not differ significantly in their coherence judgments with respect to the other two domains. Part-buildings were considered lowest of the three domains with respect to coherence by each of the three groups, but not to a striking degree. While the nature domain was the most coherent for the College students, it was the building complexes for the Architects. These two domains were tied for the Landscape Architecture students.

The pattern that emerges from these results is one of strong differences in preference as a function of area of professional interest with an understandable preference for buildings on the part of Architects, a divided preference of buildings and landscaped settings for the Landscape Architecture students, and a strong preference for unadulterated, enclosed nature settings for the College students. At the same time, the ratings of coherence and mystery show considerable agreement despite differences in training.

The prediction of preference. To determine the relative importance of mystery and of coherence in the rating of preference, a series of partial correlations was performed using the 22 slides that define the three domains. Since it has already been shown how the subject groups differed in their ratings of these domains, and how the domains differed in their relative position, it seemed pertinent to examine the interrelations among the different ratings for these items.

The prediction of preference based on coherence when the effect of mystery is partialled out ($r_{pc.m}$) and the prediction of preference based on mystery when coherence is partialled out ($r_{pm.c}$) are both very strong (.67 and .86 for the entire sample, respectively). For the Architects, the coherence rating is without a doubt the more important determinant (partial r's of .89 vs. .66), while for the College students the opposite is the case (partial correlations of .72 vs. .93) where mystery, independent of coherence, is the stronger predictor. For the Landscape Architecture students both coherence and mystery are almost equally effective (partial r's of .72 and .80, respectively) in predicting preference.

There is an interesting further question that needs to be answered, especially given such highly significant correlations. And that is the issue of the independence of the two predictor variables, coherence and mystery. It could be argued that since the same subjects produced all three ratings a "halo effect" was operative, or a response set, or some other biasing influence. In other words, a subject who liked a particular slide may have rated it high in mystery and coherence as well, since he considered these to be "good" or favorable qualities. However, when relating the ratings of these two variables, independently of (partialling out) the preference rating, the resulting correlation is in all cases negative (-.32 for the Landscape Architect sample, -.42 for the Architecture sample, -.58 for the College sample, and -.40 for the sample taken as a whole). Only the College sample's partial r is significant at $p < .05$. Clearly the subjects did not simply rate slides as consistently high or low on all ratings, but made discriminating judgments with respect to each of the ratings. Furthermore, the coherence and mystery ratings, showing partial correlations that account for only between one-tenth and one-third of the common variance, can be considered relatively independent of each other.

It would seem then that coherence and mystery are both strikingly important factors in understanding preference of a variety of physical environments. For Architects, coherence makes more difference, while for the untrained eye of the College student, mystery plays a more important role in determining preference. Further, within the ranges represented by the environmental displays used in this study, the relationships are clearly linear. And finally, coherence and mystery can both be present or absent, relatively speaking, within the same display.

Discussion

The picture that emerges here with respect to the two factors that bear on environmental preference is highly promising. The results are perhaps even stronger and more intriguing than those reported for complexity. Coherence, as one component of Kaplan's "legibility" dimension, is clearly an important factor in predicting preference. (The other legibility component, "identifiability," while not tested in this study, seemed to have some bearing on the ratings. Many of the slides that did not load on the domains analyzed here were rated as lower on identifiability by several of the people making the judgments that affected which slides were selected for the study.)

The "predicted information" dimension Kaplan proposed included both complexity and mystery. The mystery ratings were powerful in predicting preference for all three groups, but particularly for the College sample.

"Mystery" has not received as much mention in the literature as have complexity and legibility (13). Cullen's (15) "here and there" concept, especially in his instances "with a known here and an unknown there," is a closely related idea. There is a promise of more information if only one could get to that better vantage point, but the picture must communicate a feeling that there is a better vantage point to be had.

This is important. A picture that has great depth (as in the "deep" judgments discussed in terms of the organizational qualities of our slides) is not necessarily high in mystery. One can see far into the scene without feeling that a different vantage point would provide further information. A "shallow" picture, by contrast, is not necessarily lacking in mystery. The very object that blocks one from seeing beyond the foreground defines the vantage point that would change one's view. Comparably, both open and enclosed displays can provide the sense of mystery.

"Mystery" is perhaps not the best term for this notion. Our subjects however had no difficulty making the ratings and understood what was meant by this concept given the explanation on the instruction sheet. Cullen's use of "anticipation" as one case in his larger category, comes very close to the meaning intended here. He writes, "We now turn to those aspects of here and there in which the here is known but the beyond is unknown, is infinite, mysterious, or is hidden inside a black maw. First among these cases is anticipation. These two pictures clearly arouse one's curiosity as to what scene will meet our eyes upon reaching the end of the street" (p.49).

The importance of coherence and mystery in preference have direct implications for design. Both are important and both need to be present for optimal effects. Thus while a playground should look like a playground, it should also not reveal itself completely right from the start. Surprises are welcome and desired, especially when the necessities of having one's bearing and feeling comfortable in the setting have been taken care of. Frequently these two factors are operative at different distances. This is particularly true when the immediate stretch is clear and coherent. In such instances the next stretch is more inviting if we don't know completely what awaits us. This is much like the familiar urge to continue a walk to the next bend in the path.

The strong differences among the subject samples also have direct implications for designers. While this study was intended to ascertain the degree to which people in design professions show certain preferences, it was not designed to determine whether any differences found were because of training or because of prior dispositions. There is no doubt that both factors are pertinent and one would expect that designers trained at different schools would show somewhat different patterns of preference. But the preferences of those who have not self-selected themselves to be designers and have not been trained in these skills, is clearly different. The enclosed woodsy feeling of a natural setting was strongly preferred for this group, and even though the part-building displays were all in a landscaped setting, that "kind of nature" did not receive the same response. Understandably, it is the Landscape Architecture students who have the greatest appreciation for the juxtaposition of buildings and their settings, as well as appreciating both architectural renditions of building complexes and unspoiled natural settings.

These discrepancies are interesting and they are understandable. But when the time comes that the professional deals with the client or the public, they have further implications. While the designer is accustomed to being a "taste-setter" and can impose his preferences on the public, this does not necessarily lead to the happiest consequences. If designers and the public are to work together to some degree in design decisions, a mutual recognition of these differences is essential.

Notes and References

1. Craik, K. H. Environmental psychology. In New directions in psychology 4. New York: Holt, 1970.

2. Rapaport, A. Designing for complexity. Architectural Association Quarterly, 1971 (1), 3, 29-33.

3. Wohlwill, J. F. The emerging discipline of environmental psychology. American Psychologist, 1970, 25, 303-312.

4. Kaplan, S., Kaplan, R. & Wendt, J. S. Rated preference and complexity for natural and urban visual material. Perception and Psychophysics, 1972, 12, 354-356.

5. Kaplan, S. & Wendt, J. S. Preference and the visual environment: Complexity and some alternatives. In W. J. Mitchell (Ed.), Environmental design: Research and practice. Proceedings of the Environmental Design Research Association Conference Three, Los Angeles, 1972.

6. Lynch, K. The image of the city. Cambridge, Mass.: Harvard Press, 1960.

7. Craik, K. H. Environmental psychology. In New directions in psychology 4. New York: Holt, 1970.

8. Kaplan, R. The dimensions of the visual environment: Methodological considerations. In W. J. Mitchell (Ed.), Environmental design: Research and practice. Proceedings of the Environmental Design Research Association Conference Three, Los Angeles, 1972.

9. Lingoes, J. L. An IBM-7090 program for Guttman-Lingoes Smallest Space Analysis III. Behavioral Science, 1966, 11, 75-76.

10. Lingoes, J. L. Non-metric factor analysis: A rank-reducing alternative to linear factor analysis. Multivariate Behavioral Research, 1967, 2, 485-505.

11. Kulik, J. S., Revelle, W. R. & Kulik, C-L. C. Scale construction by hierarchical cluster analysis. Unpublished paper. University of Michigan. 1970.

12. Kaplan, S., Kaplan, R. & Wendt, J. S. Rated preference and complexity for natural and urban visual material. Perception and Psychophysics, 1972, 12, 354-356.

13. The term "mystery" was in fact used in the context of Landscape Architecture some time ago by Hubbard and Kimball (14), although their use of it extended beyond the restricted sense intended here. Thus, for example, they wrote of the "inability to see the landscape with any distinctness, as for instance when the scene is shrouded in haze or in a snow storm or in darkness" which would not be an instance of mystery in the sense of more information from an altered vantage point. Or, "the sheer multiplicity of detail prevents our clear comprehension of the landscape, as when we look at the misty leaves and branches of a thick deciduous wood in early spring" which would be a better example of "complexity" as used here. Another of their examples, "the foreground is clearly seen, but that an important part of the landscape known to be present is nevertheless concealed, as where a river or a road winds out of sight behind some intervening barrier," corresponds to the "mystery" concept in this paper. Thus, I agree with them that "the effect of mystery is the result of impossibility of complete perception," but not all instances of incomplete perception qualify for mystery. The strong role mystery played in predicting preference in the present study supports their assertion that "...it is a pleasant challenge to the imagination which sets the observer to trying to determine for himself by closer investigation what is concealed from his first glance, or if this be impossible, to filling in and completing the unseen landscape according to the play of his own fancy" (p.82).

14. Hubbard, H. V. & Kimball, T. An introduction to the study of landscape design. New York: Macmillan, 1917.

15. Cullen, G. Townscape. New York: Reinhold, 1961.

The work discussed here was supported by the Institute for Environmental Quality, University of Michigan, and by the Forest Service, USDA. Howard Deardorff of the Department of Landscape Architecture contributed in many ways -- conceptually, pragmatically, and through his continuing interest and enthusiasm in this project. Christine Hill took the photographs and had primary responsibility for finding suitable graphics. I am also grateful to Hillorie Applebaum for her help throughout this study.

COGNITIVE MAPS, HUMAN NEEDS AND THE DESIGNED ENVIRONMENT   5.4

Stephen Kaplan

Department of Psychology
University of Michigan

Abstract

An integrative framework is proposed which deals with those human needs underlying environmental preference through an extension of cognitive map theory. It is argued on evolutionary grounds that the basic human information processes which the cognitive map makes possible -- recognition, prediction, evaluation, and action -- must be human needs as well as human capacities. The environment which would support such needs is one that meets three essential requirements: It is (1) possible to make sense of, (2) novel, challenging, uncertain, and (3) permitting of choice. Both variety and coherence are shown to be essential for each of these requirements.

Designers look to behavioral scientists not only for information about how people experience the physical environment, but also about the sorts of human needs that the physical environment must satisfy. They would like an integrated conception of these issues to guide them in their decisions. The design professions are under frequent fire for the intuitive, rule-of-thumb, and sometimes even dehumanizing solutions they have offered. But at the same time, as has been stated repeatedly, that needed integrated framework is lacking. The practitioner thus has little choice then to act as he does, to draw conclusions from scattered empirical results, and to proceed with the faith that his intuitive expertise is better than nothing.

In the area of environmental perception and cognition, the cognitive map theory has been proposed as a model of how people experience and know the environment. The purpose of this paper is to extend the concept of cognitive maps to the area of environmental preference. In other words, the argument will be presented that the same informational processes that the cognitive map makes possible exist as essential human needs that require environmental support. The paper thus lays a groundwork for a theory of those human needs that are particularly pertinent to the designed environment, that is, those needs involved with the taking in and processing of environmental information.

Cognitive maps: Basic processes as human needs

The cognitive map is a construct that has been proposed to explain how individuals know their environment. It assumes that people store information about their environment in simplified form and in relation to other information they already have. It further assumes that this information is coded in a structure which people carry around in their heads, and that this structure corresponds, at least to a reasonable degree, to the environment it represents (cf. 1, 2, 3). It is as if an individual carried around a map or model of the environment in his head. The map is far from a cartographer's map, however. It is schematic, sketchy, incomplete, distorted, and otherwise simplified and idiosyncratic. It is, after all, a product of experience, not of precise measurement.

There remains the question of the sorts of information that would necessarily be contained in a cognitive map, the basic information processing categories that comprise knowledge of the environment. Kaplan (4, 5, 6) has described four domains of knowledge that must be included in one's mental map: Recognition (knowing where you are, being able to identify the common objects of your environment); Prediction (knowing what might happen next, being familiar with what leads to what); Evaluation (knowing whether these next things are good or bad, being able to anticipate whether alternative actions have favorable or unfavorable probable outcomes); and Action (knowing what to do, being able to think of effective alternatives). Through these processes man structures his uncertain environment and makes it livable.

The extension of the cognitive map position to environmental preference involves the assertion that these basic information processing domains also represent powerful biases or, in other words, human needs. Viewed as human needs these four domains can be described as follows:

Recognition: This includes the bias towards making sense out of the perceived environment, and the bias towards interpreting new events in familiar terms. There is thus a bias towards simplification built in here.

Prediction: The enjoyment involved in guessing about possible outcomes in uncertain circumstances. The interest in extending one's knowledge of what leads to what.

Evaluation: The delight in dividing up the world into good guys and bad guys. The discomfort generated by ambivalence.

Action: The exercise of skill, to act in such a way as to have predictable results. The concern to make a difference. The possibility of exercising choice from among alternatives, of being decisive. The knowledge that the environment is responsive, at least to a degree, to actions one could take.

Thus the human animal not only performs these categories of processes with speed, facility, and decisiveness; at the same time, he cares and cares deeply about each of them. These biases play dual roles: they insure the extensive practice necessary for the skillful performance of these processes. In addition, as is argued in the next section, they insure that the individual will have extensive and reasonably systematic commerce with his environment.

5. ENVIRONMENTAL COGNITION / 277

Space, time, and uncertainty

It seems not unreasonable to look to the environment for the satisfaction of human process needs, since it was the environment that, in the first place, made these needs necessary. The evolutionary environment may have been different in many respects from the contemporary one, but it is that one that shaped man's genetic structure, and thus, to a substantial extent, his current needs. And in terms of the basic issues of space, time, and uncertainty the contemporary environment is by no means totally dissimilar. An analysis of that primal environment is thus an essential first step in understanding contemporary man-environment relationships.

Humans, like other animals, operate in a spatial world. This world is large relative to the size of the individual and the scope of his sensory capacity. Thus at any point in time he can only know what is going on in a small portion of this environment. At the same time, environments beyond his sensory capacity and the events occurring there are of potentially great interest to him. To behave only in the here and now, as if the more distant environment did not exist, would be folly indeed. Granted this is a luxury some animals can enjoy. They found their niche relatively early in the evolutionary sequence. They can sit on their favorite leaf and munch on it and give no thought to environments distant in time and space. But man was a late arrival in the evolutionary scheme of things; his niche is neither singular nor well defined. He lives on leftovers; he is either an opportunist or a failure. His niche is composed of the scraps and odd corners not worked by other species, or not at the moment, at any rate. Thus man evolved as a far-ranging organism, able to relate (and thus take advantage of) environments dispersed in space and thus never experienced all at one time. To behave effectively with respect to extended space, especially when different places are interesting at different times, for different reasons, and never for sure, requires an organized approach. To visit all spaces every day would be impossible, and even to visit a random sample every day would be highly inefficient. Rather it would be necessary to have an overall conception of the layout of the spatial environment, and of the distribution of the assets and the dangers. The capacity to read the signs or indicators of assets and dangers would increase efficiency still further.

It thus appears that a well structured memory, a cognitive map of the spatial environment, would be essential for survival under circumstances of this kind. A cognitive map is, however, an outcome of experience, and there is no assurance that random or unmotivated experience would lead to a cognitive map that is either extensive or well structured. The occasional allegation that "you can't get there from here" would be common indeed if people always travelled the easy paths and the direct routes. A cognitive map is only an approximation to continuity. With suitable effort it can be a rather good approximation. Without such effort it is likely to be highly incomplete, full of intransitivities and dead ends. Such a map is unsuitable when one is on the chase, or, for that matter, being chased.

If the quality of cognitive map were related to the probability of survival, then those who survived would have been those who loved to explore, who craved to know, whose restlessness and eagerness for new sights constantly led them to map-extending experiences. There are, granted, limits to the adaptiveness of exploration. If man depended on knowledge for survival then being too far from well-known terrain could put him at a distinct disadvantage. Thus he would ideally station himself along the shifting fringe between the known and the unknown. As what was once fringe became known, the fringe would shift, carrying with it our knowledge-crazed organism. From this perspective one would expect humans to be both curious and fearful of the strange at the same time -- a condition Hebb (7) referred to as "man's ambivalent nature."

The data on curiosity in both man and other animals are now extensive, and the concept is widely accepted. It is not, however, a particularly well-formed or discriminating concept. Man is not randomly curious as much as curious with respect to what he already knows something about, and with respect to what he might need to know something about. He is not only curious in the sense of enjoying to see new things (given, of course, that they are not too new); he also likes to see how things connect and what leads to what and what predicts what in a time sequence. He likes to know what is good and bad, what he likes and does not like. He likes to learn how to do things and to learn about how other people did things whether he would ever be faced with the circumstances they faced or not. In brief, man is a motivated, dedicated, addicted, if you will, builder of cognitive maps.

This then is a rather rough outline of a theory of human informational needs. It can perhaps be summarized most effectively in terms of the sources of pleasure in people's lives. Although eagerly sought, primary pleasures are of limited pertinence to the designer since only a small part of the designed environment is directly concerned with such joys as food, beverage, and sex. Besides, these are often only of limited duration. Most of the time, people are dependent on process rather than content for maintaining a reasonable level of pleasure -- or at least, absence of pain. The process that feels good is the process that is most adaptive from an evolutionary point of view, that is, going along making sense out of things, anticipating, acting appropriately, and exploring new things. These activities have in common a focus on knowledge, on the acquisition, maintainence, and use of an individual's cognitive map of the environment. The central argument of this paper is that an environment that enhances the cognitive-map related processes will be an environment most suitable to the human condition.

Coherence, variety, and choice

We come, then, to the issue of the properties of the environment that would allow the expression of these basic informational needs; an environment that we might characterize, following White (8), as competence-supporting. Based on the previous discussion, three requirements for such an environment seem particularly vital. Two of these are so closely related to each other that they may even seem contradictory. In order to show their relationship and resolve the apparent contradiction, they are discussed as a pair. The third requirement, concerning choice, is then presented as a logical outcome of the circumstances satisfying the first two requirements.

1. _It must be an environment one can make sense of_. Making sense, finding order, uncovering rules and relationships are after all the very essence of environmental knowledge, of the cognitive map by which an individual relates himself to his world.

2. _It must offer novelty, challenge, and uncertainty_. As the unknown becomes known, the frontier tamed, the individual is driven to new ground to practice his powerful processes. There must always be new domains to be comprehended, new problems to be solved, new insights to be won.

It might at first appear on intuitive grounds that these two categories stand in direct conflict with each other. After all, is not what is sensible also dull? And does not uncertainty stand in the way or order, of sense? These are indeed widely held feelings and it is important to examine them with some care. In fact, the argument fails to stand up to a thorough analysis, and has been disputed empirically as well.

It is not the case that the environment that one can most readily make sense of is an environment that is neat, linear, and sterile. Granted that a housing development composed of row on row of identical white houses facilitates prediction in a limited fashion. Given any particular white house one can usually predict what will come next. But such an environment does not support the development of a cognitive map. To begin with, it is exceptionally difficult to ascertain where one is. If one cannot enter one's map, it is exceedingly difficult to use or develop it. Further, such an environment does not provide material for the identification of essential components of a cognitive map, the landmarks, differentiated regions, and the like that Lynch (9) has shown to be vital to legibility. On the contrary, an environment that fulfills the requirements of an information-processing, cognitive-map developing person, far from being dull and overly tidy, must be rich and varied.

In addition to variety, the construction and use of cognitive maps is dependent on that complex of factors that go under the heading of legibility. Of particular interest in the present discussion is coherence, that factor which deals with the way a setting "hangs together." In other words, there must be a degree of pattern, of order, running through the variety. The arrangement of the components is not random, but follows some set of rules. Coherence both facilitates recognition and makes prediction possible. Consider, for example, an architectural style. Any given style is not distinguished from all others by a single feature but by a variety of different aspects. It is thus possible to recognize a style in many alternate ways. Likewise, a region of a city may be defined by certain styles. The variety from place to place defines a region, the coherence makes it more readily recognized.

Coherence can be achieved in a number of different ways. Besides the use of multiple features to aid differentiation, there is the repetition of a given element, allowing an individual viewing part of a scene to predict at least a portion of the content of the remaining area. A third aspect of coherence is the structural basis underlying the arrangement of elements. This permits prediction not through repetition but through the expression of some underlying rule. Some rules, like the layout patterns of different cities studied by Appleyard (10), are strong on imagery, thus making them easier to use and to remember. (Appleyard's research also shows how idiosyncratic this rule system can be, demonstrating how many viable solutions are available in comprehending a rich and varied environment.)

Thus the possibility of making sense out of an environment does not depend on extreme simplicity, but on coherence and variety. When we turn to the second proposed category of factors defining a competence - supporting environment, the necessity of offering novelty, challenge, and uncertainty, we find the rules are quite similar. Here too both coherence and variety are necessary for the requirement to be met. Coherence makes possible the definition of a region as being "somewhere else," as having a distinctive character from every other region. This means there exist different, and thus novel, patterns and relationships to explore when we have mastered the region at home. If the variety were random instead of coherent, it would defeat human competence-striving on two counts. First, the environment would be homogenized. One's home area would be as likely to have one of everything as anywhere else and thus there would be no point in going anywhere else in search of novelty or challenge. The other flaw in such an arrangement would be the ultimate lack of order, the absence of any underlying rule system. The joy of exploration, after all, comes from the gradual discovery and comprehension of order, from the growth of competence in the new setting. To lack a discoverable order is to insure eventual frustration and disappointment. Such an environment does not support effective human action, and thus could hardly be expected to enhance human satisfaction.

In other words, the kind of uncertainty most favored by humans is temporary. Uncertainty, as far as humans are concerned, exists to be resolved, and challenge, to be overcome. Thus this requirement of the humane environment depends on order and rules every bit as much as the first one does. There must, of course, be different rules and different arrangements in different places.

It is important to realize that these two basic requirements for environmental preference are not intended to be unitary dimensions. There may in fact be a number of different dimensions that contribute to the extent one can make sense out of a given environment. Likewise, the category of factors contributing to novelty, challenge and uncertainty is almost certain to be multidimensional. Thus this theoretical perspective constitutes less a solution than a direction, a framework for further analysis and further research.

A beginning in this direction has been made by Kaplan and Wendt (11) in their study of preference for slides of outdoor environments. Their tentative framework, developed essentially on empirical grounds, includes two proposed dimensions for each of these categories. Their legibility category includes both coherence and identifiability, the extent to which a scene can be readily recognized for what it is. The variety or diversity category they called "information promised" to indicate that people's preference for uncertainty is a function of the possibility of learning something new, of extending the cognitive map. The two dimensions they included in this category were the familiar one of complexity and the unfamiliar one of mystery. By the latter they meant a setting that suggests that more information would be available if the observer were to take up a vantage point farther into the scene. The bend-in-the-road, the partially obscured field, and the other side of the mountain are familiar examples of this dimension. In a subsequent study, R. Kaplan (12) tested a portion of this dimensional scheme by asking subjects to rate "coherence" and "mystery" as well as preference. Although there were wide differences between groups of subjects as far as the influence of these two dimensions on preference was concerned, the composite prediction taking them both together was very substantial for all groups.

We turn finally to the last requirement of the competence-supporting environment:

3. <u>It must permit choice</u>. The bias to act, to be decisive, to make choices is a profound one. So profound, as a matter of fact, that people prefer an alternative they have chosen themselves to one chosen for them, even when the latter would otherwise be most preferred. People wish to be "origins," not "pawns" (13). The possibility of choice is of course necessary on the practical grounds that different people prefer different levels of sense and of novelty at different times. But in addition to that, there is the powerful bias on the part of humans to make their own choices, a bias that is all too easy to ignore when doing "good" on the behalf of needy others.

The requirement of choice does not impose a further burden on the designer, since the foregoing analysis necessarily provides room for choice. The designer is not called on to make every neighborhood novel and challenging. People do not require a disneyland outside their doorstep. They desire that novelty be available, not inescapable. Thus the existence of differentiated patterns and differentiated regions meets all three environmental requirements. It allows for a sensible environment, the option of novelty, and the opportunity to choose.

<u>Conclusion</u>

It turns out, then, that the complex organism we are dealing with will not be satisfied with simple solutions for his designed environment. Lynch is certainly right that humans prefer clarity and coherence. Rapaport and Kantor (14), Wohlwill (15), Craik (16), and others are also correct in their assertion that humans prefer some complexity in their environment. To the extent, however, that Rapaport and Kantor extend the complexity concept to include ambiguity, and advocate ambiguity in opposition to legibility, they diverge sharply from the position proposed here. The complexity emphasis that has so heavily influenced work in this area, may in the final analysis turn out to be a half-truth, albeit an attractive one. Complexity may be organized or it may be gibberish. The organization and the distribution of the elements are at least as important as their number. Design in terms of complexity alone could readily become a computer-generated or random number table-based activity. The requirement of order, of coherence as well as variety, the requirement that the result ultimately be comprehensible by humans, makes design necessarily not only a human but an artistic activity as well.

As we consider these informational needs it becomes increasingly obvious that curiosity is not a hobby, that such design dimensions as variety and coherence are not decorations. The designer must unavoidably deal with factors that touch deep and ancient human concerns. His role, in terms of this framework, is neither to dazzle nor to create ambiguity, but to respect these concerns through designs that develop and enhance a sense of place.

It is hard to escape the conclusions that variety can only be appreciated in the context of order and that order is lifeless and useless without such variety. These considerations apply to an internal model of the environment, to a cognitive structure of how the world works, in very much the same way as they apply to the environment itself. Given the difficult task he faces, the designer in particular needs a map of the domain he is struggling with, a model of the processes with which he must contend. It is hoped that the theoretical approach presented here will help him build such a model for himself.

Notes and References

1. Tolman, E. C. Cognitive maps in rats and men. Psychological Review, 1948, 55, 189-208.

2. Lee, T. R. Do we need a theory. In D. V. Canter (Ed.) Architectural psychology. Proceedings of the conference at Dalandhui, University of Strathclyde, 1969.

3. Downs, R. M. and Stea, D. (Eds.) Cognitive mapping: Images of spatial environments. Chicago: Aldine. 1972 (in press)

4. Kaplan, S. The challenge of environmental psychology: A proposal for a new functionalism. American Psychologist, 1972, 27, 140-143. (a)

5. Kaplan, S. Cognitive maps in perception and thought. In R. M. Downs and D. Stea (Eds.) Cognitive mapping: Images of spatial environments. Chicago: Aldine. 1972 (in press) (b)

6. Kaplan, S. Knowing Man: Towards a humane environment. et al., 1972, in press. (c)

7. Hebb, D. O. A textbook of psychology (2nd edition). Philadelphia: Saunders. 1966.

8. White, R. W. Motivation reconsidered: The concept of competence. Psychological Review, 1959, 66, 297-333.

9. Lynch, K. The image of the city. Cambridge: MIT Press. 1960.

10. Appleyard, D. Styles and methods of structuring a city. Environment and Behavior, 1970, 2, 100-116.

11. Kaplan, S. and Wendt, J. S. Preference and the visual environment: Complexity and some alternatives. In W. J. Mitchell (Ed.) Environmental design: Research and practice. Proceedings of the Environmental Design Research Association Conference Three, Los Angeles, 1972.

12. Kaplan, R. Predictors of environmental preference: Designers and "Clients." In W. F. E. Preiser (Ed.) Environmental Design Research, (EDRA 4). 1973.

13. deCharms, R. Personal causation: The internal affective determinants of behavior. New York: Academic. 1968.

14. Rapaport, A. and Kantor, R. E. Complexity and ambiguity in environmental design. Journal of American Institute of Planners, 1967, 33, 210-221.

15. Wohlwill, J. F. The emerging discipline of environmental psychology. American Psychologist, 1970, 25, 303-312.

16. Craik, K. H. Environmental psychology. In New direction in psychology 4. New York: Holt. 1970.

This work was supported in part by the Forest Service, USDA.

CONSTRUING THE PHYSICAL ENVIRONMENT: DIFFERENCES BETWEEN
ENVIRONMENTAL PROFESSIONALS AND LAY PERSONS

H. Stephen Leff

Laboratory of Social Psychiatry
Harvard Medical School

Paul S. Deutsch

Harvard Graduate School of Design

Abstract
A pilot study using a modification of Kelly's Rep Test that asked subjects to apply a set of their own constructs to a set of environments known to them investigated how graduate students in architecture, urban studies or planning and graduate students in other fields construed physical environments. Significant organizational and content differences were found between the environmental verbal construct systems of the two groups and between the results of this study and semantic differential studies. Methodological and theoretical issues and practical implications for lay-professional interactions are discussed.

Introduction
Environments designed by architects, planners and designers shape the lives of those who use them. For this reason it is important that they be designed to meet the needs of their users. The design of environments that better fit the needs of their users requires active collaboration between environmental professionals and lay persons. However, if, as Alexander (1) has suggested, professionals are socialized by their educations to think about environments quite differently than non-professionals, then this cognitive gap could be a major impediment to such collaboration. For cognitive differences translate readily into interactional problems when groups or individuals who think differently about a problem are called upon to work together to solve it.

Nevertheless, such problems may be avoided or reduced if relevant cognitive differences are understood and anticipated. The pilot research reported below was conducted to generate some tentative information about differences between the environmental cognitions of environmental professionals and lay persons that would be useful at a practical level in facilitating interactions between the two groups and at a theoretical one in increasing our understanding of how persons cognize or think about environments.

Few studies similar in intention have been carried out to date. One notable study was conducted by Hershberger (2), who used the semantic differential method to compare the environmental cognitions of students in architecture and other fields. Although this study showed some differences between the two groups, the nature of these differences is clouded by the facts that the semantic differential method employed required all persons to use the same set of scales provided by the experimenter in rating environments and pooled individuals' data for the purpose of analysis. This method leaves open the possibilities that forcing persons to use the same scales masked important inter-individual differences and that associations between variables found at the aggregate level did not necessarily reflect relationships existing at the individual level (3).

To address the above problems a "grid method" (4) was employed in the present study which asked individuals to use their own verbal categories to characterize environments and allowed for individual by individual analyses. Grid methods are adaptations of George Kelly's (5) Role Construct Reportory Test and derive their theoretical underpinnings from Kelly's Personal Construct Theory.

A grid method was deemed appropriate for use in this study because it was reasoned that people know environments in the same way that Kelly theorized they know any other object domain (6,7). More specifically, it was postulated that: 1) individuals know environments in terms of verbal constructs which they create out of their own experiences and attribute to environments (The term verbal constructs is used here because, although theoretically constructs can be verbal or non-verbal entities, those elicited in this study were ones that persons could verbally express); 2) persons' repertoires of environmental verbal constructs can be conceptualized as verbal construct systems having various organizational and content properties; 3) differences between the environmental cognitions of different groups will manifest themselves, to the degree that they exist, as differences in the properties of their environmental verbal construct systems when these are elicited by grid methods.

Concretely the grid method employed in this study asked persons to generate for a set of environments as many of their own constructs as they could, couched in their own language, and to indicate whether each construct was true or not true of each environment. The verbal construct system organizational properties investigated were derived from the numbers of constructs given by subjects and the correlations between them. The content properties were determined by applying content codes to the constructs themselves.

Table 1 summarizes the specific properties in terms of which the environmental cognitions of the groups were compared. In addition, this table indicates the basis for the assessment of each property and the experiential implication postulated for each. For example, Row 1 of Table 1 shows that the most basic organizational property investigated was Differentiation, or the numbers of constructs employed by persons. This row also indicates that this property was postulated to be an index of the number of concrete things about environments of concern to persons. The remainder of this table should be read in a similar fashion.

Only Rows 4 and 6 of Table 1 require elaboration. The "dimensions of meaning" referred to in Row 4 were isolated by means of factor analysis. A factor analysis may be thought of in non-statistical terms as a cluster of constructs that can be shown to be interrelated. The dimension of meaning represented by a factor is that theme or referrent that the interrelated constructs seem to share. Row 6

TABLE 1

Summary of Verbal Construct System Organizational and Content Properties Investigated, Bases for Assessment and Experiential Implications Postulated

Organizational Property	Assessed on the Basis of:	Experiential Implications
1. Degree of Differentiation	Number of constructs	Number of concrete things of concern about environments
2. Degree of Discrimination	Degree to which constructs applied differently to different environments	Sensitivity to environmental differences
3. Degree of Relatedness	Extent to which constructs correlated one with another	Extent to which environmental qualities are experienced as predictable
4. Degree of Dimensionality	Number of different dimensions of meaning or factors contained in a verbal construct system	Predisposition to experience environments in terms of more abstract dimensions of meaning
5. Degree of Focalization	Degree to which organization in a verbal construct system is built around most prominent dimension of meaning or factor	Predisposition to experience environments in one-dimensional manner
Content Properties		
6. Orientation	Proportions of constructs and construct clusters referring to either physical-spatial, ethno-demographic, socio-psychological, behavioral, emotional or evaluative aspects of environments	Relative importance of various aspects of environments
7. Generality	Similarities between construct clusters found in present study and Hershberger's list of factors commonly found in semantic differential studies	Not a measure of a cognitive process, therefore no experiential implications

indicates that constructs and dimensions of meaning were classified as to their orientation on the basis of their referents. Verbal constructs and dimensions of meaning referring to the physical or spatial qualities of environments were classified as physical-spatial; ones referring to the ethnic backgrounds, racial characteristics, ages or life styles of the persons inhabiting environments as ethno-demographic; ones referring to the social or psychological functions or qualities of environments as socio-psychological; ones referring to behaviors preferred in, ocurring in or being caused by environments as behavioral; ones referring to feelings elicited by environments as emotional; and ones referring to evaluations of environments as evaluative.

Methods

Subjects (Ss) The Ss for this study were a sample of graduate students in architecture, urban design or planning (the environmental professional group) and a sample of graduate students in other fields unrelated to environmental design (the lay person group). Because the study was a pilot one only 10 Ss were employed in each group. IQ (8) and socioeconomic (9) measures administered to most but not all Ss showed that the two groups did not differ in verbal IQ, quantitative IQ or social class. By chance half of the professionals had backgrounds in or were specializing in architecture and half were not architects. While the primary comparisons made were between the professional and lay groups, significant differences between architects and non-architects will also be discussed.

The Grid Method The grid method used in this study is discussed in detail in Deutsch (10). Briefly Ss were first asked to generate a list of 20 environments known to them (a house, an airport, a suburb, etc.). Ss' verbal constructs were then elicited by asking them to form as many different groups of three environments as they could, in which two environments were alike in an important way and the third differed from the other two along the same dimension. Ss' statements as to how the two environments were alike and the third different were recorded verbatim and taken as their verbal constructs. When Ss had generated all the groups that they thought important, they were asked to rate each environment by each construct, indicating whether they would say a given construct was true of a given environment, or not true of it. Ss were also asked to rate environments in terms of three constructs supplied by the experimenter. These were "I like this place," "I would judge this place to be good," and "an easy-to-utilize place." If a S had spontaneously used one of these constructs it was not added.

Results

Organizational Properties Table 2 summarizes the findings of this study for the organizational properties investigated.

Degree of Differentiation Column 1 of Table 2 shows that Ss in the professional group bracketed those in the lay group to a significant degree. In other words, Ss in the professional group tended to use either more or fewer constructs than those in the lay group. This finding suggested that there were two types of Ss in the professional group, ones who construed environments in terms of more qualities than lay persons and ones who construed environments in terms of fewer qualities.

TABLE 2
Summary of Findings for Organizational Properties[a]

Groups	Differentiation[b]	Discrimination[c]	Relatedness[d]	Dimensionality[e]	Focalization[f]
Environmental Professional Group Median	14.5	.897	.177	4.5	37.3%
Range	9-32	.775-.929	.061-.436	3-7	27.4-57.8
Lay Group Median	13.0	.895	.108	5.0	32.8%
Range	9-17	.783-.962	.008-.333	3-7	23.1-51.8
Difference Significance	Prof. > Lay Person $p < .002$	n.s.	Prof. > Lay Person $p < .05$	n.s.	Prof. > Lay Person $p < .20$

a   The measures of central tendency presented in this table are medians. The significance levels for all but the Differentiation variable are based on two-tailed, Mann-Whitney U tests.

b   Number of constructs. The significance level for the Differentiation variable is based on the Moses Test of Extreme Reactions (11).

c   Multivariate informational analysis statistic $\hat{H}$ (average for all constructs in a person's verbal construct system). $\hat{H}$ is a measure of the extent to which responses are dispersed equally throughout a set of categories.

d   Number of significantly ($p<.05$) related construct pairs in a verbal construct system as determined by Fisher's exact test (12), divided by the number of construct pairs possible.

e   Number of factors with eigenvalues greater than 1.00 extracted by P-type factor analysis (13).

f   Percentage of variance accounted for by first factors.

An examination of the socioeconomic status and IQ scores of these two groups indicated that they did not differ with respect to these characteristics. However, it was discovered that four out of five of the professionals who had used fewer constructs than lay persons were ones who had specialized in architecture at sometime, whereas four out of five of those who had used more constructs had no special backgrounds in architecture.

The finding that professionals bracket lay persons suggests that one potential, cognitive source of interactional problems between the two groups is that they are concerned about different numbers of environmental aspects. Furthermore, the finding that architects use fewer constructs than lay persons and non-architect, environmental professionals more, suggests that when professionals with backgrounds in architecture interact with lay persons, most probably conflict will occur because the latter are not concerned about as many environmental aspects as the former; while when non-architect professionals interact with lay persons most probably conflict will occur because the latter are concerned about more environmental aspects than the former. The finding also suggests that architects will have problems interacting with environmental planners and designers.

This finding also probably reflects an interaction between the personal characteristics of people who go into the different environmental professions and the different types of socialization they receive. Thus, both the environmental design program and the planning program at the university at which our study was carried out are intended to attract students with wide-ranging interests in the environment and to educate them to approach environments from a number of perspectives. This is less true of the architecture program, although this program is far from a parochial one. It may also be that persons attracted to architecture are less verbally oriented to environments than are those specializing in environmental design or planning.

This discussion points to a consideration not made explicit earlier. In the introduction we indicate that we are investigating the effects of professional socialization. Actually, our findings almost certainly also reflect differences in personal characteristics that result in a person choosing one profession rather than another. To sort out these effects, further studies like Hershberger's (14) are necessary that longitudinally or cross-sectionally explore the effects of environmental planning educations.

<u>Degree of Discrimination</u> Column 2 of Table 2 shows that professionals and lay persons did not differ with respect to the degree to which they used constructs to discriminate between environments. The median $\hat{H}$ values for each group further suggest that both used constructs in a fairly discriminating way.

<u>Degree of Relatedness</u> Column 3 of Table 2 shows that the verbal construct systems of $\underline{S}s$ in the professional group contained proportionately more pairs of significantly related constructs than did those of $\underline{S}s$ in the lay group and that this difference was significant at the .05 level. These data suggest that professionals experience the qualities of environments as being more predictable than do lay persons. Such a finding, if replicable, has two important implications. The first is that, since unpredictability generates anxiety, lay persons are apt to be

more anxious about environments than professionals. If professionals do not deal with this anxiety, interactional problems may arise because the lay persons feel that the professionals are not sufficiently concerned about their apprehensions. The second implication is that if professionals assume that lay persons see as many relationships between environmental qualities as they do, they may assume that they are communicating more by implication when they interact with lay persons than they actually are, a situation which raises the probability of mutual misunderstanding.

Degree of Dimensionality   Column 4 of Table 2 shows that the professional group did not differ from the lay one in the number of factors or dimensions of meaning their verbal construct systems contained. This was true both for absolute numbers of factors and for the ratios of factors to constructs.

Within the professional group the verbal construct systems of those without backgrounds in architecture tended to contain more dimensions than those of persons more identified with architecture. However, when ratios of dimensions to constructs were analyzed this difference disappeared. This leaves open two possibilities. The first is that persons who used more constructs also organized them into more dimensions. The second is that the two groups did not differ with respect to real dimensions since the probability of finding chance dimensions increases with the number of constructs, if the number of environments is held constant. The resolution of this issue will require additional studies of the two professional groups using more sophisticated statistical techniques.

In any case, neither of the sub-groups, taken separately, differed significantly from the lay group in numbers of dimensions of meaning or in ratios of dimensions to constructs.

Degree of Focalization   Column 5 of Table 2 shows that the first factors of $\underline{Ss}$ in the professional group accounted for more of the variance in their verbal construct systems than did those of $\underline{Ss}$ in the lay group. These data suggest that professionals tend to experience environments more in terms of one particularly salient complex or abstract dimension of meaning than do lay persons. This finding is consistent with Alexander's (15) view that professionalization increases the tendency to employ fewer, more abstract conceptualizations in dealing with environments. Viewed in terms of its implications for inter-group interactions, it suggests that interactional problems may occur because lay persons may be more concerned about more different dimensions of environments than are professionals and may feel that professionals are too "one-dimensional" in their appraoches to design problems.

Content Analysis

Orientation   Table 3 summarizes the results of classifying constructs according to their referents. The percentages on the basis of which medians were calculated were computed by dividing the total number of constructs given by a $\underline{S}$ into the number of constructs wholly or partly classified as having a given type of referent. The reason that some constructs "partly" fell into categories is that a substantial minority of constructs were coded as having two or more types of referents. Consequently the percentages on the basis of which Table 3 was

TABLE 3
Median Percentages of Constructs Wholly or Partially Falling into the Various Referent Categories

Group	Physical-Spatial	Ethno-Demographic	Socio-Psychological	Behavioral	Emotional	Evaluative
All Subjects Combined						
Median	41.5%	4.5%	9%	36%	11.5%	5.5%
Range	19%–75%	0%–24%	0%–33%	0%–50%	0%–44%	0%–33%
Environmental Professional Group Median	44%	0%	12%	36%	12.5%	6.5%
Range	19%–57%	0%–11%	4%–33%	17%–78%	0%–44%	0%–33%
Lay Group Median	36.5%	8%	7%	36%	11%	3%
Range	20%–75%	0%–24%	0%–25%	0%–50%	0%–38%	0%–20%
Difference		LayPersons > Prof.				
Significance		p < .02				

constructed usually summed to more than 100 percent, a fact which is reflected in the table. In considering the data in this table it should be borne in mind that the method used for classifying constructs was an impressionistic one that did not include procedures for assessing the reliability of the procedure. Consequently the data presented should be employed for their heuristic value for further research rather than taken as definitive findings.

Looking first at Row 1, Table 3 shows that the constructs employed by $\underline{Ss}$ in this study most often wholly or partly referred to physical or spatial aspects of environments. The category into which constructs fell with the next greatest frequency was the behavioral one. The median percentage for the remaining categories, as Row 1 shows, are substantially lower than those for either the physical-spatial category or the behavioral one, the least used category being the ethno-demographic one.

Turning now to the comparison of the professional and lay groups, Table 3 shows only one significant inter-group difference: lay persons were more likely to use constructs wholly or partly classified as ethno-demographic than were professionals. What is particularly striking about this difference is the fact that 7 out of the 10 professionals used no construct in this category. This finding suggests that the ethnic, socioeconomic, generational and life style characteristics of the persons who populate environments are more likely to be construed as important aspects of those environments by lay persons than by professionals. This cognitive difference can cause a wide variety of interactional problems as Gans (16) and others have pointed out, since it brings professionals and lay persons into conflict about basic questions such as "what is important about environments," and "when should environments be preserved or destroyed."

Dimensions of meaning as well as individual constructs were also categorized according to their referents. Table 4 summarizes these results. The numbers in this table refer to the percentages of all dimensions employed by the lay and professional groups falling into each of the categories. Two facts should be borne in mind in considering these data. The first is that every factor was given a name that appeared to reflect the dimension of meaning shared by its constituent constructs and then categorized according to its apparent referent. The second is that we did not develop a rigorous procedure for naming or categorizing factors. Therefore, we cannot present reliability data for our codings. Since the problems of naming and categorizing factors are extremely difficult ones, we want to emphasize again that these data should be treated with extreme caution and be taken only as heuristic aides, not definitive findings.

Table 4 shows that when the two samples are combined the most commonly occurring dimensions of meaning are those that refer to physical-spatial aspects of environments (33%). These are followed in frequency of occurrence by dimensions referring to environmental evaluations (19%); behavioral aspects of environments (16%); socio-psychological aspects of environments (16%); emotional aspects of environments (12%); and ethno-demographic aspects (6%).

Table 4 also shows that the most commonly occurring specific dimensions of meaning for the combined samples are: Naturalism (10%); Responsiveness (9%); Utility (10%);

TABLE 4
Summary of Percentages of Lay Persons (LP) and Environmental Professionals (EP) Dimensions of Meaning Falling into the Various Categories of Dimensions

Dimension Categories and Constituent Dimensions of Meaning	% of LP	% of EP	LP + EP	Dimension Categories and Constituent Dimensions of Meaning	% of LP	% of EP	LP + EP
A. Physical-Spatial Category				D. Socio-Psychological Category			
1. Accessibility	2	0	1	16. Responsiveness	4	15	9
2. Age	6	6	6	17. Institutional	2	0	1
3. Complexity	0	2	1	18. Privateness	6	6	6
4. Height	2	0	1	$\mathcal{E}$:	12	21	16
5. Neatness	2	0	1	E. Emotional Category			
6. Openness (vs. Closedness)	5	4	4	19. Confusingness	0	2	1
7. Ordinariness	2	0	1	20. Fulfillingness	0	2	1
8. Organization	0	2	1	21. Likeableness	8	6	7
9. Naturalism	10	10	10	22. Friendliness	0	4	2
10. Urbanism	10	5	7	23. Troublesomeness	0	2	1
$\mathcal{E}$:	40	29	33	$\mathcal{E}$:	8	16	12
B. Behavioral Category				F. Evaluative Category			
11. Rate of Activity	4	11	7	24. Goodness (vs. Badness)	10	4	7
12. Type of Activity				25. Interestingness	0	4	2
12.1 Transportation	2	2	2	26. Utility	10	10	10
12.2 Commercialism	10	0	5	$\mathcal{E}$:	20	18	19
12.3 Possibility of Self-Expression	0	4	2	G. Not Categorizable Dimensions	2	4	3
$\mathcal{E}$:	16	17	16				
C. Ethno-Demographic Category							
13. Upper Class	6	0	3				
14. Working Class	2	0	1				
15. Youth	4	0	2				
$\mathcal{E}$:	12	0	6				

Urbanism (7%); and Rate of Activity (7%).

With respect to inter-group differences, Table 4 shows that at the category level verbal construct systems of lay persons tended to contain more dimensions of meaning referring to physical-spatial aspects of environments than did those of professionals and were the only ones containing ethno-demographic dimensions. On the other hand, those of professionals tended to contain more dimensions referring to socio-psychological aspects of environments and more emotional dimensions than did those of lay persons.

These findings are further detailed by looking at the data for individual dimensions of meaning. At this level of analysis Table 4 shows that the difference between the two groups with respect to the physical-spatial category was the product of small differences for almost all of the individual dimensions of meaning in this classification. Table 4 also indicates that similar situations obtained in the case of the ethno-demographic category and in the case of the emotional category, where the inter-group difference found reflected small differences for all but the "likeableness" dimension.

Within the remaining categories of dimensions of meaning, however, inter-group differences were due to large differences with respect to particular dimensions, as Table 4 shows. First, within the group of dimensions categorized as behavioral, the rate of activity dimension appeared in the verbal construct systems of more professionals than lay persons and the commercialism dimension in the verbal construct systems of fewer. Second, within the group of dimensions categorized as socio-psychological, the responsive dimension was the only one that appeared much more frequently for professionals than for lay persons. Finally, within the group of dimensions categorized as evaluative the "goodness" dimension appeared more frequently for lay persons than professionals.

Our findings with respect to categories of factors can be very tentatively interpreted as reflecting the fact that professionals are trained to go beyond the more perceptable qualities of environments (e.g., physical-spatial, and ethno-demographic ones) to which the layman is attuned and to organize their thinking about environments in terms of more abstract or intangible categories of dimensions (e.g., socio-psychological and emotional ones). And since the dimensions of rate of activity and responsiveness can be construed as being more abstract than commercialism, a similar interpretation is possible for our findings for these individual dimensions of meaning. The finding with respect to the goodness dimension, however, does not fit the anticipated pattern. On the one hand, it is a more abstract dimension, although it is found more frequently for lay persons than for professionals. On the other, it is commonly thought that laymen are more certain than experts about what they "like" in a given domain and less certain about what is "good" or "bad". In our study lay persons and professionals did not differ in their tendencies to "like" environments, but lay persons tended to view them as "good" or "bad" more frequently than professionals. Perhaps the explanation lies in the professionals' socialization to avoid "blatant value-judgments."

In any case, it is obvious that individuals with different ideas about what specific environmental dimensions and categories of dimensions are important will have difficulties understanding and accommodating to each other's perceived self-interests and priorities unless these differences can be made explicit so that they can be discussed.

Generality An attempt was made to match the dimensions of meaning found in this study with the list compiled by Hershberger (17) of 20 factors frequently found in semantic differential studies of environmental meaning. Once again our method was impressionistic and without a known degree of reliability. Hence, again, our results should be taken as stimuli for future research rather than findings in and of themselves.

Our attempt at matching showed that nine of Hershberger's 20 factors corresponded to factors found in our study. Our factors and the corresponding factors from Hershberger's list are presented below. There were no differences between lay persons and professionals with respect to use of factors corresponding to factors in Hershberger's list.

Hershberger	Leff-Deutsch	% All Factors	Hershberger	Leff-Deutsch	% All Factors
Time	Age	6	Rigidity	Responsiveness	10
Size	Height	1	Privacy	Privacy	6
Neatness	Neatness	1	Friendliness	Friendliness	2
Space	Openness	4	Utility	Utility	10
Organization	Organization	1			

These findings indicate that although there is overlap in the findings of semantic differential and grid methods, each method produces numerous results not produced by the other. Although this does not invalidate either method for certain purposes it certainly casts doubt on the phenomenological validity of the semantic differential if one grants that grid methods elicit data that are closer to individuals' "natural" cognitions.

Summary
The major findings of this study were: 1) the environmental verbal construct systems of architects were less differentiated than those of either lay persons or non-architect professionals, while those of non-architect professionals were more differentiated than those of lay persons; 2) the environmental verbal construct systems of professionals were more organized and more focused than those of lay persons; 3) lay persons used more verbal constructs referring to the types of persons peopling environments than professionals; 4) results of this study differed to a marked degree from those of semantic differential studies.

These findings suggest that grid methods have a unique contribution to make in investigations of environmental cognition. They also point to cognitive differences that might make it difficult for environmental professionals and lay persons to understand and accommodate to each other's perceived self-interests. More specifically, they suggest: that environmental professionals might be concerned about fewer or more aspects of environments than lay persons; that lay persons might experience environments as being more unpredictable and therefore more anxiety producing than professionals; that professionals might believe they are communicating more to lay persons by implication than they actually are; that lay persons might feel that professionals are too one-dimensional in their approach to environmental problems; and that lay persons might feel that professionals do not think about environments enough in terms of the people who inhabit them.

Because of this study's small sample size, the unknown reliability of its coding methods and the unrefined factor analytic procedures used, its findings should be used heuristically rather than taken as definitive. Future research suggested by this study would be: exploring and refining methods for eliciting grid data and content and factor analyzing it; testing the relationships hypothesized between intergroup cognitive differences and interactive behavior; and investigating the utility of grid methods as tools for diagnosing areas of potential conflict prior to actual lay-professional collaboration.

NOTES

1. Alexander, C., Notes on the Synthesis of Form, Cambridge, Harvard University Press, 1964.

2. Hershberger, R., "A Study of Meaning in Architecture," in Sanoff, H., ed., EDRA 1, 1969.

3. Robinson, A., "Ecological Correlations and the Behavior of Individuals," American Sociological Review, 1950, 15:251.

4. Bannister, D., and Mair, J., The Evaluation of Personal Constructs, London, Academic Press, 1968.

5. Kelly, G., The Psychology of Personal Constructs, New York, W.W. Norton and Co., 1955.

6. Honikman, B., "An Investigation of the Relationship Between the Construing of the Environment and its Physical Form," in Mitchell, J., ed., Environmental Design: Research and Practice, University of California at Los Angeles, 1972.

7. Harrison J., and Sarre, P., "Personal Construct Theory in the Measurement of Environmental Images: Problems and Methods," Environment and Behavior, 1971, 3:4.

8. Shipley, W., "A Self-Administering Scale for Measuring Intellectual Impairment and Deterioration," The Journal of Psychology, 1940, 9:371.

9. Hollingshead, A., The Two Factor Index of Social Position, unpublished manuscript, New Haven, 1965.

10. Deutsch, P. S., Meaning in the Environment as Construed by Environmental Designers and Laymen, Senior Thesis, Massachusetts Institute of Technology, 1972.

11. Siegel, S., Nonparametric Statistics for the Behavioral Sciences, New York, McGraw-Hill, 1956.

12. Siegel, S., Nonparametric Statistics for the Behavioral Sciences, New York, McGraw-Hill, 1956.

13. Cattell, R. B., "The Three Basic Factor-analytic Research Designs - Their Interrelations and Derivatives," Psychological Bulletin, 1952, 49:449.

14. Hershberger, R., "A Study of Meaning in Architecture," in Sanoff, H., ed., EDRA 1, 1969.

15. Alexander, C., Notes on the Synthesis of Form, Cambridge, Harvard University Press, 1964.

16. Gans, Herbert J., People and Plans, New York, Basic Books, 1968.

17. Hershberger, R., "Toward a Set of Semantic Scales to Measure Meaning of Architectural Environments," in Mitchell, J., ed., Environmental Design: Research and Practice, University of California at Los Angeles, 1972.

SIX QUANTITATIVE TECHNIQUES IN ENVIRONMENTAL ANALYSIS

Chairman: Daniel H. Carson, Architecture, Univ. of Wisconsin, Milwaukee

Panelists: John Collins, Off. of Academic Planning, Univ. of Brit. Columbia
Richard Hobson, Div. of Man-Environ. Relations, Penn State Univ.
Joseph Sgro, Psychology, V.P.I. & S.U., Blacksburg, Va.
Rachel Kaplan, Psychology, Univ. of Michigan, Ann Arbor

Authors: David Seader, "Studying Community Renewal Data Relationships with Multiple Factor Analysis
Eric Schweitzer, Gwen Bell, John Daily, "A Bi-Racial Comparison of Density Preferences in Housing in Two Cities"
Trevor Denton, J. McCollum, C. Peter Ind, Richard Stutsman, "Types of User Building Evaluation"
Francisco N. Arumi, "Demodynamics - A Statistical Theory of Demographic Equilibrium"

QUANTITATIVE TECHNIQUES IN ENVIRONMENTAL ANALYSIS: INTRODUCTION   6.0

Daniel H. Carson, Session Chairman
University of Wisconsin, Milwaukee

Early quantification is exploratory, illustrative, and often demonstrates only the obvious. Quantification in Environmental Design Research has proceeded through different stages, even though there have been a number of established fields dealing with similar problems for some time. Environmental Design Research has now caught up, and although it still borrows methods and techniques, it is no longer tentative and exploratory. The four papers in this section are representative of a wide range of quantitative research that is only marginally self-conscious. They are more than mere applications of techniques; they are more than merely illustrative.

Of the four papers, two employ factor analytic techniques, another uses cluster analysis, and the fourth is basically theoretical and approaches its problem axiomatically. This order of introduction to the papers appears most logical and serves to lead the reader through the most familiar approaches first.

In the first paper, Seader casts a number of social and physical variables into a factor analysis with the aim of eliciting a more simplified structure to enhance decisions about community renewal. In order to reach decisions about relative deterioration of community sites, a great variety of data along many dimensions is collected and analyzed. However, no combinatorial procedure enjoys wide acceptance for clarity, incisiveness and consistency. Factor analysis, it is argued, will not only aid in data reduction, but can also be a technique for concept generation that provides an hierarchy of structures to replace many diverse data. Of course, factor analysis like any other correlational technique demands that data for each variable be collected from the same set of entities, e.g., people areas, or the like. Moreover, the analysis used here demanded metric data. These demands constrained Seader to data primarily from census tracts, with all the difficulties attendant. Apart from problems of updating and sampling inherent in such sources, the useful grain or possible detail in such data is limited. Nevertheless, this rather straightforward use of factor analysis was able to separate groups of social and physical attributes of city areas which might not otherwise be expected to group together so neatly.

Schweitzer and Bell also use factor analysis in their study of attributes of density preference across sex, race and location. Theirs is a more complex analytic procedure which begins by using bipolar descriptive attribute scales and proceeds to evaluate differential populations by factor scores. Factor scores reflect an overall pattern of response for single subjects by separating out the variables according to their factor loadings and assigning these weighted values to subjects. Such descriptive information about each subject may now be analyzed by other characteristics of the group, such as sex, race and location. Schweitzer and Bell treat their new data as indicators of differential perception among the subgroups which show preferences for residential structures representing different densities. Clear differences in perceptions

of attributes of different densities emerge from the analysis. Especially
strong differences are noted for location (two different cities) and race
(black and white), but much weaker differences are noted for male-female
subgoups. These final comparisons use a univariate approach where the use of
a multivariate analysis of variance might have been more satisfactory. Even
so, this extension of factor analysis is both logical and fruitful and exposes
the potential of the technique.

In the tradition of individual differences, the study by Denton, et al, points
out that groups which are otherwise treated as homogeneous may exhibit important heterogeneous evaluations of specific environments. The implications for
design are clear. If we are to develop performance standards, criteria for
evaluation or the like, they suggest that we may have to be satisfied with
merely minimum efforts, and even these may evaporate when different subgroups
are detected. The authors use a multivariate preference technique where the
scales are converted to binary choices and used as the data for a iterative
cluster analysis. A correlational tool with a significance criterion is used
to form clusters of individuals by similarities in preferences. The new groups
are passed iteratively through this procedure until a single group of all people
is reached, when the process ends. By going back through the groups at each
iteration, the authors were able to describe different subgroups by the
clustering of their preferences. The groups so clustered yield different
evaluation profiles of users for a given environment. The grain of their scales
is a bit gross for highly discriminatory judgments, but this limitation is
not inherent in the method.

The final paper is by Arumi. It is in the tradition of Stewart's "Social Physics,"
Zipf's "Law of Least Effort," and some of the descriptive work of the geographers, notably the exponential population description of urban areas given
by Berry. It departs from the above papers in being theoretical rather than
empirical, although Arumi provides some empirical tests for the theory. His
fundamental theorem states that at equilibrium, the distribution of land over
a population will be such that the excess land (that above some minimum needed
for social functioning) will give the highest entropy, disorder or uncertainty.
In short, some individuals will occupy no extra land and some will have large
amounts of extra land and this distribution will show the largest uncertainty
or widest variance. His axioms are derived from statistical mechanics and are
applicable only to very large populations, since local fluctuations increase
the errors of prediction in small populations. However, he tests the accuracy
of predictions with large human and other animal populations and shows rather
good fits. Since the data and theoretical curves are in logarithmic coordinates,
some rather large fluctuations can be masked. Still remarkably good fits are
shown and their implications are discussed. His results suggest that we may
now be very close to the equilibrium condition, but this implication requires
time parameters which are not available in the report.

This collection of papers is encouraging in its legitimate and skillful uses
of advanced techniques and the potential of these techniques for Environmental
Design Research.

STUDYING COMMUNITY RENEWAL DATA RELATIONSHIPS WITH MULTIPLE FACTOR ANALYSIS   6.1

David Seader

Assistant Professor
Division of Urban Planning
School of Architecture
Columbia University

Abstract

The emphasis on analyzing physical conditions in the study of the need for renewal has helped to limit the scope of proposed solutions to urban blight. Data restrictions and methodology constraints have contributed to the constriction of community renewal analysis. This paper explores the expansion of renewal analysis through the use of multiple factor analysis (1). Factor analysis is a technique for assimilating diverse data by producing synthetic variables which condense large amounts of data to manageable proportions. The study presented reveals that multitudinous heterogeneous data can be managed with effective analysis, and that the inclusion of a broadened data base places physical conditions in the proper perspective for community renewal and blight analysis.

Adequate measurements of comparative deterioration and blight have been eluding universal acceptance and utilization for over twenty years. Yet these measures form key elements of community renewal and urban renewal activities. The traditional emphasis has been only on physical expressions of building and environmental conditions (2)(3). Variables such as overcrowding, building conditions, poor facilities and services, dwelling unit densities, building age, community facilities and incompatible uses dominate the analysis of deterioration. Methodologies have been limited to subjective windshield surveys and consultations with Census data, once again emphasizing physical condition of building, yard, and street.

It is a truism of planning to state that urban deterioration is not synonymous with physical blight. However, the broadening of analysis to include the complex interactions of social, economic, physical and political variables pose enormous problems. One major problem is data availability and definition--what should be measured and how? Another dilemma is methodology--how can sense be made of the data, especially within the existing resource limitations? These twin problems have stymied planners for more than a score of years, even when the needs for more refined analysis have been acknowledged.

6. QUANTITATIVE TECHNIQUES / 303

Community renewal planning demands a sensitive awareness of a city's life. It is important to know how the city works (or doesn't), looks and reacts. Development and renewal strategies are premised upon some criteria of blight or need, and those criteria, in turn, are based upon statistical aggregates exposes the "gestalt" of sub-areas of the city.

In order to characterize the sub-areas of the city, large amounts of diverse data must be assimilated, digested and analyzed to obtain a statistical measurement of the forces operating in the city. The larger the amount of data, the more clumsy and less detailed the analysis usually becomes, limited by both the analytical prejudices of the professional staff (what they consider important) and sheer proliferation of numbers.

There have been traditional ways to try to discern the differences and similarities between various sub-areas of the city. Usually some aggregate indicator is utilized, based upon a priori judgment of important variables and their relative weightings. These indicators are limited by the bounds of tradition, the sophistication of the analyst and the apparent trends of the available data. A second way of tackling the problem of characterizing sub-areas in summary form is by some form of least squares regression analysis, usually step-wise multilinear. The difficulty with regression analysis is that dependencies between data series must be established and rationalized prior to the implementation of the analysis.

A third and surprisingly simple way (with the advent of high-speed electronic computers) is some form of cluster analysis which attempts to group variables together into statistically significant bundles and then summarizes the bundles with some synthetic variable composed of combinations of the ingredient variables. Factor analysis is such a clustering technique, and will be demonstrated in this paper. Although the geographic scale of the study herein is larger than the usual community renewal planning scale, the emphasis of the paper is on methodology, not on fine grain analysis. Two conclusions of the study are clear:

1. that large amounts of hetergeneous data can be systematically assimilated for local area planning, and
2. that when widely diverse data are introduced into planning analysis, physical problems are placed in the proper context of larger, more intricate urban problems.

Factor analysis is one of an emerging set of tools that should bring sophisticated analysis to the operational level. Greater familiarity with such tools should vastly improve the quality of planning analysis.

Methodology

The method of cluster of taxonomic analysis used is factor analysis. Factor analysis, a type of multi-variant analysis, was originated to aid in the explanation of psychological tests, but has found use in other fields, especially sociology and geography. No attempt will be made here to provide the theoretical and technical framework of factor analysis, as this is available in several basic works (4)(5).

Factor analysis assumes that a large number of intercorrelated variables can more easily be explained by a small set of synthetic variables, called factors. These factors provide the "dimensions" of the variability of the raw data. Just as any point in a plane can be identified by two dimensions, the factor analysis model posits a small set of dimensions to measure any number of variables. In vector algebra terms, the space defined by n variables has the dimensionality of m factors (where m is significantly less than n). The m dimensions are orthogonal, independent directions. Factor analysis is, then, a parsimonius explanation of the variability between cases (in this example, census tracts) based upon the intercorrelations of the variable set defining each case.

Rotation of the derived factors is employed to sharpen the factors' explanation of the set of variables. Rotation is like changing the direction of a "best straight line" until the variability about that line is minimized. Factors are ranked in significance according to which the factor accounts for the common variance among the variables. The significant factors are then rotated to best fit the data.

The output of the analysis is a factor loading matrix. The factor loading measures how much each variable contributes to the explanation of a factor. Thus, the significant factor loadings show the significant component variables of the factor loading matrix, inferences can be made about the clustering of variables.

Another output is the list of "scores" of each case (census tract) for each factor. The scores indicate which cases most prominently display the dimension or grouping indicated by each factor. Thus, if a factor were an "affluence" trait, the tract with the highest score would be most affluent.

The higher the amount of tract to tract variability explained by the raw selected variables, the better the analysis. If the intercorrelations of the raw variables were low, one would have to assume that each variable was controlled by unique forces, and that no groupings were realizable. The measure of the portion of a variable's total variance attributable to the other variables in the set called the communality. The greater the communalities, the greater the interrelatedness of the set of variables. Generally, the more variables included, the higher the communalities, due to better and better explanations of the variability of data from place to place.

A factor analysis of the type outlined was performed on a set of variables for the twenty-five Census tracts of a city of 90,000 to observe the differences among sub-areas of the city.

6. QUANTITATIVE TECHNIQUES / 305

Selected Variables

A large number of variables was selected that adequately represented the characteristics of the city. Variables were chosen from as wide a range as possible to avoid missing new and interesting correlations that would imply further study. Demographic, socio-economic, structural, environmental, educational, health, crime and welfare data were all used in the factor analysis.

There were two major limitations on the number of variables. Firstly, all data had to be taken from uniform geographic areas. Certainly, the most convenient data base in the census tract. Large amounts of data are collected and stored by census tract. However desirable it may have been to analyze other, finer grained areas, it was not feasible to use any other than the tract. The available data was simply not finer grained in any quantity. The second limitation was the availability of data for the chosen geographic sub-unit.

Wherever possible, data were expressed as percentages of fractions of totals to eliminate the effects of size of tract per se. Three absolute measures of population were used, however, to relate other data series to actual size criteria. In fact, because the factor analysis is used to describe variations between census tracts, the absolute amounts and uniformity of collection date are of secondary impor tance compared to the relationships between the tracts. The suitability of the data series used should be judged, therefore, in the context of their adequacy of measuring differences between census tracts that currently obtain.

The following is a listing of the data series used in the analysis, together with comments on the reason for their inclusion:

A. Demographic Variables

    Variable 1 - Total Population
          Measure of tract size and an indicator of activity volumes.
    Variable 2 - Youth Population (21 years old and less)
          Special sub-class of population with special proglems and needs.
    Variable 3 - Population 55 Years Old and Over
          Convenient measure of elderly population, a sub-class of interest.
    Variable 4 - Percentage Decrease in Population, 1960 to 1970
          Important indicator of mobility and shifts in needs and problems.

B. Socio-economic Variables

    Variable 5 - Population per Household
    Variable 6 - Median Age
    Variable 7 - Percentage Non-white Population
           An important sub-class of population with special needs.
    Variable 8 - Median Family Income
          Measure of economic stability.

Variable 9 - Percentage White Collar Workers of Total Labor Force
        Indicator of economic health, mobility, status and opportunity.
Variable 10- Percentage Unemployed of Male Civilian Labor Force
        An indicator of economic distress, lack of opportunity and mobility.

C. <u>Property Value and Physical Condition</u>

Variable 11 - Percentage of Substandard Residential Structures
Variable 12 - Percentage of Substandard Non-residential Structures
        Closely related to the previous variable, but more related to economic and commercial activity than living conditions.
Variable 13 - Mean Age of Dwelling Units
        An indicator of housing conditions.
Variable 14 - Mean Age of All Structures
        Similar to Variable 13, but including commercial, industrial and institutional structures.
Variable 15 - Total Assessed Value Per Square Foot
        An indicator of both value and associated valuation criteria, such as neighborhood quality and status.
Variable 16 - Assessed Land Value Per Square Foot
        Similar to Variable 15, but concentrating on unimproved land values; especially important for development.

D. <u>Environmental Variables</u>

Variable 17 - Percentage of Open Land (Vacant land + Open Space)
        Measure of level of development, density and amenity.
Variable 18 - Environmental Survey Score
        An excellent summary indicator of neighborhood conditions and amenities including greenery, street condition, house paint and yards (the windshield survey).
Variable 19 - PPM Sulphur Dioxide Air Pollution
        A seldom used indicator of environmental quality with interesting correlation possibilities.
Variable 20 - $\mu g/m^3$ Suspended Particulate Air Pollution
        Similar to Variable 19.

E. <u>Education Variables</u>

Variable 21 - Dropout Rate Per 1,000 Youths
        A measure of social pathology related to education programs and expectations.
Variable 22 - Median Number of School Years Completed
        Standard Indicator of educational level of the population.

F. Social Pathology Indicators

    Variable 23 - Delinquency Rate Per 1,000 Youths
    Variable 24 - Illegitimate Birth Rate Per 1,000 Youths
    Variable 25 - Percentage of Total Population on Welfare
    Variable 26 - Percentage of Youths on Welfare

The preceding variables cover a wide range of planning and development areas of concern. Unfortunately, some areas were not covered due to lack of data, such as densities, health and commercial activity, although weak sur rogate measures are buried in the data series used.

Analysis Results

The variables were coded and combined with a computer program (Bio-Medical Data program BMDX72 produced by UCLA) to complete a factor analysis. The results were groupings of variables into factors. Each factor is derived in order of statistical explanation of variability. Thus, the first factor, on General Factor is the most significant collection of variables. The other four factors follow in descending order of significance. The importance of the result is both in the order of factors and the groupings of variables within each factor.

The following is a list of the factors, the significant component variables of each in order of importance, and appropriate explanatory comments.

Factor 1: (The General Factor)

    Variable 21 - Dropout Rate
    Variable 24 - Illegitimate Birth Rate
    Variable 25 - Percent on Welfare
    Variable 7 - Percent Non-white Population
    Variable 8 - (negative) Median Family Income (this means the lower the income,
                   the more significant the contribution to the Factor)
    Variable 26 - Percent Children on Welfare
    Variable 10 - Percent Male Unemployment
    Variable 23 - Delinquency Rate
    Variable 22 - (negative) Median School Years Completed

The General Factor, Factor 1, which explains the bulk of the variability between census tract might aptly be called "Social Deterioration". While not surprising in its composition, it is remarkable how clearly and closely related all of the social pathology indicators are, even to the extent of years in school and family income.

Factor 2:

    Variable 20 - (negative) Suspended Particle Air Pollution
    Variable 5 - (negative) Population per Household
    Variable 12 - Percent Substandard Non-residential Structures
    Variable 6 - Median Age
    Variable 3 - Population 55 Years and Older
    Variable 14 - Mean Age of All Structures

This Factor shows a curious combination of variables on the surface but, except for variable 20, is readily understandable. Variables 5 (negative), 6 and 3 describe the elderly population. They are the tracts with older people living alone or couples certainly without children. Variables 12 and 14 combine to show a degree of physical deterioration. This means that the elderly tend to live in the more rundown sections of the city. Strangely, the variables related to dwelling units do not appear, meaning that the elderly do not generally occupy the worst housing. Yet they live in the older part of town. Census tracts scoring high on this Factor probably are the older commercial centers, possibly with residence hotels, or older residential areas that have been abondoned by commercial and industrial activities due to shifts in the younger, more affluent and employable population. It is of some minor comfort to know that the elderly tend to live in areas of low air pollution (negative variable 20) because of their respiratory health problems. The pollution variable is also an indicator of the absence of industrial and commercial activity, adding to the conclusion that the old are left-overs of shifts in population and activity. This Factor might be called "Inadequate Elderly Accomodations".

Factor 3:

    Variable 1 - Total Population
    Variable 2 - Youth Population
    Variable 3 - Population 55 Years and Older
    Variable 5 - Population Per Household
    Variable 19 - (negative) Sulphur Dioxide Air Pollution
    Variable 8 - Median Family Income
    Variable 9 - (negative) Median Age

Factor 3 is the only Factor in which demographic variables are most significant. The presence of variables 2, 5 and 6 (negative) show that these tracts contain the younger residential and larger, younger families. Coupled with the measures of size, these variables indicate high levels of activities and high levels of demand for services. The influence of variable 8 also indicates that the higher the score of the census tract, the greater the volume of activity, particularly commercial. One would suspect that the high scoring tracts would contain residents who are aggresive and perhaps troublesome. The negative presence of sulphur dioxide air pollution hints at the fact that little industrial (especially low grade fuel burning) use is mixed in with what would probably be overwhelmingly residential areas. This Factor might be called "Size/Activity".

Factor 4:

    Variable 15 - (negative) Total Assessed Value Per Square Foot
    Variable 17 - Percent Open Land
    Variable 16 - (negative) Assessed Land Value Per Square Foot
    Variable 13 - (negative) Mean Age of Dwelling Units
    Variable 22 - Median School Years Completed
    Variable 14 - (negative) Mean Age of All Structures
    Variable 3 - (negative) Population 55 Years and Older

This is the first mention of property values in the Factor explanations, and both land and total assessments are negatively loaded (variables 15 and 16). When coupled with the positively correlated open land percentage, the inverse relationship makes sense--open land has low assessed value generally and will depress the square foot value of total assessments. However, variables 13 and 14 are also negatively correlated, meaning that the newer structures are on open, low valued tracts. This could only mean areas in transition where abandonment and demolition depress values and increase open land, yet rebuilding activities bring younger structures. The socio-economic variables associated with the Factor 3 (negative) and 22 indicate younger, better educated people are in high score tracts. This confirms the dynamic nature of the transition, because the usual resident population of low valued areas is not the young and educated, while the resident group associated with rebuilding is. This seemingly paradoxical Factor might be called "Transition/Upheaval".

Factor 5:

    Variable 19 - Sulphur Dioxide Air Pollution
    Variable 18 - (negative) Environmental Survey Score
    Variable 4 - Percent Decrease in Population
    Variable 22 - (negative) Median School Years Completed
    Variable 11 - Percent Substandard Residential Structures
    Variable 16 - Median Age

This Factor is of great importance because of the appearance of air pollution, negative environmental score and sub-standard residential structures. These three variables together (19, negative 18 and 11) define various measures of environmental deterioration. Tracts with high scores on this Factor have poor environmental quality. Interestingly, variables 6 and 22 show who inhabits those areas--the more poorly educated and the elderly (as shown also in Factor 2). The presence of variable 4 traces the dynamic of decline of environmentally deficient census tracts. This Factor could aptly be called "Environmental Deterioration".

Conclusions

The type of speculation generated in the previous section by factor analysis is intricate and requires a sensitive analyst, familiar with both the component data series and the geographic areas. This is especially true with the factors after the first few, which uncover more and more subtle relationships between data.

Factor analysis gives a wealth of clues for further study, and can statistically confirm those commonly held but unprovable notions. Factor analysis also allows for a look at the big picture. In this analysis, the Factors, in order or importance, were:

1. Social Deterioration
2. Inadequare Elderly Accomodations
3. Size/Activity
4. Environmental Deterioration

The absence of an explicit physical deterioration factor and the prime importance of the social deterioration factor lead to the convlusion that the most significant impact to be made is in the realm of improved social services, not physical improvements. Where physical improvements are most important, they must impact on Factor 2, that is, facilities (specifically housing) for the elderly. If these two priorities are assigned and implemented, significant reductions in the variability between tracts would accrue. In simpler terms, real equality of opportunity and quality of life would result.

Secondary priorities from Factor 3 and Factor 5 would be improvements in general services that depend on population size and activity (such as schools, fire protection, shopping, jobs), environemntal improvements to public places and rights of way, aid to residents to improve their properties and rehabilitation of structures.

These conclusions are based solely upon the present analysis. This demonstrates the power of the factor analysis tool, providing inferences for priorities, impact, improvement policies and the like, all from statistical analysis. While it would be foolish to allow factor analysis to dictate final conclusions and study outputs, it is equally indefensible to ignore the powerful planning assistance it provides.

Notice that the conclusions emphasize non-physical aspects of deterioration and blight. These results would certainly not have been possible if the analysis h ad been limited *a priori* to physical blight variables. Analysis methodologies which severly limit the scope of consideration do nothing to pinpoint solutions to urban problems. In fact, the avoidance of analyzing whole urban systems can lead to the intensification of some problems by the overemphasis on others. The history of large-scale public housing projects as solutions to slums is ample testimony for this conclusion. Narrow analyses, such as those which dwell on physical improvements, do a disservice to those who look to rational analysis to overcome urban ills.

Certainly, with the availability of broader analysis tools such as factor analysis, and the development of simulation modeling, standard analysis techniques must also be widened. The measurement of blight or deterioriation can and must be broadened to include a variety of indicators beyond those limited to physical measures. An analysis like the foregoing demonstrates the limits of searching for physical solutions to complex urban problems. The way exists to use more refined analysis as standard operating procedure. Where there is a way, there should be a will.

Notes and References

1. The study in this paper was commissioned by Schwartz, Fichtner, Bick in conjunction with the Niagara Falls Community Renewal Program to go beyond the traditional statistical analyses used for redevelopment purposes. The data used in this paper were provided by the City of Niagara Falls. The Author gratefully ackowledges the substantial support of the Niagara Falls Planning and Redevelopment Department, especially Robin Stein and the Director, Harvey Albond, as well as the guidance of Harvey Schwartz of Schwartz, Fichtner, Bick.

2. Federal Housing Administration, A Handbook On Urban Development For Cities In the United States, Washington, D.C., 1941.

3. U.S. Department of Housing and Urban Development, Urban Renewal Handbook, Washington, D.C., 1968.

4. Harman, Harry H., Modern Factor Analysis, University of Chicago Press, Chicago, 1960.

5. Thurstone, L.L., Multiple Factor Analysis, University of Chicago Press, 1947.

A BI-RACIAL COMPARISON OF DENSITY
PREFERENCES IN HOUSING IN TWO CITIES

Eric Schweitzer

City Planning Commission
Richmond, Virginia

Gwen Bell

Department of Urban Affairs
University of Pittsburgh

John Daily

Political Science Department
Georgia Southern College

Abstract

This research is designed to develop a methodology that can be used to determine the preferences of a population or subgroup within a population toward the attributes of different residential densities. The subgroups were selected on the bases of the stable characteristics of race, sex, and geographical location. Respondents reacted to groups of black and white photographs and recorded their reactions on a questionnaire presenting scaled values of various attributes of each housing density. The data was analyzed through the use of factor analysis and significance tests. This gave the underlying dimensions present in the respondent's perception of different density environments depicted and the significant differences in perceptions for the different subgroups of the population itself.

Introduction

The respondent's perception of what he sees becomes important when photographs are used as a stimulus. The photographs provide a controlable number of specific items appropriate to the density environment depicted. Some of these are then used by the respondents as cues in evaluating the environment. In a simplified example, the person interested in neighborhood safety would search for cues indicating the characteristics of the neighborhood which seem to him to be predictive of a safe place to live. While other depicted items in the environment are excluded very often a person is concerned with more than one cue-category at a time e.g. safety, privacy and spaciousness. (1)

However, a semantic differential questionnaire allows the researcher to focus the respondent's attention to one cue-category or variable at a time. The factor analysis procedure then reduces the many cue inputs to their underlying dimensions. It is these dimensions which should and can be considered in the design process.

The significance tests allow the researcher to determine the differences between various subgroups within the population. These subgroups can be selected based on the stable demographic characteristics of the population, or any other identifiable variable.

This research process should give designers and planners a tool through which they can measure the environmental disposition of different populations. Environmental disposition has been defined as "information about a person's orientation toward and attitudes about the physical environment." (2)

The link between environmental disposition and behavior was provided by Asch when he argued:
> We act and choose on the bases of what we see, feel, and believe; meanings and values are part and parcel of our actions. When we are mistaken about things, we act in terms of our erroneous motives, not in terms of things as they are. (3)

Therefore, if the environmental disposition of a clientele group is understood by the designer/planner, the housing preferences of that group can be better understood. This allows the designer or planner to create a more livable environment in terms of clientele's preferences.

The Research Instrument

Photographs: Past research has used photography as a tool to evaluate residential neighborhoods and landscape preferences. (4) These studies usually used the method of ranking or rating a series of photographs. In this research no single photograph was considered to be representative, thus the stimulus used was a group or representative sample of six black and white photographs for each housing type. This minimizes photographic bias and eliminates the color preference variable.

The final group of photographs were selected from over 700 photographs of various types of dwelling units. It was considered important that the respondent be unable to recognize any dwelling unit as a local entity. Thus, all photographs depicted housing outside of the study areas. The original 700 photographs were reduced to 50 by eliminating those that did not meet standards of quality and content. The group of 50 were then divided according to the density type of the dwelling unit. Three groups, high density multi-family housing, medium density single-family attached housing and low density single-family detached housing resulted. These groups were then pretested, where respondents were required to rank the photographs by the order of preference. The high and low ranking photographs were then eliminated until 18 photographs - six in each density group - remained. The photographs depicted housing that is homogeneous with respect to condition, age and quality of architecture, although their designs vary from modern to traditional.

The Measuring Instrument: To measure the responses to the different stimuli, semantic differential scales were developed for selected variables describing housing attributes. The scaled variables applied to each stimulus group as follows:

attractiveness	private outdoor space
neighborhood density	architectural diversity
proximity to social activity	amount of greenery
proximity to employment	pollution level

neighborhood safety
interior spaciousness
neighborhood racial integration
convenience
location of housing type
acoustical privacy

The demographic characteristics of each respondent were also collected. These were sex, race and geographical location.

Testing was done in groups of 10 to 50, with photographs spaced so that all respondents had a clear view. The respondents were instructed to consider each group of photographs independently of the other groups and with the assumption in mind that they would be able to afford to buy or rent any of the dwelling units depicted. This eliminated the economic factor, and focused attention to the design aspects of the dwelling units.

<u>The Sample</u>: The sample consists of 701 respondents. Of these, 340 are college students from the Pittsburgh area and were interviewed during the fall of 1970. The remaining 361 are college students from the Savannah area and were interviewed during the fall of 1971. Our sample further divides into 323 men and 378 women and 325 blacks and 376 whites. The age distribution ranges from 17 to 51 years with a median of 20.

<u>Semantic Differential</u>: The measuring device used in this study is the Semantic Differential developed by Osgood to measure the common meanings of key concepts. (5) In its common form the Semantic Differential consists of a stated stimulus item (a concept, a person, an object, or a picture or set of pictures) followed by a set of bi-polar adjectives at the extreme end of a five, seven or nine point scale. The respondent is **asked** to keep the stated stimulus item in mind and then mark the point on the scale between each set of adjectives that best describes his view or feeling toward the stimulus object.

One of the more advantageous aspects of the Semantic Differential is the variety of adjective pairs and stimulus items that can be combined into a single questionnaire. By using the same set of adjective pairs to describe different stimulus objects, areas of common meaning between stimulus items can be discovered (as well as areas of disagreement). By using different adjective pairs to describe the same stimulus objects the scope of meaning can be identified. An additional strength of the Semantic Differential comes from the variety of ways that the data can be analyzed. Each stimulus item can be analyzed by finding the mean response to the item on each adjective pair, or pairs of stimulus items can be studied by comparing the mean responses on each adjective pair. Additionally, groups of respondents can be compared in terms of their mean responses on each adjective pair for each stimulus. Or, the responses to a single stimulus item can be studied by reducing the responses to a set of underlying dimensions through the use of factor analysis, or distance cluster analysis. (6) Likewise, multiple stimulus items can be compared by observing similarities in the factor structure or clusters obtained from each set of responses. It is also possible to study patterns of a single individual over multiple stimulus items. Finally, it is possible to combine some of the above methods into complex analysis. Such a combination of analysis methods is described in the following section.

6. QUANTITATIVE TECHNIQUES / 315

Research Procedures

Data analysis of the 701 responses to the 14 semantic differential items for each of the three stimulus picture sets was carried out in the following steps:

(1) Each of the 14 semantic differential items for each picture set (concept) was intercorrelated with every other item using a standard product moment correlation. This resulted in a correlation matrix of the order 14 x 14 with a total of 91 unique intercorrelations. (7)

(2) The correlation matrix obtained in step one was factor analyzed to reduce it to its basic underlying dimensions. Straight principle axis factor analysis was performed and all factors with eigenvalues greater than +1.00 were extracted and subjected to a varimax rotation. (8) Four factors were obtained for each of the three picture sets.

(3) The factors obtained in step 2 were then used to generate factor scores for each of the 701 respondents on each of the extracted factors. Factor scores consist of the weighted summated original response of each respondent. (9) The weighting coefficients are obtained according to a procedure suggested by Charles Spearman (10), where the weight of each variable on each factor loading of the variable is divided by one minus the factor loading squared. This procedure yielded weights of highest value for those variables with the highest loading of the factor, and which are therefore more characteristic of the factor. Thus, the final product, factor score for each individual on each factor, is considerably richer than any simple summated index for it is a summation of raw scores based on the weights assigned empirically to the variables involved. (11)

(4) Analysis of the now considerably reduced data proceeded by breaking down the sample into the subsamples of the three independent variables, geographical location, sex and race. The mean of each subsample was then compared with other relevant subsamples to discover significant differences. (12) Either t-tests based on pooled or separate estimates of variances were used depending on whether or not there were significant differences in the variances of the subsamples. (13)

Factor Interpretation: The factoring of the 14 semantic differential items for each of the three picture sets yielded rather constant factors. Each factoring process resulted in four factors and the similarity between factors across picture sets speaks to the consistent patterns of response to the three types of housing. There are, however, differences in mean response for each factor between picture sets. Table I shows how the respondents generally reacted to the different variables that constitute each factor. The total factor results are displayed on Tables II, III, IV.

Factor One - Environmental Space: The variables that loaded significantly on this factor deal with the physical space and environmental conditions related to each density group. For the high density housing this factor is composed of variables measuring the amount of outdoor space, amount of greenery, neighborhood density, level of pollution and the location within the city. For the low density housing, all the same variables except location, plus the amount of interior space, and the

TABLE I

Total Sample Mean Responses by Factor by Picture Group

Factor	Group I High Density	Group II Low Density	Group III Medium Density
1	low	high	medium
2	medium	medium	medium
3	low	medium	medium
4	medium	high	medium

These classifications are based on the total sample means for each of the variables defining the factor. Low indicates that sample means for the variables defining the factor fell between 1 - 3, medium between 4 - 6, and high between 7 - 9.

attractiveness and architectural diversity of the neighborhood structures define the factor. The medium density housing is defined with the same variables as the second picture set but with the addition of the safety variable and the reappearance of the location variable. Thus, the purely physical and environmental variables of the amount of outdoor space, amount of greenery, neighborhood density and pollution level remain constant in defining the factors across density levels, but, for the low density housing the more structure-specific variables of interior space, attractiveness and architectural diversity within the neighborhood become part of the physical evaluative dimension. At the same time the location factor present in the high density housing joins the more socially oriented factor four for the low density environment. Finally, for the medium density housing both the purely physical environmental attributes and the more structure-specific variables define the factor but with the additional variables of safety and location lose their social meaning and become more physical in meaning.

Factor Two - Personal Activity - Convenience: This is the most constant of all factors. The same three variables define this factor regardless of the stimulus pictures. The factor is defined by proximity to employment, proximity to social activity and general overall convenience.

Factor Three - Acoustical Privacy: This third factor is defined by the single variable acoustical privacy. Only in the case of high density housing do other variables join with acoustical privacy to define the factor: these are the structure-specific variables, interior space and architectural diversity. Thus, as we move from the high density housing to the low and medium density units the variables interior space and architectural diversity seem to lose the more personal meaning they have in the high density housing and become part of the environmental space dimension.

Factor Four - Integration - Safety: The variable that seems to define this factor best is racial integration. Additionally in two of the three picture sets the integration variable is joined with other variables. For the high density

TABLE II

Factor Matrix for Picture Group I - High Density Housing

| | Factor Loadings | | | |
Variable	Factor I	Factor II	Factor III	Factor IV
Outdoor Space	.697	-.001	.232	.210
Greenery	.692	.138	.044	.128
Neighborhood Density	.661	-.019	.049	-.206
Pollution	.631	-.202	.088	-.180
Location	.474	-.351	.171	-.168
Social Activity	-.130	.673	-.070	-.038
Convenience	.245	.668	-.185	.033
Employment	-.287	.601	.222	.204
Acoustical Privacy	.065	.192	-.704	-.032
Indoor Space	.249	.294	.608	-.265
Arch. Diversity	.217	-.077	.481	.074
Racial Integration	.115	.233	.108	.781
Safety	.442	.206	.171	-.576
Attractiveness	.327	.322	.339	.055
Cumulative Proportion of Total Variance	21%	35%	43%	50%

housing the integration-safety factor consists of views toward the level of integration, the amount of safety, with a negative relation between these two defining variables (indicating that respondents feel that as integration goes up safety goes down or vice versa). For low density housing the negative relationship between safety and integration is maintained together with location. Finally, for the medium density housing, only integration defines the factor, while both safety and location take on meaning within the environmental space dimension.

Results

The following paragraphs attempt to present in summary form the results of the methodology when applied to a specific sample. These results are not meant to be complete nor has the data at hand been exhausted. Rather, the discussion of results is intended to be suggestive of the kinds of results that can be obtained by the application of the methodology. The results discussed here revolve around one main set of comparisons. Ideally, the comparisons should continue, including the first and second order interactions of the three independent variables. Analysis of variance would of course be the proper statistical technique for the discovery of significant higher order interactions. Unfortunately, we did not have access to com-

TABLE III

Factor Matrix for Picture Group II - Low Density Housing

Variable	Factor Loadings			
	Factor I	Factor II	Factor III	Factor IV
Outdoor Space	.749	-.042	-.164	.130
Neighborhood Density	.663	-.053	-.041	.204
Greenery	.662	-.022	-.148	.029
Interior Space	.633	.042	.002	.177
Attractiveness	.603	.126	.160	-.055
Pollution	.459	-.035	.022	.418
Arch. Diversity	.412	.047	.342	-.001
Social Activity	-.036	.802	-.069	.002
Employment	-.088	.790	-.047	-.105
Convenience	.231	.563	.200	.102
Acoustical Privacy	-.276	-.014	.751	.051
Safety	.163	.235	.308	.649
Racial Integration	.136	.002	.490	-.585
Location	.284	.286	-.089	.553
Cumulative Proportion of Total Variance	22%	35%	43%	51%

puter routines that would do analysis of variance for our particular data set (the main problem being that most analysis of variance programs require equal cell size, which we do not have). While we have done t-test for all of the higher order interactions the variety of significant combinations make the identification of patterns nearly impossible and a presentation of the results of the larger study must await additional analysis.

The perceptions of Pittsburgh respondents are compared with Savannah respondents towards (1) high density housing, (2) low density housing, and (3) medium density housing. (14) Similar comparisons are also made for the black and white, and male and female subgroups. (Table V)

<u>Pittsburgh-Savannah Comparisons</u>: The high density stimulus picture set produced significant differences between the Pittsburgh and Savannah groups in their perceptions of each of the major dimensions. The Pittsburgh group gives significantly higher means to the variables defining the environmental space factor, although the means on these variables are generally low for all groups. Thus, in relation to the Savannah sample, Pittsburghers see the high density housing as more adequate in an environmental space context. In terms of personal activity/convenience again the cities differ: Pittsburghers see high density housing less con-

TABLE IV

Factor Matrix for Picture Group III - Medium Density Housing

Variable	Factor Loadings			
	Factor I	Factor II	Factor III	Factor IV
Outdoor Space	.727	.078	-.147	-.171
Pollution	.688	-.165	.115	.123
Neighborhood Density	.686	.114	.242	-.161
Interior Space	.644	.276	.237	-.134
Greenery	.641	-.101	-.292	-.106
Safety	.576	.230	.434	.143
Location	.518	-.187	.262	.246
Attractiveness	.505	.309	.082	.233
Arch. Diversity	.466	-.016	-.119	-.147
Employment	-.129	.836	-.078	-.090
Social Activity	-.017	.774	.061	.036
Convenience	.389	.569	.016	.193
Racial Integration	.027	.038	-.855	.090
Acoustical Privacy	-.184	.061	-.080	.878
Cumulative Proportion of Total Variance	27%	40%	49%	56%

venient than the Savannah respondents. The acoustical privacy dimension showed that the Savannah group viewed the high density units as less private than the Pittsburgh group. Finally, for the integration/safety factor, clearly different perceptions were manifested. Pittsburghers perceive the high density housing as less integrated and more safe than the Savannah residents. It would seem that the housing patterns in the two cities account for this different view of the social milieu. Pittsburgh respondents may be thinking in terms of the high rise luxury apartments that can be found in some areas of that city, while Savannah respondents may be equating high density housing with public (and thus integrated, i.e. black) housing.

For low density housing there are few marked differences between the Pittsburgh and Savannah samples. The only factor on which the groups differ significantly is the personal activity/convenience dimension. Pittsburgh respondents find single family houses less convenient than the Savannah respondents, although the entire sample viewed low density as more convenient than either high or medium density housing. (Table V)

The perception of the medium density housing resulted in significant differences between the Pittsburgh and Savannah samples on two factors. Savannah respon-

TABLE V

Significant t Scores for Each Picture Group by Factor by Stable Characteristic

Picture Group	Stable Characteristic	Factor I	Factor II	Factor III	Factor IV
I	Location	-3.31	3.08	2.09	7.02
I	Sex	--	--	-4.65	4.68
I	Race	6.50*	-4.19	--	--
II	Location	--	8.03	--	--
II	Sex	3.95*	--	--	--
II	Race	6.68*	-4.32*	3.80*	-2.28*
III	Location	2.01	--	--	4.35
III	Sex	--	--	--	--
III	Race	5.13*	-3.67	-1.97*	-4.26*

*These t scores are based on separate estimates of variance, all others are based on pooled estimates of variance.

Note: All t scores are significant to the .05 level.

dents view the environmental space of these houses in more positive terms than do the Pittsburgh respondents. However, in regard to acoustical privacy, Savannah respondents see these housing units as less private than do the Pittsburgh respondents.

Moving from high to low density housing, the degree of significant differences between responses in the two cities is reduced. This could be attributed to regional/cultural variations or to the fact that in Pittsburgh, a much larger city, there is a greater amount and variety of high density housing.

Black-White Differences: Racial differences are not as pronounced for high density housing as locational differences. In responding to photos of high density housing, the black and white populations differ on only two factors. Black respondents view high density housing more positively in terms of environmental space than do white respondents, but feel that such housing is less convenient for personal activities than do white respondents.

The low density housing stimulus pictures resulted in significant differences between racial groups on all four factors. Black respondents give higher environmental space scores, but lower personal activity/convenience scores and lower acoustical privacy scores than whites. White respondents located low density housing in environments that are less integrated and safer than do blacks. (However, the entire sample views these housing units as being segregated, safe, and suburban.)

The pattern of differences that emerged is repeated for medium density housing. Racial groups differ significantly on all four factors. Black respondents again give higher environmental space scores than white respondents. However, unlike the pattern in low density responses, whites see the medium density units as more convenient for personal activities than do blacks. Whites also see the medium density units as less private than the black respondents. Finally, whites perceive that these units are located in a more segregated area than do blacks.

Male-Female Differences: The final category for comparison is sex. This category produces fewer significant differences than either of the other categories. In responding to the high density housing photos men and women show significant difference only on factors three and four. Woman think such housing is more acoustically private than do men, and women see the high density housing as more integrated and less safe than the men. The only other significant differences between men and women occurs on the environmental space dimension for low density housing. Here women respond in a more positive direction than the men. It is curious and perhaps fortunate that men and women share common perceptions of the housing environment. Pittsburghers and Savannahians never have to live together in the same housing environment. Blacks and whites seldom live in the same housing environment and according to our results it would be most difficult to find a common design base for the development of integrated neighborhoods. Yet men and women most always live in the same housing situation and it is at least convenient for the designers that they share common perceptions.

Evaluation of the Methodology

The results show that the combination of groups and photographs and a semantic differential questionnaire can be used to present a specific environment and elicit meaningful responses to predetermined variables relating to characteristics of the environment.

The clarity of the clusters developed by the factor analysis and their consistency over different housing densities shows that the research methodology does uncover common cognitive structures of the respondents. Since the original goal was to develop a methodology for evaluating different housing environments, only a small number of variables was used. This is the reason for the factors only accounting for 50% to 56% of the total variance of each density group. An expansion of the variable list should increase the completeness of the factor analysis. The significance tests show that the structure of the data allows for determining the perceptions of one subgroup when compared with another.

Improvements: Before reliability of the methodology can be certain it should be replicated several times with varying stimulus and respondents. (15) Several other improvements that should be made include developing a more complete list of housing attribute variables, increasing the number of stable characteristics collected on the respondents to include past housing experience, past family income, ages, etc. Also obtaining the preferences of the respondents to different density housing by defining different life styles on semantic differential scales. This should include the life styles of those who are single; married with or without children; devoted to careerism, consumerism or familism; experimenting with communal living; the elderly; and the peripatetic.

Too often studies such as this continue to focus on the short stage in the life cycle during which the "ideal family unit occurs - the young couple with their perpetually school-age children." (16) This study in itself dealt with three housing densities by four factors by six population groups, and the data became very complex - the edge of the problem was merely isolated. Yet if we are to build cities, neighborhoods, and housing groups that satisfy the difersity of needs rather than homogenize society, such studies will have to be carried out to feed into policy making from the level of the nation to the design of each unit.

Notes

1. Peter Warr and Christopher Knapper, THE PERCEPTION OF PEOPLE AND EVENTS. (London; John Wiley and Sons, 1968).

2. Gary Coates, "Residential Behavior Patterns," DESIGN AND COMMUNITY - A STUDENT PUBLICATION OF THE SCHOOL OF DESIGN, ed. by David Alpaugh, Vol. 19, No. 2 (Raleigh, North Carolina; School of Design, North Carolina State University) p. 279.

3. S.E. Asch, SOCIAL PSYCHOLOGY (Englewood Cliffs, N.J.; Prentice Hall, 1952), p. 646.

4. George Peterson, "Measuring Visual Preferences of Residential Neighborhoods," EKISTICS, March 1967; Henry Sanoff, "House Form and Preferences," EDRA II, ed. by Eastman and Archea; Roger M. Downs, "The Cognitive Sturcture of an Urban Shopping Center," ENVIRONMENT AND BEHAVIOR, Vol. 2, No. 1, June 1970; Raymond M. Craun, Jr., "Visual Determinants of Preference for Dwelling Environs," EDRA I, ed. by Sanoff and Coher.

5. C. Osgood, G. Suci and P. Tannenbaum, THE MEASUREMENT OF MEANING (Urbana, Ill.; University of Illinois Press, 1957).

6. Fred N. Kerlinger, FOUNDATIONS OF BEHAVIORAL RESEARCH (New York; Holt, Rinehart and Winston, Inc., 1967), p. 573.

7. Reliability of the semantic differential scales was tested with a retest of 20% of the Pittsburgh sample five weeks after initial test administration. The test-retest correlation was .872.

8. R.J. Rummel, "Understanding Factor Analysis," JOURNAL OF CONFLICT RESOLUTION, Vol. XI, No. 4, December 1967, pages 444-480; Benjamin Fruchter, INTRODUCTION TO FACTOR ANALYSIS (Princeton, N.J.; Van Nostrand Co., Inc., 1954); Paul Horst, FACTOR ANALYSIS OF DATA MATRICES (New York; Holt, Rinehart and Winston, Inc., 1965), Harry H. Harman, MODERN FACTOR ANALYSIS, 2nd edition (Chicago; The University of Chicago Press, 1967).

9. Benjamin Fruchter and Earl Jennings, "Factor Analysis," COMPUTER APPLICATIONS IN THE BEHAVIORAL SCIENCES, ed. Harold Borko (Englewood Cliffs: Prentice Hall, Inc., 1962), p. 262.

10. Charles Spearman, THE ABILITIES OF MAN (New York; Macmillan and Co., 1927), p. xix.

11. W.J. Dixon, BMD BIOMEDICAL COMPUTER PROGRAMS (Berkeley; University of California Press, 1971).

12. W.J. Dixon, BMD BIOMEDICAL COMPUTER PROGRAMS, X-SERIES SUPPLEMENT (Berkeley; University of California Press, 1969).

13. Hubert Blalock, SOCIAL STATISTICS (New York; McGraw-Hill, 1960), pages 169-179.

14. All significances discussed in this section have a $p = .05$.

15. For a more complete analysis of the Pittsburgh results the reader is referred to: Robert J. Diaiso, David M. Friedman, Lester C. Mitchel and Eric A. Schweitzer, PERCEPTION OF THE HOUSING ENVIRONMENT: A COMPARISON OF RACIAL AND DENSITY PREFERENCES (Pittsburgh; Graduate School of Public and International Affairs, University of Pittsburgh, 1971).

16. Margaret Mead, Delos Symposium, EKISTICS, October 1972.

TYPES OF USER BUILDING EVALUATION(1)   6.3

Trevor Denton

Department of Sociology and
Urban Studies Institute
Brock University

J. McCollum

School of Architecture
Carleton University

C. Peter Ind

Director of Planning
Brock University

Richard Stutsman

St. Catharines, Ontario

Abstract

The aim of this paper is to outline an approach to user building evaluation which is grounded in the notion of diversity of user needs rather than a mythical average user. It is suggested that user evaluations may be sorted into types via cluster analysis, and that the members of these different types of evaluation profiles may seize on different features of a building in making their evaluation. Several implications for building design and user evaluation generally emerge from this approach.

Problem

Planning for diversity is much like the weather - everybody talks about it but it is hard to do much about it. Flexible environments which can meet a variety of user needs are as often an ideal as a reality, and depend more on the designer's intuition and imagination than on objective planning strategy. The result tends to be the creation of "average" (or inappropriate) buildings which cater to the needs and interests of a mythical "average" user - the average office worker, the average apartment dweller.

The aim of this paper is to outline a formal approach to user building satisfaction and evaluation; one which is grounded in the notion of diversity or types of users. Rather than treating the totality of users of a building as a group, it is suggested here that it is conceptually possible, indeed preferable, to sort user needs into different types via cluster analysis. Using data from a user evaluation of a university student residence it will be shown that different types seize on different features of a building in making their evaluation.

The results of this approach have serious implications for building design and user evaluation generally.

Case Study

Setting. To illustrate this approach let us turn to an examination of a student evaluation of the Brock University student residence.(2) Brock University is situated in St. Catharines, Ontario, a city of some 110,000 persons. The university has about 2,200 full-time students and the residence houses about 400 of them. The residence is arranged so that eight rooms (four singles, four doubles) form a "floor", sharing lounges, a washroom and a kitchenette. Access to these floors is by a stairwell. Four "floors" stacked one on top of the other constitute a "house". The residence is composed of two groups of five houses, plus supporting dining, administrative, recreational and circulation spaces.

A sample of forty students was drawn and data were collected on attitudes toward eleven varieties of persons and places judged important in a resident's life - attitudes to room, floor, house, residence, university, city, people on floor, people in house, people in residence, people in university and people in city. These attitudinal data were collected on a six point scale which was later collapsed to yield two categories, 1 being average or unfavourable and 2 being favourable.

Method. In numerical taxonomy at least three conceptual problems typically are involved - selection of a domain of characteristics on which objects (residence students here) are to be typed, selection of a suitable index of relatedness and selection of an appropriate algorithm creating groups of homogeneous membership. The domain was outlined above.(3) Because of the possibility of two-state binary characters, Sokal and Michener's similarity coefficient (number of matches divided by number of matches plus number of mismatches) was used.(4) Finally, a hierarchical clustering algorithm was selected.(5)

Results. Figure 1 displays the dendrogram showing how samples - residence students in this case - are merged into successively larger groups until the tenth cycle when all are merged into one group. The sample numbers are the identification numbers assigned to the students and were arbitrarily selected.

Figure 1 also displays the dendrogram in a form showing significant characteristics of a type at a given cycle. "Significant" characteristics means those eleven attitudinal preferences significant at the .05 level or less, using a one sample two-tailed binominal test sequentially on each attitude. For each type a row of numbers is presented from 1 to 11. These numbers stand, respectively, for attitudes to city, people on the floor, people in the house, people in residence, people in the university, people in the city, room, floor, house, residence, university. If the attitude is significantly distributed at the .05 level (two-tailed test) or less, it has a circle around it. If there is a significantly large number of 1's (unfavourable attitude) the character number is circled with an X inside the circle. If there is a significantly large number of 2's (favourable attitude) there is no X. If the character is not significantly distributed it is not circled. Thus in cycle 9 the profile ①2③ ④5⑥7 8⑨10 ⑪ indicate that members of this type have the following statistically significant attitudes: unfavourable to city, unfavourable to people in the house, unfavourable to people in the residence, unfavourable to people in the city, unfavourable to the house, and favourable to the university.

Now let us turn to the selection of types. At cycle 9 there are two types - likers and dislikers. Six people like people on the floor, people in the university, their room and the floor. Thirty-four people dislike the city, people in the house, people in the residence, people in the city and the house, but like the university. At cycle 8 this latter group separates into two types. One group of eight students dislikes the people in the house but likes people in the city. The other group dislikes the city, people in the house, people in the residence, people in the city and the house, but likes the university.

Conceptually, there is good reason to accept three types of attitude preferences one group of "likers" and two groups of dislikers - "mild dislikers" and "strong dislikers." The first group of likers (six people) like many places and people, as with people on the floor, people in the university, their room and the floor. The second group of dislikers (eight people) seem oriented toward the city, disliking people in the house and liking people in the city. The third group of strong dislikers (26 people) dislike many people and places - the city, people in the house, people in the residence, people in the city and the house. However, it is interesting to note that they like the university. At cycle seven these 26 people break down into two groups of five and 21 members. The group of five people cannot be tested at the .05 level because the numbers are not sufficient for a two-tailed binomial test. Perfect agreement among the five people yields a .062 probability. If for the moment we accept this level for the purposes of comparing the 21 and five members of the strong dislikers, the major difference is that the first likes the people on the floor while the second dislikes the people on the floor but likes the floor itself. Findings reported elsewhere(6) indicate that it is the floor which is the meaningful social unit in the residence. This evidence makes it doubtful whether a floor - people on the floor distinction is meaningful since both involve commitment to the floor. In addition, the small numbers of members in types make finer interpretations somewhat unrealistic.(7)

Discussion

Typal Approach to User Building Satisfaction. Through the application of a hierarchical taxonomic algorithm it has been demonstrated that different patterns of attitudinal preferences to people and places exist among students in the Brock University residence. In particular, three conceptually different patterns were suggested - persons tending to like things generally, persons tending to dislike the people in the house but who like the people in the city, and people disliking many things but who like the university and who seem to have a commitment to the floor.

6. QUANTITATIVE TECHNIQUES / 327

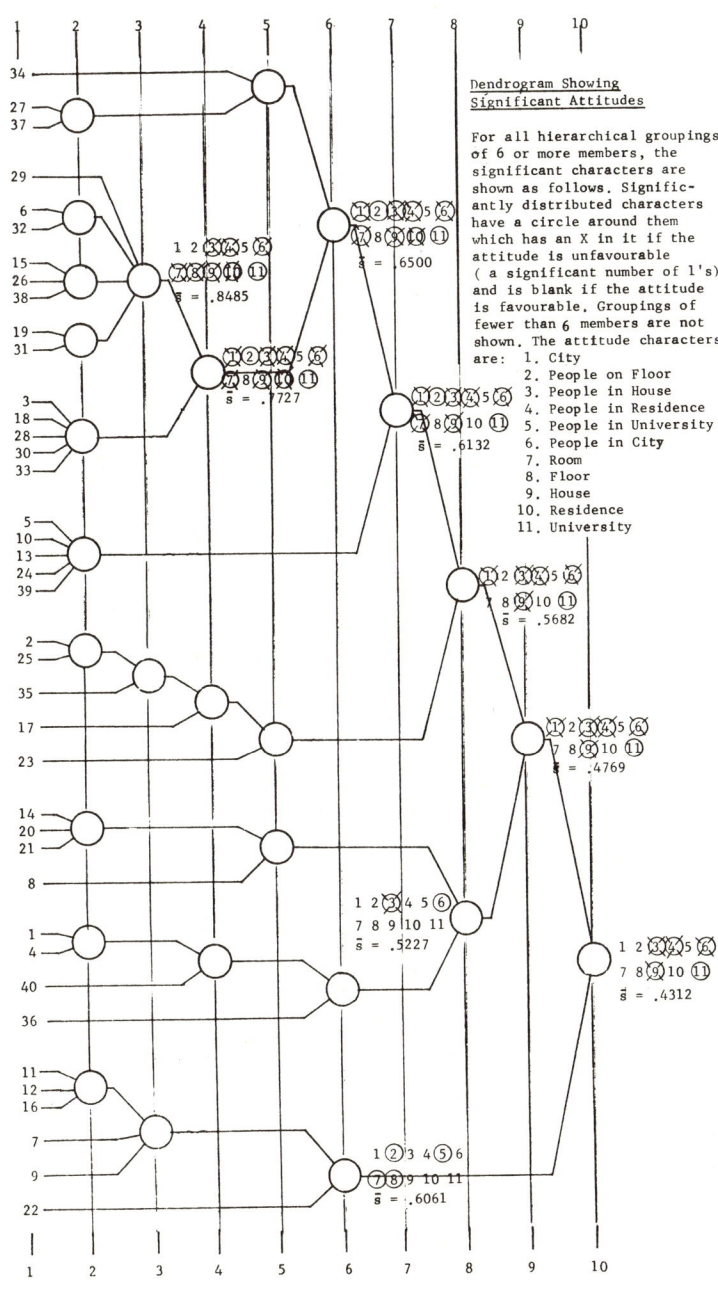

Figure 1.

Because types of users exist who have discernibly different evaluation profiles it conceptually does not make sense to visualize a mythical "average" user. Different types exist and must be recognized as such.

The selection of the domain of items on which users are to be typed is all important in yielding useful results. In the residence study outlined above attitudes to people and places were used. Other studies might wish to consider only physical environmental preferences, depending on the problem at hand.

The weighting of items on which people are to be typed may be equally important.(8) In the university residence study all items were treated as being of equal importance to users. It may be useful to weight individual items according to the importance which users attach to them.

Weighting can also enter into the creation of rational as opposed to naturally occurring types.(9) The examples used so far have been naturally occurring types in that they are the outcome only of users' characteristics. Rational types may be created when the analyst wishes to add some criterion external to the subjects. While this might be done for theoretical reasons arising out of a theory of human nature (eg. basic environmental needs), operational reasons arising out of practical necessities are likely to occur more frequently. For example, suppose it is required to type people on office building preferences on the basis of preferred lighting patterns, geographic location and office size. If the purpose is to establish user preference types which can in fact be satisfied, it would only make good sense to weight geographic location as being several times more important in the creation of types than the other two items because one type with different geographic preferences could not practically be satisfied. It might even be preferable to demand that each type created have perfect internal agreement on this item.

If the domain and weighting of items on which types are to be created have been satisfactorily (from the analyst's perspective) selected, there is little reason to be concerned with items in a type for which there is disagreement (i.e. no significant distribution). Cluster analysis can be assumed to have minimized intra-type differences and to have maximized inter-type differences.

There are other questions which bear on the selection of types. Depending on the nature of the research problem, different theoretical or operational concerns may make some similarity indices and some clustering algorithms much more appropriate than others. However, a detailed discussion of these is beyond the scope of this paper.

In all the above areas alternative methodological procedures exist. In deciding which to use one must ask which are most useful in meeting the needs of the particular design or other operational problem at hand.

<u>Social Determinants of Types</u>. A typal approach to user building satisfaction may be visualized in causal terms to discover what it is that "causes" the different types.(10) Presumably a building cannot cause types of people.

Rather than visualizing aspects of the building itself as the determinants, it makes better sense here to look for social determinants - the characteristics of the users themselves.

Let us return to the case study of university residence evaluation types. In order to discover the influence of independent variables the three evaluation profile types mentioned above were related via a two-tailed binomial test to each of a set of socio-economic variables. The aim here was to discover independent variables capable of being viewed as possible "causes" of the types. A .05 significance level was used. The results are displayed in Table 1.(11) The strong dislikers are in first year, do not own cars, and are in Arts courses. The likers are males.

TABLE 1

Type		Course Type 1 = Arts 2 = Science	Car Owner 1 = no 2 = yes	Father's Occupation 1 = blue collar or deceased 2 = white collar or professional	Sex 1 = male 2 = female	Year 1 = 1st year 2 = 2, 3 or 4	Age 1 = 20 or less 2 = more than 20	Room Type 1 = double 2 = single
	Strong Dislikers (n=26)	ⓔ21	ⓔ23	13	18	ⓔ19	16	16
	Mild Dislikers (n=8)	5	7	2	4	3	5	3
	Likers (n=6)	5	4	3	ⓔ6	3	2	5
	All Students (n=40)	ⓔ31	ⓔ34	18	ⓔ28	25	23	24

The table shows frequencies of 1's.
Frequencies significant at the .05 level are circled (two-tailed test).

The variables associated with the strong dislikers seem to be reasonably consistent with the type. The type members are first year Arts students restricted to the residence by lack of easy transportation, who are enthusiastic about life on the floor and the university, but otherwise critical of their surroundings. The other types are less "reasonably explained" by the independent variables significantly related to them. The small numbers involved in these types may have masked relationships with year, age, sex, father's occupation, room type, et cetera. In particular, it is hypothesized that a larger sample size would show that the likers or mild dislikers were upper year students who had returned to the residence even though they had made friends in the university during the previous year or years.

Practical Implications. There are clear planning implications for building design and user evaluation generally which arise from the finding that 1) user evaluations may be sorted out into different types, and 2) the members of these different types may seize on different features of a building and life inside it in making their evaluation.

Diverse patterns of user evaluation will be the norm rather than the exception, and it cannot be hoped reasonably to satisfy everybody unless these variations are taken into account.

The approach to user evaluation outlined here will be especially applicable in cases where it is deemed useful to plan to accommodate sub-groups of a population of users. If it is possible to specify in advance with reasonable accuracy who the population of users of a building (a university residence, an office building) will be, the following planning algorithm might be adopted. 1) Establish via cluster analysis types of relevant user building needs from a sample of representative users (or the population if possible). 2) Design building environments based on the satisfaction of these user profile patterns. These can be used to create environments which closely match specific types of users, or to enhance the fit in more flexible design solutions, or both where appropriate. 3) Wherever feasible, make it possible for users to be located in environments consistent with their profile patterns. This might also be achieved on the basis of user scores on independent "causal" variables where these have been adequately specified and if, for whatever reason, it is not appropriate to use profile type scores.

A fourth step might be added to this algorithm. Once users are located in environments via the above procedures, it would be helpful to re-evaluate the fit between users and environments. This would serve the purpose of validating the environments created and, if done over time, would point to any changing user need patterns.

A typal approach to user building needs has several advantages. It can provide a fit between users and their environments which, on the basis of probability, has a good chance of proving satisfactory. It can be used either in the design of new buildings or in the redesign of existing building environments. In either of these cases, however, it must be recognized that "environment" includes not only physical structures but also norms and regulations governing user behaviour which may yield design solutions which do not utilize physical structures.

From Preference to Use. While the examples given in this paper have been ones of attitude or preference, there is no reason why the algorithm can not be extended to include use. Measures of use are concerned with what people actually do as opposed to stated preference. However, it cannot be said that types created out of use characteristics are better or worse than those of preference. The choice between the two should be made on the basis of which is conceptually most likely to yield results which meet the needs of the particular design problem at hand. In practice, types constructed out of both use and preference characteristics are likely to be relevant to a wide range of design problems.

Planning for diversity is a complex business. While the algorithm suggested above is not the only means of coping with the problem of diversity, and the study of cluster analysis as a statistical technique is only in its infancy,(11) the perspective outlined in this paper appears to be one which should not be ignored.

Notes

1. This paper is a shared collaborative one and the authors are therefore listed in alphabetical order.

2. The ideas and data presented in this paper emerged out of a larger study of the Brock University residence reported in Brock University Planning Department and Trevor Denton. An exploratory study of the Brock University residence. Volume 1. Social life in the residence. St. Catharines: Brock University Planning Department, 1972.

3. The domain here includes references to both places and people in the residence because both were felt to be important aspects of users' overall evaluation of the building. Other studies utilizing the evaluation approach outlined in this paper might find it preferable to use only parts of the building in question. The domain of characteristics is important and will be discussed again later in the paper.

4. Sokal, R.R., and Sneath, P.H.A., Principles of numerical taxonomy. San Francisco: Freeman, 1963, 133.

5. Here is the algorithm. It draws on the hierarchical algorithms of R.R. Sokal and P.H.A. Sneath (Principles of numerical taxonomy. San Francisco: Freeman, 1963), on the write-up for computer programme CSD 113 from the University of Guelph, and on the writers' own contribution. In Sokal and Sneath's terminology the algorithm is a "weighted" technique. Initially, a matrix of similarity coefficients is computed for all pairs of samples (persons). Samples are progressively linked into groups at successively lower and lower levels of similarity. The procedure is iterative, with new values of similarity being calculated between groups and ungrouped samples at each cycle. At every iteration those samples (or groups of samples) whose coefficients are highest in both row and column of the matrix are linked together. In this way several linkages can take place at a single iteration. At each cycle new values of similarity are calculated from the latest array of similarity coefficients, not from the original array. Cluster cycles are continued until all samples and groups of samples have been linked together into a single large group.

As stated so far, in the case of ties in similarity coefficients, the order of pair formation would presumably depend on the order in which data were stored in the array. The algorithm was therefore altered to permit merging of reciprocal strings - i.e. sets of types in which each member reciprocates with at least one other member in the set.

Obviously, this is not the only algorithm which could be used. Many others are available but this one was chosen because it allows the analyst flexibility in the selection of final types.

6. Brock University Planning Department and Denton, Trevor. An exploratory study of the Brock University residence. Vol. 1. Social life in the residence. St. Catharines: Brock University Planning Department, 1972.

7. At this point it is appropriate to comment here on the selection of types in the dendrogram. It is possible to set out some criteria which can be optimized as a means of deciding on final types from among the many available in the dendrogram:

- internal consistency of types (designated by $\bar{s}$ in Figure 1)
- conceptual sense
- numbers of persons per type
- number of types

Generally speaking, the higher up the dendrogram the more members per type, the fewer the number of types and the less the internal consistency or homogeneity of type members. Conceptual sense should always be a criterion; here, it and number of members per type were the only ones used. Nevertheless, the number of optimizing principles used, and the relative importance given each, may vary as the context of the individual research project and the use to which it is to be put (infra).

8. For a discussion of different techniques of weighting items on which people or objects are to be typed, see R.R. Sokal and P.H.A. Sneath. Principles of numerical taxonomy. San Francisco: Freeman, 1963, Chapter Six.

9. The rational versus naturally occurring type distinction follows the terminology of R.C. Tryon and D.E. Bailey. Cluster Analysis. New York: McGraw-Hill, 1970.

10. This approach obviously differs from building evaluation studies such as Mary C. Avery, Gerald Davis and Ronald Roizen. Architectural determinants of student satisfaction in college residence halls. San Diego: TEAG - The Environmental Analysis Group, University of California, San Diego, 1970, which have attempted to create a model of user building satisfaction in which all users are treated as a single group.

11. Admittedly, the causal argument outlined here is tentative because the small numbers of subjects did not permit controlled multivariate causal modelling. The number of subjects was set at 40 because of considerations in-

volved in other aspects of the research project from which the data came. Nevertheless the variables cited here as "independent" are plausible ones and the analysis may be viewed as illustrative rather than definitive. Further work in this area should make more use of predictions based on theoretical formulations rather than simply the demographic variables used here.

Although three variables in Table 1 (sex, car ownership and course type) are significantly distributed for the entire 40 students, this is not seen as influencing the distributions of independent variables for the three types. From the perspective of this paper, the population is an artificial entity which is simply the summation of all type members.

12. Borgen, F.H., and Weiss, D.J. Cluster analysis and counseling research. Journal of Counseling Psychology, 1971, 18, 590.

DEMODYNAMICS - A STATISTICAL THEORY OF DEMOGRAPHIC EQUILIBRIUM

Francisco N. Arumí

School of Architecture and Planning, and Physics Department
The University of Texas at Austin
Austin, Texas 78712

Abstract

A model describing the land distribution patterns of a system in demographic equilibrium is developed. The model is derived from three postulates: a) the existence of a quantum of individual area; b) the inevitable occupation of all habitable land; and c) the fundamental distinguishability among individual members of a large population, and indistinguishability of the quanta of individual area. The fundamental theorem of this model states that the available land will be distributed among the members of the population so as to maximize the degree of disorder (entropy) in the system. Real systems that have not achieved demographic equilibrium will proceed in time seeking the state of greatest disorder. Land occupation patterns in the USA are matched successfully with theoretical predictions. Demographic stabilization of two previously demographically isolated systems is described. Some implications of this process to urban planning and development are discussed. Within the tolerance of the predicted fluctuations when the model is applied to small systems, comparisons are made of the equilibrium conditions with the observed clustering effect of small animal groups. An extension of the theorem of greatest disorder predicts an optimum population size, and this prediction is compared with the observed population leveling of confined clusters of animals.

6. QUANTITATIVE TECHNIQUES / 335

Introduction

The behavior of social systems, regardless of species must conform to some natural laws in the same way as non-social systems.

The work presented here represents an effort to apply the well-established methods of statistical thermodynamics to demographic systems. The immediate purpose of the investigation is to determine the potential of these methods to help elucidate the rules of land distribution within a given population. The total inhabited land is not evenly distributed among the members of the population. The pattern of distribution becomes evident when large samples of demographic data are analyzed. This paper develops the argument of a macroscopic theory capable of describing these observed patterns of land distribution.

The Statistical Mechanical approach of Thermodynamics (Ref. 1) contains the description of the energy distribution of particles. The solutions advanced by Boltzman and later by Bose and Einstein, and by Fermi and Dirac enjoy universal recognition within their stated range of applicability. The statistical nature of the method was exploited by Shannon in his development of Information Theory (Ref. 2). The resounding success of these works with their reliance on proper statistical counting and their exploitation of the entropy function have profoundly influenced our perception of nature and our philosophy of knowledge.

Demodynamics, the resultant model presented here, is a statistical theory. Its predictions, therefore, are not deterministic but probabilistic. It predicts the most likely values to be observed and the concomitant uncertainty of their realization. This uncertainty decreases as the size of the statistical sample increases. For large samples, expectation values become virtual predictions. The smaller the sample, the more important the statistical fluctuations become. For samples for which statistical fluctuations are important, but where the expected values can still be differentiated from the statistical noise, the predictions become qualitative guides to the patterns we are likely to observe. When the sample is so small that the noise arising from statistical fluctuations drowns the signal from the expected value, then we have reached the microscopic level and no meaningful information can be extracted from the theory.

Instead of viewing this as a failure of the theory, we should view it as a manifestation of its merits. The theory contains no information about close neighbor interactions, and the growing uncertainty of its predictions as we apply it to smaller systems is its way of reminding us that it has no way of knowing what is going on at the microscopic level.

Demodynamics can be applied to macroscopic systems without intervening in the description of their microscopic structure. The identification of macroscopic rules of behavior, however, may help elucidate rules of microscopic behavior, just as Statistical Thermodynamics helped steer the ever more precise formulation of atomic theory of matter.

The fundamental Theorem of Demodynamics states that the equilibrium distribution of a population in their inhabited land area seeks the configuration that yields the greatest demographic disorder. The validity of the theory rests with its ability to describe observed phenomena.

Three independent empirical tests are presented as verification: (a) quantitative agreement with land distribution patterns of the urban population in the USA, (b) qualitative agreement with the clustering tendency of small populations, and (c) qualitative agreement with the space-limited optimum population effect.

Postulates

The theory is derived from the following postulates:

1) <u>There exists a fundamental quantum of individual area</u>. The quantum of area is taken to be a fundamental quantity that specifies the land required for the social functioning of each member of society. This cell of land need not be contiguous. Its defining properties are: a) a person may occupy more than one cell; b) no cell can be partially occupied; c) no cell can be occupied by more than one person. Knowledge of the magnitude of the cell is not required to develop the theory. Its size may be adjusted to fit empirical data.

2) <u>All additional quanta of area are occupied</u>. The standing room only state occurs when the number of people is equal to the number of cells available. In this case, every member of the population is in the ground level. As additional cells become available, this postulate, through its dynamic role, requires them to become occupied.

3) <u>People are statistically distinguishable while the quanta of area are statistically indistinguishable</u>. Although each person is treated equally to everyone else, the intrinsic distinguishability of each one is preserved by the counting procedure required by this postulate.

Definitions

A demographic system is defined here as an isolated, steady population residing in a given and fixed land area. A system in demographic isolation cannot exchange population or land with other demographic systems. The state of a demographic system is defined by specifying the total population and the total inhabited land. Specification of the total inhabited land in turn specifies the total number of quanta of individual area available to the population. A configuration of the state is any possible distribution of the additional quanta among the population members. A micro-configuration in turn is any distinguishable arrangement of individual members of the population that yields a given configuration. A micro-state is the generic term for all microconfigurations of a state regardless of the corresponding configuration. The total number of micro-states is the sum of all the microconfigurations of all the possible configurations of the system. The disorder of a configuration is defined as the number of possible microconfigurations. The disorder of the state is defined as the number of possible microstates of the system. Entropy is defined, as in statistical thermodynamics and information theory, as the natural logarithm of disorder.

Example

A system with three individuals, call them A, B, and C, have six quanta to distribute among themselves. Since each one must have one quantum, there are three additional quanta left. The state is defined by the two respective numbers, (3,3). The three additional quanta can be distributed in three configurations: a) all three quanta are claimed by one person, and the other two persons claim none; b) two quanta are claimed by one person, one quantum is claimed by another person, and the third person claims none; and c) each person claims one quantum. Since the individuals are distinguishable, exchanging two individuals with different number of quanta creates a new distinguishable arrangement. On the other hand, since the quanta are indistinguishable, exchanging any two cells does not give rise to a new distinguishable arrangement. Therefore, the possible microconfigurations are:

for configuration a)

```
         A B C      A B C      A B C
         x              x          x
         x              x          x
         x              x          x
        (3,0,0)    (0,3,0)    (0,0,3)
```

for configuration b)

```
         A B C    A B C    A B C    A B C    A B C    A B C
         x x      x x      x x      x x      x   x    x   x
         x          x        x          x    x          x
        (2,1,0)  (1,2,0)  (0,2,1)  (0,1,2)  (2,0,1)  (1,0,2)
```

for configuration c)

```
              A B C
              x x x
             (1,1,1)
```

The number of microconfigurations for configurations A, B, C are respectively 3,6,1. The total number of microstates, therefore, is 10 (=3+6+1). The corresponding configurational disorder and entropy are respectively again 3,6,1 and $S_1=\ln(3)$, $S_2=\ln(6)$, and $S_3=\ln(1)$. Likewise, the state's disorder and entropy are 10, and $S=\ln(10)$.

Fundamental Theorem

The formal development of the theory proceeds to determine the number of microstates and the average configuration solely from the knowledge of the state of the system. The above example illustrates the counting procedure to determine the number of microstates. The average configuration for the above example can be determined by counting the total number of ways in which an individual may end up with 0,1,2 or 3 extra quanta out of all the possible microstates the state may have. In configuration a), each microconfiguration has two individuals with no extra cells, and since there are three microconfigurations, then this configuration has six possible ways for an individual to have no extra cells. In configuration b), each microconfiguration has one individual with no extra cells, and since there are six microconfigurations, then this configuration has six possible ways for an individual to have no extra cells. In configuration c) there is no way for an individual to have no extra cells. Therefore, out of the 10 microstates of the system, there are 12 possible ways for a person to end up with no extra cells. On the average, therefore, there will be 1.2 persons occupying only their mandatory cell and none of the extra ones. The average number of individuals occupying two, three, and four cells (the mandatory cell plus 1,2, and 3 extra cells) can be calculated in a similar way to be respectively: .9,.6, and .3. Note, of course, that individuals and cells are conserved, ie, 1.2+.9+.6+.3=3 individuals, and 1.2x1+.9x2+.6x3+.3x4=6 cells.

The average configuration is also called the expected distribution, or simply the density distribution. If the cell is $1m^2$, a person occupying two cells is in a density regime of 1/2 person per m^2. In this example, the density distribution is: 1.2 persons at 1 person per m^2; .9 persons at 1/2 person per m^2; .6 persons at 1/3 person per m^2; and .3 persons at 1/4 person per m^2.

The fundamental theorem of demodynamics states that the most likely density distribution of a demographic system corresponds to the configuration with the greatest disorder - allowing the existence of the probabilistic fractional individual. For very large systems, this disorder or entropy corresponds to the state's disorder.

6. QUANTITATIVE TECHNIQUES / 339

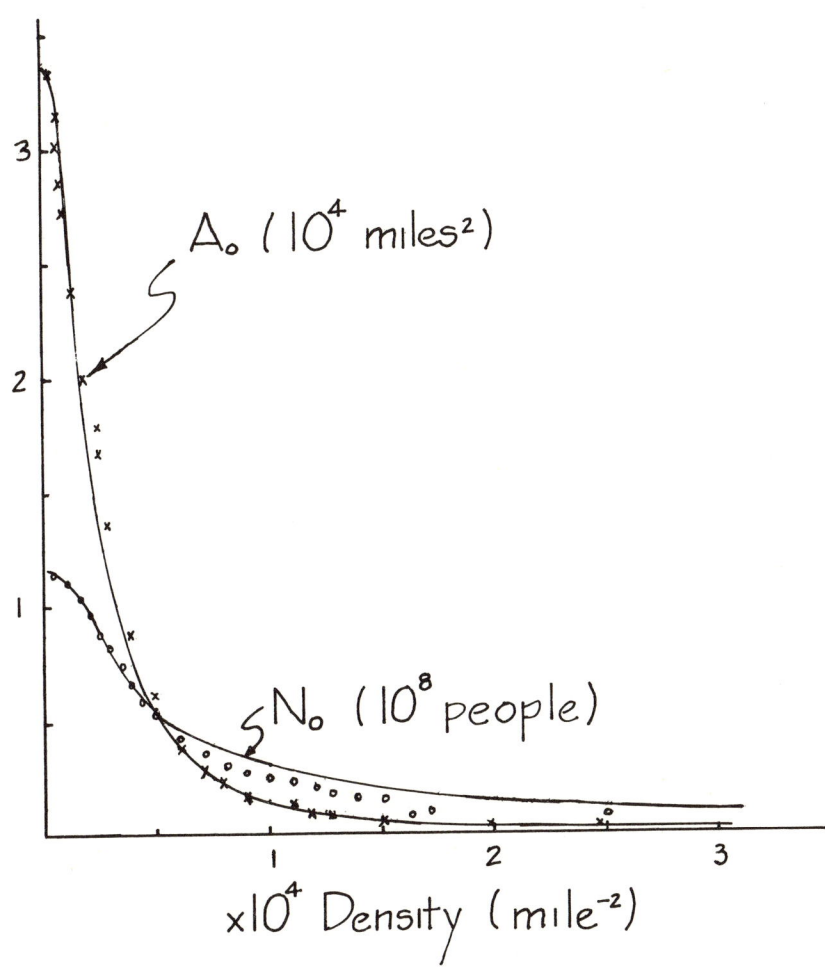

Figure 1. Empirical area (x) and population (o) as a function of density from the urban population data of the 1960 US Census. Solid curves represent the theoretical predictions. See the text for interpretation.

Rigorously speaking, this is an idealized equilibrium distribution. Since real systems are not likely to be found in the equilibrium state, the distribution predicted by demodynamics represents that state towards which real systems are evolving in time.

Empirical Test

In order to test empirically the density distribution predictions, data was collected from the 1960 Census Data (Ref. 3). The population, land area, and density figures for each of the 5409 towns and cities in the USA with a population greater than or equal to 1000 people were collected. Each town was then treated as a statistical data point totally disregarding its internal structure. These figures were then arranged in descending order of density and the accumulative population and area for each density level were calculated, and displayed graphically as a function of population density.

Figure 1 shows the resulting curves. Both cumulative population and area are displayed on the same graph. There are 1.1×10^8 people living above densities greater than 0 persons per square mile, and they occupy a total land area of about 3.5×10^4 square miles. As we go up in the density scale, we find, for instance, that there are 5×10^7 people living with density levels greater than 5000 persons per square mile, and occupying a total land area of about 5000 square miles, etc. The total population and land area at a given density are given by the corresponding density derivatives. The solid curves are the corresponding theoretical predictions of demodynamics. Their agreement with empirical data is within 1% through most of the density range, and never more than 5%.

Figure 2 displays the fractional population as a function of the fractionally inhabited land. Twenty percent of the land ($A/A_o=.2$) is occupied by fifty-five percent of the people ($N/N_o=.55$); fifty percent of the land ($A/A_o=.5$) is occupied by eighty percent of the population ($N/N_o=.85$). In this graph, again the points represent the empirical data, and the solid curve represents the theoretical prediction, and the two are again within 1% agreement through most of the range, and never more than 5%.

Other Predictions and Their Potential Significance

  Cross-Cultural Validity. The postulates and density distribution of population are enunciated without regard to cultural, economic, or political conditions of the demographic system. This implies, then, that precisely the same rules of land distribution should be detectable in any one demographic system in this world. The only variation allowed within the formulation of the theory is the magnitude of the quantum of area. Two possibilities arise. Every demographic system in the world

6. QUANTITATIVE TECHNIQUES / 341

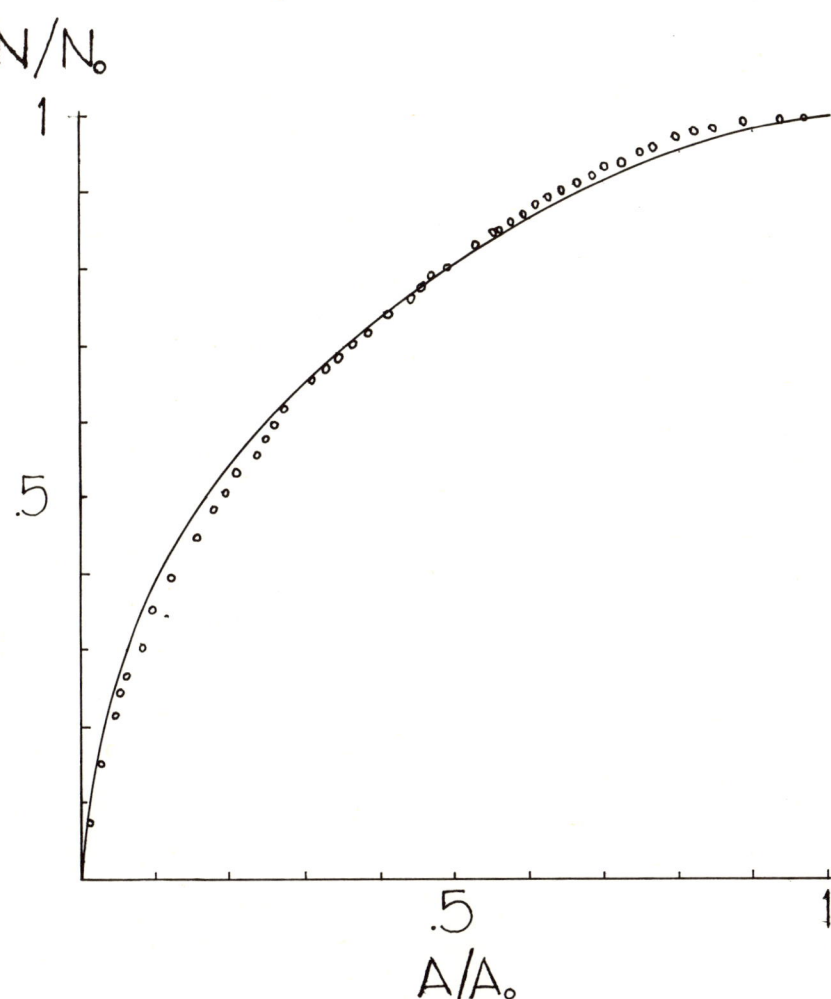

Figure 2. Fractional population that occupies a given fraction of the land. Empirical points derived from US Census data 1960. Solid curve is the theoretical prediction. See the text for interpretation.

exhibits the same fundamental magnitude for the quantum of area; which in turn implies the existence of a universal constant that determines the social development of man; or the magnitude of this quantum is parametrically dependent on local cultural, economic or political conditions, in which case, we would have discovered a quantitative gauge of cultural differentiation.

It should be emphasized, however, that the relationship between the concept of individual quantum of area and that of personal space is not clear at this point. Whether they are equivalent concepts or not, it appears certain that they will complement each other. The nature of this complementarity and their relationship, if they exist, will of course be subject to the empirical evidence that can be uncovered.

<u>Migration of Population Between Density Levels</u>. The fundamental theorem of the theory states that the population will seek to distribute itself so as to attain the greatest possible disorder. A demographic system that has not achieved this state, exhibits pressure gradients in the direction necessary to achieve this state. Comparison, therefore, between achieved and equilibrium distribution provides a predictive tool for future migrations of population between density levels.

A high density housing development, for instance, that reduces the entropy of the system within which it was built, will inevitably give rise to an unstable configuration and the population of the system will redistribute itself so as to bring the entropy to its maximum value again. In the process, of course, the original purpose of the housing development will have been defeated.

<u>Demographic Contact and Clustering</u>. When two previously isolated systems, each in demographic equilibrium, come into contact, they blend into each other seeking a common and different configuration unique to the resulting system. According to the theorem of greatest disorder, we expect the most favorable composition of a demographic system to be the one that yields the greatest disorder. A single cluster has a greater disorder than two linearly super-imposed sub-clusters. This result can be used to trace, and subsequently predict the development of urban subsystems such as suburbs and twin cities.

For small systems, of course, no single and unique configuration can be identified; instead, oscillations and fluctuations from one configuration to another are expected. The only trait that this theory can identify is a general tendency for small populations to cluster into single groups. Empirical observations to this effect have been made by McBride (Ref. 4) in his controlled observations of chickens.

<u>Space Limiting of Population Growth</u>. Generalizing the fundamental theorem to read: The size and distribution of the population will seek the state that yields the greatest possible disorder, introduces

a space limitation on the population growth. When a given land area has no population, its entropy or disorder is zero. On the other hand, when the number of quanta in this land area is equal to the number of people, the land is fully occupied and there is only one distinguishable way in which cells can be distributed among the people. In this case, again, the disorder of the system is zero. In going from zero to standing room only population, the entropy function went from zero to zero. Analyzing the entropy function, we find that for a system with a fixed land area, a population half of the standing room only value yields the greatest possible entropy. This means that the population of a system will increase or decrease, always seeking that value that yields its greatest entropy. The significance of this result is that it predicts a leveling off of the population at a value equal to half of the standing room only value.

This phenomenon has been observed by Calhoun (Ref. 4). He reports that under controlled conditions, a mice population whose standing room only value were to be 4000 stopped growing at about 2000. This result and its agreement with demodynamics is remarkable. Here again, however, the relationship between the two uses of the expression "standing room only value" remains to be determined. It is not possible, of course, to carry out Calhoun's experiments with humans. Demodynamics, on the other hand, suggests ways in which we can determine if this phenomenon is present in human populations without having to resort to controlled experiments. Available census data can be used for this purpose.

Conclusions

Demodynamics arrives at the above predictions from very general arguments. Far from entering into conflict with microscopic studies that seek to explain the same phenomena by determining close-neighbor interactions, it complements those efforts by providing a unified formalism at the macroscopic level.

At the time of the writing of this report, tests are being organized to seek empirical verification of the predictions outlined above.

Further tests could exploit the concept of demographic pressure to seek correlations with pathological social behavior. The permeability of international boundaries in presence of severe pressure differences could also be studied and its old established correlation with conflict cast into the fresh light of a unified formalism.

The fundamental appeal of demodynamics rests with the economy of its formalism and the wealth of phenomena it can explain. Its potential contributions may include its policy implications in urban and regional planning and design; its contribution to mathematical demography; its macroscopic modeling of species distribution in the ecological chain; and last but not least, the pleasure one derives from seeing an abstract thought process compatible with the ways of nature.

References

1. Huang, Kerson, Statistical Mechanics, John Wiley & Sons, Inc., 1963.
2. Brillouin, Leon, Science and Information Theory, Academic Press, Inc., 1962.
3. Census of Population, 1960, Vol. I, "Characteristics of the Population Part A," US Bureau of the Census.
4. Esser, A.H. (ed.), The Use of Space by Animals and Man, Plenum Press 1971.

SEVEN   DECISION MAKING TOOLS

Chairman: Donald P. Grant, Arch. & Environmental
              Design, Cal. Poly. State Univ., San
              Luis Obispo

Panelists: William Miller, Design Methods Century City,
                 Los Angeles
              Jean-Pierre Protzen, Architecture, Univ. of
                 California, Berkeley
              Morton Rubinger, Architecture, Nova Scotia
                 Technical College, Halifax
              Thomas Thomson, Architecture, Washington
                 Univ., St. Louis, Mo.

Authors: John C. Simon, "The Value of User Evaluation
                 Studies to the Design Process and
                 Formula Financing"
              Volker Hartkopf, "An Economy Model for
                 Generating, Evaluating, and Selecting
                 Architectural Design Alternatives"
              Robert E. David, "Proposal for a
                 Diagrammatic Language for Design"
              Leon A. Pastalan, Robert K. Mantz II, John
                 Merrill, "The Simulation of Age Related
                 Sensory Losses: A New Approach to the
                 Study of Environmental Barriers"

DECISION MAKING TOOLS: INTRODUCTION   7.0

Donald P. Grant, Session Chairman

School of Architecture and Environmental Design,
California Polytechnic Institute, San Luis Obispo

Richard Bellman was one of the inventors of dynamic programming. Speaking to a group of architectural educators a few years ago, he gave a humorous description of the derivation of the term "dynamic programming" for his sequential, adaptive decision making technique, then observed that its great benefit might not be in the form of solutions for specific problems, but rather in the provision of a new, non-static frame of mind for problem formulation. Horst Rittel has similarly observed that the great benefit of systematic design methods may not be in the form of specific solutions arrived at by their application, but rather in the process of educating and informing the designer that takes place in the course of working with systematic design methods. The title of this session, "decision making tools," is not a precise fit to the content of the papers included, at least in the sense of "tools" as implements for direct application to tasks. However, if one takes a broader view and accepts as "tools" procedures that educate and inform designers, then some worthwhile directions are indicated in these papers.

A growing concern that planning be viewed as a process for dealing with problems as they arise in an ever-changing contest, rather than as the production of static diagrams for future reference and conformity, is often voiced in debates over master plans for communities and educational systems. Simon proposes an educational planning and programming procedure in which "...planning is a process rather than a finite act...." The procedure proposed deals with the characteristic difficulty of combining evaluations of quantitative with qualitative considerations in a three-step process. First, planning problems are formulated as issues; then programming is carried out in the form of establishing a hierarchy of physically achievable objectives; and finally, these are translated into specific performance requirements against which plans and in-use facilities are evaluated. In this paper, planning is viewed in what might be termed a "dynamic programming frame of mind," rather than as the production of a fixed plan for long-term development.

Purely economic measures of performance of program and plan evaluation have fallen into well-deserved disrepute. There are many stories about cost-benefit analyses in which significant human values have been excluded or violated, or in which generally unacceptable underlying worldviews are concealed. Hartkopf describes an economic decision making procedure for architectural design that includes a variety of interesting variety generation and variety reduction activities. He does not, however, deal with the problem of how to quantify the messier end of the spectrum of human needs and desires, nor does he propose ways in which "...it will be possible to select a higher cost alternative which, considering the intangibles, creates a better value." For this approach to constitute what the author describes as "...a model suitable for decision-making in architectural design...," these problems will have to be dealt with. The example in this paper works largely

7. DECISION MAKING TOOLS / 347

because it is a decidedly non-wicked problem, that of efficiently storing and accessing goods in large blocks. Interesting features of the approach are the combination of an exhaustive enumeration of possibilities, in the manner of the morphological approach, with variety reduction using cost-benefit analysis; and the author's attempting to deal with the problem of different configurations implied by optimizing against different criteria. Economic considerations and measures of performance are not going to go away; we may or may not be able to look forward to a Keynsian "age of abundance" in which we might get on with decent human pursuits, but for the moment we must deal with the dismal science. Conflicts of interest and opportunity and scarcity of resources are facts of human life, and we must seek some way in which to integrate economic considerations into the processes of design and planning, along with not-so-quantifiable considerations. An approach that succeeds in this essential integration is still lacking.

David's paper perhaps comes closest to the session title, "decision making tools." He surveys systems of notation in several fields, and draws together threads from the Shannon-Weaver model of communication and botryology (clustering, that is) to present the rudiments of a language for use in the process of designing. His preliminary formulations elicit immediate comparisons with Alexander's pattern language.

Pastalan, Mautz and Merrill describe a simulation in which graduate architecture students undergo sensory deprivation over a long period as a simulation of the sensory situation of aged persons. Related work several years ago by K. Izumi simulated the sensory situation of mentally ill persons by use of the drug LSD. A game, "End of the Line," played last year at the Berkeley gaming club and now being developed there, simulates sensory and motor deprivations increasing with age in the course of playing a role in which tasks and transactions must be carried out. Sensory deprivation experiments have long tempted architects to use them as ammunition in arguments for greater variety and stimulation in the environment, on the assumption that the extreme disorientations and malfunctions occuring in laboratory settings with extreme deprivation must have less extreme but still undesirable counterparts in less extreme situations of deprivation, like drab housing and workplaces. Perhaps in work such as this paper reports sensory deprivation data can be useful in two ways: the compiling of specific data observed in the simulation, and the education and information of designers about the overall life situations of their client-users. The experience of walking in the moccasins of one's clients might be useful in several situations in which the clients' life situations differ greatly from the designer's. The next step in the sort of work reported here might logically be its incorporation into a more comprehensive gaming situation, with roles, tasks and transactions. There has been a reaction against design and planning methods that go too far in the direction of precise definition and quantified representation of human experience. Gaming simulations and empathic simulations like this one might provide a means for understanding situations not readily modelled in the explicitly defined and highly quantified techniques of the first generation in design methods.

THE VALUE OF USER EVALUATION STUDIES TO THE DESIGN PROCESS AND   7.1
FORMULA FINANCING

Joan C. Simon

Department of Consumer Studies,
College of Family and Consumer Studies,
University of Guelph

Abstract

The objective was to investigate whether qualitative as well as quantitative environmental factors can be:

1. Assessed in use
2. Integrated into formula financing
3. Blended into a form which is readily useable to architect and the client in the building design process.

It was concluded that only by revamping the building design process can the introduction of qualitative factors be assured. Architectural planning must be perceived as a continuing process rather than the production of a finite product. Therefore, the paper focuses on development of such a process.

7. DECISION MAKING TOOLS / 349

Introduction

In 1965 the Government of Ontario created a system of Colleges of Applied Arts and Technology (C.A.A.T.) to provide an alternative to university education at the post-secondary level. The Colleges were intended "...for full-time and part-time students, in day and evening courses, and planned to meet the relevant needs of all adults within a community, at all socio-economic levels, of all kinds of interests and aptitudes, and at all stages of educational achievement" (1) The Colleges were allowed considerable latitude for individual college development but the enabling legislation did direct that "...some features will be common to all programs; they will be occupation oriented, for the most part; they will be designed to meet the needs of the local community - and they will be 'commuter' colleges." (2)

Barely sixteen months after the initial statement in the legislature by the Minister of Education, nineteen (later increased to twenty) Colleges opened for classes in September, 1967. The form of development of the Colleges has been strikingly similar. All Colleges have a new primary campus location and vestibular campuses dotted throughout the region that they serve. The vestibular campuses range in types from major new purpose built buildings to converted factors, 'shop fronts' and even two trap-lines. Eighteen of the primary campuses were designed for the new institutions, the other two were newly constructed Ontario Vocational Institutes which were incorporated into the college system. The form of construction on the primary campuses represents the development of a new building prototype - an enclosed campus complex.

To date in excess of $200 million in capital funds have been authorized for construction and equipment purchases. It was necessary to devise governmental procedures for allocating funds which would assist the orderly development of the necessary facilities without hampering either the rate of growth or imposing preconceived limitations on the type of shelter appropriate to these new institutions. The relationship between Department of Education officials and college administrators was characterized by mutual trust and respect. The five years of construction reviewed in this study were undertaken in what could be called a pioneering period. In October, 1971, a governmental reorganization created a new Department of College and Universities.

C.A.A.T. Planning Procedures

The original means devised for reviewing college requests for capital funds was a series of five forms, the first of which (C.A.A.T.I.) prescribed the development of a Master Plan.

Document (3) Typical sections of this report were:

1. Educational Philosophy
2. Educational Programs
3. Educational Specifications
4. Community Analysis
5. Site Feasibility Studies
6. Enrolment Projections for Ten Years
7. A physical Master Plan for the Primary Campus

The documents produced are typical of college and university master plans in North America in the late 1960's. The deficiency in the planning methodology was that it veered from "motherhood" statements about education, such as, "Man cannot live by bread alone, and all students irrespective of the specific training they are receiving need to enrich themselves culturally, if they are to become whole persons and truly mature citizens," (H2-22) to specific requirements given a priori, for example, "language lab positions should be increased from the present fifteen to at least thirty-five. Reading skill lab positions should similarly be increased to thirty-five," (4) Planners, clients and governmental agencies accepted these documents because they looked impressive. This attitude was not unusual for the time. Richard Saul Wurman recorded his feelings "Until then I had always fooled myself. If something looked good, it was good... If a map or planning report looked good, it was good, and I had not asked myself two very simple questions: did I understand what I was seeing and could I tell somebody about it." (5)

Government asked for a master plan in the form of a finite document rather than a continuous process. The approach was accepted by the Colleges and their professional advisors. With hind sight, the attempt to produce a definitive statement about the initial ten years of the development of a college which was a new type of educational institution without precedence in Ontario seems almost naive. What would public reaction be to these Colleges? How might changing governmental policies, technological events, or student attitudes effect the development of a particular college campus? The C.A.A.T.I. required an assessment that was new to educational planning in Ontario and given the limitations of the static approach, it is surprising how much of the reports are still valid in an area of educational philosophy and program structure.

At the operational level of student enrolment, organizational structure and operating finances the Master Plan documents rapidly became outdated. The capital finance section of the C.A.A.T.1 form was never used by the Department of Education for authorizing funds. Despite this there was little reassessment of the physical planning proposals. The Connect/Campus Computer Simulation Model was introduced to try to make planning more responsive to changing conditions. The model attempts to project space needs according to twenty-one categories of use. Despite three years of work on the model, for various reasons it is not yet utilized for physical planning by either of the colleges studied.

The original intention of this research project was to evaluate the Colleges on the basis of the Master Plan proposals. A review of the available information showed that this approach was not feasible because they contained no logical development from objectives to activity criteria to performance standards. It was not even possible to evaluate actual versus planned growth on a per square foot and per student basis because neither the Master Plans or the simulation model recognized that physical space has a geographical attribute. Hence a number of critical projections lumped together several distinct campus locations.

Planning by Issues/Programming by Goals

In keeping with the viewpoint that planning is a process rather than a finite act, it is suggested that governmental planning methodology should also be subject to continual reassessment. As has been noted, during the last five years government has encouraged the Colleges to try first a finite Master Plan and then a computer simulation model. What is proposed is that a goal oriented approach to physical planning be initiated and that the goals form part of the basis for capital formula financing. Goals would also form the criteria against which the performance of the built form could be evaluated. In turn the performance evaluation should lead to an improved planning base for subsequent building projects.

This approach would complement the computer simulation model by using the quantitative data and providing the qualitative input lacking in computer analysis and blending these inputs into a readily usable form. It would reduce the paper work involved in submitting requests for capital expenditures because the working and approval format would be identical. The procedure outlined would allow the College to retain individual forms of administration and operation procedures in the area of physical planning but it would also encourage the development of expertise in these matters.

It is suggested that the initial stage in the building approval procedure call for the submission of a Tactical Working Document which

1. Outlines the methods to be used by the College to generate a plan. Knowledge of the process should enable a co-ordinated participation by all the various people, groups, agencies and specialities involved.
2. Examines every problem which comes within the planning sphere in terms of issues. Issues cover administrative, financial, procedural and motivational concerns. Issues should be analysed in a standard manner and such analysis should be kept to one page. (More complex presentations have been shown to be the result of compound issues which should be segmented for analysis.) The analysis should take the form of

- issue statement
- source of available information
- current Assumption/Policy/Situation
- comments on current Assumption/Policy/Situation
- recommendations re Further Analysis of Current Assumption/Policy/Situation (6)

3. Formulates a hierarchy of physically achievable goals. Issues relate directly to functional concerns, they clearly delineate the nature, extent and direction of the work to be done. They are the working brief to the architect and the basis of in-use evaluation. They could provide the means for introducing environmental concerns into formula financing. Goals are arranged in a hierarchical order so that if conflicts between goal attainments are unearthed during the design process it is clear which environmental consideration takes precedence. The College may change, add, delete, or revise the order of the goals during the subsequent design stage, but it should be clearly understood that these modifications constitute a revised brief to the architect and the College is responsible for any ensuing delays.

Program Goals and Building Design

One of the conceptual difficulties which arises in planning for both architects and their clients is understanding the difference between a building program and a building form. Far too many architects engaged in planning and programming too quickly extrapolate the input data into three dimensional shapes. Similarly, many clients want to see what the building will look like rather than diagrams of growth strategy, movement, organization and relationships. This is the core weakness in the original C.A.A.T.1 documents. One direct means of overcoming premature conceptualization is not to employ the same professionals for programming and design, but the benefit of continuity is also apparent because by preparing the program the architect will have developed an understanding of the client's problems and characteristics which no document can fully communicate. The review of programming goals by the Department of College and University Affairs prior to the approval of funds for design fees should encourage the rationalization of the programming process and prevent the program from becoming a justification of design forms.

Programming goals establish the performance criteria for the building. They should be concerned with the kind of things which will happen or should be allowed to happen in the building as the result of the planning issues. The hierarchy of goals should enable the architect to make an appropriate physical response based on the knowledge of the activities that will take place.

The establishment of program goals will enable the client to evaluate the design phase on a rational basis. All too frequently the consideration of major building projects by the client focuses on relatively insignificant details revolving around personal preferences and ignores the basic goals of the scheme.

Goal statements should avoid presetting the form of the design solution. This will afford the architect more latitude in prescribing environmental solutions for novel problems and the opportunity to reconsider historical solutions (such as the traditional classroom or lecture theatre) in light of present conditions and technology.

Goal statements should not be a catalogue of specific requirements. This type of

7. DECISION MAKING TOOLS / 353

catalogue programming tends to repeat past solutions without evaluating this present validity. Since these types of electrical outlet, lab bench etc. listings are usually compiled by or with particular reference to an individual faculty member they presuppose a particular teaching format. The survey conducted showed that the most costly building and equipment reappraisals occurred in spaces purposely designed in detail for specific faculty members who by the time of completion were no longer involved in activities in those spaces.

An example of the differences between issues, goals and specific requirements is illustrated by the following hypothetical case.

Issue: What is the College's policy regarding community use of physical facilities? The resolution of this issue would entail policy decisions about the College's community vote, the basis of admission of public - groups or individuals, to community developed programs or to college developed programs, scheduling, fee acquirements if any. Financial considerations might include payment of the facilities entirely through funds for college construction, possibility of grants from local authorities, agencies or groups, private subscription, or special mortgage arrangements.

Goals

1. The gymnasium should be available for use by organized community groups on weekends and holidays when the College is not in operation.
2. The gym should be conveniently located for informal use by students during their unscheduled hours of the day.
3. The gym will provide scheduled classroom space for students in the recreation department and therefore should directly relate to the other classrooms and faculty offices in that department.

It is evident that these three goals relate directly to the location characteristics of the gym and might, given the limitations of a specific site result in conflict. The hierarchy of goals established by the College would tell the architect which is the clients primary locational consideration.

Specific requirement

The gym floor should be hardwood with an epoxy finish.

Relationship of Goals to Formula Financing

Since April 1, 1969, capital monies have been allocated to Ontario universities based on a devise known as Capital Formula Financing. "The formula is based upon involvement projections - Recognizing that certain types of students require more space than others, the enrolment projections are weighted by course of study and year level..A unit of space is then applied to each weighted unit of enrolment. Thus a total cumulative space need is determined for any particular year. From this Total Cumulative Space Need is subtracted the Existing Space in order to calculate the Additional Space Required. A unit cost is applied to the Additional Space Required and a cumulative dollar entitlement is calculated for any one year." (7)

To date, the Colleges of Applied Arts and Technology have not been financed under this system but since the reorganization of government combined the responsibilities for universities and colleges into a single department speculation has existed about the extension of the formula theory.

Direct application of the university formula to the Colleges ignores their unique form of development and their mandate for community involvement and relevancy. The formula was devised at a time of increasing student enrolments. If enrolments become static or decrease the formula theory in effect cuts the institution off from capital expenditure funds even though the space available may not be compatible with current institutional aims and policies. Cyclical renewal funds equal to 1% of the allocation inventory plus 1% of the cumulative cash flow since April 1, 1969, are provided for annually. (8)

The present cost per unit of space (for universities) is a fixed allowance of $55. per net assignable square foot of the building project. For many of the vestibular campuses this is an overly generous figure. Some of this space is leased to maximise flexibility by easily eliminating it from the College inventory when needs change. Clearly the cost allowance is inappropriate to remodelling and equipment needs.

The only truly manipulable factor in the formula is the enrolment weightings. There are five weights for university students ranging from 1.0 for Arts & Science undergraduates to 4.0 for Ph.D's in science. Refinements are presently being considered. Even though there are allowances for part-time students the College's heavy involvement in evening courses, special sessions, study groups etc. would be difficult to accommodate to this concept.

Enrolment projections are evaluated annually by a high level committee both from the total provincial standpoint and from the individual institutions' standpoint.

What I am suggesting is that the institutions environmental goals be assessed by a select committee and that the goals from the primary manipulable factor in the formula. This would allow for the provision of facilities not directly related to expanding enrolments and projects not involving total new construction. Also considerations of cost overtime could be taken into account.

The initial weighting of goals can be established through reference to historical data. User Evaluation Studies could lead to refinements and sophistication in goal weighting.

Goal/Evaluation Studies

A study of the in-use operation of two college primary campuses was undertaken. The aim was to scale data which would be useful both for the client in establishing the building program and the architect in the design of the next phase.

One of the major limitations to the establishment of a feedback cycle between evaluation and design decisions is that the type of information presently being generated by environmental researchers is not the type of information used in the architectural design decision process. As Perrin has so well recorded, the architects education does not predispose him towards the systematic refinement of design solutions nor does it provide him with an understanding of how to interpret sociological, psychological or even statistical data. (9) However, there are a number of architects searching for data which they can use, but they become confused and disillusioned by much of the material they read or hear at conferences (10).

Much environmental research seems to be arrived at establishing norms or a range of norms that the author presumes will have a wide application. This presupposes that the environmental response sequence must be space man and, indeed, a non-variable man response which denies a cultural dynamism. The search for these finite norm ranges may be intellectually interesting but the relevance of this type of research to architectural design seems dubious. For a building to have validity over time the response sequence needs to be man space. Thus instead of designing for historical norms the architect needs to build spaces in which the user can create the environmental response appropriate to changing cultural, etc. circumstances.

There needs to be a recognition that there is a grain of data beyond the scope of architectural consideration. This is true of both psychological data and physical information. An architect is not concerned with the molecular composition of steel but the strength of two particular steel sections. Those supplying the architect with input need an understanding of the appropriate data grain useful for design. For example, faced with Barkers goal, program, deviation countering and vetoing circuits, the architect is perplexed. As an educational planner Barkers conclusions about school size related to the behavior of inhabitants of undermanned and optically manned behavior settings has utility.

Goals are composed of sets of considerations. Psychological factors are only one set type. Economical (both capital and operating), structural, political and aesthetic aspects of goals must be balanced along with the psychological. Sets other than psychological may be more significant to both client and architect.

Examples of Goal Evaluation and Utilization from Pilot Study

Data Collection

Psychological data series were assembled by six paid researchers who were familiar with market research techniques and had some previous experience in environment analysis surveys. The researchers were instructed to answer questions about what they were doing directly and honestly. They were to be friendly towards students and staff, but avoid involvement in college activities.

Classrooms were assessed through observations made by the team. The form covered teacher and student activity, equipment usage and classroom atmosphere as they related to movement, perceived noise level, physical arrangement of furniture, clustering of students and proximity of faculty to students. Four-hundred-and-fifty classroom observations were made a College A and seven-hundred-and-twenty at College B. (The difference in the number of observations is primarily due to the open plan concept of College A.)

Individual reactions to the environment were ascertained by questionnaires supplemented by a small number of in-depth interviews. Questionnaires were distributed to students in classes to obtain a balanced distribution through divisions. In order to disassociate the research work from the College administration questionnaires were returned at various drop boxes 355 questionnaires were returned at College A and 310 at College B. This procedure resulted in an inbalance of return by division with those students in labs being three times as likely as those in lectures or seminars to return the questionnaires. Some items on the questionnaire provided checks on the equipment usage data collected in observations. Library usage and informal behaviour in the communal spaces - lounges, eating areas and circulation spaces where students lingered (surrogate lounges) were recorded photographically and graphically by coding on floor plans. In keeping with the hypothesis of the study, that have grain behaviour data is not of primary significance in architectural design decisions. Undisguised photography proved satisfactory. The research team found that when the subjects had assumed lounging positions, activity from the point-of-view of this study was not affected. Subjects in motion (and therefore not of interest) did modify their behaviour - There was a tendency to move out of the way of the camera. Photographs were taken every twenty minutes during one day from 8:00 a.m. to 5:00 p.m. from each of twenty-three viewing stations at College A and nineteen at College B. Viewing stations were established from previous observations of areas which attracted groups of stationary students graphic coding supplemented photographic areas by picking up behaviour between stations and provided a cross check on the photographic material. Instructed interviews amplified data files. Twenty-eight activity maps were compiled of all public circulation and lounge spaces in each college and six-hundred and forty-four photographs of College A and five-hundred and thirty-two of College B. The photography also proved to be a source of fine grain data about small group interaction.

Economic data series were assembled from records, budgets and interviews with college staff. Mechanical and structural data series were obtained from interviews with college personnel, architects and consultants and review of working drawings.

7. DECISION MAKING TOOLS / 357

Summary Examples of Goal Evaluation

1. Issue: The College should be involved in the community.
   Goal: Pedestrians strolling through the park will be attracted into the College.
   Evaluation Procedure: During a one-week period all visitors to the College were asked reasons for visit.
   Evaluation Finding: Every outside visitor had a specific reason for coming to the College and the journey involved a special journey.
   Goal/Evaluation Correlation: None
   Correlation Analysis: Poorly formulated goal which ignored readily available psychological input data about pedestrian behaviour.

2. Issue: Students are non-academically oriented.
   Goal: English teaching will be dialogue oriented.
       A recreation room atmosphere is wanted to encourage rap sessions.
   Evaluation Procedure: Classroom observation data.
   Evaluation Finding: Students were in more than one cluster 58% of the time. This compared with 34% at the other College with more "conventional English classes". The scale of activity showed little difference between campuses and this evaluation base is being reassessed. The activity pattern of faculty showed a marked difference with faculty talking to students at the control college in 66% of the observations, while at the goal test college faculty were doing something other than talking to students 78% (of which 31% of the time they were observed listening to students).

3. Issue: In response to provincial budgetary restraints on educational spending a program of self-directed learning utilizing learning objectives conveyed through audio-visual media will be staged.
   Goal: The business division will teach typing and shorthand using a system which utilized slides and cassettes.
   Evaluation Procedure: Questionnaires, classroom observations.
   Evaluation: 43% of the typing classes used were using the A.V.T. system at the time of observation. 73% of secretarial students reported that they had used the A.V.T. system. Of those reporting use 67% found it very useful, 8% found it of some use.
   Correlation: Economic data required.
   Correlation Analysis: Economy was the essential factor. Economic analysis is being prepared by Business Division. Method of evaluation inadequate to assess goal correlation.

4. Issue: Because of the difficulty of defining the role of colleges in post-secondary education over an extended period of time and the changing nature of post-secondary education as a whole, college space should be flexible.
   Goal: Classroom and laboratory space should be designed for changing class sizes, class types and teaching techniques.
   Evaluation Procedure: It was decided that when activity patterns and space provision were grossly mismatched, renovation resulted. Renovation costs could provide the assessment parameter. The costs of renovations were assessed on a per square foot, per student and percent of construction cost basis. Because the

initial provision of flexible space tends to involve more initial investment than fixed feature space the latter was used to establish a correlation scale.

Evaluation: College A financed all renovation work for five years from their operating budget. They had purchased two portable classrooms which might be interpreted as reducing renovation costs. It was decided that these were disposable assets and should not be classified with renovation costs. College B had requested funds for renovations (amounts are confidential at this time).

Goal Correlation: A reverse scale of ten was established giving College A a score of 0. A score of 1 would indicate remodelling costs per year were 1% of construction costs, etc. Very significant differences were noted between the two scores.

Goal Analysis: The remodelling costs incurred by College B were the result of a poor definition of the parameters of flexibility. The parameters did not include mechanical and plumbing installations.

Formula Financing Significance: Costs overtime are one type of factor which could receive a formula goal weighting and produce a considerable net saving for the tax payer.

References

1. Ontario Department of Education, Colleges of Applied Arts and Technology, Basic Documents, Toronto, 1967, page 8.

2. Ibid., page 12.

3. Ontario Department of Education, Guidelines for Planning Colleges of Applied Arts and Technology Including Approval Procedures, Toronto, 1967.

4. Humber College of Applied Arts and Technology, "Request for Approval, Colleges of Applied Arts and Technology, 1", June 28, 1968, page 2:6

5. Wurman, Richard Saul, "The Invisible City" Forum, Volume 136, Number 4, May, 1972, page 41.

6. Scott, David H., Compilation of Papers on Programming, Guelph, Ontario, 1971. Also see Royal Institute of British Architects Manual Sections 3.520 to 3.525.

7. Ontario Department of Colleges and Universities, Capital Formula, Toronto, 1971, page 1.

8. Ibid. page 12.

9. Perrin, Constance, With Man in Mind.

10. Campbell, Sheila, "Environmental Design Research Association Conference, has Angles, January 1972", Architectural Design, Volume XLII, June 1972, page 386-7.

11. Barker, Roger G., Ecological Psychology, Standford University Press, California, 1968.

AN ECONOMY MODEL FOR GENERATING, EVALUATING,
AND SELECTING ARCHITECTURAL DESIGN ALTERNATIVES

7.2

Volker Hartkopf

Assistant Professor
Department of Architecture
Carnegie-Mellon University
Pittsburgh, Pennsylvania

Abstract
The case study as described in the paper exemplifies a method utilizing economic analysis to generate and evaluate architectural design alternatives early in the design process. This is achieved by means of present worth analysis and a computer program which was developed for the case study. Alternatives which under specified conditions provide for the best value can be selected for further development. Initial and future investments necessary for the construction, maintenance and operation (wages and machinery for example) of each design alternative are included in the process. The effect of various economic environments on the economic ranking of the alternatives is tested using sensibility analysis. Confidence limits are established within which certain alternatives provide for the best value. Implications for environment design are discussed.

Acknowledgements
Precursory work was started at the University of Texas at Austin. Much of the work, however, was conducted at North Dakota State University at Fargo. The author would like to thank his students at this institution, as well as Mr. K. Nagarasan, Fargo, and Mr. Kelly McAdams, Austin, who were invaluable with their profound knowledge of computer programming. I am most grateful to Mr. Gary D. Frank, Sales Engineer at Clark Equipment Company, for his generous assistance establishing the costs of the storage systems.

Introduction
For the purpose of this paper architectural design is seen as making explicit the needs of client and user and translating these into physical form. Needs can be met in many ways. In order to choose between possible approaches to a given problem the architect has to evaluate the various alternatives at hand. This should be achieved early in the process. In many cas s it is not feasible to fully develop designs from all conceivable alternate concepts. It is therefore important to separate the parameters of prime importance from those of lesser importance. In the author's opinion, economic analysis as applied in this case study is a suitable tool to evaluate the initial and future economic performance of a range of conceivable solutions. Together with other concerns such as values and code restrictions, the model can be used in decision making. It is in this sense that the model is seen in a decision context rather than an analytical context.

The Case Study

The objectives of the case study are:

to develop an approach by which design concepts can be generated systematically.
to generate and select a range of suitable design alternatives.
to define the variables involved and their interrelationship over the full lifetime of each alternative.
to subject the range of selected alternatives to financial evaluation.
to test the sensitivity of these alternatives to various economic parameters.
to select the alternative which is most likely to produce the best value under the identified conditions and possible future conditions.

The case study is based on the following problem statement: a producer of canned goods (soups and vegetables) needs storage facilities for 10,000 pallets. (A pallet is an assembly of cases for transportation and storage purposes, each of which contains a given number of cans. All pallets have equal dimensions: 4' x 4' x 4'. There are three different kinds of goods: A pallet of item A, weighing 3000 pounds, accounts for 60% of the inventory. A pallet of item B, weighing 4000 pounds, accounts for 32% of the inventory. And finally a pallet of item C, weighing 1000 pounds, accounts for 8% of the inventory.

Heuristics in Concept Uncovering

Instead of engaging in the impossible task of investigating every possible state of the system, the designer identifies those parts of the problem which may have far-reaching implication. He then forms concepts, and, as more and more constraints are identified, eliminates those concepts which cannot be developed within the established constraints. The objective of the heuristic approach is to establish promising concepts and to eliminate lesser ones. Once a range of possible solutions is found a formalized approach is used in which all the promising courses of action are evaluated to find that one which, within economic constraints, holds the greatest promise.

This relatively simple task, the storage of 10,000 pallets, could be solved in many ways. Each pattern has implications for the equipment used, and the labor force involved, as well as the building shape. What are the aspects of the problem which have the most far-reaching implications in the context of the problem statement?

The Logic of the Search for Alternatives

Objective: to formulate alternate concepts which satisfy the objective of storing 10,000 pallets and handling 800 transactions daily.

To limit the search for alternatives it is best to look for that part of the system which most restricts the range of possible solutions.

The case study addresses itself to a specific task in a general environment; no specific site is chosen. Therefore, the limitations usually imposed by a specific location such as codes, labor situation, height restrictions for buildings, and dimensions of the piece of land, are not used in generating possible alternatives. Within the general framework, it becomes evident that material handling equipment is a most limiting factor to possible solutions. Forklift trucks can store pallets in stacks 20-25 feet above the floor level. In our case with 4' high pallets and the necessary margins for shelving, we can stack 5 pallets on top of each other. Presently built storage systems which utilize stackers and transfer cars (see Figure 1) can reach up to 100 feet above floor level. Considering our pallet height again and the margins necessary for shelving, we see that we can stack a maximum of 20 pallets.

At this point we may establish a range of 20 different heights, 1-20. Since it is not reasonable to stack only one pallet high using only 4' of an 8' minimum acceptable ceiling height, we are left with a range of possible heights from 2-20 pallets. For all this range of stacking heights there are hundreds of thousands of plan arrangements. The two main factors to be considered while searching for reasonable layouts are: the efficient use of equipment and labor, and the cost of the related building shape.

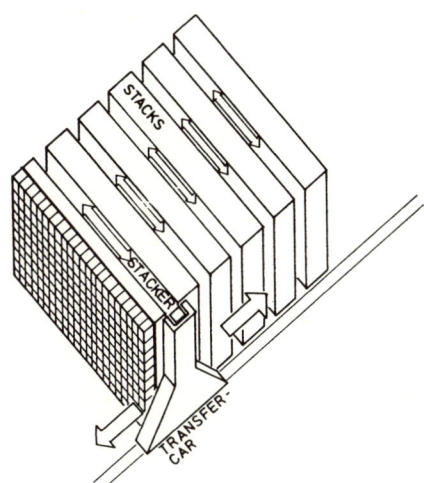

Figure 1

A good indicator for building economy is the ratio of enclosure built to volume created. This is especially true in respect to warehouses. Since it is the skin which costs money (initial cost, cost of maintenance as well as cost for heating and cooling) and the sheltered volume is the usable item, it is clear that the task is to create a maximum volume with a minimum total surface area. The solid with the highest ratio of volume to surface area is the sphere. Presently available handling equipment and shelving systems provide for rectangularly organized storage systems. Therefore, the occupied space within a spherical warehouse would leave considerable unusable space and suggests the most complete utilization of the building volume lies in the use of rectangular building shapes. (See Figure 1) For each of the identified possible stacking heights, the building shape becomes optimal in terms of construction, maintenance, and heating and cooling costs, when the perimeter is minimal. Under the previously established condition of rectangularity, the perimeter of a given area is minimal for a square.

The efficiency of equipment use is another important consideration for the layout of the warehouse. Part of my work establishes that the mean distance to all storage locations is minimal when the length of the building is twice the width. At this point we do not know which factor is more influential for the total cost of a given alternative, the building cost or the efficiency of layout.

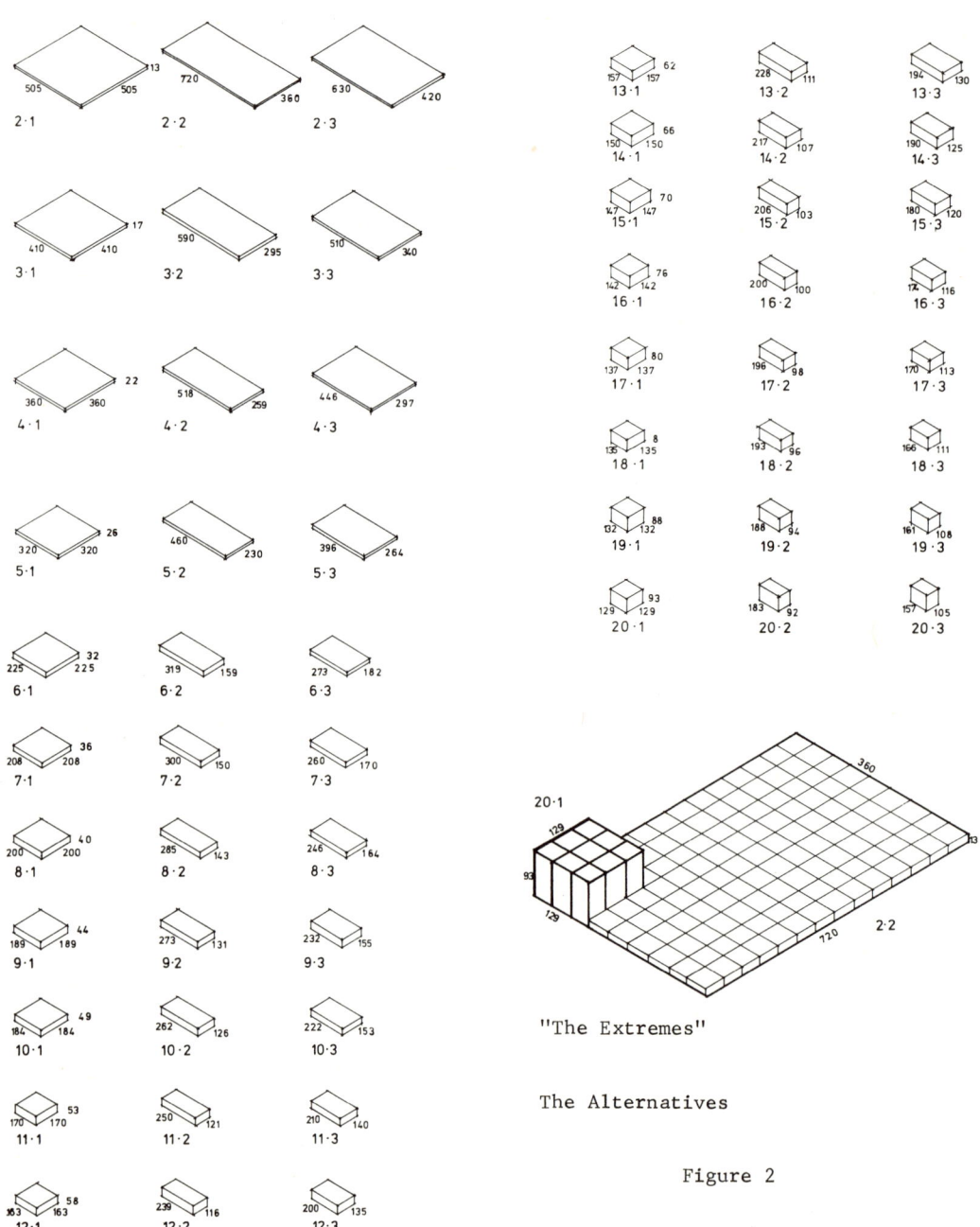

"The Extremes"

The Alternatives

Figure 2

Therefore, a third alternative for each stacking height was established. It provides for the same area, but the length of the building is the arithmetic mean of the length of the "square" alternative and the length of the "rectangular" alternative. There are now 19 different stacking heights, and three alternatives for each. Thus we have 59 alternatives.

The floor area required for each alternative depends on two factors: 1) the number of pallets "stored" on the ground floor; 2) the equipment used to stack the required height.

Once the stacking height, layout, and required area are identified, the configuration of each of the 59 alternatives is determined. (See Figure 2)

The Formalist Approach

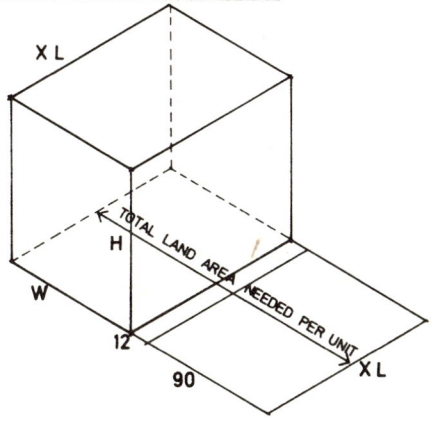

Figure 3

The design alternatives were generated by the heuristic approach described above. They are now evaluated by a formalistic approach with the objective of expressing the anticipated costs and benefits as a direct function of the building dimensions. (See Figure 3) The following is a list of costs directly dependent on the building dimensions: The total land cost, total cost of site preparation, cost of paving, the total building cost and all of its parts with the exception of structural cost and foundation cost, the maintenance cost of building and site, the heating and cooling cost.

Some costs could not be expressed as a direct function of the building dimensions, and have to be separately established. They are the costs of: the structure (see Figure 4), the foundation, the heating and cooling equipment, the shelving system, the material handling equipment, the labor, the maintenance of equipment, the replacement of equipment and parts.

The Computer Program

Once the specific cost data for each alternative were found, and the equations describing the direct relationship between building dimensions and respective costs were established, a computer program was written. The purpose of the program was to provide an economic base for the evaluation of all alternatives subjected to financial analysis. Specifically, the program was to facilitate the consideration of a number of different economic environments so that the sensitivity of the alternatives to different tax situations, rates of return, land costs, and lives of the investments, etc., could be tested.

The alternatives differ in the costs of their parts. At different times in the life of the building investments have to be made to account for these costs. Therefore, a method is required making these cost comparisons possible. The method selected by which all alternatives are compared is the present cost analysis. Money has a time value. If a cost or a benefit occurs at a future date, the sum of money needed or gained at that date could be equaled by investing a certain lesser sum now. Present cost analysis accounts for this fact. Rather than assuming a revenue, it was decided to only use the costs and benefits of each solution for comparison.

The present cost (PC) of all future costs (C) and benefits (B) is:

$$PC = \frac{C_o - B_o}{(1+R)^o} + \frac{C_1 - B_1}{(1+R)^1} + \frac{C_2 - B_2}{(1+R)^2} + \ldots = \frac{C_n - B_n}{(1-R)^n} \quad (1)$$

Where C is the initial investment, B is the benefit such as tax savings for example, during year O, R is the expected return on capital, or opportunity cost. The subscripts of C and B and the power of the expression in the denominator are the years in which the costs or the benefits occur.

It can easily be recognized that a high return will make future investment benefits less important than a low return. With a high rate of return the initial investment has more weight in terms of the total cost than with a low rate of return.

The benefits considered in this case study are equal to the taxes saved. There are basically two tax situations considered: 1) No taxes, as is the case for government operations; 2) 50% tax bracket, as is the case for most corporations. In the case of the 50% corporatate tax, half of all noncapitalized expenses occuring during the year are deductible from the taxes to be paid. Therefore, half of the maintenance costs, labor costs, noncapitalized repair costs, etc., are saved and recovered. The depreciation allowance granted by the Internal Revenue Service is another source for tax savings. There are three methods most generally used by which depreciation can be computed. The straight line method, the declining balance, and the sum of the years digit method. Presently only the straight line method can be used for the type of problem discussed in this case study. Consequently only that method was considered in the computer program. The procedure to compute the yearly depreciation allowance is quite simple. "The depreciation for each year is determined by dividing the ad- (2) justed basis of the property, less salvage value, by the remaining useful life."

For the purpose of this study three alternate lives of the warehouses are considered: 20, 40, and 60 years. Forklift trucks are assumed to have a useful life of five years, and the stacker cranes are assumed to have a useful life of twenty years. Some parts of the storage systems are assumed to have a life of only five or ten years, the replacement cost of these parts then is capitalized and also used for depreciation purposes. The heating and cooling equipment is assumed to have a life of ten years.

After the costs and benefits and their occurrence over time were determined, the computer program provided for the kind of output discussed in the following paragraphs.

The Output
The computer print-out shows the results of the cost analysis. For each of the various economic environments (inputs), i.e., the rate of return, cost of the land, site preparations, climate, tax situation, etc., the computer program provides one set of print-outs. Each set has two major parts--the tables showing the results of the analysis of the individual alternatives and the comparative ranking of the alternatives according to their final cost.

The yearly costs and tax benefits are listed for each alternative. For the years 1-4, 6-9, 11-14, 16-19, and so forth, the costs are assumed to be constant. They consist primarily of the labor cost, the equipment operating cost, the cost of heating and maintenance, and typical equipment replacement which account for marked increases in cost. Every 10 years the heating equipment is assumed to be replaced. In the case of an alternative in which stacker cranes are used, the storage system is assumed to need replacement every 20 years.

The costs and benefits lead to the present cost for each individual year when the present cost analysis method is applied. The final cost is the sum of all present costs occurring during the life of the investment.

Parameter Analysis
Figure 4 shows the land area, total enclosure area, roof area, and wall area, for each of the alternatives required. These parameters are good indicators of building economy in terms of initial and future costs. Naturally, the smaller the areas required per alternative, the lower the construction cost, and the lower the future maintenance, repair, and heating and cooling costs. With increased stacking height the areas of total enclosure, roof and land decrease, whereas wall area increases.

The graphs on the right hand side of Figure   show the cost of the building, the annual heating cost, and structural costs are directly dependent on the total enclosure area.

Every building design affects construction and maintenance costs, as well as the cost of operating within the structure. The warehouse alternatives have different operational costs. The relationship of alternative to initial operational investment is also shown in Figure 4.

Several observations can be made from the typical graphs presented. As stacking height increases, the cost of the building decreases, even though the wall area increases. The decrease in total building cost, with increasing stacking height, results from marked decrease in roof area, floor area, and structural cost. However, with increased stacking height the operational investment necessary increased dramatically.

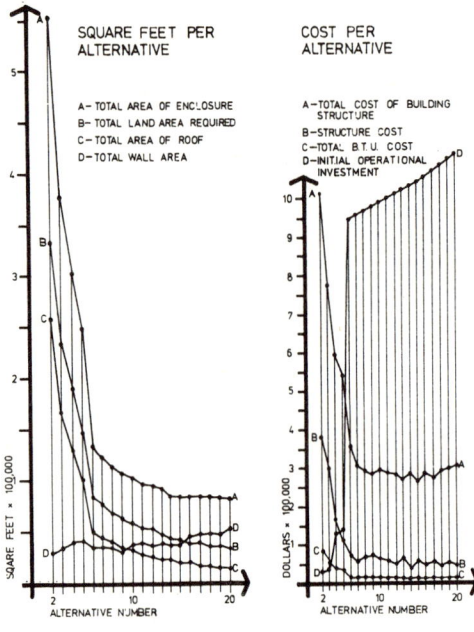

Observations based on graphs in Figure 4 indicate that alternative 5-1 might be the most economical solution to the problem, as stated. However, the question arises: how does the final cost of this alternative compare with the cost of other alternatives in various economic environments?

Figure 4

Sensitivity Analysis

Sensitivity refers to the relative magnitude of the change in one or more elements of an engineering economy problem that will reverse a decision among alternatives. Thus, if one particular element can be varied over a wide range of values without affecting the decision, the decision under consideration is said not to be sensitive to uncertainties regarding that particular element. On the other hand, if a small change in the estimate of one element will alter the decision, the decision is said to be very sensitive to changes in the estimates of that element. (3)

With a constant basic set of data, varying one factor at a time, considering only 6 major cost factors, e.g., cost of land, expected return, tax situation, the life of the venture, site preparation cost, and cost per square foot of wall, with 3 different values for each, produces 18 economic environments. These same changes made in combination produce 3^6 = 729 economic environments.

While the possiblities were numerous, I decided to investigate effects on comparative rankings by varying the cost factors listed before one at a time, thus assuring that the computing time remained reasonable. This part of the case study demonstrates the sensitivity of the model, rather than the effect many conceivable economic environments have on it.

Wall Panel Cost PSF*	$1.00	$3.50	$3.50
Annual Repair	$ .10	$ .00	$ .00
U-Factor Heat Transfer	$ .17	$ .13	$ .13
Rate of Return	20%	20%	30%

Ranking	Alternative No.	Present Cost	Alternative No.	Present Cost	Alternative No.	Present Cost
1	5-1	1,077,995	5-1	1,145,462	5-1	1,039,718
2	5-3	1,110,769	5-3	1,180,393	5-3	1,073,592
3	5-2	1,140,102	5-2	1,213,079	5-2	1,105,528
4	4-3	1,216,100	4-3	1,280,591	4-3	1,165,336
5	4-1	1,230,986	4-1	1,293,420	4-1	1,180,528
6	4-2	1,246,405	4-2	1,314,084	4-2	1,197,865
7	3-1	1,352,453	3-1	1,404,389	3-1	1,281,107
8	3-3	1,408,701	3-3	1,462,577	3-3	1,337,303
9	3-2	1,433,132	3-2	1,489,582	3-2	1,363,589
10	12-1	1,572,624	12-1	1,654,285	12-1	1,550,228
11	13-1	1,573,231	13-1	1,657,462	13-1	1,553,838
12	12-3	1,578,539	9-1	1,662,199	9-1	1,555,896
13	14-1	1,578,768	12-3	1,662,462	7-1	1,557,537
14	15-1	1,582,622	14-1	1,664,579	12-3	1,557,824
15	13-3	1,582,842	7-1	1,665,927	14-1	1,561,646
16	15-3	1,590,573	11-1	1,669,718	13-3	1,565,542
17	9-1	1,591,055	13-3	1,669,757	11-1	1,565,556
18	15-2	1,591,848	15-1	1,671,912	15-1	1,568,999
19	11-1	1,592,091	11-3	1,673,979	11-3	1,569,101
20	13-2	1,592,999	10-1	1,675,852	10-1	1,569,847
21	11-3	1,594,059	9-2	1,681,667	9-2	1,574,444
22	12-2	1,597,454	15-3	1,681,699	15-3	1,578,474
23	10-1	1,598,448	13-2	1,683,989	13-2	1,579,320
24	14-3	1,599,377	15-2	1,685,757	10-3	1,579,952
25	7-1	1,602,485	10-3	1,686,312	12-2	1,581,156
26	16-1	1,602,488	12-2	1,686,428	9-3	1,582,076
27	11-2	1,603,989	9-3	1,688,534	15-2	1,582,242
28	9-2	1,605,511	11-2	1,688,747	11-2	1,583,015
29	10-3	1,607,415	14-3	1,689,460	7-3	1,583,481
30	14-2	1,609,562	7-3	1,692,409	14-3	1,585,764
31	16-3	1,610,508	10-2	1,695,416	10-2	1,588,940
32	10-2	1,613,674	16-1	1,696,274	7-2	1,593,240
33	9-3	1,615,671	14-2	1,702,278	16-1	1,593,866
34	17-1	1,617,464	7-2	1,702,932	14-2	1,598,526
35	16-2	1,619,580	16-3	1,706,285	16-3	1,603,553

*PSF equals per square foot

TABLE 1: SENSITIVITY ANALYSIS

Table 1 shows, in Columns 1 and 2, the effects a lower price wall panel, with consequently higher repair costs, and a larger U-Factor, has on the comparative ranking of the alternatives, and the respective total costs. In Column 3 the influence of a higher rate of return on the ranking is demonstrated. Again, the decision to select Alternative 5-1 is not reversed. It is interesting to note that once Alternative 5-1 is chosen its total cost can be lowered by further refinement of the design. As this case suggests, the ranking of alternatives would not be altered by applying the same improvements to the other alternatives.

GRAPH NO.	1	2	3	4	5	6
Site Preparation Cost - PSF*	$7.50	$5.00	$3.00	$ .50	$7.00	$7.00
Rate of Return	5%	5%	5%	5%	10%	20%
Ranking	Alternative	Alternative	Alternative	Alternative	Alternative	Alternative
1	5-1	5-1	5-1	5-1	14-1	15-1
2	14-1	5-3	5-3	5-3	15-1	14-1
3	15-1	5-2	5-2	5-2	13-1	13-1
4	16-1	14-1	4-1	4-1	16-1	16-1
5	17-1	15-1	4-3	4-3	17-1	17-1
6	13-1	13-1	4-2	4-2	12-1	12-1
7	12-1	16-1	14-1	3-1	15-3	15-3
8	15-3	12-1	13-1	3-3	5-1	18-1
9	18-1	17-1	15-1	3-2	18-1	16-3
10	19-1	15-3	12-1	12-1	16-3	19-1
11	16-3	13-3	16-1	13-1	19-1	15-2
12	17-3	15-2	15-3	14-1	15-2	13-3
13	15-2	16-3	13-3	15-1	17-3	17-3
14	13-3	11-1	12-3	12-3	13-3	12-3
15	20-1	18-1	11-1	11-1	11-1	11-1
16	19-3	12-3	17-1	13-3	14-3	14-3
17	18-3	14-3	15-2	15-3	12-3	19-3
18	14-3	17-3	14-3	9-1	19-3	18-3
19	20-3	19-1	16-3	11-3	18-3	20-1
20	11-1	18-3	11-3	15-2	20-1	20-3
21	12-3	16-2	13-2	16-1	20-3	17-2
22	16-2	19-3	17-3	14-3	16-2	16-2
23	17-2	17-2	18-1	13-2	17-2	5-1
24	5-3	11-3	9-1	10-1	14-2	14-2
25	14-2	14-2	14-2	12-2	13-2	13-2
26	13-2	13-2	16-2	7-1	11-3	11-3
27	18-2	20-3	10-1	16-3	18-2	18-2
28	11-3	20-1	17-2	17-1	19-2	10-1
29	19-2	9-1	19-1	14-2	10-1	19-2
30	20-2	10-1	18-3	11-2	9-1	9-1
31	10-1	12-2	12-2	10-3	20-2	12-2
32	9-1	18-2	19-3	9-2	12-2	20-2
33	12-2	4-1	20-3	9-3	5-3	11-2
34	11-2	19-2	11-2	16-2	11-2	10-3
35	10-3	11-2	10-3	10-2	10-3	9-3
36	5-2	20-2	20-1	17-3	9-3	7-1
37	9-3	10-3	9-3	17-2	10-2	10-2
38	10-2	9-3	18-2	18-1	7-1	5-3

*PSF equals per square foot.

TABLE 2: SENSITIVITY ANALYSIS

Table 2 shows the effects changes in site preparation costs and interest rates have on the comparative ranking of the alternatives. See also Graphs 1-6 in Figure 5. When the site preparation cost reaches $7.50 as presented in Column 1, Alternative 5-1 still leads the field, followed very closely by the high-stacking Alternative 14-1. This alternative requires considerably less land - 37,800 sq. ft. as compared to 135,000 sq. ft. in Alternative 5-1. In Columns 2-4, site preparation cost decreases and Alternative 14-1 drops from the second rank to place 7 in Column 3 and place 12 in Column 4. It is interesting to note that an increased interest rate--from 5% to 10%, as seen in Column 5--has a strong effect on the ranking. Alternative 5-1 drops from place 1 to place 8, even with a slightly lower site preparation cost--$7.00 in Column 5 as compared to $7.50 in Column 1. If the interest rate is increased to 20%, as in Column 6, Alternative 5-1 drops to place 23.

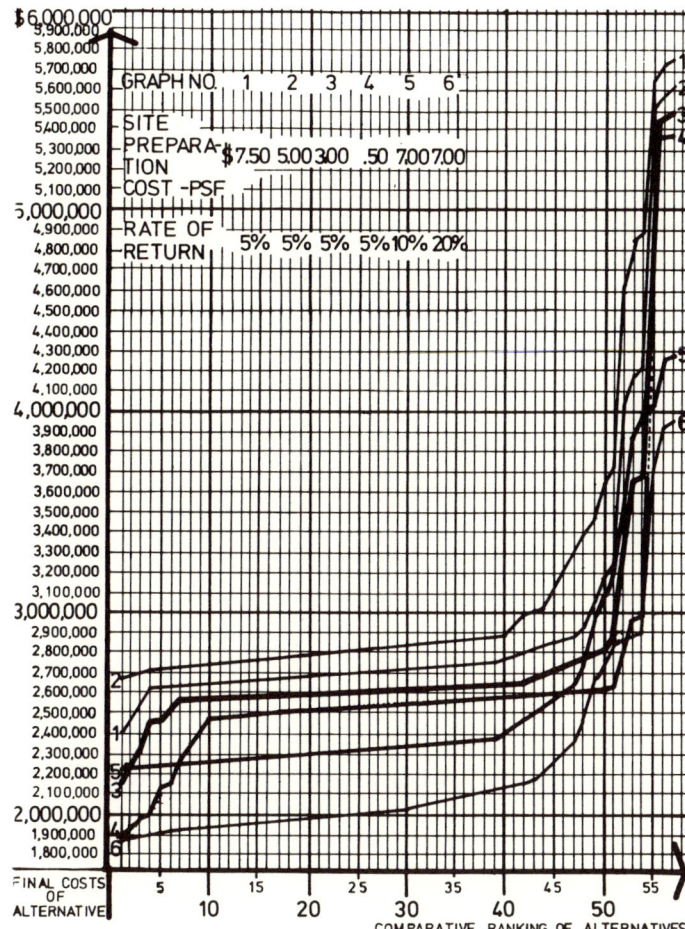

Figure 5

Only high land costs and/or high site preparation cost were found to affect the comparative ranking of alternatives severely. Alternative 5-1 would not be the first choice, should the sum of land and site preparation cost per square foot exceed a limit of $7.50.

Conclusion
The case study describes a model suited for decision-making in architectural design. The approaches taken in the generation, and evaluation of possible concepts and finally in the selection of the most promising alternative, seem to be applicable to larger and more complex problems particularly where the total building can be thought of as consisting of a sum of units (modules) as in office buildings.

Providing for the needs arising from a problem in the best way possible is the crucial task in environmental design. Many aspects of the needs, especially those which concern themselves with amenity and our value system, i.e., the nonphysical environments, escape quantification. However, these nonquantifiable aspects are often the real essence of a problem and, therefore, need to be used in the evaluation process. Alternate courses of action proposed to solve a problem are different means to achieve the same end. Alternatives usually vary to a great degree in their provision for the nonphysical and physical aspects of the needs. To generate an alternative which under identified conditions promises to provide for the least monetary cost does not necessarily mean that this same alternative provides for the best value. This depends very much on the context in which a problem is seen. Value strictly seen as monetary value for the owner or as value and amenity to a community, for example, will have strong bearings on the proposed evaluation. In any case, the approach as developed in this thesis is capable of establishing the financial outcome of design decisions; it allows for testing the relative feasibility of alternative courses of action. Often it will be possible to select a higher cost alternative which, considering the intangibles, creates a better value. (Decision context versus analytical context.)

One of the major advantages of the model is its capability to select the most suitable alternative early in the design process, making it possible to focus all effort into developing and perfecting this alternative.

Notes

1. Küsgen, Horst, Planungsökonomie-Was Kosten Planungsentscheidungen? Karl Krämer Verlag, 1970, p. 44.

2. Internal Revenue Service, Depreciation, Amortization, and Depletion, Publication 534 (2-72), p. 6.

3. Grant, Eugene L. and Ireson, W. Grant, Principles of Engineering Economy, Fifth Edition, The Ronald Press Company, 1970.

PROPOSAL FOR A DIAGRAMMATIC LANGUAGE FOR DESIGN

Robert E. David

Design Division
Kansas City Art Institute
Kansas City, Missouri

Abstract

This paper proposes a notational language of diagrammatic elements which can provide the designer with a communication tool that permits him to visualize basic design ideas in a consistent manor at a high level of abstraction. The primitive elements of this language represent a set of ideas that, in various combinations, have been found to recurrently make up the basic entities of various problems in environmental design.

For the purposes of this paper we will consider the design process to consist of two major (but not necessarily distinct) phases: the analytic phase and the synthetic phase. The analytic phase begins with a definition of the problem and expands into a search of pertinent literature and existing solutions to similar problems. The intent of this phase is to produce what the architectural profession calls the "program," a delineation of the objectives or requirements of the final design solution. The medium of expression in the analytic phase is, for the most part, the designer's vernacular and thus is essentially verbal. The synthetic phase of the design process hopefully will produce a working drawing and/or scale model description of an acceptable solution to the design problem. These working drawings are commonly preceded by several stages of "preliminary" drawings of varying degrees of completion which are, in turn, preceded by many stages of "sketches." The medium of expression in the synthetic phase is, for the most part, scale drawings of varying degrees of crudeness and thus is essentially visual.

This brings us to that stage of the design process which can be considered the middle zone of the process, where the verbal information of the analytic phase is evolved into the visual information of the synthetic phase. This transformation from verbal to visual information is characterized by series after series of sketchy expressions best described as "doodles" (i.e. bubble diagrams, etc.). If we consider design as a communication activity, these doodles, their verbal predecessors, and their visual successors are tools of communication which the designer uses to continuously record developing ideas. In the case of the single designer, these records are maintained so that ideas may be communicated accurately back to himself. In the case of a team of designers working on a common problem, the records serve as verbal and visual conversational aids among the designers involved.

Conceivably, the sophistication of the design solution depends to a certain extent on the sophistication of the communication tools used. The tools in current use in the initial verbal stages and the final visual stages of the design process are reasonably well developed (i.e. the vernacular and the working drawing systems. However, the efficient handling of ideas generated as doodles is often difficult for a single designer because of his lack of recall, and it proves to be impossible for a team of designers since each member of the team will have a different doodle "style." Actually, collaboration at this stage of the design process is most likely not even considered. It follows that there may exist a need for a notation system which would allow design ideas to be expressed and recorded in a consistent way during this doodle stage of the process. Hereinafter these doodles will be termed diagrams, and their conceivable organization will be termed a diagrammatic language.

My interest in a diagrammatic language for design stems from some experience in the use of decomposition computer programs for the analysis of design problem structure and the use of constructive diagrams as part of this design method (1).

It is my contention that there exists a set of diagrammatic elements which designers subconsciously use as part of their thought patterns which can be identified and formalized into a grammatical structure as a diagrammatic language for design. Research to date has been concerned with the following five stages:

1. Design methods and the design process in general.
2. Diagram systems in current use in other disciplines.
3. Artificial languages and their compilers.
4. Identification of the primitives and grammar of the design language.
5. Delineation of the possible visual form of the notation elements.

The next stage of the project will be:

6. Experiments in the use of the language.

Full development of the language will be an evolutionary process resulting from its use. This paper, therefore, constitutes a preliminary report on a continuing project and is presented at this time to draw comments and criticism from designers and educators. This project was initially undertaken as this author's thesis done in partial fulfillment of the requirements for the degree of Master of Science of Product Design at the Institute of Design of Illinois Institute of Technology.

The Design Process

It may be helpful at this point to look at some conceptual models of the design process in order to see where this diagrammatic language could be used. Figure 1 shows a very simplistic and familiar conception of the design process. The starting point is the definition of the problem, and the endpoint is its solution. The area of divergence is essentially verbal (phase 1), and the area of convergence is essentially visual (phase 2). The area at the center is the realm of diagrams which supports the transition from verbal to visual.

Figure 2 is a more complex model adapted from Shannon's model of communication (2). It can be considered in two ways: first, as an overall model of the design process in the sense of problem definition (input), designer activity (process), and solution (output); second, as a model to be duplicated end-to-end as many times as there are identifiable stages of communication activities in the design process. For

7. DECISION MAKING TOOLS / 373

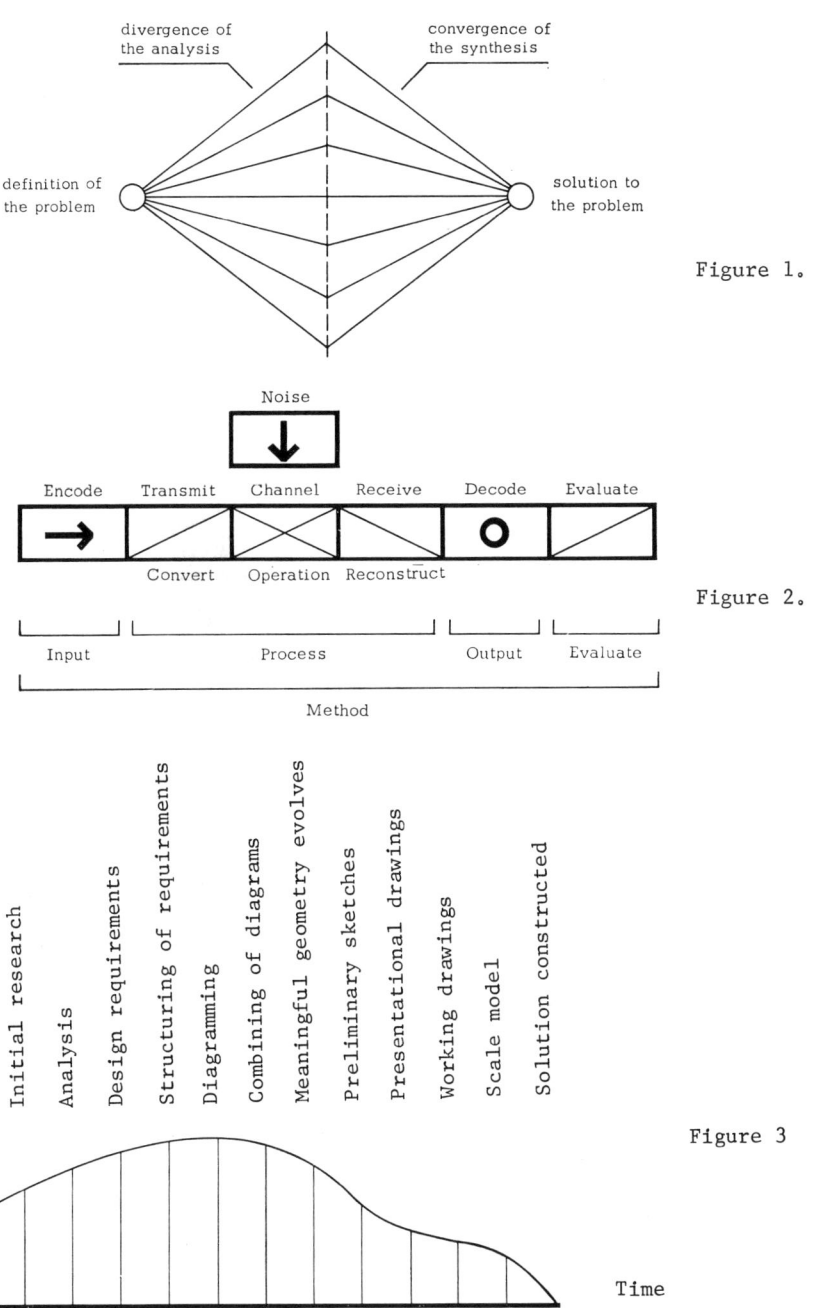

Figure 1.

Figure 2.

Figure 3

instance, the designer may "encode" his thoughts and transmit them via a "channel" of a sketch on paper to be "received" by a design partner or consultant or himself at a later time. "Noise" could be anything that would in any way deteriorate the quality of the intended message, which could be anything from wrinkling the paper to the designer's lack of talent as a draftsman. Relative to diagrams as a communication activity we must consider such questions as: What is the encoding and transmitting process? What kind of channel is used? What is the receiving and decoding process? What kind of noise will be present during the process? How is the result evaluated? For the sake of simplicity the discussion will presume a handwritten notational system using suitable two-dimensional surfaces as "channels."

Figure 3 shows a still more complex model. It is a graph of the overall elapsed time of the design process divided into a sequence of the various modes of expression which constitute the periodic output of the design process. The vertical dimension of the graph indicates relative degrees of abstraction. For example, preliminary sketches are more abstract than a scale model. Zero abstraction is, of course, the final solution fully constructed and ready for use. We can say that the area of the curve to the left of the peak is the analytic phase of the process, and the area to the right is the synthetic phase. The important issue is that the higher the degree of abstraction of a mode of expression, the more manipulatable it is in terms of efficient use of human time and energy relative to its effect on the form of the final design solution. For example, a single line as an element of a diagram may represent the basic configuration of an entire architectural space, whereas a single line as an element of a working drawing may only represent the surface of a single wall. Thus, in terms of designer time and energy, the removal of this single line in a diagram might remove the entire wing of, say, a school building, whereas this same amount of line removed from a working drawing might only mean the removal of a single partition.

The reason for the for the different modes of expression for the various stages of the design process is that each of these modes is best suited to record the type of human thought being generated at that stage of the design process (3). Herbert Simon refers to the elements of thoughts as "chunks," and the human mind generates and receives these chunks at reasonably constant intervals (4). If the nature of a chunk is different at one stage of the design process than at another, the mode of expression (notation) used to record these chunks as they are generated must be capable of recording at a rate commensurate with the interarrival time of the chunks. If the lag between chunk speed and notation speed becomes great, an important chunk may be forgotten before it can be notated. A diagrammatic language must provide a notation system that can cope with the relational type of thought produced at this middle zone of the design process (5).

Diagrams and Language Systems

If we are to devise a diagrammatic language for design, it would seem appropriate to survey diagrammatic systems in use in other disciplines. Space does not permit the discussion of any of these here, but the following are worthy of mention:

pictographs (6)   shorthand

alphabet (7)
music notation (8)
Gantt and PERT charts (10)
Therbligs (12)
graph theory (13)
bubble diagrams
Polyominoes (14)
Motation (16)
Proxemic notation (18)
aUI: The Language of Space (20)

nomograms
Labanotation, dance notation (9)
process charts (11)
flowcharts
Venn diagrams
engineering schematics
Sequence Experience Notation (15)
Kinesic notation (17)
Blissymbolics (19)
Sign Language (21)

All of the above are visual systems containing a limited number of symbolic elements which in various combinations and formats are capable of describing a limitless number of situations (22). In addition to the field of visual communication, traditionally associated with graphic design (23), there is new interest in the use of visual languages as design tools (24). The more firmly established world of non-visual linguistics (25) functions with a few elements representing everything the human voice is capable of uttering. Basic English reduces to 800 the number of English words necessary for most conversations (26). Inter-disciplinary research in environmental design has spurred an interest in formulating a set of environmental descriptors for an agreement on terms in speaking about natural and man-made environments (27).

Diagrams as a Language System in Design

The design process is an evolutionary progression; an idea is refined continuously until the resulting form satisfies the initial objectives. It is a process that proceeds from the general to the specific and from the abstract to the real. It is a process of leveling and sharpening and reorganization (28). Once the analytic phase of the process is complete and the design objectives defined, the diagrammatic language can be employed to set down in a presentational visual way (as opposed to the sequential verbal methods of the analytic phase) the desirable relationships between the conceptual components of the problem. These relationships will evolve into more specific relationships that are defined by type and degree. The diagrammatic language must be both visual to show relationships, and numerical to show degree of relationship (29).

Primitives and Grammar of a Design Language

Since the initial diagrams in the design process would express the desirable relationships specified in the design requirements, it seemed logical that the place to begin a search for the basic elements, or primitives, of the design language would be among available collections of design requirements that had been compiled by various designers for particular problems. Consider the following three design requirements which are relevant to three different design problems:

1. Urban mass transportation system: Any pollution created by the transportation system must be contained and not be allowed to enter the environment

external to the transit system.

2. Elementary educational facility: Distraction factors external to an academic setting (such as a classroom) must be controlled so as to not disturb the students' concentration.

3. Residence: Children in their play activities should not be subject to contact with vehicular traffic.

The notion to be abstracted from these three requirements is a common one: "protection." Protection is the desirable relationship that should exist between the pairs of parties named in each of the three requirements.

In a similar way, the accumulated design requirements of seven different documented environmental design problems were searched. The average number of requirements per problem was about 100; approximately 150 notions were abstracted from these 700 requirements. Some notions occurred only once or twice, whereas notions such as "protection," "proximity," "access," and "control" occurred 10 to 30 times each. Redundant notions (such as "control" and "supervision") were eliminated until approximately 100 notions remained. A positive or negative relationship was identified between every possible pair of notions and a decomposition computer program (30) was used to identify subsets of highly interrelated notions. Analysis of the subsets revealed further synonymous ideas which were eliminated to produce a list of 49 notions (see Table 1), which may be considered as an initial attempt to see what the primitives of the design language might be.

Table 1. Notions Abstracted from Design Requirements.

1. adjacency	18. feedback	35. privacy
2. barrier-access	19. format	36. process
3. behavior setting	20. gate	37. property
4. boundary	21. generate-terminate	38. protection
5. communication channel	22. guidance	39. proximity
6. comparison	23. implicit	40. queue
7. compliance	24. information	41. relation
8. conceptual	25. input-output	42. replace
9. content	26. media	43. seize-release
10. context	27. mobile-stabile	44. sequence
11. dependency	28. motion (speed)	45. similarity
12. disipation	29. negative-positive	46. static
13. dynamic	30. parts	47. tool
14. enclosure	31. people	48. supervision
16. enter-leave	32. physical	49. vehicle
17. explicit	33. point of contact	
18. facility	34. position	

These 49 primitives were, in turn, processed through the decomposition program and simplified further by observation and experiment to produce the structure shown in Figure 4. An initial attempt at defining the visual elements of the language is

7. DECISION MAKING TOOLS / 377

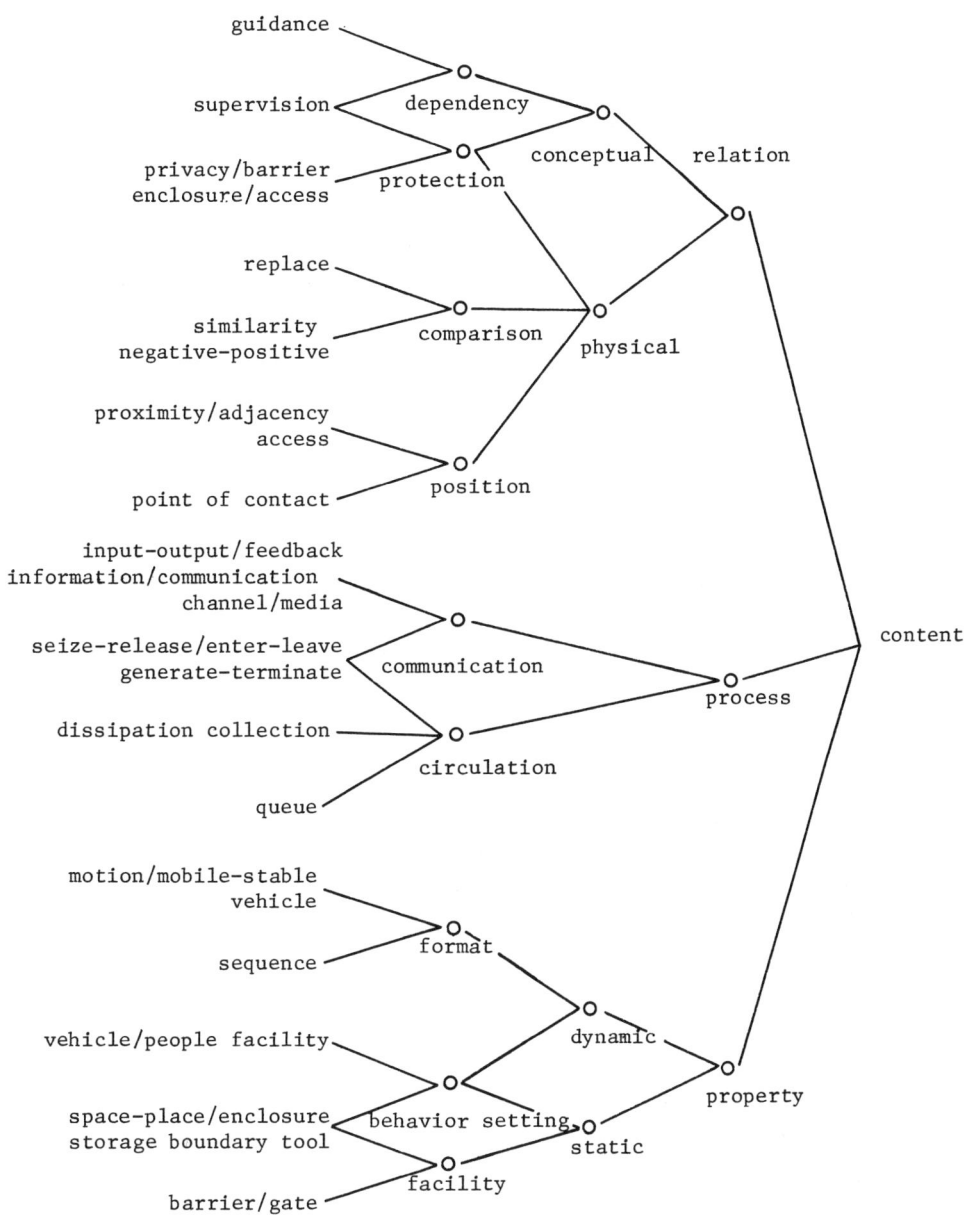

Figure 4

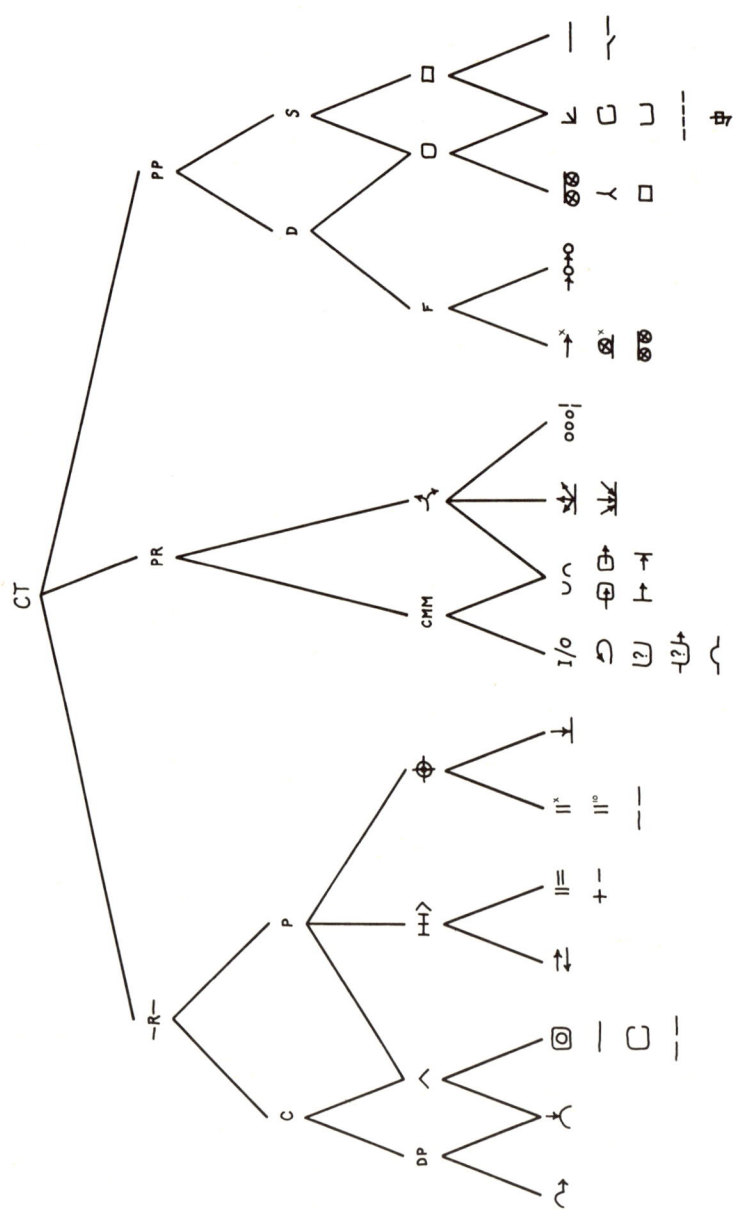

Figure 5.

shown in Figure 5, where symbols and mnemonics have been substituted for the words in the structure of Figure 4.

The structure of this "grammar" proceeds downward from the general to the more specific. This is analogous to the leveling and sharpening nature of the design process. This structure is in need of further refinement, but it will provide a reasonable vehicle for illustrating the manner in which it might be used.

Application

As an illustration, let us consider the design of the architectural students' academic work station -- the studio. For the sake of simplicity let us consider two of the many design requirements which might be produced by the analytic phase of the design process as applied to this problem:

1. Security from theft: The typical design school layout must permit continuous student access to the student studio areas. This open situation also permits access by strangers whose intention might be theft of the students' work tools.

2. Studio teaching routine: Instructors typically come to the student studios for individual consultation with each student. However, there is often need of group discussion between faculty and students in the studio intermittent with individual consultation.

Consider the structure in Figure 5 and keep in mind that it will be used in a top-to-bottom process. Beginning with requirement 1, we have a group of four student work stations. These are entities which are identifiable by their property which is both dynamic and static as a behavior setting. Let us place these in a behavior setting drawn as an enclosure (Figure 6A). In Figure 6B we add another entity whose property is dynamic as a person and is considered to be an intruder. This person has a certain relation "R" with each behavior setting, and the desirable relation is protection (Figure 6C) which can be accomplished by the addition of an additional enclosure (Figure 6D) about each behavior setting.

Continuing with requirement 2, we have the same four behavior settings in their studio enclosure (Figure 7A). In Figure 7B a new entity is added as a person who is the instructor, and this person assumes a relationship with each behavior setting and the person at each of these behavior settings. The problem presented is that of group discussion, which might be solved by the instructor being able to assume a position of equal proximity to each of the persons in the individual behavior settings. This might require a rearrangement of the behavior settings as in Figure 7C.

The superior figures ("4" and "6") used with the person symbols in the diagrams of Figure 7 refer to the relative roles each of the persons depicted in the diagram play in the behavior setting. The "role" in question is that of relative centrality or "level of penetration" on a scale of 1 to 6 as devised by Barker in his theory of behavior settings (31). Thus, the person with the "6" has a more central role in the behavior setting than those with the "4." Superior figures or mnemonics might also be used along with the behavior settings themselves in order to give an

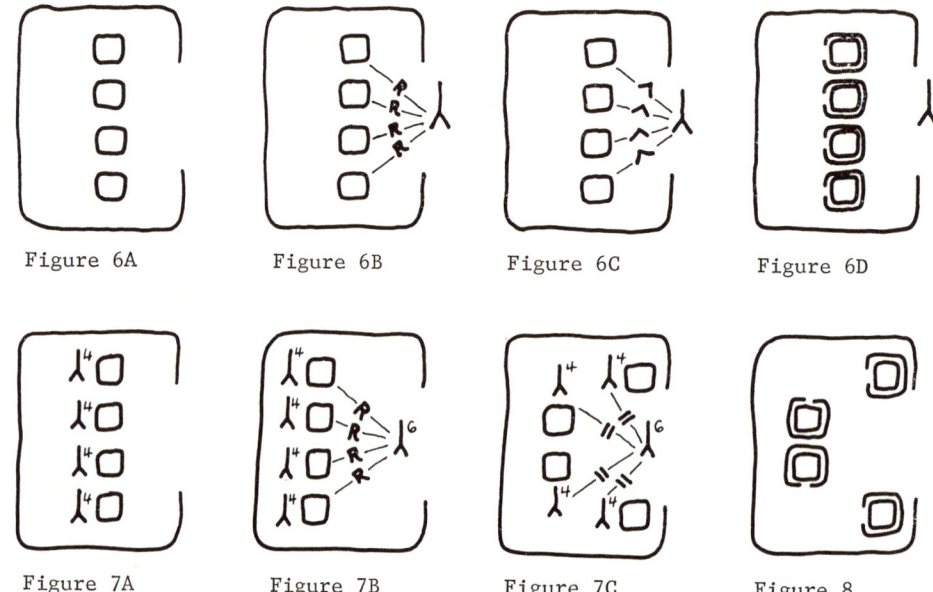

Figure 6A   Figure 6B   Figure 6C   Figure 6D

Figure 7A   Figure 7B   Figure 7C   Figure 8

indication of what type of behavior setting is present.

The combination of Figure 6C and 7C is shown in Figure 8. The physical form which will ultimately satisfy the diagram of Figure 8 might be individual work stations arranged in a circular configuration to satisfy the discussion requirement. The theft problem might be solved by the use of a roll-top structure which could cover the entire work surface in one easy operation.

Conclusion

The example above was, of course, overly simplified; many more design requirements would have come into play in the normal design situation and the diagram would have become quite complex. However, the elements for one approach to a basic theory of diagrammatic language is presented. The grammatical structure as presented here needs to be tested and the visual elements have to be refined. The design requirements themselves will, no doubt, have to be written in a special format to make their translation into diagrams as efficient as possible. The next stage of this research effort will be experiments in the use of the language, carried out both in academic and professional environments. The form of the language will undoubtedly change in response to these experiments. Its ultimate efficient use will require, as with any language system, that the language be learned.

Notes

1. Serge Chermayeff and Christopher Alexander. Community and Privacy, (New York: Doubleday, 1963); Christopher Alexander. Notes on the Synthesis of Form, (Cambridge: Harvard, 1964).
2. Claude Shannon and Warren Weaver. The Mathematical Theory of Communication (Urbana: University of Illinois, 1949).
3. Charles Rush and Stuart Silverstone. "The Medium Is Not the Solution," AIA Journal (December, 1967).
4. Herbert Simon. The Sciences of the Artificial (Cambridge: M.I.T., 1969).
5. Christopher Alexander. "From a Set of Forces to a Form," The Man Made Object (Gyorgy Kepes, Editor), (New York: Braxiller, 1966).
6. Otto Neurath. International Picture Language: The First Rules of Isotype (London: Kegan Paul, 1936).
7. David Diringer and Reinhold Regensburger. The Alphabet: A Key to the History of Mankind (New York: Funk and Wagnalls, 1971).
8. Gardner Read. Music Notation (Boston: Allen and Bacon, 1969).
9. Ann Hutchinson. Labanotation (New York: New Directions, 1954).
10. Federal Electric Corporation. PERT (New York: Wiley, 1963); Mel Levinson. PERT for Designers (Chicago: Illinois Institute of Technology, 1968.
11. Ralph M. Barnes. Motion and Time Study: Design and Measurement of Work (New York: Wiley, 1968).
12. American Society of Mechanical Engineering. Operation and Flow Process Charts (New York: A.S.M.E. Standard 101).
13. Oystein Ore. Graphs and their Uses (New York: Random House, 1963).
14. Solomon Goulomb. Polyominoes (New York: Scribners, 1965).
15. Philip Thiel. "A Sequence-Experience Notation," Town Planning Review (April, 1961); Donald Appelyard, Kevin Lynch, and John Myer. The View from the Road (Cambridge: M.I.T., 1964).
16. Lawrence Halprin. "Motation," Progressive Architecture (July,1965).
17. Raymond Birdwhistell. Kinesics and Context (Philadelphia: University of Pennsylvania Press, 1970).
18. Edward T. Hall. "A System of Notation for Proximic Behavior," American Anthropologist (October, 1963).
19. Charles Bliss. Semantography (Blissymbolics) (Sydney, Australia: Semantography Publications, 1966).
20. John Weilgart. aUI: The Language of Space (Decorah, Iowa: Cosmic Communication Company, 1962).
21. Bornstein, Hamilton, and Kannadell. Signs for Instructional Purposes (Washington, D.C.: Gallaudet College Press, 1969).
22. Leonard Bernstein. The Infinite Variety of Music (New York: Simon and Schuster, 1966); Frank Heacock. Graphical Solutions of Technical Problems (Ann Arbor, Michigan: University of Michigan Press, 1964); Henry Dreyfuss. Symbol Sourcebook (New York: Wiley, 1972).
23. Paul Klee. The Thinking Eye (New York: Wittenborn, 1961); Arthur Lockwood. Diagrams (New York: Watson-Guptil, 1969); Emil Ruder. Typography: Manual for Design (New York: Hastings House, 1968); Anton Stankowsky. Graphical Presentation of Invisible Processes (New York: Hastings House, 1966).
24. Constance Perin. With Man in Mind (Cambridge: M.I.T., 1970).

25. Francis Dineen. An Introduction to General Linguistics (New York: Holt, Rinehart and Winston, 1967).
26. Julia Johnsen. Basic English (Bronx, N.Y.: H.W. Wilson Comapny, 1944); Charles Ogden. Basic English (London: Paul, Trench, Trubner, 1933).
27. Joyce Kasmar. "The Development of a Usable Lexicon of Environmental Descriptors," Environment and Behavior (September, 1970).
28. Charles Rusch. "On the Use of Leveling and Sharpening as an Analytic Tool in the Study of Artistic Behavior," Proceedings (Washington, D.C.: American Psychological Association, 1969).
29. Christopher Alexander. Ten-Year Program for Research on Environmental Design (Berkeley: University of California, 1964).
30. Charles Owen. An Algorithm for the Decomposition of Non-Directed Graphs (Chicago: Illinois Institute of Technology, 1968).
31. Roger Barker. Ecological Psychology (Stanford: Stanford University, 1968).

Additional References

Christopher Alexander, Sara Ishikawa and Murray Silverstein. A Pattern Language which Generates Multi-Service Centers (Berkeley: Center for Environmental Structure, 1968).
D.C. Engelbart. Augmenting Human Intellect (Stanford: Stanford Research Institute, 1962).
IBM. General Purpose Simulation System/360 Introductory User's Manual (White Plains, N.Y.: 1969).
Marvin Mannheim. Hierarchical Structure: A Model of Design and Planning Processes (Cambridge: M.I.T., 1966).
Murray Milne and Charles Rusch. A Method of Systematic Design in Architecture (Los Angeles: UCLA, 1969).
Jane Sammet. Programming Languages (Englewood Cliffs, N.J.: Prentice-Hall, 1969).
Alan C. Shaw. The Formal Description and Parsing of Pictures (Stanford: Stanford Linear Accelerator Center, 1968).

THE SIMULATION OF AGE RELATED SENSORY LOSSES: A
NEW APPROACH TO THE STUDY OF ENVIRONMENTAL BARRIERS

Leon A. Pastalan
Robert K. Mautz II
John Merrill

University of Michigan

Abstract

The purpose of this study was to explore the feasibility of developing a new research approach which would enable researchers to simulate relevant environmental experiences of an elderly population. Sets of simple mechanical appliances such as specially prepared lenses, ear plugs, a masking device to decrease olfaction and a fixative to temporarily desensitize the tactile sense were developed and assembled. Four doctoral students in architecture specializing in environmental problems of the elderly wore these appliances for approximately one hour a day over a period of one year in three standardized settings--a dwelling unit, a multi-purpose center and a shopping center. Each of the participants kept an on-going written account of their experiences over the course of the one year duration of the study.

Comparative evaluation of participants' reports, derivation of generalizations and implications for future efforts involving the use of this model as a research technique will be discussed.

The purpose of this study was to construct a simulation model which would duplicate relevant environmental experiences of an elderly population suffering from sensory deficits. Such an approach would provide a unique tool to discover and evaluate the nature of environmental barriers which the elderly and other physically vulnerable people face daily in their homes, neighborhoods and communities.

Since the organism can respond directly only to those aspects of the environment experienced through sense organs, age changes in sensory and perceptual mechanisms affect very real environmental changes in the world in which the aging individual lives. There has been an impressive accumulation of literature regarding the relationship between age-related sensory losses, environmental experiences and behavior. (1)

Because of the availability of this kind of basic data it was possible to simulate certain types of sensory decrements such as increased opacity of the lens, increased rigidity of the middle ear or presbycusis and diminished tactile and olfactory sensitivity by mechanical means. Sets of simple mechanical appliances such as specially coated lenses, ear plugs, a masking device to decrease olfaction and a fixative to temporarily desensitize the tactile sense were developed and assembled.

A basic strategy regarding the empathic model was to develop devices which simulated only "normal" loss and to steer clear of pathologies at least until some baseline information was established for normal losses. Thus our visual, auditory, olfactory and tactile appliances simulate only losses that occur within the context of the normal aging process. An attempt was made to simulate the condition of a person in his late seventies in the belief that this would represent a kind of average or mid-point in terms of this progressive condition, keeping in mind the wide range of variation within the elderly population.

The visual loss which was simulated was the problem of light scatter or glare. A technical description of this condition will not be recounted here; suffice it to say that for all practical purposes the lens of the human eye typically begins to lose its elasticity and gradually starts to become opaque from about the mid-fifties and the condition continues to progress with age. The preparation and coating of the lenses was done under the supervision of Dr. Byron C. Floyd, a practicing optometrist.

Hearing loss for the elderly typically occurs above the 2,000 cycles range and decibel loss averages around 30 for those 65 years of age or older. (2) Through the cooperation of Ronald Rogers, a speech and hearing specialist with The University of Michigan, a material was tested which when used in the form of ear plugs, simulated the above loss very precisely. Each of the researchers had individualized ear plugs made from this material and each were tested for their hearing beforehand.

The literature is rather sketchy in the area of age-related olfactory loss and since there was no reliable guidelines or practical instrumentation available to establish the magnitude of loss, it was felt that simple cotton wading introduced into the anterior of the nasal passages would reduce olfactory stimulation sufficiently so that it would give the researchers at least some idea of the kinds of environmental messages lost from the inability to smell acutely. This was the most primitive of the devices and perhaps the least successful. Certainly the challenge of refining this simulation is a priority task for the future.

There was also very little information in the literature regarding age-related tactile loss. Our research team developed its own instrumentation, standardized the losses with an aged population in the community and experimented with a number of liquid and spray fixatives until we established the appropriate fixative with the proper coating procedures. (3) The procedure involved coating the finger tips with a liquid fixative until the proper thickness was secured to elicit the necessary desensitivity.

Once the devices were developed and assembled the research team which consisted of four doctoral students in architecture specializing in environmental problems of the elderly initiated the field work. The researchers not only had several years of experience in professional practice but had advanced graduate work in the behavioral sciences including the physiology, psychology and sociology of aging. It was felt this team of researchers was uniquely suited for the undertaking.

Procedures. The procedures regarding the execution of the research consisted of each person experiencing 3 standardized settings for at least one hour each day. (4) That is, one setting would be experienced for one hour on one day, the next day the second setting would be experienced, and so on. The entire cycle would take on the average three days and then the cycle would begin again. The experiment was carried out on a trial and error basis for a number of weeks to work out a standardized routine and was then pursued on a full scale basis for approximately six months. The three settings consisted of a dwelling unit, a multi-purpose senior center and a shopping center. Each of the researchers kept a daily record of his experiences for the duration of the study. During the course of the study periodic meetings with the research team were convened and discussions were conducted comparing experiences and reactions. (5) Ultimately, general categories of experiences were derived from these periodic reviews which served as bases for the first tentative summary of experiences.

Summary of Experiences. Gradual accommodation to the simulated sensory deficits was part of the research design. There is considerable shock value associated with "instant" sensory deprivation and certainly produces an immediate empathy. However, it is how one begins to perceive the environment after ceasing to be pre-occupied with the deficits that one begins to concentrate on evaluation. Then too, it seemed more natural this way since the elderly adjust to these losses over time and learn to make effective accommodations.

A number of experiences were readily apparent to all members of the research team and elicited unanimous agreement. It will be these experiences that will be summarized below. Those experiences which seemed more idiosyncratic will be the focus of further work and will not be discussed at this time.

The experiences will be summarized under the four senses tested: the visual, auditory, olfactory and tactile.

Visual. a) Glare from uncontrolled natural light and from unbalanced artificial light sources was the single most ubiquitous difficulty encountered. For instance, when walking up an aisle toward the front of a supermarket the typical vast expanse of plate glass across the front of the store on a bright day serves to obliterate most of the detail in surrounding objects. If only a single intense artificial light source is used for illumination rather than several, the chances of inducing uncomfortable glare is increased.

b) Colors all tended to fade, the cool colors such as green and blue faded most while red faded the least.

c) Contouring was a difficult problem. One example of contouring involves the capacity to perceive the boundary between two contrasting surfaces. The problem was most apparent when two intense colors such as red and green bounded each other. The boundary becomes visually unstable because the intensity of the colors seem to overlap and as one focuses on the boundary it appears to shift. This becomes a real hazard when an elderly person has to negotiate stairs or distinguish floor from wall surfaces.

d) The opposite problem from unstable boundaries is the disappearance of boundaries. Closely related colors such as blues and greens tend to fade and blend into each other. This also creates problems in distinguishing wall and floor surfaces. For example, a light green wall and a blue-green carpet becomes virtually impossible to distinguish, and stumbling into walls is common.

e) Depth perception is affected. Frequently it is difficult to judge risers and treads going down a flight of stairs particularly when stairs are carpeted with a floral print carpet or painted the same color.

f) There was difficulty in eye recovery when moving from a lighted area to a dark area or vice versa. The abrupt movement from an area having too much light to an area having too little should be avoided or mitigated with transitional lighting arrangements.

g) Dark wall surfaces bounded immediately by windows admitting bright sunlight make it difficult to see objects located near the walls. Again, the extreme in contrast needs to be reduced.

h) Ability to discriminate fine visual detail was seriously impaired. The reading of printed information such as names on people's doors, directional signing in hallways of public buildings, hospitals, stores and the like were continual burdens.

<u>Audition</u>. a) Inability to hear conversation clearly with background noise such as noise from appliances, air conditioning units, or when people congregate together and talk such as at parties, theaters, lecture rooms, etc.

b) Parts of words in a conversation are frequently unintelligible. This apparently occurs when a part of the word sound goes above the 2,000 cycle frequency. Thus it is not only a matter of loudness but even if the sound is loud enough part of the sound can be filtered out if the frequency is high enough.

c) Difficult to locate and identify sounds. For example, noises from down the hall sounded much like noises only a few feet away.

d) Some combinations of carpeting, acoustical ceiling and draperies absorb too much sound and make functional hearing even more problematic.

<u>Olfaction</u>. a) The single most dramatic experience was the drop off in the taste of food and the pleasure of eating. Appetite and interest in food was reduced.

b) Odors associated with various rooms in the dwelling unit which are used to aid environmental coding such as cooking and food smells in the kitchen, the smell of deodorants and bathing paraphernalia in the toilet, were missing. Street smells such as exhaust fumes, bakery smells, freshly mown grass, the scent of flowers were all significantly reduced and affected the richness of environmental information.

<u>Tactile</u>. a) Difficulty with fine muscle control in eye-hand coordination tasks such as unfolding napkins, adjusting dials, turning pages of newspapers, magazines and books, adjusting pressure in gripping objects.

b) Making fine discriminations in temperature differences such as in dish and bath water.

c) Problems in identifying subtle differences in textures.

<u>Design Implications and Other Uses</u>. The study has demonstrated that age-related sensory decrements can seriously constrain a person from freely using buildings, facilities and other environments as presently designed. While it is apparently impossible to forestall age-related sensory losses, this experience suggests that through appropriately programmed environmental stimuli, the environment can be made to function as a more effective support network and mitigate the consequences of sensory losses.

While there are a number of specific design implications that the model helped make apparent, it is not our purpose here to recite a shopping list of design problems along with suggested solutions but rather to suggest organizing principles which call attention to whole classes of design solutions and in that way enlist the creative talents of interested designers and others to develop their own specific sets of solutions.

There are two basic principles that emerge from our study: 1) Organized space as stimulus and 2) Organized space as orientation.

Organized space as stimulus involves the principles of getting the message or environmental cue across through stimulation. The environment is organized as intricately and systematically as any spoken language. It has a system of cues that tells us how to respond to specific situations. However, the environment communicates meaningfully only to the degree that the cues which are sent out can be received and perceived by an individual. If the sensory modalities such as the auditory or visual are deteriorated to the point where the message comes through only very weakly, then the person is in obvious danger of responding inappropriately. Organized space as stimulus involves a design concept called redundant cuing. Redundant cuing means beaming the same message through more than one sensory modality. An example of this kind of redundancy is getting the cue that one is in the kitchen by hearing the clatter of pots and pans, the aroma of food cooking and seeing kitchen appliances. This brings three senses into action and at the same time all three are saying the same thing, "you are in the kitchen."

Organized space as orientation is a design concept which seeks to organize spaces for its predictive value. The idea is that in general, a space should have a singular and unambiguous definition and use. Again the purpose is to compensate with environmental arrangements for lessened sensory acuity. The spaces should be cued with landmarks which act as focal points for functionally different spaces. For example, color coding surfaces to signal functionally different spaces in terms of visual perception, textured surfaces for the tactile sense, and so forth. The purpose is to sensorally load the spaces so that they may more effectively serve as points of reference and avoid ambiguous messages. It should be kept firmly in mind that the changes in sensory acuity and other important physiological factors of this population is such that the usual subtle and complex architectural statements are not only largely unappreciated but are dysfunctional as well.

It is felt that the kind of organizing principles for design suggested by the model is only the beginning and that a systematic, longitudinal effort will surely yield even greater dividents.

In addition to suggesting the above organizing principles for design, the model holds great promise as a unique research tool since it makes it possible for the researcher to be the experimenter and subject simultaneously. It also examines the total context of the situation rather than examining relationships between a limited number of variables.

The model has proved to be a very powerful training tool for designers and others who work with physically vulnerable people. It has been used extensively for this purpose with planning and design professionals, educators, social services personnel, housing managers and community service personnel such as firemen, policemen and telephone operators who frequently deal with the elderly in emergency situations. Graduate students representing a large number of disciplines have also experienced the model.

<u>The Future</u>. The future calls for further refinement of the simulation appliances, field testing the design concepts, and continuation of developing the potential the model has for teaching and/or training purposes.

Notes

1. See References.

2. Morrisett, Farr and others (See References)

3. The instrument involved making comparative judgements regarding the 10 grades of 3M sandpaper. Each of the 10 grades were mounted on 4"x 4"x 1" blocks and subjects were asked to feel these various grades of sandpaper with their fingers and make comparative judgements as to whether the different blocks were coarser, finer, or the same. The subjects were not allowed to see the blocks but had to rely entirely on the sense of touch. Authors will be glad to supply further information on this procedure.

4. Experiencing here means something other than observation and reportage, it is more akin to discovery and evaluation as almost a simultaneous process.

5. Members of the first research team consisted of Uriel Cohen, Edward Steinfeld, Gerald Weisman and Paul G. Windley.

References

Birren, J.E., Schapiro, H.B., and Miller, J.H. 1950 The Effect of Salicylate Upon Pain Sensitivity. Journal of Pharmacology and Experimental Therapy, 100, 67-71

Chapanis, A. 1950. Relationships Between Age, Visual Acuity and Color Vision. Human Biology, 22, 1-31.

Chapman, W.P. 1944. Measurements of Pain Sensitivity in Normal Control Subjects and in Psychoneutiotic Patients. Psychosomatic Medicine, 6, 252-55.

Crouch, C.D. July, 1967. Lighting Needs for Older Eyes. Journal of American Geriatrics Society, 15, 685-8.

Douek, E.E. April, 1967. Smell - Recent Theories and Their Clinical Application. Journal of Laryngology and Otology, 81, 431-9.

Farr, L.E. 1967. Medical Consequences of Environmental Home Noises. Journal of the American Medical Association, 202, 171-4, Oct. 16, 1967.

Fisher, R.F. 1968. The Variations of the Peripheral Visual Fields With Age. Documented Ophthalmology, 24, 41-67.

Geldard, F.A., and Crockett, W.B. 1930. The Binocular Acuity Relations as a Function of Age. Journal of Genetic Psychology, 27, 139-45

Gilbert, Jeanne G. 1957. Age Changes in Color Matching. Journal of Gerontology, 12, 210-15.

Hilger, J.A., Glorig, A., and Mueller, W. 1954. The Facts About Hearing Aid Fitting. Transactions of the American Academy of Ophthalmology, 59, 617-29.

Hofstetter, H.W. 1944. A Comparison of Duane's and Donder's Tables of the Amplitude of Accommodation. American Journal of Optometry and Archives of the American Academy of Optometry, 21, 345-64.

Jonouskova, K. 1955. (Color Vision and Age.) Ceskoslovenska Oftalmologie, 1955, 11, 37-48 (Ophthalmology Literature, Vol. 9, No. 134 (1955).

Kleemeier, R.W. 1952. The Relationship Between Ortho-Rater Tests of Acuity and Color Vision in a Senescent Group. Journal of Applied Psychology, 36, 114-16.

Kleemeier, R.W., and Justiss, W.A. 1955. Adjustment to Hearing Loss and to Hearing Aids in Old Age. In I. L. Webber (ed.), Aging and Retirement, pp. 34-48. Gainesville: University of Florida Press.

Mesolella, V. 1934. L'ofatto Nelle Diverse Eta. Archivio Italiano di Othologia, Rinologia e Laringologia, 46, 43-62.

Moncrief, R.W. 1951. The Chemical Senses. London: Leonard Hill, Ltd.

Morgan, M.W. 1958. Changes in Refraction Over a Period of Twenty Years in a Non-visually Selected Sample. American Journal of Optometry and Archives of the American Academy of Optometry.

Morrissett, L.E. 1950. Plight of the Nerve-Deaf Patient. The uselessness of all present therapy, the practical usefulness of aural rehabilitation. Archives of Otolaryngology, 51, 1-24.

Obi, S. (Senile Change in Colour Sense). Acta Societatis Ophthalmological Japonicae, 58, 451-54. (Ophthalmology Literature, Vol. 8, No. 969 (1954).

Pastalan, L.A., and D.H. Carson, Spatial Behavior of Older People, University of Michigan, Ann Arbor, 1970.

Pastalan, L.A.,"Privacy as a Behavioral Concept," Social Science, Vol. 45, No. 2, April 1970.

Pastalan, L.A., Shimizu, K., and R. Yaste, Study for Change and Growth: Chelsea Methodist Home, Detroit Annual Conference, Ann Arbor, Michigan, 1971.

Ronge, H. 1943a. Altersveranderungen des Beruhrungssinnes. I. Druckpunkt-schwellen und Druckpunktfrequenz. Acta Psychiatrica Scandinavica, 6, 343-52.

Schenk, H., and Pfeifer, H. 1957. Untersuchungen des Gesichtsfeldes und der Dunkel adaption bei seniler Pigmententartung der Netzhaut. Albretch von Graefes

Archio fur Ophthalomolgie, 158,326-33.

Slataper, F.J. 1950. Age Norms of Refraction and Vision. Archives of Ophthalmology, 43, 466-81.

Tiffin, J., and Kuhn, Hedwig S. 1942. Color Discrimination in Industry. Archives of Ophthalmology, 28, 851-59.

Vaschide, N. 1904. L'etat de la sensibilite olfactive dans la vieillesse. Bulletin de Laryngologie, Othologie et Rhinologie, 7, 323-33.

Walton, W.G., Jr. 1950. Refractive Changes in the Eye Over a Period of Years. American Journal of Optometry and Archives of the American Academy of Optometry, 27, 267-86.

EIGHT SPACE PLANNING TECHNIQUES

Chairman: Weldon E. Clark, Souder, Clark & Griffin Assoc., Encino, California

Panelists: Jerry Finrow, Center for Environ. Research, Univ. of Oregon, Eugene
Alan Hershdorfer, Civil Engineering, MIT, Cambridge
Robert Mattox, Caudill, Rowlet & Scott, Houston
L. Stephen Windheim, Leo A. Daly, Co., San Francisco

Authors: Charles M. Eastman, "Requirements for Man-Machine Collaboration in Design"
Elliott E. Dudnik and Robert Krawczyk, "An Evaluation of Space Planning Methodologies"
Christos I. Yessios, "Syntactic Structures for Site Planning"
Richard Hobson and Imre Kohn, "Utility Ratings and 0-1 Programming in Housing Design"
Margaret A. Frederking and Alton J. Penz, "An Integrated Methodology for Office Building Elevator Design"

SPACE PLANNING TECHNIQUES: INTRODUCTION 8.0

Welden E. Clark, Session Chairman

Souder, Clark, Griffin and Associates, Inc.
Consultants for Health Care Planning and Architecture
16633 Ventura Boulevard, Encino, California 91316

The five papers of this session show some indication of the breadth and ambiguity of the topic area. Eastman and Dudnik are both concerned with assessment and reporting of the state of the art. Yessios offers a conceptual extension to the art. Hobson and Kohn, and also Frederking and Penz, illustrate the application of known techniques to a design problem. The examples chosen as illustrations similarly cover a broad spectrum.

The perspectives of the different authors are glimpsed in their papers, from a proper academic concern with theory of design to an equally proper pragmatic concern with costs and benefits in commercial building. An attempt to review the state of an art (and science) so broad and ill-defined is beyond the skills of this chairman, so a few remarks on the papers and a short statement on some other problems that might have been aired in this session will have to suffice.

Eastman's thesis is that although the successful designer is not often able to describe or teach the details of his skills in creative synthesis we can learn much from studying the processes of present and developing computer design systems. The notion is fascinating, and is, of course, also being pursued in other fields of computer use. He makes a good case, but he, like many others of us, is committed to a rational view of design and to the ultimate ability to measure goodness of solutions by quantitative (and thus programmable) measures.

Dudnik is concerned, in his paper, with the lack of available comparisons among different computer-based space planning techniques. He compares a number of techniques (and some extensions of them) on a common problem and reports his findings. He concludes that methods exist for deriving good and efficient architectural plans, that the form of the solution is greatly affected by the choice of method, and that the resulting availability of multiple good solutions could greatly aid planning investigations.

Yessios presents a portion of a linguistic approach to computer-aided design, using syntactic relationships to describe patterns and allowable combinations. His system, as applied to site planning problems, provides for three levels of aggregation: a module level, a pattern neighborhood level, and a diagrammatic plan level. The approach, currently being programmed, offers much promise.

Hobson and Kohn illustrate the use of a linear programming approach with utility values determined by a panel of experts as raters. Their example, the choice of bathroom fixtures in tract housing, does not introduce complexities to mask the presentation of method, although they note that their assumptions of additivity of utility values may limit usefulness of the approach.

Frederking and Penz attack a complex problem, the design of elevator systems for modern commercial buildings. They present a cost/benefit analysis approach and discuss thoroughly many of the factors entering into choice of elevator systems.

These five papers attack important problems and gaps in development and application of rational design processes to the design of our man-made environment. One wonders, however, if the widespread application of these directives does not hinge as much on new capabilities for determining requirements and criteria, as on better algorithms. Perhaps the pattern language of Yessios and the cost/benefit analyses of Frederking and Penz can be used in the manner suggested by Eastman to show human designers and their clients how they might restructure functions, time schedules, and processes to accommodate designs that are simpler, more effective, and more conserving of resources.

REQUIREMENTS FOR MEN-MACHINE COLLABORATION IN DESIGN

Charles M. Eastman

Associate Professor of Architecture, Computer Science and Urban Planning
Institute of Physical Planning
Carnegie-Mellon University
Pittsburgh, PA   15213

Abstract

This paper examines the intelligence required by a computer to successfully collaborate in design. It focuses in two areas where design augmentation is justified, in the production of working drawings and in the storing and accessing of design related information. The semantics of line drawings, operations for locating an element, the sequencing of elements to be located, planning considerations, and graphical communication are reviewed as necessary intellectual capabilities for an automated draftsman. The capabilities of an extended architectural information system are suggested by an example, and some of its desired structural properties are suggested.

Introduction

Almost all building today would be vastly improved if the time available for planning and design were doubled. The implication is that architects already have more to consider than they have time to resolve in an integrated way. New information is being made available continuously to guide design decisionmaking. As information proliferates, such as in the behavioral sciences, and the opportunity grows to plan facilities which adapt to different uses over their lifetime, the gap can only increase between the potential of design and what will be realizable within the professional timeframe.

One obvious tool to enhance the information processing capabilities of designers is the computer. Many different design contributions of the computer have been proposed and are various stages of development. I wish to focus on two of them. The organization, storing, and easy accessing of design related information via information retrieval systems is one obvious application that responds to the above problem. Another application is the generation of detailed designs from preliminary drawings. Over forty percent of an architect's fee is devoted to the task of producing working drawings[1]. While this effort is necessary, it makes little contribution to the quality of the overall design result. Expediting working drawings is mandatory if the designer is to address more strategic design issues.

Significant effort has gone into applying the computer in the above two ways. The limited results suggest that a mismatch exists between the practice of architecture

---

* The work presented here is supported by a grant from the National Science Foundation, GJ31188.

and the structure and organization of computer tools developed to date. The current development of computer tools seems comparable to a jeweler wishing to use electric tools to expedite his work. He surely would be disillusioned if he could only obtain heavy equipment, such as 1/2" drills, turret lathes and arch welders. Even though these tools are useful in other applications, they do not respond to the jeweler's problems. Similarly, the computer tools developed to date in other areas do not fit the problems of architectural design. The issue is more than one of the size and crudeness of the tools. In each area where an attempt has been made to develop a design collaboration with the computer, serious technical problems have arisen. These problems concern the fundamental nature of the design process, both at the psychological level of perception, cognition, and communication, and at the organizational level regarding the structuring and management of decisions. The mismatch between computer tools and architectural practice will only be resolved by incorporating the requisite "design intelligence" into these tools.

A perspective which clarifies the issues regarding the type of design intelligence requisite for computer collaboration is to consider the qualities required for human design collaboration. Consider first the design consultant who contributes special knowledge and evaluation skills. The consultant who provides useful design information is expected to know the important design variables and vocabulary of the architect. His job is delineate the relations between design and performance variables. For example, the sociologist consultant should be able to relate office layout decisions with employee satisfaction or productivity; the management consultant should be able to relate the type of space being offered to market conditions. Both the vocabulary, the relations organized within it, and the ability to evaluate design decisions in terms of these and other performance variables, are critical determinants of the consultant's value. Now consider the architectural draftsman. He is expected to take roughly sketched details and to complete them, applying such standard considerations as structural rigidity, moisture control, construction sequences, etc. He is also expected to complete partially drawn designs. In both cases, he must know the standard sizes of materials, the tolerances required for different temperature ranges, and to have the appropriate synthesizing capabilities necessary to generate details that meet the criteria involved. Moreover, he must be able to communicate easily and if required he should be prepared to revise his work in light of new information provided him.

If we consider the capabilities expected of their human counterparts, similar criteria can be generated for other computer-aided design tools. Judging from the results to date, the intelligence required for successful design collaboration, by man or machine, has received insufficient attention.

In this paper, I review current knowledge regarding design intelligence, with particular emphasis placed on the tasks of drafting and information retrieval. I attempt to generalize our current understanding about the requirements for machine collaboration in design, as gained from a variety of computer and design methodology research. I attempt to review what is known, identify what is not known and should be, and indicate some promising directions for future research.

Automated Drafting

If an architect hired someone to do drafting, it would be mandatory the person have some knowledge about buildings and the way they are constructed. While several programs and languages exist for computer drafting, most simply translate a symbolic description of a drawing into plotter commands which create the drawing. They incorporate no knowledge about the class of objects drawn[2].

Certain fundamental definitions are generally recognized and should be responded to in the syntax of any design description. For example, a plotter drawn rectangle will have at least two meanings. It can depict either an object or a window in an object. Most current drafting languages are not able to remove this ambiguity and are thus unable to remove hidden lines, or apply properties to the elements that are being represented. The problem is that objects are not represented, only lines. This ambiguity is easily resolved. First, it is necessary to recognize that each line of a drawing depicts the border of an object. That is, it separates space with one characteristic from space with another. It is possible to make the rectangle disambiguous by adding a semantic definition to all lines. Suppose that each line is directed, such that the inside of the object it bounds is on the right of the line. Thus the rectangle drawn clockwise denotes an object and drawn counterclockwise denotes a window. Other definitions can also remove the ambiguity.

The above line definition allows distinguishing occupied from unoccupied space. Two objects side by side create two overlapping lines with opposite directions. Using these distinctions, it is also possible to define all unfilled areas, that is, the complement of the solids. This is another important semantic feature of orthographic drawings. They depict empty as well as filled space.

Computer drafting languages which do not explicitly depict empty space are not able to quickly tell if the location of an object is feasible (not overlapping with another object). Two approaches may be used to determine location feasibility. A check can be made if the location is completely inside the empty space or, alternatively, if the location overlaps with the shape of any other object. The second method of testing is very expensive. Without an explicit representation of empty space, similar difficulties arise if one wishes to measure the size of a particular empty space[3].

It is surprising that none of the available production-oriented drafting languages incorporate these two obvious semantic features. They were conceived as languages for representing drawings, not objects, and are of little value to the architectural draftsman. It should be noted that several experimental programs have been written which include these features and they could be rewritten for production uses[4].

Object Location

Several experimental programs for automated design use the above features in generating arrangements of objects which satisfy simple criteria[5]. Given the above capabilities, a fundamental task in generating arrangements is the adding or changing of the location of an element so that the location satisfies a set of criteria. Consider for example, the refrigerator placement problem shown in Figure One. Given the arrangement shown, find a location for the refrigerator so that it satisfies the

8. SPACE PLANNING TECHNIQUES / 399

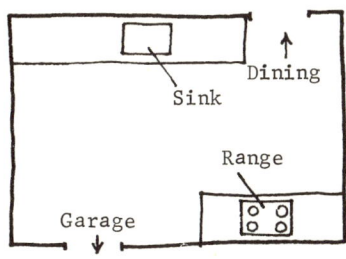

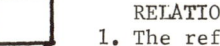
Refrigerator

RELATIONS:
1. The refrigerator door should not block the counter front.
2. The back of the refrigerator should be against the wall.
3. Its door should open so, that the hinge is on the side away from the stove and sink.
4. Its door should not block circulation.
5. The distance between the stove, refrigerator, and sink should all be small.

FIGURE ONE

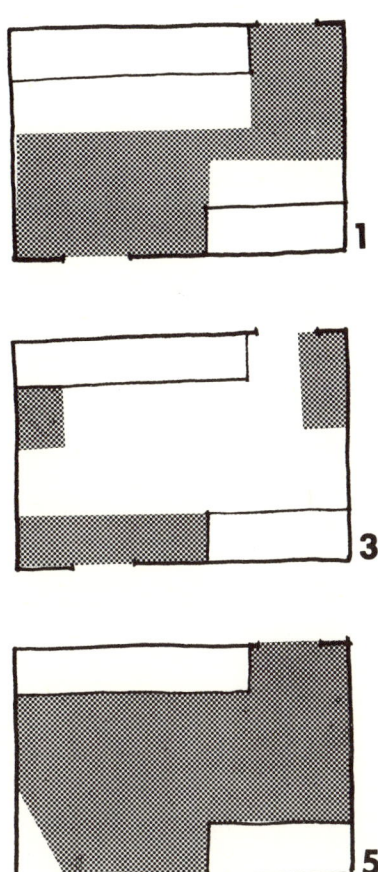

FIGURE TWO - Areas of acceptable location for each of the Relations.

list of Relations[6]. The reader is encouraged to resolve this task, without the aid of paper.

In solving this problem, it is likely that the reader used a problem solving process well known to researchers. Specifically, it is most likely that he used one of the Relations shown at the right of the Figure to generate alternative locations. Each of the trial locations was then tested to see if it satisfied the other Relations. That location was chosen which passes the tests defined by the other Relations. This strategy for finding locations within a partial arrangement seems to be the easiest one for humans. It does not require extensive use of short term memory, requiring the problem solver only to remember the operation for generating trial locations and the sequencing of relations that are used as tests. This location method is known as <u>generate-and-test</u>[7].

The above method of solving location problems is so trivial that it is easily ignored by a human designer, but incorporating the intelligence it involves into a computer requires that it be understood explicitly. Several automated design programs find locations by generate-and-test. It is much more efficient than the earlier approaches which relied on random locations or exhaustive trial and error[8]. But in recognizing that this method seems preferred by human designers because it requires little memory, an even more efficient technique recently has been developed that relies on the unique capacities of the computer.

In generate-and test a Relation is used to define a set of locations that were acceptable to it. But the interaction between locations and Relations holds not only for one location, but for all of them. More precisely, for each Relation, one or more spatial domains can be defined inside of which any location is acceptable but outside of which none are. See Figure Two. In generate-and test, only one of these domains is considered. With the aid of tracing paper, one could generate all such domains as in Figure Two. The desired location of the element, of course, is at the intersection of the domains, eg. the common area defined by overlaying all the individual tracings. This location method can be easily automated. We call it <u>Projective Location Generation</u>.

Projective Location Generation allows quick generation of a small empty domain inside of which any location for an element is acceptable. It does so without any trial and error and integrates our understanding about the interaction of locations with Relations that goes beyond standard human design practices. An automated drafting system is now being designed at Carnegie-Mellon University which incorporates the capability of projecting the domains defined by an object's spatial Relations and of automatically defining their intersection. We expect Projective Location Generation to be useful for both interactive and automated drafting and opens up opportunities for still other kinds of efficiencies.

<u>Sequencing of Objects</u>
The locating of one object is simply a single step in the larger sequence of locating a set of objects. The sequence is not linear. Indeed, the assumptions of Projective Location Generation guarantee that the generation of an arrangement is an iterative process. The relevant assumption is that all previously placed objects are located correctly. If the earlier locations were not in the correct part of the projected

8. SPACE PLANNING TECHNIQUES / 401

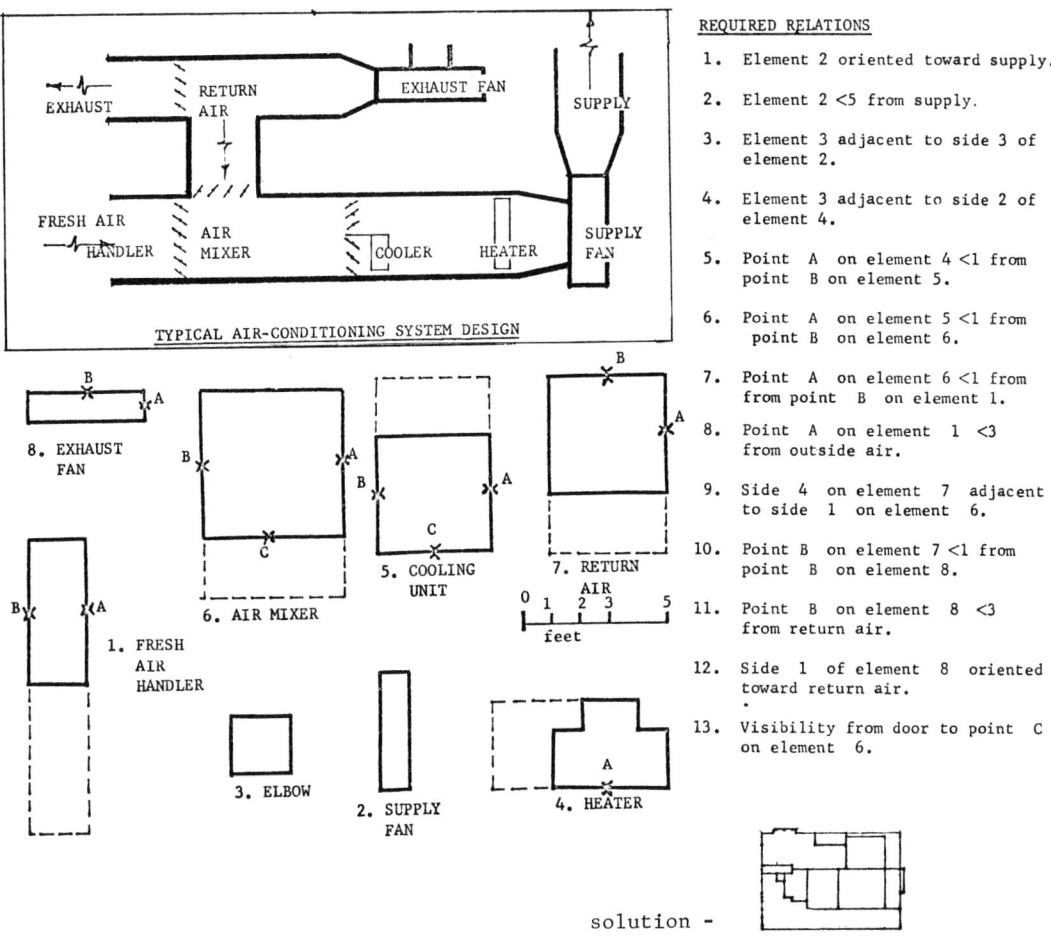

FIGURE THREE

intersection, then the current arrangement may not be correct either. Unfortunately, there is no means to eliminate this possibility. Thus an arrangement is inevitably the search for a set of locations for each element and the result is a combinatorial tree (9).

In generating locations for an object, an important prior question arises. In what order should the objects be added to the arrangement? This question has been answered in various ways by space planning programs and by the earlier space allocation programs(10). Projective Location Generation provides a useful, if not final, answer.

A general rule of broad importance to the processing of many types of tasks defines the most efficient order for sequencing a set of Boolean tests (11). The rule states that a sequence of conjunctive tests should be made in ascending order of a function α_i, where

$$\alpha_i = \frac{\text{cost of test i}}{p(\text{failure of test i})}$$

The logic behind this rule can be understood intuitively through an example. Suppose that a person has a set of criteria by which to select a car for cross-country travel. It must be able to tow his trailer over six percent grades, get twenty miles to the gallon, have at least a thirty gallon fuel tank, plus other criteria. Not believing salesmen, suppose that this person is going to make the tests for these criteria himself. (Among the cars meeting his criteria, he will take the cheapest.) The question is how should he deal with these tests so that he can select the correct car as efficiently as possible.

Of these three tests listed above, suppose that he also knows that most cars he is interested in offer the large fuel tank, some but not many can meet his towing test, and even fewer will get the required mileage. If all tests required the same amount of effort, what order should he apply the tests? The answer is mileage-trailer-fuel tank. If the first test fails, he is saved from doing the others, thus he should make the most-likely-to-fail test first.

Suppose instead that all cars were equally likely to pass each test. He could ask the salesman about the fuel capacity easily while the other tests would take about a half days work apiece. It should be obvious that the fuel capacity question should be settled first, so that he won't waste any time with unnecessary testing. In general, easy tests with high probability of failure should be made early in any sequence. These result in low values of α.

Applied to the automatic generation of arrangements, we can consider the placement of a set of objects as a set of tests, of the form - is there a location for this object such that the whole arrangement is satisfactory? The cost of testing each object can be derived empirically from cpu time on the computer. The probability of failure requires that certain assumptions be made. The best assumption to date is that the probability of failure of an object is proportional to the area of the object, divided by the total restriction on that element resulting from the Relations which are associated to it. That is, the likelihood that a location does not exist is proportional to the area of the object divided by the area of the projected intersection. This assumption ignores shape considerations; that is why it is defined probabilistically. In general, big objects go early in the arrangement sequence, as do those which are highly restricted. This sequencing test has been found to buildup arrangements very efficiently, minimizing the need to backtrack, that is to reconsider prior decisions (12).

Together, Projective Location Generation and the above object sequencing rule allows simple arrangement problems to be resolved quickly. An example problem and solution is shown in Figure Three. This problem was solved in 77 seconds of cpu time, using an IBM 360/67. These techniques provide another set of capabilities which will eventually allow an automated drafting system to intelligently collaborate with a designer.

Planning - by Simplifying Structure

Planning is generally defined as the organizing of actions or operations. Since the process of design involves many operations, a design strategy can appropriately be called a plan for generating a design.

Since Alexander's monograph on the structure of design problems and their decomposition into more manageable subproblems, many students have attempted to use graphs to represent problem structure and to use decomposition in the planning of large design projects. Alexander's proposal was to use a boolean graph, with design criteria as nodes. His definition of an edge of the graph was intuitive. For a period, many different techniques for decomposing and then reassembling a problem were proposed(13). The underlying and unresolved problem with this planning technique was the lack of a workable definition of the units which make up a problem; both nodes and the Relations between them were ambiguously defined.

The measure of restriction, generated from Projective Location Generation, provides a metric leading to new insights into the structure of simple design problems. Any Relation between a set of objects can be defined in terms of Relations between pairs of objects (14). A Relation between two objects can be characterized by the restriction it places on one object if the other is located. Thus any arrangement problem can be characterized by a graph of restrictions. Shown in Figure Four, for example, is the graph of the Relations shown in Figure Two. Each node in the graph is an object in the final arrangement. An edge from node A to node B gives the restriction on object B if A is already located. Such a graph can be called a <u>Constraint Graph</u>.

In order to find the restriction on any object, given some arrangement, one need only find all nodes from the located objects to unlocated ones. For each unlocated object, the intersection is computed of the restrictions having the unlocated object as its head and a located element as its tail.

The Constraint Graph and relations defined in automated design programs make operational Alexander's intuitive concepts. Pfefferkorn's program, for example, includes a procedure for decomposing modified Constraint Graphs and uses it to simplify large problems. In solving equipment arrangement problems within a room, his program decomposes the Constraint Graph representing the total problem into smaller graphs which are internally highly restricted. No subgraph may include more than five elements. For each subgraph, a separate arrangement of elements is developed. The arrangements are then combined. Constraint Graphs have been used as an integral planning tool in the operation of three automated design programs (15).

Planning-by Adding Structure

Programs which use a Constraint Graph to lay out rooms in a building or equipment in a room predominantly use what is called a buildup approach. That is, they start with an empty space and add elements or set of elements one at a time, adjusting already placed ones as the arrangement is built up.

Many problems are made up of a large number of elements, yet their Constraint Graph, as defined above, does not break down into well-defined subsets. Good examples are found in residential site planning. Such problems may be solved using a Constraint Graph and decomposition, but two fundamental difficulties arise. First, the site requirements, including access, drainage, views, and circulation apply equally to all units; the most meaningful Relations are not those between one unit and another, but between units and the site and are dependent upon particular topographical features. These features are not easily incorporated into a Constraint Graph. Second, there are a very large number of ways that the elements (roads, buildings, yards) may be built up. It is very tedious (and expensive) for the computer to base an arrangement on a Constraint Graph because much of its time is spent building up partial arrangements that are later rejected. It turns out that this type of problem is difficult because it is so loosely defined. The Constraint Graph provides too little structure to guide the program to good solutions. In such problems, an alternative approach to decomposition has been found useful. Rather than eliminating Relations so that the Constraint Graph can be broken down, Relations are added in a pattern which call a planning diagram. The Relations add organization which facilitate finding a solution quickly.

In site planning, Yessios has explored diagrams which structure access and group residential units (16). The diagram is a set of lines which depicts the topological features of the circulation pattern. See Figure Five. The diagram is first matched to the site so that slope, groundcover and other criteria are met for circulation. After this initial phase, the pattern is used to guide the placement of individual units so that views, access, and yard considerations are most likely to be satisfied. Currently, the diagrams are selected and composed by the program user.

When planning diagrams are used, design alternatives are not explored in the rearranging of individual elements, but in selecting and fitting of one or more diagrams to a context and then in fitting design elements to the diagram. It seems likely that in large loosely structured design problems, the imposing of a planning diagram is <u>necessary</u> to achieve good results. In many design problems now solved by the computer, adding relations simplifies the problem.

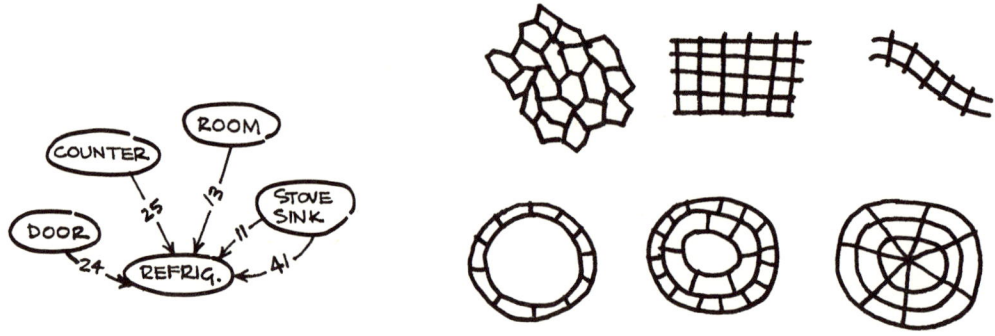

FIGURE FOUR

FIGURE FIVE

To me, this experience with automated and interactive design programs parallels the experience with more traditional design processes. Most complex building types have evolved prototype solutions. Airports are a good example. Two well known prototypes are the "finger" and "feeder" plans. O'Hare Airport in Chicago is the largest example of the finger plan and Dulles Airport by Saarinan is the only major example of the feeder plan. In the terms of this paper, both airports can be abstracted into planning diagrams. The abstract patterns derived from them provide an organizing structure of Relations that simplify the large but unstructured set initially defined for any particular airport problem.

New planning diagrams are an important source of design innovation. It seems reasonable to expect that at some time in the forseeable future, the Relations defining different sets of design problems and the solutions for them may be systematically examined in a way similar to Alexander's proposal regarding environmental structure. Indeed, this may become a central function of architects. Efforts in other areas have already examined some existing patterns, in both office buildings and bridge design (17). A lesser kind of design intelligence, but one of more importance to a draftsman is the knowledge of and ability to apply a variety of patterns to a particular context.

Variable Shaped Objects

Many of the Objects which are manipulated and arranged in an architectural problem are not of fixed shape. Kitchen counters, pipes and ducts, walls, rooms, and buildings and parking lots in site planning, are a few examples of design elements which are fabricated to accommodate different contexts. The treatment of these types of elements has not been adequately studied.

Two approaches have been identified thus far for dealing with variable shapes. One approach was introduced at the very beginning of computer graphics by Ivan Sutherland and has not received much attention since that time. It might appropriately be named the rubber object method (18). Shapes are normally defined in computer graphic languages in terms of ordered points which denote the corners of an object's perimeter. In the rubber object method, an object's shape is specified in terms of a set of constraint equations defining the necessary relations between one point on the perimeter and others. Distance, relative placement, and angle of connection are typical relations that might be specified. These equations internal to the object provide the "substance" which causes the object to take a particular form. Other objects may displace a point or line on a variable object. This in turn requires the internal constraints to reconverge on a feasible assignment of locations. Reconvergence was done sequentially using a least-squares-means-fit technique. The effect is that any time an object is nudged, the program attempts to reform it in a satisfactory shape.

An alternative approach is the generative method (19). Instead of constantly adjusting the shape of an object as it is imposed upon during the generation of an arrangement, an object's shape is generated once or only a few times in a controlled sequence. The basic operation generates one change in the object's form. Specifically, it redefines one line on the object's perimeter. This operation is called iteratively, with internal checks guaranteeing that the partially generated form leads toward a desired one. A great variety of shapes which fulfill different objectives are thus allowed.

To provide an example of how the generative method can be applied, I outline one program now being implemented for laying out departments within an office building. All previous programs essentially assign departments to modules of the building; the task is considered an assignment problem (20). New spaces of undefined area cannot be considered; thus circulation space cannot be located as a separate entity; circulation is usually treated as a proportion of each department's area. Departments are also specified in terms of a fixed number of modules which must be located contiguously. No shape constraints can be imposed. Circulation distances are approximated by the straight line distance between the centers of two departments or as the sum of the X and Y coordinate distances. Because of these limitations, programs for space allocation are only useful in generating schematic layouts; they require extensive modification before they can be of practical use.

Our objective has been to develop a procedure that directly generates a working layout of an office floorplan. Circulation is to be explicitly represented and departments shapes are to respond to dimensions, area and other spatial attributes.

In the generative approach to office planning, the basic expansion is first organized into two higher level processes. One generates circulation space and another expands the shapes of rooms. The circulation generation process takes as input two specified points on different departments. The process connects them with a pathway of defined width which is made up of a set of rectangles connected end to end. To achieve this, the basic expansion is called iteratively, each call expanding one part of the path in a direction leading to the element to be connected. When some blocking element is encountered, the shortest path around it is taken and the program proceeds. The path generation process fails whenever it finds that it must cross its own path.

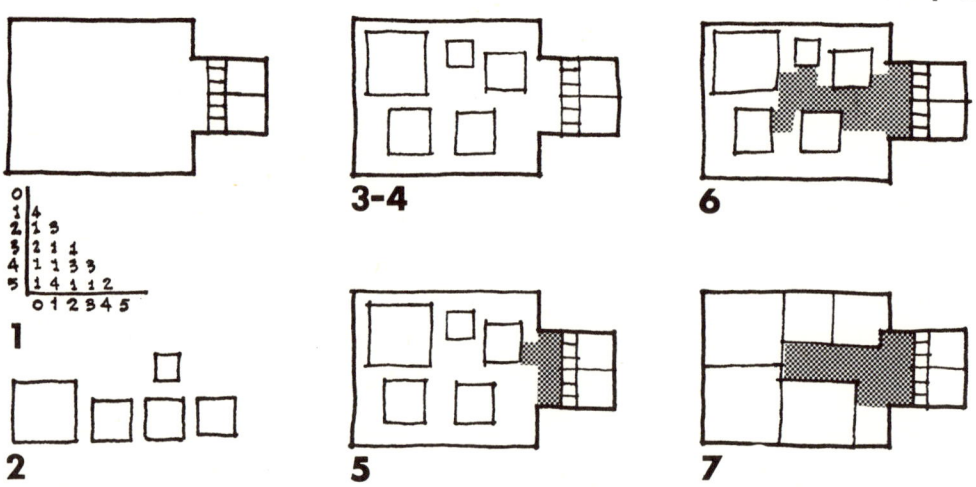

FIGURE SIX

The room expansion process iteratively calls for an expansion of a small "core" of the room, checking at each step that it: (a) quickly fills out to existing barriers, (b) satisfies the internal relations as to the department's form, (c) and takes the smallest form that satisfies these criteria.

With these two processes, the department layout program generates a layout directly, with little trial and error. Inputs to the program consist of department criteria, including area, shape, and environmental criteria. A circulation interaction matrix is included, along with the building envelope. The program's steps for generating a departmental layout are listed below; each step is shown graphically in Figure Six.

1. the inputs are made to the program to describe the particular problem at hand;
2. the minimum dimensions of each department are used to define the "core" of each. Area criteria are ignored at this stage;
3. the sum of the areas of each minimal shape is compared with the area available and a function of the ratio of these two values defined;
4. each department core is located, separated from other spaces by a function defined in step 3. Departments are located sequentially, those with the highest interaction with fixed or already located ones being placed first;
5. the path connector constructs a link between the two department cores or with a department and a fixed element with the highest interaction;
6. other departments are connected to the nearest point on an existing pathway. The order is in terms of the original order in which the departments were included in the arrangement;
7. After all circulation has been defined by 5 and 6 above, the cores of the departments are expanded to their required areas and shapes. The sequence of department expansions takes those in the center of the building envelope first.

The details of these steps are described elsewhere (21). They are outlined here to suggest how relatively simple generative operations can be organized sequentially and hierarchically to allow treatment of a variety of variable shaped objects. Other processes have been developed for the generation of counters, pipes and ducts. An intelligent draftsman must know how to manipulate shapes of an object so as to respond to different contexts. Both of the above methods seem promising for the achievement of these ends.

Design Communication

Before leaving the intelligence known to be required for drafting, the issue of communication should be mentioned. We require easy and "natural" communication with design collaborators. Exploration has been made of programs which accept "sloppy" sketches and interpret them (22). To date, the techniques developed have been general and do not focus on any particular design application. Other work has translated drawings of one form to another.

Let us consider how these techniques might be extended to allow a user to roughly sketch a detail and have the computer interpret, correct, and extend it according

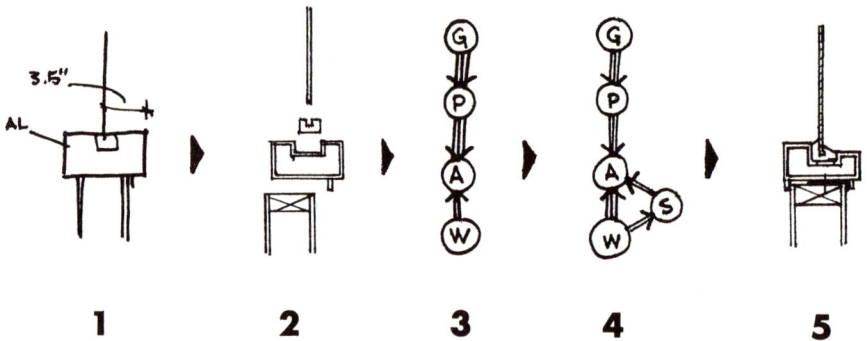

FIGURE SEVEN

to the computer's knowledge of the problem. See Figure Seven. The first step of the task is simply for the computer to recognize the different sketched forms. It would do so by matching elements of the sketch with stored models of the objects. The models would recognize different shapes. The organization of elements would then be analyzed and the topological characteristics defined by the person making the sketch would be identified. The types of properties identified at this level would include adjaciencies, spaces for thermal expansion, critical dimensions, etc. An important preliminary program which does analysis of this sort is Evans' (23). The program could then match features of the sketch with stored information about construction details and retrieve other information pertinent about details of the same type. With this added information to guide it, the program would then modify and complete the detail so as to satisfy the considerations implicitly addressed by the sketch and also by the knowledge available to it. Several programs have undertaken parts of this task. More effort, though, is required before they can be put together.

Other work on drafting applications has been undertaken in the area of formal languages. Instead of a sketch, a description of the drawing is given in algebraic form. Preliminary results show that not only are quick drawings possible, but that the drawings may reflect some intelligence about their composition (24).

Design Information Systems

Given a large and complex design problem, all the information required to solve it is not likely to be readily available. Unless a design team is very familiar with the building type and the client, much of their time initially will be spent understanding the functions of the organization, the technology involved and the problem context.

Many proposals have been made for the development of design information systems. Most of those proposed or implemented thus far rely on a set theoretic and taxonomic organization of data. Building components and construction technology have been emphasized (25).

Computer science has for some time considered the following type of information system an appropriate long term research objective. They have proposed the construction of a data input system capable of reading and storing written text. The information read would be parsed and organized not only along set theoretic and taxonomic lines, but also according to the kinds of relations of importance to the particular area of discourse. In general, such organization of data is said to be associatively structured. The system could then be interrogated about its information in a subset of natural English and it could answer about anything that it had read. For the short term, such systems are likely to be feasible for only limited areas of knowledge. A good survey of research in this area is Minsky's (26).

Consider the following discussion between an architect and a geriatics consultant.

A - "What special considerations apply to the residential units?"

C - "Older people's balance and muscle control are weaker than younger persons. Safety is an important consideration in all spaces. It is recommended that all surfaces be non-slip. Handrails should also be available on walls near any activity. Because of muscle control, all switches, latches and other storage closures should be simple, so that even a person with arthritis can operate them."

A - "What other special considerations are required for kitchens?"

C - "Because of balance, there should be no high or hard to reach storage. Memory is also poor, suggesting automatic controls for the oven and other cooking."

This straightforward conversation provides much design information which can be considered a supplement to the architect's own experience. It suggests the contents an architectural information system should include as well as its associative structures.

One kind of information about which architects require information is obviously the users of a space. The characteristics of the users and the schedules and distribution of their activities are important design inputs. In the above discussion, balance, muscle control, memory, and incidence of arthritis are examples of user characteristics. Another kind of information is the definition and organization of generic facilities. In the above discussion the consultant was told to consider residences. Within that category, he knew something about kitchens, storage, doors and latches and surfaces. The consultant's most useful insights were derived from relations he made between user's characteristics and the parts of a residence. In other situations, it has also been found that the technology available is another important category of information used in design.

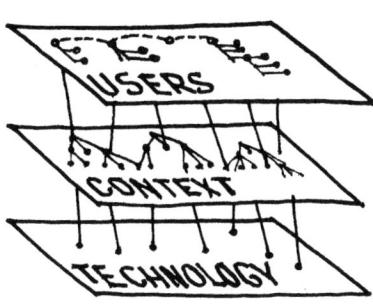

FIGURE EIGHT

Together, this example and other analyses made of similar hypothetical discussions with a consultant suggest that three tiers of information required for an intelligent architectural information system. See Figure Eight. The first level would include information about user's of facilities and would include the following kinds of relations:

>   (user group) - (characteristics)
>   (user group) - (activity)
>   (activity) - (context & time)

Several of these relations and a data base to store them already have been developed (27).

The second level of the data base would include generic object descriptions and generic design descriptions (called objects below). It would include the following relations:

>   (object name) - (component object name)
>   (object name) - (generic description)
>   (object name) - (system name)
>   (system name) - (generic description)

Information systems with capabilities for representing this type of information have been developed by Myer, among others (28). The last level of the system would provide specific product information, as requested in terms of object names with delimited object descriptors. That is, information at the third level would be primarily accessed through the second. Linkages between the first and second levels would also be crucial. They would be structured by routines which found relations according to generic types of considerations. For example, one kind of consideration might be safety. For each characteristic of the users, this routine would scan the descriptors of generic objects and list those which should be delimited to respond to this issue. Similarly, a convenience operator could suggest groupings of objects, i.e. spaces, reflecting the sequences of activities found in the user description. Other operators could be defined for specifying relations between different levels of the data base. Similar systems are being explored in other areas of discourse (29).

While many of the details of such a system have not been resolved, a system with the design outlined does have the potential for replicating the types of discussion presented earlier. It seems an appropriate objective for research in architectural information systems and incorporates an expanded but realizable design intelligence.

Conclusion

The computer offers the opportunity of reducing many mundane design tasks. In so doing, it allows widening the considerations going into any design project. It is assumed that architecture has more to offer society than simply the production of drawings for buildings that don't fall down. I assume that by reducing the cost of production documents, most of the savings will transfer to other activities further improving design.

The cost for accepting the computer as design collaborator is the same as for any other collaborator. The Architect will not immediately know all the considerations that resulted in the final drafting detail or recommendation. Of course, in any office involved in large projects the same situation holds today. In the current case, no one person has complete knowledge of the design; complete knowledge is only held by the design team. Any member may be questioned about his decision so that the criteria from which they derived can be made explicit. There is no reason the same could not apply to a computer program; for any decision made, the criteria which led to it could be stored for later retrieval.

The emphasis throughout has been on a review and expansion of the stock of ideas regarding design intelligence. Some have suggested that computer-augmented design has not made progress or lived up to its potential. If this is so, it is because of the scarcity of effort going into this area. Hopefully, the directions posed here will suggest further directions and the motivation will be generated to realize them.

Notes

1. "The Economics of Architectural Practice" report by the American Institute of Architects, Washington, D.C., 1968.

2. Frank, Amalie J. "B-Line: Bell Line Drawing Language" Proceedings, 1968 Fall Joint Computer Conference, Vol. 1 (1968). Harris, Herbert R. and Smith, Dale Autodraft - A Language and Processor for Design and Drafting" Design Automation Workshop, Atlantic City, N.J., 1965.

3. Eastman, Charles "Representations for Space Planning" Communications of the Association for Computing Machinary, 13:4 (April, 1970), pp. 242-250.

4. Eastman, Charles, "GSP: A System for Computer Assisted Space Planning." Proceedings of the Eighth Design Automation Workshop, Atlantic City, N.J. 1971; Johnson, Timothy, Image: An Interactive Graphies Based Computer System for Multi-Constrained Spatial Synthesis," Department of Architecture, M.I.T. (September, 1970); a review of several other programs is offered by Kasmierczak, H. "Image" Processing and Pattern Recognition," Information Processing 68, North Holland Publishing Co., Amsterdam (1969) pp. 1056-1071.

5. See (4) above.

6. In this paper I use Relation (capitalized) to denot a constraint between one design variable and another. A Relation may be defined algebraically, in functional form.

7. Newell, Allen and Simon, H.S., Human Problem Solving, Prentice-Hall, Englewood Cliffs, N.J. (1971); Eastman, Charles, "Problem Solving Strategies in Design", EDRA I: Proceedings of the First Environmental Design Research Association

Conference, Chapel Hill, N.C., H. Sanoff and S. Cohn (eds.), (1969).

8. Fromboluti, C. "The Site Plan Idea Generation System", Proceedings of the Kentucky Workshop on Computer Application to Environmental Design, (1970); Lee, Robert and Moore, J., "Corelap-Computerized Relationship Layout Planning" Journal of Industrial Engineering, 18:3 (March, 1967).

9. Eastman (1969) op.cit.; Nilsson, N.J. Problem Solving Methods in Artificial Intelligence, McGraw-Hill, New York, (1971).

10. Lee and Moore (1967), op.cit.; Pfefferkorn, Charles, "Computer Design of Equipment Layouts Using the Design Problem Solvers", unpublished Ph.D. dissertation, Department of Computer Science, Carnegie-Mellon University, (May, 1971); Whitehead, B. and Elders, M.2. "The Planning of Single-Story Layouts", Building Science, Vol. 1 (Sept., 1965)pp. 127-139.

11. Slagle, J. "An Efficient Algorithm for Finding Certain Minimum Cost Procedures for Making Binary Decisions", Journal of the Association for Computing Machinary (1964) pp. 253-264.

12. Eastman, C. "Automated Space Planning", unpublished paper, Institute of Physical Planning, Carnegie-Mellon University, (September, 1972).

13. Alexander, C., Notes on the Synthesis of Form, Harvard University Press, Cambridge, (1964); Alexander, C., HIDESC 3: Four Computer Programs for the Hierarchical Decomposition of Systems whib have an Associated Linear Graph, Civil Engineering Systems Laboratory, M.I.T. Cambridge, (1963); A good Survey of other decomposition algorithms is provided in Moore, G. (ed.) Emerging Methods in Environmental Design and Planning M.I.T. Press, Cambridge, (1970).

14. Eastman, C. (1972) op. cit.

15. Eastman (1971) op.cit.; Pfeffercorn (1971) op.cit.; Weinzapfel, Grey "Image: An Interactive Graphics Based Computer System for Multi-Constrained Spatial Synthesis", Proceedings of the Eight Design Automation Workshop, Atlantic City, N.J. (June, 1971).

16. Yessios, Christos "Syntactic Structures for Site Planning" this volume.

17. Alexander, C., Ishikawa, S. and Silverstein, M. "A Pattern Language Which Generates Multi-Service Centers," Center for Environmental Structure, Berkeley, (1968); Wong, A.C.K. and Au, T., "A Parallel Processing Approach to Bridge Planning" Journal of the ASCE (Structural Division), 1970.

18. Sutherland, I.E., "Sketchpad: A Man Machine Graphic Communication System", Proceedings Spring Joint Computer Conference, Vol. 23, pp. 329-346 (1963).

19. Eastman, Charles and Schwartz, Michael "A Generative Method for Creating Variable Sized Objects in Design" Institute for Physical Planning Research

Report No. 34, Carnegie-Mellon University, Pittsburgh, (January, 1973).

20. Nugent, Christopher, et.al., "An Experimental Comparison of Techniques for the Assignment of Facilities to Locations", Operations Research, 16:0 (1968).

21. Eastman and Schwartz (1973) op.cit.

22. Negroponte, N., Groisser L., Taggert, J. "Hunch: An Experiment in Sketch Recognition", Environmental Design: Research and Practice, Vol. 2, EDRA 3 Conference, UCLA, (January 1972).

23. Evans, Thomas "A Program for the Solution of Geometric-Analogy Intelligence Test Questions", in Semantic Information Processing, Marvin Minsky (ed.), M.I.T. Press, Cambridge, (1968).

24. Yessios Christos, "Fosplan: A Formal Space Planning Language" in Environmental Design: Research and Practice, Vol. 2, EDRA 3 Conference, UCLA, (January, 1972).

25. See the papers in Session Six in EDRA TWO: Proceedings of the Second Environmental Design Research Associatation Conference, J. Archea and C. Eastman (eds.) Pittsburgh (1970).

26. (1968) op.cit.

27. Estes, Mark "Data Management Techniques Applied to People/Activity Relationships Within the Built Environment", in Environmental Design Research and Practice, Vol. 2, EDRA 3 Conference, UCLA, (January 1972).

28. Myer, Theodore "An Information System for Component Building" in EDRA TWO: Proceedings of the Second Environmental Design Research Association Conference J. Archea and C. Eastman (eds.), Pittsburgh, (1970).

29. Lederberg, J., et.al. "A Heuristic Program for Solving Scientific Inference Problems: Summary of Motivation and Implementation", in Theoretical Approaches to Non-numerical Problem Solving, R. Banerji and M. Mesarovic (eds.), Springer-Verlag, New York (1970).

AN EVALUATION OF SPACE PLANNING METHODOLOGIES

Elliott E. Dudnik

Assistant Professor of Architecture
University of Illinois at Chicago Circle
Department of Architecture
Chicago, Illinois 60680

Robert Krawczyk

C. F. Murphy and Associates
Chicago, Illinois

Abstract
While numerous algorithms and approaches have been suggested and applied to the problem of space-planning there has not been any systematic comparison or evaluation of these methods. Such an evaluation is important because the spatial configurations produced by such methods vary considerably. In this paper the major space planning techniques, as well as several new approaches and modifications to existing methods, are described, compared and evaluated. The use and consequences of certain options and/or modifications to the basic methods are included and evaluated. The results for each method using an identical problem are given, indicating the comparative speed and optimality of each approach, as well as the final plans generated. The implications of these results and their relationship to the design process are discussed.

Introduction
The past decade has seen a variety of techniques for evaluating and/or automating the space planning process. To a great extent, these allocation techniques have become very familiar to many designers and their utilization has become more widespread. Design literature consistently includes applications, extensions, or computer program descriptions which rely on any number of different space allocation algorithms. The resulting spatial configurations or floor plans are generally presented with discussion reserved only for the end product. Usually, a particular algorithm has been implemented, with certain other related decisions inherent to the chosen approach, and the system or solutions built up from this initial decision. There is little explanation, however, as to why the particular technique has been chosen nor why certain secondary decisions related to the implementation of this approach have been made. The results of any of the many different techniques are not, however, the same. Even the secondary decisions that arise as a consequence of any algorithm yield widely varying solutions. It has, however, rarely been the case where the results or consequences of using each of the different approaches have been carefully evaluated, analyzed or compared.

This paper takes each of the many space allocation techniques available plus several new approaches and compares and evaluates them with respect to one another.

Identical example problems are used to clearly demonstrate the differences in results that occur in each case. Comparisons are made of computing time as well as the "scores" of each resulting solution. For every technique used, all possible sub-options are explained and demonstrated to clearly illustrate the consequences of their use. Thus, when the random generating approach is discussed, the implications of a technique which randomly assigns locations to spaces is compared to one that switches locations of two randomly chosen spaces and also to one that only performs the switch if the new random location is within a chosen distance of other elements. Similarly, various approaches and options within the framework of assignment techniques are discussed and compared to one another. Thus, one may see the consequences of various approaches or rules to the spatial location and choice of each incoming element to the plan and/or its dependence toward any or all previously located elements. All approaches are evaluated and compared to one another, and not only to their related generic types.

The purpose of this paper is to both clearly define and illustrate the various techniques available for space planning, to carefully and systematically demonstrate the differences that arise from use of each method, and to evaluate these results. This paper shows that the initial choice of algorithm does have a profound affect upon the resulting design and that the consequences are such as to warrant more serious consideration of certain methods for particular situations. This paper also, in illustrating these techniques presents several new approaches, extensions and modifications of space planning algorithms, heretofore, not presented in the literature. These new techniques are clearly recognizable as logical extension of the choices or decisions that would occur during the manual design process but ignored or avoided previously in the automation of the process.

A Review of Space Planning Techniques
A number of space-planning or allocation methods have, at various times, been developed or presented. For the past ten years, design literature has contained many applications of space-planning techniques as well as extensions of these methods. Comprehensive summaries of these various approaches are given by Mitchell[1] and Eastman[2]. Mitchell not only provides an extensive bibliography of space-planning, but also a systematic summary and taxonomy of space-planning techniques which will be referred to in this paper.

With few exceptions, the bulk of space-planning techniques fall into the category of assignment techniques. Essentially, this is an approach which considers the space-planning problem as a combinatorial problem of assigning the various required spatial elements to discrete locations or modules in the available space in such a way as to satisfy a given set of constraints and to optimize some objective function. In general, this objective function relates to the distance between elements and some type of interaction function. This interaction may be expressed as one or more weighted values with either an objective (travel cost, trip volume) or subjective (relative importance, observed hierarchy) basis.

While other approaches to space-planning have begun to emerge recently, many of these are either still in conceptual form and have not been implemented successfully or may not necessarily seek to optimize or improve upon a plan. An example of the former may be found in Grason's proposal[3] for use of graph theory, which seems only successful for small planar graphs. As for the latter situation, the overlay techniques used by Grant[4] or Ward[5] locate elements on the basis of greatest "utility" or "suitability", but this in fact does not seek to improve the relationships between the parts of the plan as much as it seeks to resolve conflicts between the available space and each individual element. The work of Johnson, et al[6] appears closer to resolving the difficulty of other techniques by being best able to internally represent the relationships, the boundary conditions, and geometry of the elements and the space. At present the major difficulty appears to be the large amounts of computer storage and time required for solution and the absence of a "closed-form" problem, thereby preventing the possibility of obtaining a single "best" solution but rather producing several alternative results.

One returns, therefore, to consideration of the large group of assignment techniques. They are important for a number of reasons. First, the bulk of work and literature is centered in this area, mainly because of the relative ease of problem statement and structure. Secondly, the algorithms that can be and have been developed for obtaining solutions by these approaches are such that computation time is generally extremely fast. This has generally led to continued reliance upon and use of these methods and less tendency to seek new or further develop more complex, costlier, and more time-consuming techniques. Thirdly, despite the relatively large amount of activity and writings devoted to these techniques, little if no objective evaluation or assessment has been made of these techniques. Finally, the structure of the assignment model, with its clearly defined objective function provides the means for such an evaluation. This is not to say, however, that other approaches or methods cannot be evaluated. However, many of the other methods, some of which rely only on relationship of element to site of available space, e.g., the overlay techniques, and do not necessarily possess a set of interelemental relationship, could not be "scored" in a similar method and might necessitate a greater reliance upon subjective evaluation.

It should be noted that the general area of assignment technique comprises a wide range of methods. The generic grouping is often assigned to any approach that bears some resemblance to the model description of Brotchie[7] or possesses one or more of the attributes or characteristics of that model. It is necessary therefore to clearly define and classify the different methods within a more detailed framework.

Assignment Techniques
Within the broad classification of assignment techniques, one may first divide all approaches into two distinct classes: 1) constructive or generative procedures and 2) improvement procedures. The constructive procedures begin with an empty field and locate each element of the plan successively in accordance with some given set of rules or algorithm. The improvement procedures take an initial configuration and attempt to modify it so as to produce a better or improved configuration. While this classification provides some means of categorizing the various techniques, a further breakdown is possible by consideration of the algorithms used to improve or generate the plans. These algorithms will be considered in the context of each of the two classes.

    A.   <u>Constructive or Generative Procedures</u>. As noted above, the constructive procedures are typified by an "ex nihilo" approach to the space-planning problem.

That is, the technique begins with an empty field ("tabula rasa") and locates the elements in accordance with some algorithm. The technique for choosing which element should "enter" the plan and for determining its location provides the means for classification.

1. <u>Random Generation Method</u>. The simplest and most obvious method for generation or construction of floor plans is by random choice of elements and/or locations. The technique employed is a simple one: using a random number generator, successively generate a pair of X- and Y-coordinates for each element on the list of spaces. These coordinates will determine the location of that element on the plan. To prevent an element from being placed in a location already occupied by a previously placed element, a check is made for occupancy. If the space is not vacant, a new set of coordinates are generated. The method may be continued indefinitely with a scoring algorithm used to determine which solutions should be "saved" or printed if one wishes to avoid seeing every solution.

2. <u>Ordered Scores (Assignment Method)</u>. This method, the simplest and least sophisticated use of the affinity interaction matrix, builds the plan by choosing the elements in accordance with an ordered list based on total interaction scores. The element with the highest score enters the plan first followed by the next highest scoring element and so forth. The positions of incoming elements are tested for best score within a given radius or distance from previously placed elements. This approach is essentially that of Whitehead and Elders[8] and is essentially that used by Lee and Moore in CORELAP[9], although the latter method does not adhere strictly to the ordered interaction scores in certain cases.

3. <u>Polyomino Assembly (Assignment One)</u>. This method is based on the polymino assembly procedure described by Mitchell and Dillon[10]. The choice of elements to enter a plan is made on the basis of interaction with elements already placed and the location on the basis of adjacency to the placed elements. In particular the choice and location of the n + 1th chosen element is based upon interaction with the nth chosen element of the plan. The method may be modified somewhat by the following options:

    a. <u>Positioned Elements</u>. Choice of element is based on interaction with all elements already placed on the plan and the element with highest interaction score individually to the most of the placed elements is chosen for entry.

    b. <u>Number of Dependent Elements</u>. Positioning of an element chosen by the previous option may be restricted to only locations around the n elements with which it had the greatest interaction.

    c. <u>Interacted Elements</u>. Choice of element to be placed is again chosen on the basis of interaction with positioned elements. The one with the highest score to the most elements is next to enter the plan. Location, however, is only possible next to those elements of the plan with which the incoming element had highest interaction scores.

4. <u>Nuclear Growth (Assignment Two)</u>. This method is a new approach to the problem of space-planning. Selection of an incoming element is based upon the scores of the available elements to the cluster of elements previously positioned. That is, if three elements have been placed, the choice of a fourth element is made on the basis of the element having the highest <u>total</u> score to all three of these elements. Position is tested at all points adjacent to these elements. The method differs from the original assignment technique in that this method chooses on the basis of only interaction with <u>positioned</u> elements rather than from a list ordered by total scores. It also differs from the polyomino approach by using <u>total</u> interaction with all positioned elements rather than only with the last element(s). The method may be modified by allowing for:

    a. <u>Number of Dependent Elements</u>. Rather than interaction with <u>all</u> previously located elements, choice of element may be restricted to only the last n elements to have entered the plan.

B. <u>Improvement Procedures</u>. As noted earlier, the improvement procedures are those which systematically seek to improve upon the score of the plan constructed in the previous cycle. Elements are <u>re-positioned</u> in an effort to achieve a better score. In general, efforts that fail are disregarded or discarded and computation ceases when no further improvement is possible. Classification is made by the technique used for obtaining improvement.

1. <u>Random Switch</u>. This method takes the initial plan and seeks improvement by randomly switching any two elements of the plan matrix. To avoid switching empty space elements or a non-empty space with an empty space, the restriction is included that only if two non-empty spaces are within a given radius of the empty space will a switch be permitted. This rule also prevents a plan with "detached" spaces and thereby maintains contiguity of plan.

2. <u>Ordered Scores and Alternative Check</u>. This method utilizes the same technique as the first generative technique (Ordered Scores - Assignment). At the end of the plan generation phase, however, a systematic switch of every occupied element with every other occupied element of the space is made in an effort at improvement. A more sophisticated approach to this method uses the option:

    a. <u>Alternative Check</u>. This option performs a systematic switch of plan elements seeking improvement at each cycle of the plan generation, i.e., as each new element enters the plan, switches are performed seeking overall plan improvement.

3. <u>Single Switch</u>. This method systematically switches every element of the initial plan with every other element of the plan. Again, to avoid non-contiguity, the rules prescribed for the random switch are invoked. The method also permits K passes through the plan, where a pass is defined as every position switched with every other position (a maximum of n^2 switches for a non-empty space).

4. <u>Computerized Relative Allocation of Facilities Technique (CRAFT)</u>. This method has been thoroughly documented[11] and several extensions proposed[12,13]. Basically, elements are interchanged to achieve improvement on the basis of meeting at least one of three criteria: 1) they are the same size, 2) they have a common border, and/or 3) they border on a third element. At each cycle the interchange performed according to these rules is that producing the greatest improvement in score. The procedure ceases when no further improvement is possible.

There are, therefore, eight different assignment space-planning methods which can be modified by invoking the several options. In addition, all the generative methods may include the options:

i. <u>Value Only</u>. This option calculates interaction scores on the basis of relational value X distance, as opposed to:

ii. <u>Value-Area</u>. This option calculates interaction scores on the basis of relational value X distance. This option is useful when elements are of varied size and importance is desired for larger plan elements so that they would enter the plan initially.

<u>Evaluation of the Space-Planning Methods</u>

The various methods and options for space-planning described above were evaluated using as a test problem, a middle school (junior high school). The data for the school including room types, areas, and relationship matrix are shown in Fig. 1. The choice of this building type and its associated program for testing allows for many clearly defined functions and relationships. The number of spaces (twenty-five) permits the possibility of sufficient variation in the final plan configurations and scores which would not be possible with less complex building types or those with a smaller number of spaces.

Using the information given and a set of computer programs developed for each of the methods described above, test runs were performed to assess and evaluate each method. Since scoring was possible on the basis of value X distance in all cases, comparitive scores were obtained. In addition, computer times were obtained for each run to assess speed and cost as well. The scores and computer times for each method and option used are given in Table I.

With a major objective of any space-planning technique being the achievement of the best possible results without generating excessive costs, it is interesting to note that the variation between the final scores of the various methods is less than 5%. More importantly, the time differential between the worst and best solution is more than forty-fold! This means that for a sacrifice of only 5% in the efficiency of plan, a <u>savings</u> of nearly 20 minutes of computer time (23.2 minutes versus .52 minutes) is achieved. Interestingly enough, a solution only 2% "worse" than the "best" solution can be achieved in 1.02 minutes.

It is surprising to find even with a relatively "small" building, that the solutions generated by the various methods generate so great a variety of solutions. While certain options of each method fail to produce much, if any, change a wide range of

FIGURE I – INPUT DATA FOR SAMPLE TEST PROBLEM

TABLE I

Generative Methods

Method	Options	Score	Time
Ordered Scores			
	No options	57827	.40
	Check alternatives at completion	57779	5.38
	Check alternatives at each cycle	56664	24.20
Polyomino Assembly			
	3 dependent elements/Interacted/Area + Value	58924	.49
	3 dependent elements/Positioned/Area + Value	59218	.52
	3 dependent elements/Interacted/Value	58510	.52
	3 dependent elements/Positioned/Value	58304	.49
	6 dependent elements/Interacted/Area + Value	58578	.57
	6 dependent elements/Positioned/Area + Value	57789	1.02
	6 dependent elements/Interacted/Value	57619	1.02
	6 dependent elements/Positioned/Value	57922	1.04
	9 dependent elements/Interacted/Area + Value	58578	1.02
	9 dependent elements/Positioned/Area + Value	57809	1.09
	9 dependent elements/Interacted/Value	57671	1.05
	9 dependent elements/Positioned/Value	57671	1.05
	12 dependent elements/Interacted/Area + Value	58578	1.03
	12 dependent elements/Positioned/Area + Value	57977	1.13
	12 dependent elements/Interacted/Value	57671	1.08
	12 dependent elements/Positioned/Value	57671	1.14
	15 dependent elements/Interacted/Area + Value	58578	1.07
	15 dependent elements/Positioned/Area + Value	57977	1.23
	15 dependent elements/Interacted/Value	57671	1.11
	15 dependent elements/Positioned/Value	57671	1.20
Nuclear Growth			
	3 dependent elements/Area + Value	57734	1.27
	3 dependent elements/Value only	57771	1.24
	6 dependent elements/Area + Value	58306	1.28
	6 dependent elements/Value	57710	1.23
	9 dependent elements/Area + Value	58426	1.26
	9 dependent elements/Value	57825	1.23
	12 dependent elements/Area + Value	58426	1.25
	12 dependent elements/Value	57896	1.23
	15 dependent elements/Area + Value	58426	1.25
	15 dependent elements/Value	57920	1.22
	all dependent elements/Area + Value	58426	1.26
	all dependent elements/Value	57920	1.23

Improvement Methods

Single Random Switch

Switch	Score
0	76387
600	59210
1200	57528
1800	57018
2400	57018
3000	56919
3450	56794
3825	56782
6000	56782

Time = 37.37

Single Switch

CRAFT

Option	Score	Time
1 Pass	57899	7.42
2 Passes	56905	15.34
3 Passes	56356	23.20
4 Passes	56356	30.68
5 Passes	56356	38.49
5 x 5 Matrix	58058	23.27
6 x 6 Matrix	58103	78.78

different configurations are possible. If one carefully inspects the final solutions shown in Fig. 2, it is clear that a wide number of alternatives have been obtained using these different approaches. Methods such as the random generating routine produce a great number of plans, of course, but not many are optimal. The more sophisticated generative techniques do, however, differ in their final product. It can be seen in Fig. 3 that the growth patterns resulting from the different approaches vary sufficiently to produce these variations, but the scores tend to indicate that near optimality has been achieved.

Conclusions

It is probably difficult to denote any one of the many space-planning methods as "best". Nearly all of the methods that have been evaluated here have certain attribute that make them valuable. Even the random generating or random switching routines, both of which are seemingly based on an "unscientific" approach, are valuable to the designer in producing alternatives not foreseeable by a strictly rigorous, rational approach. Clearly, however, if certain methods do little to improve upon a final configuration, but are much more time consuming, there is some doubt as to their values and consideration should be given the more efficient approaches.

On a rational, philosophical basis, one may prefer certain methods of space-planning because the technique more closely approaches the actual design process. The second and third assignment techniques (Polyomino Assembly and Nuclear Growth) are the most sophisticated methods and come closest, in this regard, to a designers process. Despite the fact that scores were not as "good" as that produced by the switching routines, these two techniques have a rational appeal and appear to justify one's faith in the approaches taken by the speed with which they achieve solutions. Clearly, a direct route to solution is more satisfying than a trial-and-error solution.

It should be noted that the improvement techniques, particularly CRAFT, are valuable to any designer by demonstrating alternatives to the initial plan. This is particularly valuable when the initial configuration is one that has been carefully developed rather than arbitrary as in this paper. Unfortunately, however, since the lower bound on the score for a plan is unknown, these methods may perform thousands of needless operations and tests, producing no real improvements. This is the greatest disadvantage to these methods along with the associated lengthy computation times.

For the designer, the implications of these results are three-fold. First, methods exist and can be further refined that can produce good and efficient architectural plans. The adjacency of related elements into well-defined clusters seems to indicate that these methods not only provide efficient plans, but also that these plans are quite similar to those traditional design methods achieved. Secondly, the ability for certain space-planning techniques to quickly generate numerous efficient alternative solutions to a given program and set of criteria provides the possibility for a richer variety of architectural solutions available for investigation than would be possible by manual methods. Finally, and most importantly, the results of this evaluation indicate that many of the different space-planning techniques produce solutions that are all as nearly efficient despite fundamental

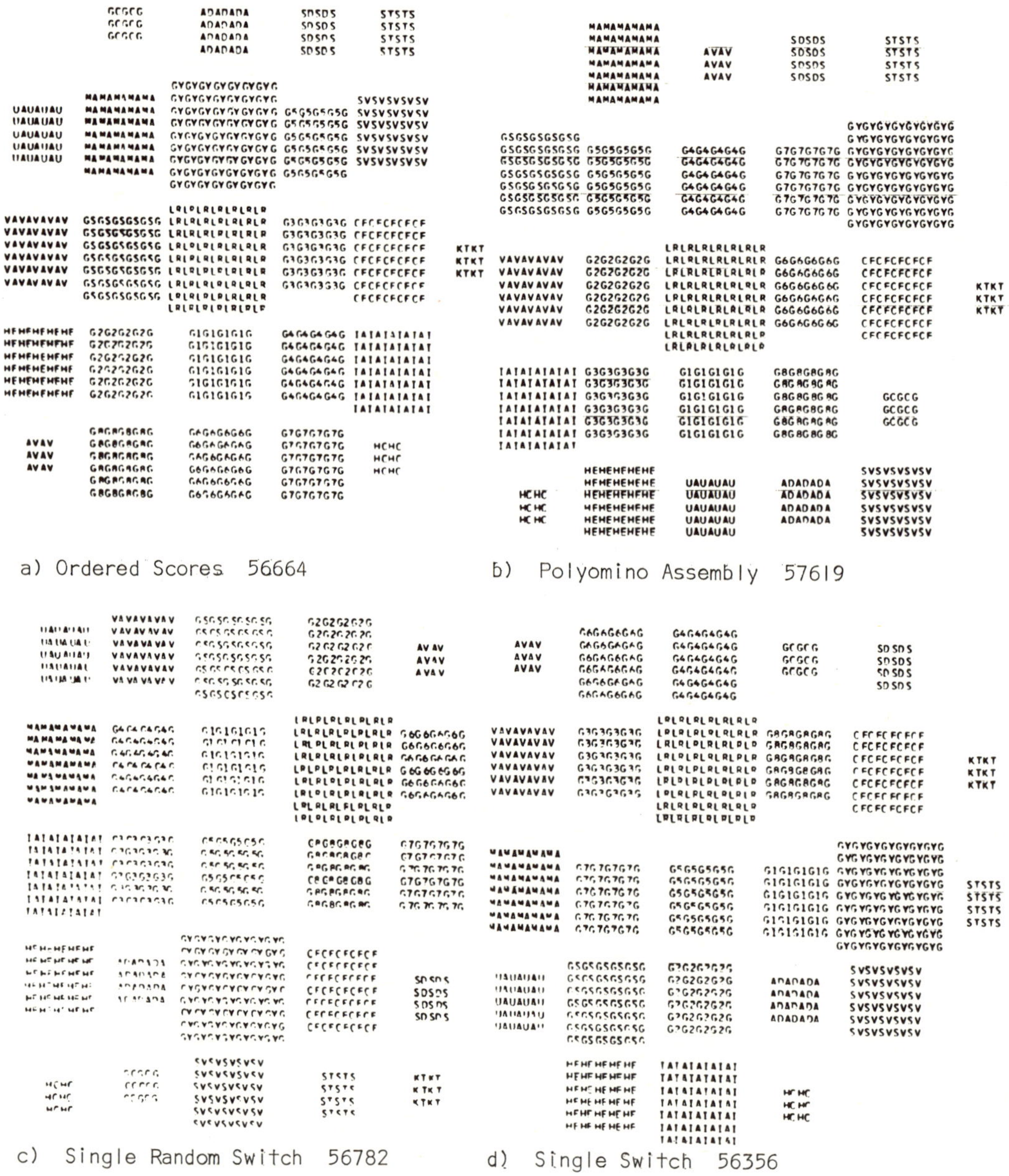

a) Ordered Scores   56664

b) Polyomino Assembly   57619

c) Single Random Switch   56782

d) Single Switch   56356

FIGURE 2 – TYPICAL FINAL CONFIGURATIONS

8. SPACE PLANNING TECHNIQUES / 425

Ordered Scores	Polyomino Assembly	Nuclear Growth

CURRENT PLAN MODIFIED WITH THE ADDITION OF THE G1 ELEMENT | CURRENT PLAN MODIFIED WITH THE ADDITION OF THE G4 ELEMENT | CURRENT PLAN MODIFIED WITH THE ADDITION OF THE G4 ELEMENT

```
                                G4                          G4
     MA  GY                G2  LR                     G2  LR
     G5  LR                G3  G1                     G3  G1
         G1
```

CURRENT PLAN CONFIGURATION HAS A TOTAL INTERACTION VALUE OF 1311   ... 1903   ... 1903

a) After 5th element entered plan

CURRENT PLAN MODIFIED WITH THE ADDITION OF THE G6 ELEMENT | CURRENT PLAN MODIFIED WITH THE ADDITION OF THE AV ELEMENT | CURRENT PLAN MODIFIED WITH THE ADDITION OF THE AV ELEMENT

```
                                    AV
     MA  GY  G5          G5  G4  G7              G5  GA  G8
     G5  LR  G3          G2  LR  G6              G2  LR  G6
     G2  G1  G4          G3  G1  G8              G3  G1  G7
         G6                                              AV
```

CURRENT PLAN CONFIGURATION HAS A TOTAL INTERACTION VALUE OF 9012   ... 9548   ... 9548

b) After 10th element entered plan

CURRENT PLAN MODIFIED WITH THE ADDITION OF THE VA ELEMENT | CURRENT PLAN MODIFIED WITH THE ADDITION OF THE HE ELEMENT | CURRENT PLAN MODIFIED WITH THE ADDITION OF THE IA ELEMENT

```
                              MA  AV                      IA
     MA  GY  G5          G5  G5  G4  G7          IA  G5  G4  G8
     VA  G5  LR  G3  CF  VA  G2  LR  G6          VA  G2  LR  G7
     G2  G1  G4  IA      IA  G3  G1  G8          MA  G3  G1  G7
     G8  G6  G7              HE                      G5  AV
```

CURRENT PLAN CONFIGURATION HAS A TOTAL INTERACTION VALUE OF 23143 ... 23478 ... 23229

c) After 15th element entered plan

CURRENT PLAN MODIFIED WITH THE ADDITION OF THE SD ELEMENT | CURRENT PLAN MODIFIED WITH THE ADDITION OF THE CF ELEMENT | CURRENT PLAN MODIFIED WITH THE ADDITION OF THE SD ELEMENT

```
         AD  SD                 MA  AV                      IA  HE  AD
     UA  MA  GY  G5  SV     G5  G5  G4  G7           UA  G5  G4  G8  GC
     VA  G5  LR  G3  CF     VA  G2  LR  G6  CF       VA  G2  LR  G6  HC
     HE  G2  G1  G4  IA     IA  G3  G1  G8  GC       MA  G3  G1  G7  SD
     G8  G6  G7             HC  HE  UA  AD              G5  AV
```

CURRENT PLAN CONFIGURATION HAS A TOTAL INTERACTION VALUE OF 39619 ... 38118 ... 37965

d) After 20th element entered plan

CURRENT PLAN MODIFIED WITH THE ADDITION OF THE KT ELEMENT | CURRENT PLAN MODIFIED WITH THE ADDITION OF THE KT ELEMENT | CURRENT PLAN MODIFIED WITH THE ADDITION OF THE ST ELEMENT

```
     ST  AD  SD  GC             MA  AV  SD  ST                IA  HE  AD
     UA  MA  GY  G5  SV     G5  G5  G4  G7  GY         UA  G5  G4  G8  GC
     VA  G5  LR  G3  CF  KT VA  G2  LR  G6  CF  KT     VA  G2  LR  G6  HC
     HE  G2  G1  G4  IA     IA  G3  G1  G8  GC         MA  G3  G1  G7  SD
     AV  G8  G6  G7  HC     HC  HE  UA  AD  SV             G5  AV  GY  SV
                                                                KT  CF
```

CURRENT PLAN CONFIGURATION HAS A TOTAL INTERACTION VALUE OF 57827 ... 57619 ... 57710

e) After 25th element entered plan

FIGURE 3 – TYPICAL GROWTH PATTERNS

differences in approach. This implies that the designer may choose any method which comes closest to his particular design approach and philosophy and know that his solutions will not suffer in comparison to any other approach.

This final point is most interesting if one looks at the process by which the methods construct the solutions. The elements entering a plan at any stage and their location are often different as a direct result of the problem criteria and the method used. Despite this fact, and independent of type of problem, the final configurations still more than adequately meet the rquirements and standards generally expected of a good architectural solution. The broad implications of this apparent independence of good design solution to individual method may well be the most important result of this evaluation and indicate the continued need for investigation of the entire process of design.

Notes

References

[1] Mitchell, William, "Notes on Approaches to Computer-Aided Space Planning", Proceedings of the Kentucky Workshop on Computer Applications to Environmental Design (ed. M. Kennedy), Lexington, Department of Architecture, University of Kentucky, 1971.

[2] Eastman, Charles, "Logical Methods of Building Design: A Synthesis and Review", Institute of Physical Planning Research Report No. 28, Carnegie-Mellon University, December 1971.

[3] Grason, John, "A Dual Linear Graph Representation for Space Filling Location Problems of the Floor Plan Type", in Emerging Methods in Envirnomental Design and Planning (ed. G. Moore), Massachusetts Institute of Technology Press, 1970.

[4] Grant, Donald, "Combining Proximity Criteria with Nature-of-Spot Criteria in Architectural and Urban Design Space Planning Problems Using a Computer-Aided Space Allocation Technique: A Proposed Technique and an Example of its Application", Proceedings of the Ninth Design Automation Workshop, Dallas, Association for Computing Machinery, 1972.

[5] Ward, W. S., D. P. Grant, and A. J. Chapman, "A PL/I Program for Computer-Aided Architectural and Planning Space Allocation", Proceedings of the Fifth Annual Urban Symposium of the Association for Computing Machinery, New York, Association for Computing Machinery, 1970.

[6] Johnson, Timothy, et al, "IMAGE: An Interactive Graphics Based Computer System for Multi-Constrained Spatial Synthesis", Massachusetts Institute of Technology, Department of Architecture, 1971.

[7] Brotchie, John, "A General Space Planning Model", Management Science, volume 16, number 3, 1969.

[8] Whitehead, B. and M. Z. Elders, "An Approach to the Optimum Layout of Single-Story Buildings", Architects Journal, June 17, 1964.

[9] Lee, R.B. and J. M. Moore, "CORELAP" Computerized Relationship Layout Planning", Journal of Industrial Engineering, volume 18, number 3, March 1967.

[10] Mitchell, William and Robert Dillon, "A Polyomino Assembly Procedure for Architectural Floor Planning", Proceedings of the Third Environmental Design Research Association Conference, University of California at Los Angeles, School of Architecture and Planning, January 1972.

[11] Armour, G. C. and E. S. Buffa, "A Heuristic Algorithm and Simulation Approach to the Relative Allocation of Facilities", Management Science, volume 9, number 2, January 1963.

[12] Lew, Paul and Peter Brown, "Evaluation and Modification of CRAFT for Architectural Methodology", in Emerging Methods in Envirnomental Design and Planning (ed. G. Moore), Massachusetts Institute of Technology Press, 1970.

[13] Rohn, Joachim, "The CRAFT Program: Improvements and Proposed Improvements", SIGSPAC Bulletin, volume 4, number 3, June 1970.

SYNTACTIC STRUCTURES FOR SITE PLANNING [1]

Christos I. Yessios

Institute of Physical Planning
Carnegie-Mellon University

Abstract

The paper focuses on Site Planning and expands on work reported earlier [2]. Syntactic structures and space grammars for different levels of composition are explored and developed. As an implementation, it relates to the Computer Implemented Site Planning System currently under development at IPP, C-MU. Some running portions of this system will be presented as illustrations. The approach emphasizes consideration of a class of local conditions which lend themselves to the development of inexpensive composition oriented procedures. The result is a language by the use of which the computer can be programmed to resolve site planning problems.

1. Introduction

The space planning problem is to derive a functional configuration given a set of physical elements and a set of conditions to be satisfied. Such problems can be at different levels and range from room layouts to land use planning. In computer implemented space planning, issues as the uniformity of the problem at the different levels, the nature of the conditions, etc., are rather crucial. There is no intention to elaborate on these issues. But a short exposition of the principles upon which the work reported here is based, will help.

Each space planning problem refers to some global space (room, lot, site) and there are basically two types of conditions: global and local. The global conditions can be dealt with only if the global space and the elements it already contains are considered. In contrast, the local conditions extend only to a limited neighborhood around the respective element and are independent of the global space. Some global conditions can be reduced to local in cases where the global space can be treated as a common element.

The local conditions tend to have some interesting properties. They are usually transitive (a condition initially satisfied for a single element, is also satisfied for the whole configuration, part of which it is); distributive (a condition satisfied for the overall configuration, is also satisfied for each and all of the constituent elements); or even associative (first transitive and then distributive). The global conditions on the other hand, tend to be intact (neither transitive nor distributive).

An implication of the above distinctions is the suggestion that each condition should be treated according to its type. The local conditions have been especially neglected. The most advanced systems to date employ a feasibility approach by which all conditions are treated as global.[3] For each individual element the required conditions are tested rather independently by the use of heuristic problem solving techniques. Such an approach tends to build up the final configuration by sequentially adding or manipulating one element at a time. Even though the feasibility approach tends to result into expensive systems, there is no doubt that it is valid and that it constitutes a significant contribution. What is in doubt is its exclusiveness.

An alternative is the syntagmatic approach, which this paper advocates and explores. This treats the local conditions as constituents of an element's definition, and taking advantage of their properties, it employs a linguistic method for the definition of functional configurations. The result is a language through which a user can program the computer to resolve a space planning problem.[4]

It is the author's belief that future computer implemented space planning systems will seek to utilize both approaches simultaneously. Whatever the general case, the space planning problem at the site level is of a syntagmatic nature.[5] Given the problem's scale, it is inconceivable to try to build up a site plan by adding one element at a time. The main concern is to compose use domains and linkages in such a way that a desired degree of accessibility, human concentration, dispersion of uses, etc. is achieved. This paper explores a class of local conditions and the respective syntagmatic structures for the site planning problem. The work relates to the Computer Implemented Site Planning System currently under development at the Institute of Physical Planning, C-MU.

The specific problem to be dealt with is as follows. Given the space diagram of some site, a set of uses and an overall or partial conceptual scheme, derive a plan of the uses for the space diagram. The space diagram is a kind of a map showing the site's borders, existing uses (roads, points of interest, natural features, etc.) and delineating the buildable vacant land. It is the result of an analytic process applied to a given site. It is assumed that the preliminary, comprehensive planning type of studies have been completed and that the uses to be located have been decided and are given. By conceptual scheme is meant the planner's "image" of the solution, which is a schematic physical interpretation of concepts like community, uninterrupted walks, (wo)man-machine interaction, or even maximum profit, full coverage, etc.

The above problem is divided into four consecutive levels of composition; namely the Lot Level, the Module Level, the Neighborhood Level and the Plan Level. At each level, we start with a set of primitives and going through a number of composition sublevels, we derive the level's top composite. An element is primitive if it is taken to consist of itself only; it is composite otherwise. An intuitive outline of the whole process can at best be given by referring to the sketches in Figure 1.1.

Assume we are given the site in 1.1(1), which is intersected by some highway. Preliminary studies, mainly slope analysis, delineates the space diagram shown in 1.1(2). The shadowed areas denote spaces which cannot be used. Initially, at the Plan Level, we seek to subdivide the usable spaces of a space diagram into as many spaces as the number of neighborhoods we wish to locate. Area constraints and accessibility relations count heavily. The space diagram (1.1(2)) in our example is already divided into three spaces all of which have access to the highway. To keep the example simple, assume that three neighborhoods are to be located and that the accessibility requirements are already satisfied. The neighborhoods, called N1, N2 and N3 are located as shown in 1.1(2). The next task is to build up each neighborhood. The configurations M1, M2 and M3 (Figure 1.1(3)) are defined at the Module Level. These consist of use spaces (domains) and roads (links) and constitute the modular patterns to be employed in deriving the respective neighborhoods. At the neighborhood Level, the Module Level composites (called modules) may further be composed and are ultimately adjusted to the respective spaces, as shown in 1.1(4). Finally, at the Lot Level, the plan is finished by locating buildings within the use domains. This can be done in a variety of ways, as shown in 1.1(5).

The above example oversimplifies some aspects of the process but does offer an overview of the general space planning model pursued. This paper presents the Module and the Neighborhood Levels only (in sections 2 and 3 respectively), in as much detail as the space available permits. The Lot Level has been reported elsewhere [6]. The Plan Level involves problems which go beyond syntactic structures and deserves a detailed presentation by itself.

Throughout the remainder of this paper, names will be used for the identification of elements. A name is a string of up to six alphanumeric characters the first of which is always a letter. Different types of names will be distinguished through their initial letters. The linkage names start with W; the domain names with T; the module names with M; the neighborhood names with N; the composite names (variables) with V; the space names with S; and the parametric names with any letter from A to L. For the definition of the different elements, a variety of parameters will be used. They will be either of a constant or of a variable value and the two types will be abbreviated as c-parameter and v-parameter. The latter is a parametric name.

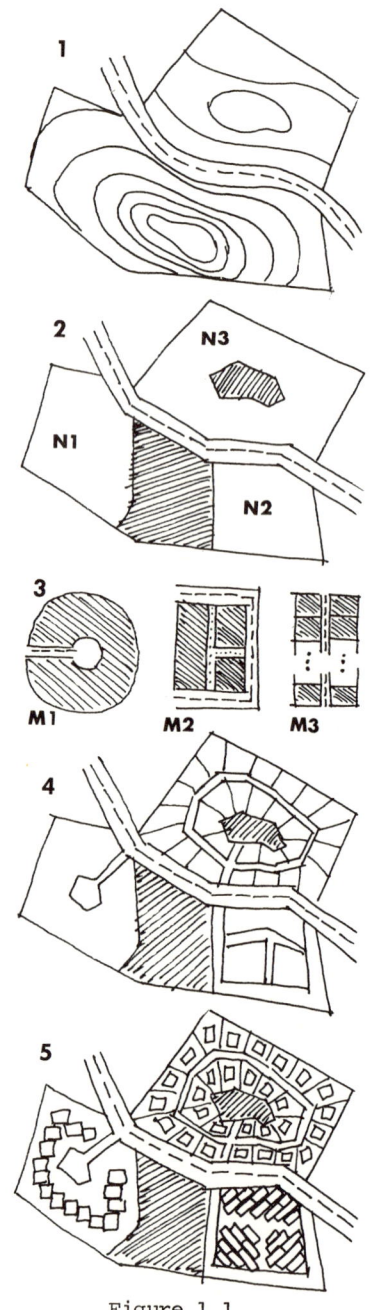

Figure 1.1

2. Module Level and Grammars

At the Module Level (M-level) there are two types of primitive elements, the <u>linkages</u> and the <u>domains</u>. These are composed deriving <u>pattern modules</u> which are this level's top composites.

Semantically, a linkage is any channel of movement of communication. Typical examples are the routes for pedestrians, cars, disposals, etc. For this paper, a linkage will be either a pedestrian route (to be indicated by a 'P') or a vehicular road (to be indicated by an 'R'). Syntactically, a linkage is a curved or straight line of predefined and significant width and of variable length. The <u>linkage specification</u> is a linkage name, followed by an equal sign, followed by a width parameter, followed by a type parameter. The parameters can be of a constant or a variable value. The constant value of the width parameter is any number; for the type parameter it is P or R. A linkage specification contains no information about its length. This is initially undefined and is assumed to vary. It takes a specific value when associated with some domain. Examples of linkage specifications with constant values, and their graphic representations are given in Figure 2.1. As shown, a linkage is represented by its axis and sides. The "color" of the axis indicates the linkage's type (dashed line for R and dotted line for P).

WA = 3 ; R   **WB = 2 ; P**

Figure 2.1

A domain can be any use or activity space such as a building lot, a garden, a parking space, a lake, etc. Graphically it is any two-dimensional physical element delineated by a closed curve (any curved line or any line segment approximation of such a line). Initially, the delineating curve is a regular polygon or the stretch equivalent of such a polygon. [7] The regularity of its shape is not necessarily preserved in the derived pattern module. Depending on the operations applied and the associations pursued the shape of a domain may be adopted to other shapes.

A <u>domain specification</u> is a domain name followed by an equal sign, followed by two dimension fields separated by a semicolon, possibly followed by a number of neighboring conditions. Each dimension field can be a single parameter or two parameters separated by a hyphen or empty. Each parameter can be of a constant or variable value. When, for a single dimension, two parameters are given, they define a range for the respective dimension. For example, T=3.5-8;10 means that the first dimension of T can range from 3.5 to 8 units. When a domain field is empty, it is left undefined and works as the length of the linkages. That is, its value is determined when associated with some other element. Either or both dimension fields can be empty. The semicolon should always be written to indicate which field is which dimension. Examples will be T1=3.5-8; , T2=;10 , and T3=; .

Each neighboring condition is a colon followed by a number of names separated by slashes. The neighboring conditions indicate the elements which are accepted as neighbors by a domain's sides. The symbols Q and Z can be used to indicate "no neighbor" and "any neighbor" respectively. A domain specification can take a variable number of neighboring conditions and, theoretically, there are no upper limits. Such limits are imposed by the implementation by which the neighboring conditions can only be one or two or four or eight. The number is significant and indicates the number of sides a domain has. One indicates a circle, two a semicircle, four a rectangle, and eight an octagon. If no neighboring conditions are given, the system defaults four sides fully permissive. That is if T=3.5-8;10 is given, the system defaults T=3.5-8;10 ::::Z, where ::::Z is a condensed notation for :Z:Z:Z:Z.

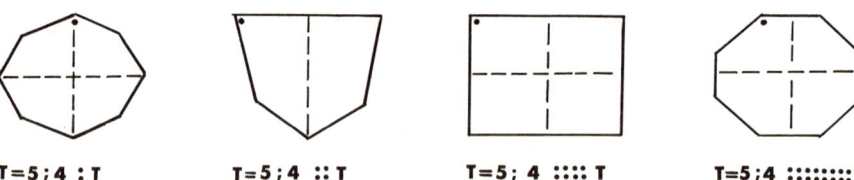

T=5;4 : T   T=5;4 :: T   T=5; 4 :::: T   T=5;4 :::::::T

Figure 2.2

Four examples of domain specifications and their respective graphic representations are given in Figure 2.2. Notice that they all have the same dimensions but different numbers of neighboring conditions. This results in different shapes. The circle is approximated by the octagon which it circumscribes. One of the corner points of a domain is designated as its origin and its sides are indexed clockwise. The sequence of the neighboring conditions corresponds to this indexing. The location of the origin is denoted by a dot in the respective corner. When the origin is at the upper left (or simply left in the case of the circle) corner, as shown in Figure 2.2, we say that the domain is in a normal position. The first dimension field of a domain specification refers to the horizontal and the second to the vertical axis of a domain, when in its normal position.

At the M-level, there are four composition operators; the junction (denoted #), the inclosure (denoted @), the envelopment (&), the string junction (#...#) and the exponentiation (*). The latter two are variations of the junction.

Envelopment (&) is the operation by which one element (the enveloped) is placed within another (the envelope). By the implementation, the envelope can only be a rectangular domain; the enveloped can be of any kind. For an example, see Figure 2.3.

Junction (#) is the operation by which two elements are joined in such a way that a side of the one becomes tangent with a side of the other. Only rectangular domains, composites and/or linkages can be joined by the junction. A non-rectangular element can be an operand to the junction, if it is first located within an envelope. An example is shown in Figure 2.4.

Inclosure (@) is the operation by which one element (the inclosed) is surrounded by another (the incloser). By the implementation, the incloser can only be a rectangular domain, composite or linkage; the incloser can be of any type. An example is given in Figure 2.5.

8. SPACE PLANNING TECHNIQUES / 433

T1 & T2
where
T1=7-10;7:**T2**
T2=8;5-8::::T1

Figure 2.3

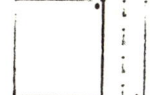

T3 # W1
where
T3=7;6-10:W1:::Q
W1=3;R

Figure 2.4

T4 @ W2
where
T4=6;5:::::::::W2
W2=1;P

Figure 2.5

The string junction is a sequence of junctions which derive a string of elements. An example is given in Figure 2.6. This is the only operation for which more than two elements are allowed in a row without the use of parentheses. Compare Figure 2.6 with the example of Figure 2.7, where the use of parentheses caused the operator # to be interpreted as a plain junction and a non-linear configuration was derived.

T5#W3#T5#W3
where
T5=5-7;7:W3:Q:W3:Q
W3=2;R

Figure 2.6

T8((T7#T8)#T7)
where
T7=4-6;4:::T7/T8
T8=4-8;2::T7::Q

Figure 2.7

T6 * 4
where
T6=7;2:T6:Q:T6:Q

Figure 2.8

When a number of identical elements are to be joined by a string junction, the exponentiation (*) can be used as a condensed notation. That is, T#T#T#T#T is equivalent with T*5. The exponentiation is a very convenient feature since it may be used in cases where we wish to allow the number of times an element is joined to itself to depend on its associations. In such cases we write T*X, where the value of X is initially undefined, and its value depends upon the element with which it is joined. An example of the exponentiation is given in Figure 2.8.

The derivation of compositions and the assignment of names to composites can be denoted by the use of <u>productions</u>. Syntactically, a production is a composite name followed by an up pointing arrow (↑), followed by a composition string. A composition string is two elements joined with a junction or an inclosure or an envelopment (e.g. T#W or TT@T or V&T), or a single element followed by an exponent (e.g. T*10 or T*X) or a number of elements joined with string junctions (e.g. T1#T2#T3#T4). Examples of productions will be V1↑T1#VV, VV↑T4*KK, V3↑T2#4↑#T0T0S, VAVA↑VILA@WAY, etc.

The composite names are more specifically called <u>variables</u>, and the primitive names <u>terminals</u>. A special type of a variable is the <u>start</u> name. This is a parametric name and is the top composite's name. A production with a start name on its lhs (left hand side) is called a <u>start</u> <u>production</u>.

Semantically, a production can be interpreted in either one of two ways. (i) As the operation by which the rhs (right hand side) composite is derived and is assigned the lhs name. Or (ii) as the operation by which the lhs variable, whenever encountered, is replaced by the rhs composition string. By (ii), a set of productions generates <u>unspecified</u> composites, which are composites represented by extended terminal composition strings. A composition string is terminal when it consists of terminals (primitives) only. It is of an extended form when, by the use of parentheses, a number of plain composition strings are joined together. For example, (((T2#T)@W)#((TT*3)&TTT)#TAT#WW). By (i) an unspecified composite is <u>specified</u> and its graphic (or some equivalent) representation is derived.

For the definition of a sequence of compositions and of the primitives involved, grammars are used. In general, a <u>grammar</u> is a <u>header</u> followed by a <u>set</u> of <u>productions</u> (the first of which is a start production always), followed by a <u>set</u> of <u>specifications</u>, followed by a <u>tail</u>.

For the M-level Grammars (M-grammars), the header is the sys (system's symbol) MODULE followed by a start name, possibly followed by a list of parameters enclosed in parentheses. The tail is the sys ENDMOD. The M-grammar specifications are linkage and domain specifications.

An M-grammar is called for execution through the <u>call command</u>, which is the sys CALL followed by a start name. The latter functions as the grammar's name also and should be accompanied by a list of constant values (c-parameters), if the header of the grammar definition was given with a list of parameters. These constant values replace the parametric names in the specifications, before the execution of the grammar proceeds.

An M-grammar is executed by three mechanically defined procedures, the <u>generator</u>, the <u>specifier</u> and the <u>tracer</u>. The generator starts with the start name and by sequentially replacing all variables acccording to the respective productions in the grammar, it derives the unspecified terminal composition string of the top composite. The specifier works in reverse. It starts at the lowest level of composition and by sequentially applying the respective specifications, it builds up specifications for the intermediate and top composites. Finally the tracer, which is a plotter oriented procedure, translates the specifications into the respective geometric shapes, it calculates a variety of statistics referring to the area percentages of the constituent elements and eliminates unnecessary partitions (partitions between elements of identical context). This completes the derivation of the top composite(s) defined by the grammar, which can then be plotted or transferred to the Neighborhood Level for further manipulations.

Examples of M-grammars as defined, called and executed by the current implementation are given in Figures 2.9, 2.10 and 2.11. The implementation accepts format free statements and blanks may be embedded freely. The symbol $ is used to signal the end of a statement. This feature allows more than one statement to be typed on the same teletype line. The system marks the beginning of a line with a ">" or a "-" sign to indicate that it is within or out of a grammar definition, respectively.

8. SPACE PLANNING TECHNIQUES / 435

```
-MODULE BBB ↑ V3#V2#V4 $
>V2 ↑ V1&T2 :Q:V4:Q:V3 $
>V3 ↑ T4#W1#T4 :V2:::Q $
>V1 ↑ T1@W1:T2$  V4 ↑ T3@W2 $
>T1 = 6-3;8-10 :W1 $
>T2 = 10;12 ::::V1 $
>T3 = 7-9;4 ::::W2 $
>T4 = 3-5;4-6 :W1:::Q $
>W1 = 1.5;R$  W2 = 2;R$
>ENDMOD   $

OK: GRAMMAR IS SYNTACTICALLY PROPER.

-CALL BBB $

((T4#W1#T4)#((T1@W1)&T2)#(T3@W2))
```

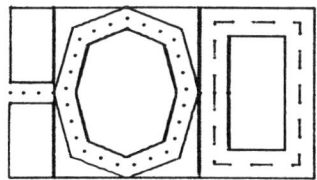

Figure 2.9

```
-MODULE CP ↑ VVV#VOV $
>VVV ↑ VV#W#VV#VV#W#VV :VOV:::Q $
>VV ↑ T1*5 :W:Q:VV:Q $
>VOV ↑ VB&T3 :VVV:::Q $
>VB ↑ VA@VC ::T3$  VC ↑ T1*X :VA:::Q$
>VA ↑ T2@W :Q:VC $
>T1=4;2-4:T1:Q:T1       :Q$
>T2=8;4:Q:W$  T3=20;10::::VB$
>W=2;R$  ENDMOD   $

OK: GRAMMAR IS SYNTACTICALLY PROPER.

-CALL CP $

(((T1*5)#W#(T1*5)#(T1*5)#W#(T1*5))#(((T2@W)@(T1*X))&T3))
```

Figure 2.10

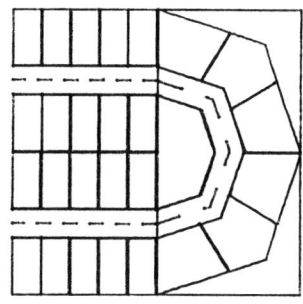

```
-MODULE DASY (AA,BB) ↑ TTT@VV $
>VV ↑ V*8 :TTT:::Q $
>V ↑ T&TT :TTT:V:TTT:V $
>T= ;AA::TT $  TT= ;BB::::T $
>TTT= AA;BB :VV$  ENDMOD       $

OK: GRAMMAR IS SYNTACTICALLY PROPER.

-CALL DASY (10,10) $

(TTT@((T&TT)*8))
```

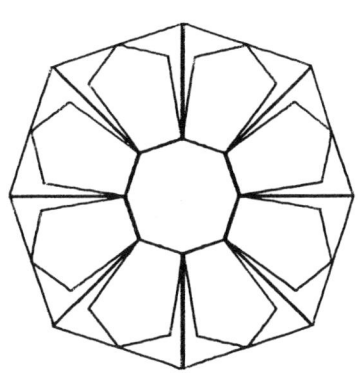

Figure 2.11

3. The Neighborhood Level and Grammars

At the M-level, the elements are composed according to locally defined syntagmatic conditions. Given that the constituent primitives are of regular shapes, the derived top composites have an overall regular shape. In contrast, the basic feature of the Neighborhood Level (N-level) is that its top composites, called neighborhoods, are referred to some given space and are ultimately adjusted to its shape.

The N-level Grammars (N-grammars) are basically as the M-grammars. The main difference is that, in addition to the linkages and domains, modules defined by M-grammars can be called from within an N-grammar, and are consequently dealt with as primitives. Syntactically, the header of an N-grammar is the sys NEHOOD (for neighborhood), followed by a start name, possibly followed by a list of parameters. The tail is the sys ENDNED. The set of productions is as before and so is the set of specifications with only the addition of the domain specifications. These are a module name, followed by an equal sign, followed by an M-grammar start name, possibly followed by a number of neighboring conditions. The neighboring conditions are as before with the difference that the symbol Y can be used to indicate that a side accepts as a neighbor itself only. An example will be MANA=ANA(2,3,4.5,6,3,R) :Y:Q: Y:Q $. In actuality, a module specification is a call for the execution of an M-grammar.

The N-level call command is followed by the indication of some space, with respect to which the defined neighborhood should be derived. For example, CALL KLP ; SPC $, where KLP the start name of some N-grammar defined without a parameter list and SPC can be any of the spaces shown in Figure 3.1.

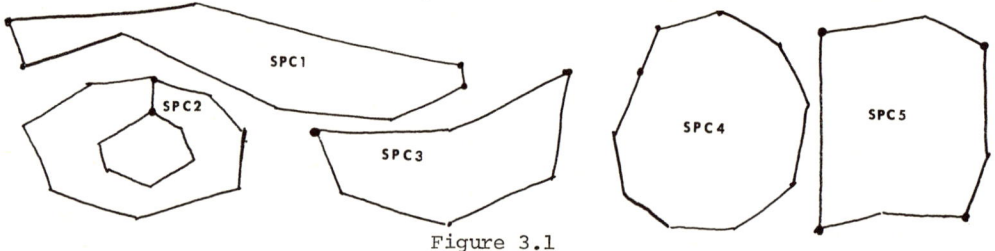

Figure 3.1

Depending upon the kind of operations applied, the types of elements composed and the mixture of compositions defined by the set of productions, the top composites derived by an N-grammar inherit an overall shape and pattern. For example, the string junction and the exponentiation imply a linear shape, the inclosure, generally but not necessarily, implies a circular pattern, etc. The referred spaces are also characterized by some predominant shape. That is, from the spaces in Figure 3.1, SPC4 can easily be taken as a circle or as an octagon, SPC5 as a rectangle, SPC3 as a semicircle, etc. It is the user's responsibility to make sure that an N-grammar is referred to the appropriate shape. An inherently circular N-grammar may well be referred to, say, SPC2, which is but circular. The result will be a neighborhood the constituent elements of which will be badly distorted.

```
-MODULE AAA↑W#VV#W  $
 VV ↑TT&TTT  :W:Q:W:Q $
 W=.5;R $
 TT=3-5;2  :TTT $
 TTT=4;2  ::::TT $
 ENDMOD $

-SPC1=::(1,3;0,0);13,7::(31,4;31,2);27,0;19,0;8,5  $

-SPC2=::(9,7;9,10);15,7;15,3;8,0;2,2;0,6;5,9
      ::(9,7;9,10);5,5;6,3;9,2;12,4;11,6  $

-NEHOOD ALFA ↑ MM*X $
 MM=AAA  :Y:Q:Y:Q $
 ENDNED $

-CALL ALFA ; SPC1 $

-CALL ALFA ; SPC2  *
```

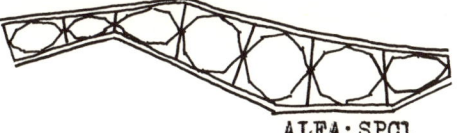
ALFA; SPC1

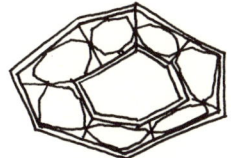
ALFA; SPC2

Figure 3.2

```
-MODULE BBB ↑ TT@W   $
 W=0.5;R $
 TT=2-3;2-3 ::::W $
 ENDMOD $

-SPC5=:,0,0:0,13;7,14:11,12;11,5;10,1;4,1 $

-NEHOOD BETA ↑ VV*X $
 VV ↑ MM*X $
 MM=BBB ::::Y $
 ENDNED $

-CALL BETA ; SPC5 $
```

BETA; SPC5

Figure 3.3

For the execution of an N-grammar, two additional procedures are applied before the tracer and after the specifier. They are the _partitioner_ and the _adjuster_. The partitioner follows the pattern of a neighborhood, as specified by the specifier, and accordingly partitions the referred space. Each constituent element in the neighborhood now corresponds to a cell in the partitioned space. The adjuster takes each element and adjusts it to the respective cell. The tracer works as before and completes the derivation of a neighborhood. Examples of N-grammars and the neighborhoods they derive are given in Figures 3.2 and 3.3. Since the implementation of the N-Level is, at the time of this write-up, still being debugged, the graphics in these examples are hand simulations of the expected outputs.

4. Concluding Remarks

There are quite a few awaiting questions that can be answered only when the whole implementation is available to play and experiment with. The underlying philosophy of the project is not to provide for automated site planning. In site planning there are too many value judgments involved and they are crucial enough not to be entrusted to a computer. The aim has been to leave the value judgments to the user and have the system do all the procedural and tedious work which will check the feasibility of the user's propositions.

The discussed structures are designed to be a problem solver in an indirect sense. They primarily constitute a language through which a user can program a solution and ask the system to derive it. Thus, as the algebraic languages are designed for the expression of algebraic relationships and operations, which, when compiled and executed in the given order, derive the solution to an algebraic problem, the suggested linguistic structures are designed for the expression of site planning relationships and operations, which when compiled and executed derive the solution to a site planning problem. Therefore, the main question is whether or not the means offered are efficient for the expression of such problems. It is an advanced belief that, even though these means may not be universal, they are general enough. But the claim is yet to be verified.

By the above, two situations are expected to occur. At first, the success of the system will depend largely on the user's capabilities and it will take a good designer for the system to derive a good site plan. Secondly, the system will have a personality and a style of its own. As discussed, it is based on the concept of patterns and unless a user agrees with such an approach, it may be of no use to him.

Notes

1. This work is supported in part by the National Science Foundation Grant GJ31188. I wish to acknowledge and thank Professor C. Eastman for his continuous advice on the project.

2. For example, see C. I. Yessios, "FOSPLAN: A Formal Space Planning Language" in Environmental Design: Research and Practice, edited by W. J. Mitchell, UCLA, 1972.

8. SPACE PLANNING TECHNIQUES / 439

3. See T. Johnson, J. Perkins and G. Wenzapfel, "IMAGE: An Interactive Computer System for Multi-Constrained Spatial Synthesis", in Proceedings of the 8th Design Automation Workshop, Atlantic City, N.J., 1971; C. Eastman, "GSP: A System for Computer Assisted Space Planning", in Proceedings of the 8th Design Automation Workshop, Atlantic City, N. J., 1971; and C. Pfeffercorn, "Computer Design of Equipment Layout Using the Design Problem Solver", Ph.D. Thesis, C-MU, 1971. For a discussion on the approach, see C. Eastman, "Heuristic Algorithms for Automated Space Planning", in Proceedings of the 2nd International Joint Conference in Artificial Intelligence, London, 1971.

4. See Chomsky's work on Transformational Grammars, In Specific, N. Chomsky, "Syntactic Structures", Mouton & Company, The Hague, 1957. For an overview of additional developments, see J. Hopcroft and J. Ullman, "Formal Languages and their Relations to Automata", Addison-Wesley Publishing Company, 1969. From an Artificial Intelligence point of view, space planning is a kind of picture processing and the work presented relates, in principle, with the generative aspects of such systems. For pointers to the related literature, see M. Clowes, "Transformational Grammars and Classification of Images," edited by Graselli, Academic Press, New York, 1969. For earlier works, see J. Feder, "Linguistic Specifications and Analysis of Classes of Patterns", Technical Report 4dd-147, Department of Electrical Engineering, New York University, 1966.

5. A related term in use is "pattern languages". In particular the Alexander team at Berkeley has regenerated interest in the use of patterns for space planning. For example, see C. Alexander, et al, "Houses generated by Patterns", Center for Environmental Structures, Berkeley, California, 1969. Alexander's work and a variety of anonymous designs of communities, primarily in the Mediterranean area advocates the appropriateness of the syntagmatic approach for site planning problems.

6. See C. I. Yessios, "Modeling the Site Planning of Homogeneous Uses", in Proceedings of the 9th Design Automation Workshop, Dallas, Texas, 1972.

7. If n the number of sides of a regular polygon, n can range from 3 to ∞ (infinity). Thus, the triangle is the lower limit and, by convention the circle is the upper limit of the countably infinite set of regular polygons. The stretch equivalent of a regular polygon is derived by an analogous stretching of its sides in such a way that its area remains the same.

UTILITY RATINGS AND 0-1 PROGRAMMING IN HOUSING DESIGN

Richard Hobson and Imre Kohn

The Pennsylvania State University

Abstract

    A decision model is presented for the inclusion or exclusion of interior household fixtures based on preferences, formulated as utilities, of a "client" population, where the clients' choices are restricted by environmental constraints. MacMillan's program for solving 0-1 linear programming problems is employed.

    An example of how the technique might be used by developers of large-scale furnished housing for a specific space--the bathroom--is provided. In this case the constraints take the form of restrictions on (1) money, (2) space available for fixtures, and (3) permissible water consumption per capita per day.

The Design Problem

    A familiar problem faced by designers, when planning for the housing needs of a client population, is to decide what features to include in homes in order to maximize the user's satisfaction, given that the users cannot have everything. The problem becomes even more difficult when decisions must be made for a community without working directly with the residents of that community. Too frequently, planners and designers select the facilities to be included in housing units, parks, and service centers on the basis of criteria presented to them by a handful of community leaders. These community spokesmen generally represent the client population in political and related environmental matters, however, when it comes to the many decisions required of a designer for optimizing interior layouts; i.e., selecting kitchen and bathroom facilities, providing adequate storage and circulation space, etc., the community leaders lack a sense of the client population's values. Further, they lack a systematic way of determining and utilizing such values.

    Thus, many household spaces requiring decisions in large-scale developments are typically designed without the aid of direct consultation with the client population; as an example of such a space, we chose the bathroom for two reasons. First, there has been little research on the basic design and functioning of bathroom fixtures, particularly when compared to the degree of attention other spaces in the home receive. Kira [1] has shown that bathroom arrangements fall far short of meeting functional and psychological hygiene requirements. Satisfactory, i.e., marketable, bathroom solutions can be generated with little difficulty, but it might be of value to test assumptions regarding what people would like in such spaces.

The second reason is that in order to examine the feasibility of applying utility theory and linear programming concepts to the planning process, we felt that it would be best to work with a space in which the various decision-making costs could be quantified, and certain assumptions, which serve to narrow the scope of the problem, could be reasonably made. For example, we treat the bathroom as a single, separate, and enclosed space in the home; we further assume that the decisions to include or exclude different fixtures can be made independently.

The present study is an effort to obtain from a sample of a user population information about what built-in fixtures and purchase options they conside more useful than others; and then to design a planning program which would allow for the inclusion of such items based on the values expressed by that population, taking into account various constraints imposed upon the designer. The "values" can be quantified as utilities [2], and the constraints can be expressed in terms of limitations on money, space, and water usage.

Problem Formulation

For any given space to be used as a "bathroom", there are a number of different fixture choices, say N, and the client population may select any number of these. The latitude of their decision, however, is constrained by (1) the amount of money they are willing to spend on hygiene items, (2) the amount of space they are willing to devote for hygienic use, and (3) the amount they are willing to pay for, or tolerate, excessive water consumption. Hence, the designer's objective is to maximize the utility of the bathroom for a large population given budget, space, and water restrictions. The procedure given below could be applied to the decision-making process of anyone wishing to program the furnishings of apartments in a large-scale setting.

We have a list of N variables: $x_1, x_2, \ldots, x_N$ which can only take on values of 0 or 1. If the ith possible bathroom item is included in the design solution, then we let $x_i = 1$; if the ith possible bathroom item is not to be included, $x_i = 0$. Suppose we let d_i be the cost of the ith item in dollars, s_i its space utilization in square inches and w_i its estimated water requirement per capita per day. We now obtain from members of the population an evaluation, for each possible bathroom item, of the utility of that item to individuals in that population, and let u_i be the utility of the ith item. (Our method for arriving at values for d_i, s_i, w_i, and u_i, $i = 1,\ldots,N$, is described below.) Then $\sum_{i=1}^{N}(u_i x_i)$ represents the sum of the utilities for the chosen bathroom items.
A reasonable objective for designers might be to maximize this sum. If we have D dollars to spend per bathroom, S square inches of space for fixtures, and W gallons of water per individual per day, then the problem can be stated in the following way:

Maximize $\sum_{i=1}^{N} u_i x_i$

st $\sum_{i=1}^{N} d_i x_i \leq D$

$\sum_{i=1}^{N} s_i x_i \leq S$

$\sum_{i=1}^{N} w_i x_i \leq W$

When more than a few variables are present, such a problem is completely unwieldy to work by hand and it becomes necessary to use a computer. We used MacMillan's FORTRAN Program for solving 0-1 linear programming problems [3]. In order to use this particular program, it was necessary to transform the problem from maximization to minimization which can be done by observing that the same values of x_i which maximize $\sum_{i=1}^{N} u_i x_i$ also minimize $\sum_{i=1}^{N} -u_i x_i$. If we let $x_i = 1 - y_i$, as MacMillan suggests, then the objective becomes

Minimize ($\sum_{i=1}^{N} u_i y_i - \sum_{i=1}^{N} u_i$).

This minimum is achieved by the same values of y_i that minimize $\sum_{i=1}^{N} u_i y_i$. The constraints were handled in a similar fashion.

Obtaining Utility Values

For purposes of trying out the technique, we let 36 second year male architecture students serve as our "client population". We selected architecture students, because they could be expected to have some experience with the design process, and therefore might offer some useful criticisms of this information gathering process.

Each student was given a list of 29 bathroom items and was asked to ascribe utility ratings to each item without conferring with anyone.
The Instructions were:

> Evaluate each of the listed bathroom items in terms of its utility to you. Presume that you presently have one bathroom with a basin, medicine cabinet, and toilet. For each item that you evaluate, ignore any considerations related to cost, available space, and water use. Let the number "1" be the utility of the least preferred item. Assign numerical values to the remaining items so that the values describe their relative importance to you. The most important item should receive your highest utility rating.

Following the assignment of utility ratings, the students were asked to check their ratings for internal consistency. Unfortunately, the most desirable procedure, that of successively comparing every ranked item, i, with up to

N-(i+1) combinations of the other items [4, p. 142] could require up to 27 + 26 + ...+1 comparisons. Instead, the students had the process of successive comparisons explained to them.. They then were asked to spend ten minutes arbitrarily checking the consistency of their preferences with their ascribed values by comparing clusters of items, and to adjust any discrepancies accordingly.

The three items mentioned in the instructions: basin, medicine cabinet, and toilet were omitted from the list since we assumed that these items constituted a "basic bathroom", and would automatically be included in every possible bathroom. The list of 29 possible bathroom items appears in Table 1.

Table 1

Utility and Constraint Coefficients

Subscript	Fixture Name	Utility	d_i Cost in Dollars	s_i Space in Sq. Inch.	w_i Water Use in Gallons
1	Bathtub	1305	164	900	18
2	Bidet	144	128	644	-4
3	Cleaning brush and holder	224	5	20	0
4	Clothes hamper	401	16	231	0
5	Concealed tank for toilet	155	54	0	0
6	Dehumidifier	319	95	252	0
7	Intercom	249	60	0	0
8	Mirror (full length)	361	13	0	0
9	Paper cup dispenser	260	4	0	0
10	Paper towel dispenser	199	4	0	0
11	Robe hook	314	2	0	0
12	Sauna bath	344	1200	2880	0
13	Scale	313	16	144	0
14	Shower (cabinet type)	1272	125	900	13
15	Slot for used blades	88	18	0	0
16	Soap dispenser	172	12	0	0
17	Spray attachm. for basin	170	39	0	0
18	Sun lamp	219	16	0	0
19	Swivel spout for basin	159	15	0	0
20	Telephone extension	176	60	0	0
21	Tissue dispenser	376	2	0	0
22	Towel rod	537	10	0	0
23	Urinal	282	199	180	-5
24	Vanity with chair	167	108	672	0
25	Wall clock	168	10	0	0
26	Wall heater	402	18	0	0
27	Wall shelf	285	11	114	0
28	Waste basket	535	7	64	0
29	Water heater (separate)	401	77	256	0

In order to transform the individual utility ratings from each student into a representation of the group's utility ratings, a procedure suggested by Fishburn [5] was used. The ratings of each individual were divided by the sum of the utility values for all items for that individual. These new values are called "normalized" utilities. For the group's utility rating of item i, we simply took the mean of the 36 normalized utility ratings for the item. This mean value (multiplied by 10,000 for ease in comparison) is represented by u_i in Table 1.

In order to get some idea of the amount of agreement among the ratings for each judge into rankings: 1, 2,..., N. Kendall's Coefficient of Concordance was calculated for these rankings of 29 items by 36 judges, yielding W=.385 ($p < .001$).

Coefficients for the Constraint Equations

The three resource constraints imposed on the planning process (budget, space, and water restrictions) were selected because the first two typically confront the designer, and the third may well become a realistic concern in the near future. The notion that water is a limited commodity is no longer novel to some sectors of the United States [6].

The fixture costs were rounded to the nearest dollar for parts [7], [8], and labor [9]. For each item, i, the costs for parts and labor were added and appear as d_i in Table 1.

Space utilization data are expressed to the nearest square inch and appear as s_i in Table 1. Space for circulation was not included in the formulation. Each fixture was measured in terms of the floor area it required. Thus for instance, a cup dispenser takes up space, but was judged not to have any floor area requirement, so that $s_9 = 0$. It is important to realize that the intention of the programming model is to provide a format for the inclusion or exclusion of environmental items. The specific spatial relationships between the items would be planned by the designer.

The water consumption coefficient, w_i, represents the amount of water used per day per person specific to the ith fixture. Water consumption data is relation to each fixture are difficult to obtain. There are a number of comprehensive studies of residential water use, and some are related to household activities, but none differentiates between wateruse activities within the bathroom. Average per day per capita "domestic" water consumption data[10] allow us to make educated guesses of water consumption for showering, toilet flushing, etc. Carson[11] is presently developing instruments which might precisely measure water use within the bathroom. In Table 1, there are a few minus coefficients, w_i, representing water consumption. These represent the water savings associated with that fixture. For example, a bidet is used for body washing and requires less water than a bathtub or shower. We calculated that a home equipped with a bidet would use, on the average, four gallons a day less than a home with no bidet due to less frequent use of the tub or shower. Likewise, a urinal requires about half the water of a toilet for each flush, so that during a day an average of five gallons might be saved. The minus coefficients thus represent a saving in daily water use over that of a "basic bathroom" with neither a bidet nor a urinal. We assume these coefficients are independent of the configuration of the rest of the bathroom. In order to compare utilities for a bathtub and for a shower we chose to include exactly one of the two in each solution. We thus ignored the possibilities of having both, or a combination bath-shower. In our formulation we therefore added the following constraints:

$$x_1 + x_{14} \leq 1, \text{ and } x_1 + x_{14} \geq 1,$$

The resource restrictions imposed in the constraints (values for D, S, and W) represent amounts over and above the consumption figures for the basic bathroom of basin, medicine cabinet, and toilet. We calculate that these basic items alone would result in a minimum cost of 251 dollars, a minimum space utilization of 1204 square inches (excluding circulation space), and a minimum of 28 gallons of water per person per day. D, S, and W would be added to the figures to obtain total amounts in a real situation.

Table 2

	Cost	Space	Water	Subscripts 1-29	Total Utility
Linear Programming Solutions	140	1200	18	(see figure)	2757
	150	1200	18		3294
	170	1200	18		4002
	190	1200	18		4480
	220	1200	18		4984
	250	1200	18		5352
	300	1200	18		5681
	350	1200	18		5846
	400	1200	18		6024
	450	1200	18		6194
0-1 Solutions	350	1200	12		3576
	400	1200	12		4930
	450	1200	12		5480
	450	3000	12		6429
	1400	3000	12		8214

Results

The solutions (those bathroom items slated for inclusion given specific values for D, S, and W) appear in Table 2. Limitations on cost, space, and water in the constraints were systematically varied, producing bathrooms with different facilities and different total utilities for the client population.

For example, if we spend 140 dollars, allot 1200 square inches to fixtures, and allow 18 gallons a day (all _in addition_ to the amounts for the "basic bathroom" given above), then the solution involves the fixtures with indices 9, 11, 14, 21, and 28. From Table 1, we see that these fixutres are: paper cup dispenser, robe hook, shower, tissue dispenser, and waste basket.

As one might expect, slight incremental changes occur as resource contraints are changed. For example, as the cost constraint figure is increased from 140 to 150 dollars, then item 22 (towel rod) enters the solution with a utility of 537, raising the total utility represented by fixtures in the bathroom from 2757 to 3294.

It is interesting to note that neither the vanity with chair nor the sauna bath ever enter any solution, and that in only one solution, the last, is the bathtub picked over the shower. These results may very likely reflect the fact that the client population consisted only of males. It is also interesting to observe the effect of decreasing W while leaving C and S the same. Many fixtures are necessarily eliminated from the solution because their use represents an excess water consumption.

Discussion

In order to use the technique we must make the rather arbitrary assumption that the utilities expressed by the clients are additive. Thus we assume that the utility of any subset of the N bathroom fixtures is the same as the sum of the utilities of the items.

Another problem is that people are not accustomed to thinking in terms of one item having a "utility" which is represented as some multiple of a "least preferred item." Furthermore, the ideal check on the rating procedure suggested by Churchman [12] of successive comparisons turned out to be too exhausting for our clients. We reported earlier, however, that there was a high degree of agreement between the subjects about the rankings of the u_i values. Thus, although the procedure may appear artificial to clients, they appear to be able to carry it out fairly consistently.

8. SPACE PLANNING TECHNIQUES / 447

References

1. Kira, A. The Bathroom, Criteria for Design, Center for Housing and Environmental Studies, Cornell University, Ithaca, New York, 1966.

2. Fishburn, P.C. Utility Theory, Management Science, Vol. 14, No. 5 (1968) pp. 335-373.

3. MacMillan, C. Mathematical Programming: An Introduction to the Design and Application of Optimal Decision Machines. John Wiley and Sons, New York 1970.

4. Churchman, C.W., Ackoff, R.L., and Arnoff, E.L., Introduction to Operations Research, John Wiley and Sons, Inc., New York, 1957.

5. Fishburn, P.C. A Normative Theory of Decisions Under risk. ONR Project--NONR --1141(11). Operations Research Group, Case Institute of Technology 1961.

6. Jansma, J.D., and Kerns, W.R., An Economic Analysis of Water Utilization in Pennsylvania, Institute for Research on Land and Water Resources, The Pennsylvania State University, Research Publication Number 54, 1968.

7. Building Construction Cost Data, 1967, 25th Annual Edition, Robert Snow Means Co., Inc., Duxbery, Massachusetts.

8. Sears, Roebuck and Co. Catalogue, 1970.

9. Building Construction Cost Data, 1967, 25th Annual Edition, Robert Snow Means Co., Inc. Duxbery, Massachusetts.

10. Jansma, J.D., and Kerns, W.R., An Economic Analysis of Water Utilization in Pennsylvania, Institute for Research on Land and Water Resources, The Pennsylvania State University, Research Publication Number 54, 1968.

11. Carson, D.H. Residential Water Usage Change Potential Related to Attitudes Toward Water and Knowledge of Usage Behavior, Unpublished manuscript, Division of MER, College of Human Development, The Pennsylvania State University.

12. Churchman, C.W., Ackoff, R.L. and Arnoff, E.L., Introduction to Operations Research, John Wiley and Sons, Inc., New York, 1957.

AN INTEGRATED METHODOLOGY FOR OFFICE BUILDING ELEVATOR DESIGN[1]

Margaret A. Frederking

Alton J. Penz

School of Urban and Public Affairs
Carnegie-Mellon University

Abstract

The authors have outlined a systematic approach to elevator system design which applies cost-benefit methodology to the design process. This approach relies on a cash flow modeling base for analysis of a total building. It then interprets the effect of elevator operation on the building's value for various traffic conditions. Elevators affect building value via direct and indirect operating costs, the opportunity cost of rentable floor area foregone for shaft space, the rental value derived from user satisfaction, and the cost of lost production time consumed in elevator travel. A computer simulation of elevator operation employing a statistical model of elevator behavior provides the measurement of elevator performance. The traffic patterns simulated reflect, via the use of probability, the nature of traffic generated by floor populations which may vary over time. The resulting methodological approach facilitates evaluation of elevators from the perspective of total building value to the client and to the users of a building.

The Problem

Each participant in the elevator design process of a high-rise office building seeks to have his own particular criteria define which elevator system would be most appropriate. The architect, for instance, concerns himself primarily with coordinating the functional aspects of a building with his aesthetic plans. Such coordination is sometimes achieved, however, only at the expense of efficiency elsewhere--in elevator service, for example. Similarly, the architect's client may consider only system cost, often with little regard for performance criteria. On the other hand, the elevator consultant may favor extreme elevator efficiency. Though this recommendation may stem from a concern for the satisfaction of employees with better than adequate elevator service, it may also be the result of reasoning which suggests that future contracts will be gained or lost on the basis of performance of his currently operating systems. Specifically, he may recommend a system which provides for a morning peak-period maximum lobby waiting time of only twenty-five seconds. The gains from a five-second reduction in waiting time over the generally recognized standard of a thirty-second maximum[2] may not compensate for the additional system costs and lost rentable area which detract from the investor's financial position.

The result of this diversity of inputs into the choice of elevator design may result in a decision in the financial interests of neither the building's future tenants nor the building owner. Of critical importance but often absent from present elevator deliberations is a systematic examination of system costs, both initial and operating; the performance of the system in terms of lobby waiting time and passenger travel time; the opportunity cost of floor space occupied by elevator shafts;[3] and rentability of the building as affected by tenant satisfaction with the quality of elevator service.

Two other factors of importance seem to be missing from the criteria for elevator system design. The first of these concerns the uncertainty with which different levels of service demand are experienced. This demand is affected by population size, the mix of occupations in the population, and the distribution of the population over the floors served by the elevators. Each of these factors affects service quality. The second concern is a quantitative consideration of the amount of time spent in the elevator system by employees. This time represents a cost to the building tenants in terms of foregone productivity. If this cost is excessive, it will represent a source of reduction in the owner's cash flow.[4]

No existing analytical technique would seem single-handedly capable of providing the investor, consultant, or architect with the needed broader perspective. The purpose of this paper is to suggest such a tool: an integration of cost-benefit analysis, principles of cash-flow modeling, and computer simulation of elevator performance to provide improved focus.

Proposed Methodology

The methodology presented here constitutes a first attempt to devise a systematic approach for analyzing a bank of a proposed elevator system via cost-benefit analysis. The procedure, which is implemented as a computer program, essentially simulates a specified elevator system under realistic, random conditions specified by the user. The program then provides means of evaluation focusing on expected performance values to aid the designer in selecting a particular elevator system. The following discussion outlines the structure of the program with the intent of demonstrating its validity for real projects.

Basically the method for analysis involves two types of calculations. (See Figure 1). In the first section, the nominal financial structure of the building and the basic mechanical characteristics of the elevator system are established from specified input data. This information is nominal in the sense that it does not include adjustments for realized elevator performance. For instance, the potential rental rate may differ from the nominal rate to the extent that elevator performance affects rentability. In the second section, performance of the elevators is simulated over the lifetime of the building for varying traffic conditions. This performance is weighted by the probability that the specific conditions arise. The financial characteristics are then altered according to the hypothesized effect imposed by elevator performance. The resulting characteristics are presented in a manner facilitating evaluation of the proposed design.

A powerful feature of the program is the ability to perform sensitivity analyses on various parameters. Since the program analyses proposed systems but does not automatically search for solutions, the ability to quickly obtain the effects of parameter changes guides the user in searching for better systems. In the search for an optimal solution, the building designer may readily explore the merits of elevator systems containing different numbers of cars, car sizes, or numbers of floors served by a bank of cars.

In addition to varying the elevator system, the values for financial characteristics such as the cost of capital and the rent rate may be changed. An analysis of a building project should not rely only on estimates of average values for rent, capital costs, and other indeterminant characteristics but should incorporate evaluation over a range of values, both high, expected, and low, so that the sensitivity of the project values to those parameters can be established.[5]

```
┌─────────────────────────────────────┐
│ INPUT                               │
│ Building Financial Character-       │
│ istics, Elevator Description,       │
│ and Traffic Conditions              │
└─────────────────┬───────────────────┘
                  ▼
┌─────────────────────────────────────┐
│ SIMULATION                          │
│ Evaluation of Elevator              │
│ Performance                         │
└─────────────────┬───────────────────┘
                  ▼
┌─────────────────────────────────────┐
│ CASH FLOW MODEL                     │
│ Adjusted for Elevator Perfor-       │
│ mance                               │
└─────────────────┬───────────────────┘
                  ▼
┌─────────────────────────────────────┐
│ OUTPUT                              │
│ Cash Flow, Return on Equity,        │
│ Lost Production Time, and           │
│ Other Evaluation Criteria           │
└─────────────────────────────────────┘
```

Figure 1

The financial structure of the model is handled very similarly to standard cash flow models.[6] The anticipated development costs and construction costs exclusive of elevator costs are given by the user. Elevator costs are provided by a detailed cost breakdown of elevators derived from a leading manufacturer's generalized cost data. The anticipated financial flows from operations for the building are initially established from data supplied by the architect's client. The rental income is adjusted to recognize the reductions in net rentable area attributed to elevator shaft space; operating costs are adjusted to include operating costs for specific elevator systems. The debt financing, depreciation, and sale or equity reversion calculations are handled in a customary fashion for cash flow modeling. The process for deriving the cash flows from initial construction through occupancy to sale of the building is illustrated in Figure 2.

The three components of cash flow--initial construction, annual operations, and building sale--are discounted to present value terms using a cost of capital provided by the client. Most importantly, this cash flow value incorporates most of the effects on the viability of the project contributed by elevator performance. The initial elevator costs, the elevator operating costs, and the loss of rentable floor space consumed by shafts are explicitly in the valuation.

Mechanics of the Elevator Simulation

The criteria for good elevator system design primarily focus   on acceptable

8. SPACE PLANNING TECHNIQUES / 451

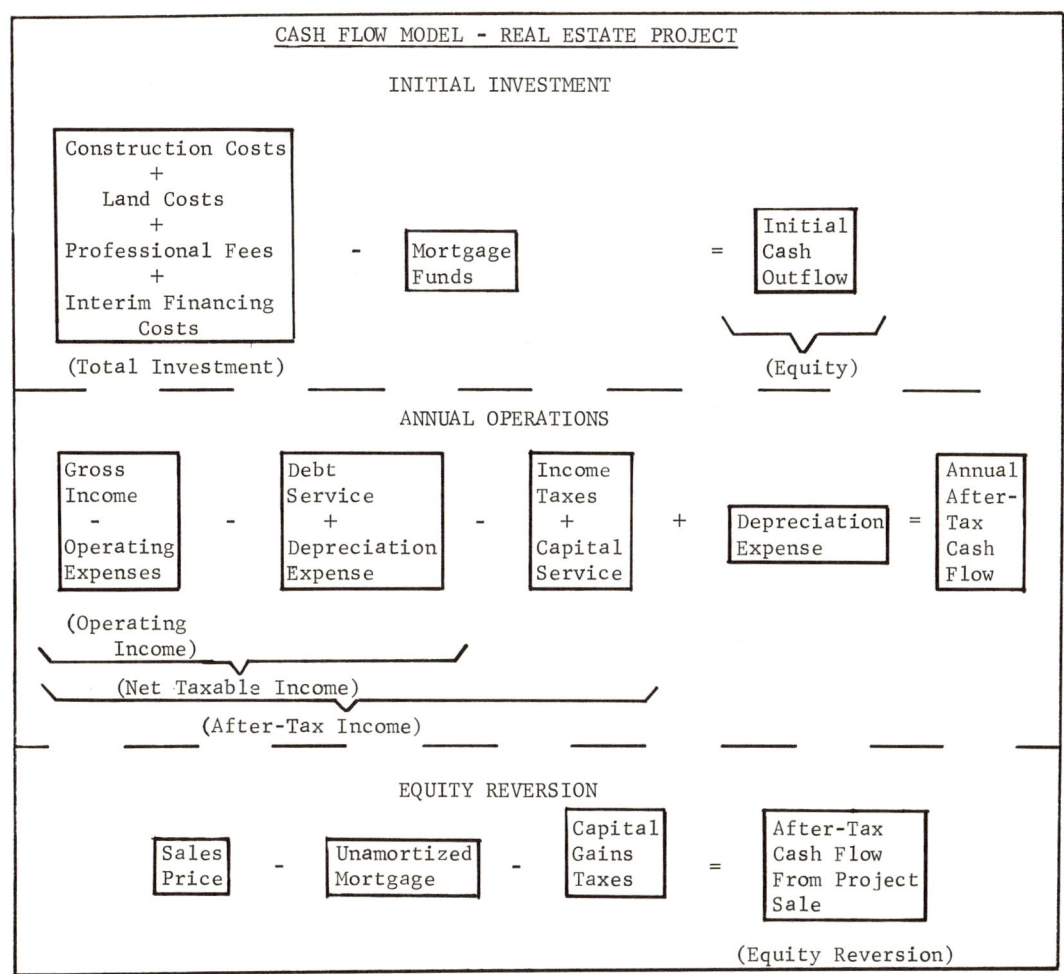

Figure 2

system performance under the heaviest traffic conditions. If service is satisfactory under the heaviest or worst conditions, then it should be satisfactory at all other times. Usually the heaviest conditions occur during periods of traffic peaks. At this time, if elevator system capacity is not greater than traffic demand, lobby areas quickly fill with dissatisfied people.

In office buildings peak demand occurs during the morning arrival period, the noon lunch hours, and the evening departure period.[7] Of these three periods, the morning period is usually studied for determining the capacity of a proposed elevator system. This traffic places strenuous demand on the elevator system and includes direct implications for lost production time as a result of lobby waiting and time spent in the elevator cars.

A further justification for analysis of the morning peak period is no efficient means currently exists for accurately studying the more complicated elevator performance during the noon and evening traffic periods. Only the morning period is included in the methodology presented here. To the extent that the effect of performance at other times can be inferred from morning elevator performance, then it can be included in the evaluation.

Although consideration of traffic demand is restricted to the morning arrival period, the analysis problem remains a complex one in that there are an infinite variety of possible morning arrival patterns. Two of these possibilities are pictured in Figure 3. The pattern on the left portrays

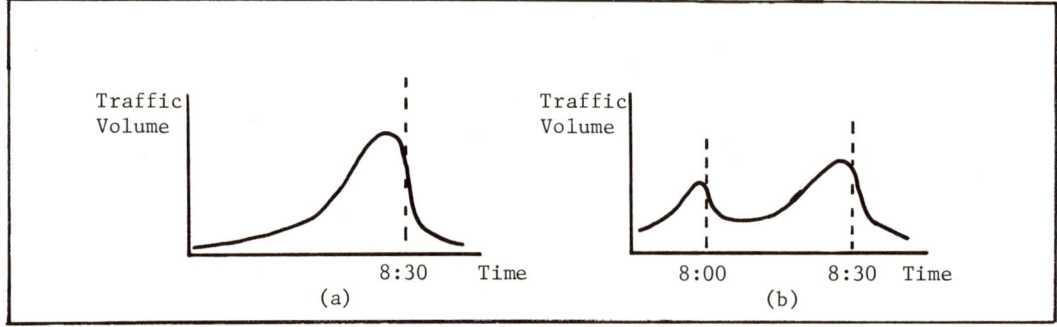

Figure 3

a morning arrival distribution which, by virtue of its higher peak, will place a greater demand on the service capabilities of the elevator system than will arrivals in the bimodal pattern of Figure 3b. As it will generally be true that any multi-modal arrival distribution will be less exacting of an elevator system than a unimodal, attention will be restricted to the unimodal case.

The mean number of people arriving per unit of time is given by the expression
$$N(p,t) = L(p,t) \, p,$$
where $L(p,t)$ is the mean arrival rate of a Poisson process [8] and p is the population of the set of floors served by the elevators. Usually, $L(p,t)$ is expressed as the percentage of the population p arriving per five minute interval. $L(p,t)$ is a function of p as well as of t because of the dependence of arrival rates on population composition by occupation. Arrival patterns for bank clerks and secretaries differ from those of executives, for example.[9] Thus, we represent $L(p,t)$ as
$$L(p,t) = A_i(p) \, L_i(t)$$
where $L_i(t)$ is the arrival rate for occupation set i at time t and $A_i(p)$ is the percent of total elevator bank population p consisting of occupation set i. That is, $L(p,t)$ is a weighted sum of all arrival rates characteristic of the occupations represented. The area under the curve $L(p,t)$ multiplied by p is the total number of building occupants likely to arrive during the morning rush period.[10]

We have presented this concept via a continuous function; typically, empirical data, based on five minute intervals, is used to establish a discrete representation of arrival patterns.

Performance of a specific elevator system is evaluated from a simulation of the elevators under various traffic conditions specified by $N(p,t)$. The simulation is constructed by calculating elevator performance at specific time intervals over the range for t and for different values of p. In each time interval the number of people to arrive is generated by a Poisson process, assuming a uniform arrival rate for the people generated and a constant mean arrival rate for the interval. At the end of the interval, the number of people in queues is compared with the capacity of the elevators. If the capacity is exceeded, the excess people are left in the queue and are assumed to wait an additional time interval. The performance of the elevators for a particular load of people is determined by statistical approximations of elevator operation.[11] From these calculations, values for average waiting time and average travel time are derived for each one minute interval of a traffic distribution. The calculations of waiting time and travel time for all passengers form the basis for estimating the value of production time lost in elevators.

The use of simulation in the model facilitates the calculation of a satisfaction value (provided a satisfaction function can be established). Two systems with the same average waiting time can have different distributions of waiting time (e.g., the number of people to wait x seconds). Only by simulating the elevator operation for realistic arrival conditions in which system capacity may be exceeded can one hypothesize the degree of satisfaction or dissatisfaction arising from elevator performance.

In addition, a simulation approach avoids a source of error in the statistical method of Gaver and Powell. As shown in Figure 4, mean waiting time remains relatively constant over large ranges of traffic volume below approximately 95 percent of elevator system capacity. Traffic levels within 5 percent of capacity cause service to deteriorate, however, because of difficulties associated with scheduling elevators to coincide with random passenger arrivals at the lobby.[12]

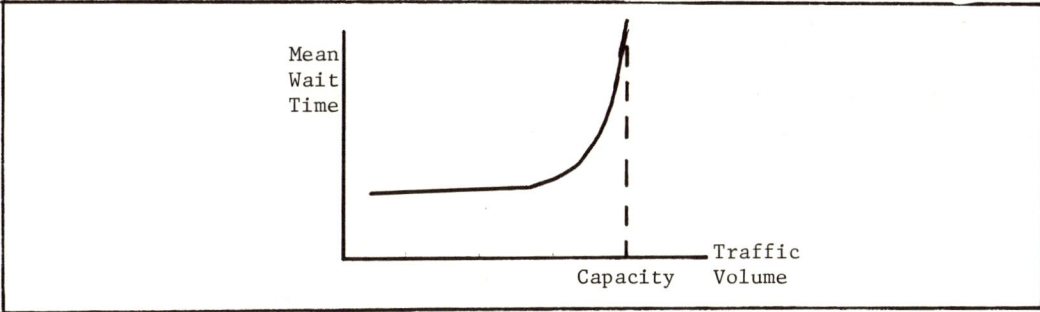

Figure 4

Implementation of the Methodology

Summarizing briefly, the proposed methodology contains three components: a statistical simulation of elevator performance, cash-flow modeling to represent the financial position of the building owner given varying expenditures for purchase and operation of an elevator system, and cost-benefit analysis to weigh changes in performance against changes in the owner's financial position.

The authors have investigated the implications of this methodology by employing it in the context of an existing thirteen floor office building of conventional structure. The building costs approximately $25 per square foot to construct in a recent year. It contains 21,500 gross square feet per floor and is served by five 3,000-pound elevators. The maximum estimated demand per five minute interval is 13 percent of the building population, and the system was estimated by its designers to provide service within a maximum wait interval of 30 seconds. Table 1 shows the values obtained for the actual system as well as for other alternative elevator systems.

As a first example of tradeoffs, let us consider the question of an elevator system with a twenty-five second maximum lobby waiting time as opposed to one with a thirty second maximum. A priori such a change is of questionable value for two reasons. First, the ability of passengers to discriminate between a maximum waiting time of twenty-five seconds versus thirty seconds is uncertain, particularly since the average waiting time is approximately fifteen seconds. Furthermore, any advertising advantage which might accrue to the building owner by virtue of exceptional elevator service may be of limited time duration. That is, should such a criterion become widely accepted, the novelty of the building's offering will cease to be a drawing card as more buildings with similar service are constructed.

In the immediate context of the example presented, however, Table 1 reveals that a system with a twenty-five second maximum waiting time (either six 2,500-pound cars or six 3,000-pound cars) increases initial system cost substantially without decreasing the value of passenger time consumed in the system by a corresponding amount. The difference must be made up in additional rent associated with increased passenger satisfaction to justify the faster system.

As yet no method has been delineated for explicitly recognizing the passenger satisfaction (or dissatisfaction) associated with lobby waiting time. To account for this aspect of elevator service, we have introduced the notion of satisfaction $S=f(WT,WT^*)$, a function of the realized waiting time WT and some criterion waiting time WT^*. Though no actual experimentation has been done concerning this concept, presumably some positive satisfaction occurs when an elevator arrives after only a brief wait, but dissatisfaction is experienced with a wait beyond some criterion time limit. In such a case, it may be that satisfaction can be represented as shown in Figure 5. If this view is appropriate, it may be possible to mathematically represent satisfaction by the following equation:

$$S = 2.0 - e^{-B(1.0 - WT/WT^*)}$$

where the constant coefficient B, $0 < B < 1$, adjusts the slope of the curve. A

8. SPACE PLANNING TECHNIQUES / 455

TABLE 1 Elevator Systems for a Thirteen Floor Office Building					
Number of Cars	4	5	5	6	6
Car Capacity (lbs.)	4000	3000	3500	2500	3000
System Capacity (% of pop/5 min)	12.2	13.0	14.3	13.6	15.8
System Cost (000)	$334.2	$384.0	$394.0	$450.0	$460.8
Max. Wait Time (sec)	42.2	29.0	31.6	21.0	24.1
Net Rent. Area (000)	222.0	221.6	220.7	220.3	219.6
Syst. Oper. Cost (000)	4.05	4.87	4.75	5.66	5.27
Project Value[1] (000)	$123.1	$ 75.5	$ 53.3	$ 0.4	$-17.7
Value Lost Time[2] (000)	$176.2	$132.4	$140.0	$111.5	$119.4

1. Net Present Value of after tax cash flows, based on $7.00 rsf, unadjusted for satisfaction, discounted at 5%.
2. Present Value of all future lost production time, valued at $5.00 per hour.

larger B generates greater satisfaction or dissatisfaction depending upon the value of WT. Again, there is no proof that satisfaction behaves in the manner described or even that 30 seconds, that point at which people waiting for elevators have been observed to become fidgety, should be the criterion time used. However, such a representation appears plausible and permits use of the performance measure WT in calculation of a satisfaction coefficient. One may verify that the coefficient equals 1.0 when WT equals WT*.

The mean value of S multiplied by the nominal net rental income represents the expected value for rental income. This value may be interpreted as a change in the level of rent which may be levied as a result of the quality of elevator service. The value for revenue as adjusted by the satisfaction coefficient is

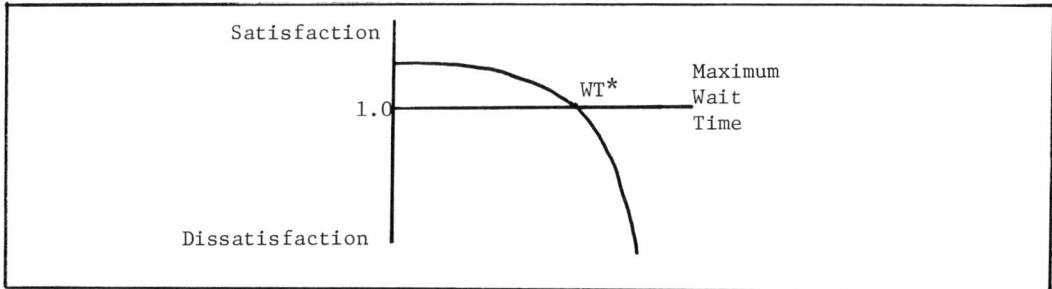

Figure 5

employed in the cash flow model and thus affects the financial evaluation of the project.

The financial evaluation and the value of lost production time derived from the simulation of specific traffic conditions are multiplied by the probabilities of occurrence associated with the traffic conditions to obtain weighted values. When these values are summed for all traffic conditions, the expected value of lost production and the expected financial value for the project are obtained. Both the expected value of lost production and the expected financial value of the project derived from cash flow modeling and adjusted for satisfaction should be used in the design process of a building project. For all other factors held constant, the best proposed design in a cost-benefit sense would be that system resulting in the greatest

$$V = f(W,P)$$

where W is the value of lost production and P is the satisfaction-adjusted project value exclusive of production time. Use of V as a criterion avoids the possibility of a decision on the basis of P alone or any component of P or W without regard for the total system. A sensitivity analysis should, of course, be performed on V to examine the effect of changes in market, financial and economic conditions on the project.[13]

The complete model has now been presented. A schematic diagram of its structure is illustrated in Figure 6. The value of V represents the expected value of a project when tenant satisfaction, employee performance, rentable floor area, and elevator costs are calculated for a specific elevator system operating under stochastic traffic conditions.

Several difficulties prevent the use of the methodology presented in an analytic context, i.e., to calculate THE value of an elevator system. Initial indications are that the interaction of satisfaction criterion, the value of lost time, and the market rental rate predominate the decision process. Unfortunately, no means currently exist for establishing the validity of a quantitative formulation of passenger satisfaction with elevator service. In addition, the passenger time saved by a more efficient elevator system may not have much value. Some studies suggest that the value of marginal time gains is negligible.[14] Finally, the authors have found no research indicating how market rental rates reflect the benefit of efficient elevators, either via improved satisfaction *per se* or via reductions in lost time.

As a result of these observations the methodology proposed here is for a decision process rather than an analytic process. In this context, alternative solutions are evaluated on the basis of known parameters, and then decisions are made according to the range of values for unknown parameters implied by each alternative. If all parameters except satisfaction are known, for example, boundary conditions can be calculated for satisfaction. A client can then choose between alternatives associated with different ranges of satisfaction according to his own judgment. In Figure 7, either solution 1, 2, or 3 is selected, depending upon whether the satisfaction parameter is judged to be less than a, greater than b, or between a and b.

8. SPACE PLANNING TECHNIQUES / 457

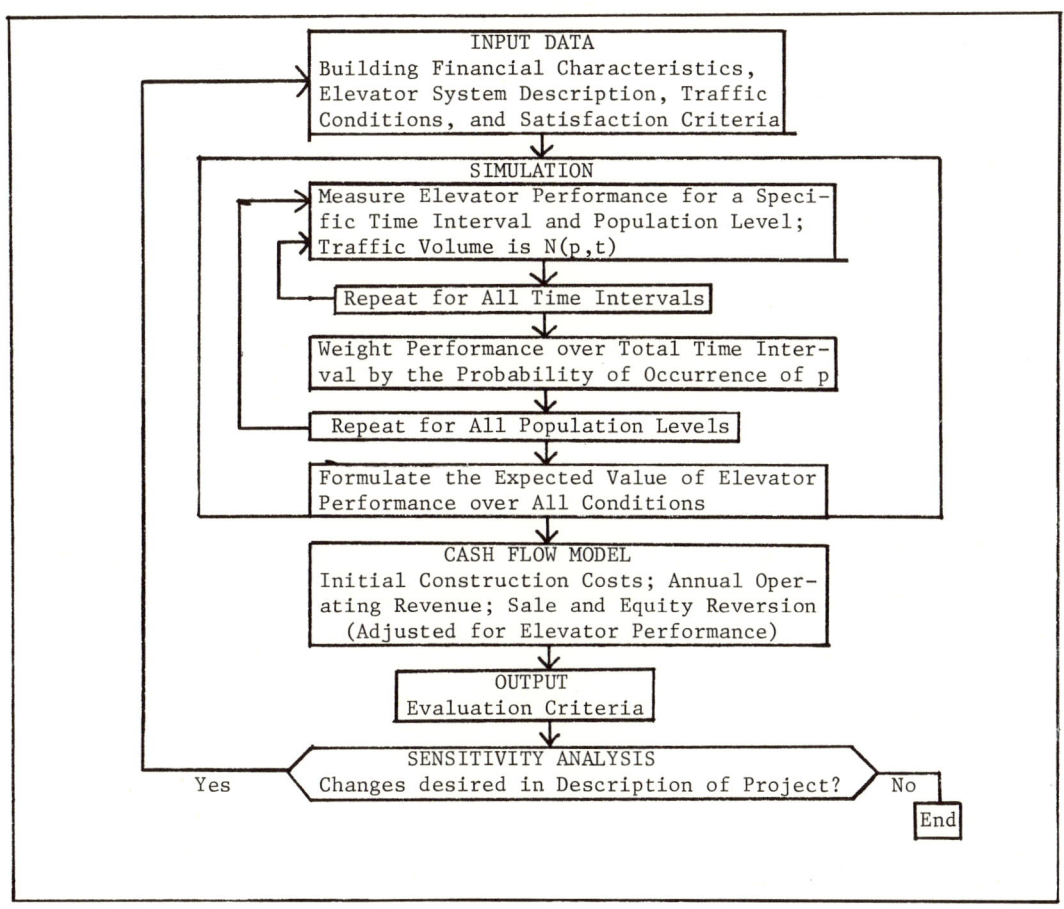

Figure 6

Figure 7

Application

The methodology suggested here represents a powerful tool for examining the performance of a system which operates in the context of a larger system. This combination of statistical simulation, cash-flow modeling, and cost-benefit analysis enables multiple objectives to be evaluated in the context of the more universal objectives of the client and users.

The proposed approach emphasizes the need for additional research to provide appropriate data for quality decision-making. As mentioned above, no appropriate procedures exist for representing the behavior of elevators during the noon and evening peaks. The noon hours are particularly difficult because demand is in both directions as opposed to only up traffic in the morning. In addition, traffic during the noon and evening periods is clustered as people travel in groups. Improved capabilities for analysis during these two periods would expand the situations of peak demand under which elevator performance could be observed.

As the quality of elevator service is so dependent on parameters of the building population, much uncertainty would be eliminated if studies were conducted of buildings near the proposed location to determine:

(1) arrival patterns characteristic of various occupations,[15]

(2) expected population densities and trends,[16] and

(3) a relationship between population or population density and the mix of occupations represented.

Other important data concern anticipated changes external to the building which may seriously affect the rates of arrival of employees. An example is the installation of a subway station within a very short distance of the office building. Though individual walking rates may disperse the crowd, dispersal will naturally be reduced as distance is shortened. It is under situations such as this that considerations of the duration of the demand peak and evaluation of satisfaction become particularly important.

Extensions

The model described in this paper was designed with computer implementation in mind. The approximations made and the assumptions contained in the model are believed to be reasonable ones with respect to actual implementation. Subsequent research should resolve these questions when a fully validated model is implemented with empirical data to support the traffic behavior. At that time its full potential can be examined by comparing the actual decisions made on projects to the recommended decisions based on V.

A crucial point in the presentation of this model is that all building design factors are either constant, maintain constant rates of change, or are influenced only by elevator design. Realistically, this situation does not hold; structure, mechanical requirements, legal codes, site and market conditions may very well affect the design more strongly. Nevertheless the conceptual approach remains the same. In the continual give and take of a design process when elevators and all other aspects of design are evaluated, changed, re-evaluated, and changed again, the form of evaluation should always be in the context of the total problem. Hopefully, as more of the evaluation process is systematized, the interaction between project components can be more effectively integrated and evaluated.

The model for evaluation has been presented for analysis of one bank of elevator cars at a time. In high rise buildings, a major problem involves the allocation of floors to several elevator banks so as to provide the best overall elevator service. In this case a dynamic allocation program as currently developed[17] could be used, with evaluation of each state handled as suggested here. This approach may very well be expensive to operate, however.

Some Broader Implications

The concepts contained in the approach delineated here are not restricted to elevator design. Many other components of an architectural project may incorporate the same considerations of human satisfaction, employee performance, and system costs (both indirect and direct) which are tied to system capacity and randomly occurring system operating conditions.

Methodology similar to that proposed here may also be applied to the design of interior environments. An example is the design of office building furnishings where concern must be given to rates of change of operating procedures and organizational structure, deterioration of the equipment, and employee efficiency and comfort.[18]

Recent work has indicated the value of adaptive capabilities for a building. The more readily a space can be transformed to satisfy changing use requirements over time, the more desirable it will be to prospective tenants. On the other hand, flexibility often requires additional expenditures. Thus the costs associated with varying degrees of flexibility must be related to the expected gains from the flexibility.

Notes

1. The authors acknowledge receiving numerous suggestions on the substance and form of this paper from Dr. Alfred Blumstein, Professor Charles Eastman, and Robert Carbone, Carnegie-Mellon University.

2. Adler, Rodney. Vertical Transportation for Buildings. (New York: American Elsevier Publishing Company, 1970), p.15.

3. Swartz, W.W. "Optimizing Space Requirements for Elevators", Architectural Record (March, 1970), pp. 133-136.

4. Strakosch, G.R. Vertical Transportation. (New York: John Wiley and Sons, Inc. 1967), p.353.

5. Penz, Alton J. "Financial Analysis in Architectural Design". Carnegie-Mellon University (1971). Submitted for publication.

6. Farrell, Paul B., Jr. "Financial Analysis of Real Estate." AIA Journal (August, 1968), pp.74-81.

7. For a discussion of traffic periods, see G.R. Strakosch, Vertical Transportation (New York: John Wiley and Sons, Inc., 1967),p.53.

8. Morning arrivals are assumed to obey a Poisson distribution.

9. Brown, J.J. and Kelly, J.J. "Simulation of Elevator System for World's Tallest Buildings." Transportation Science, 2:1 (February, 1968), p.35-36.

10. See G.R. Strakosch, Vertical Transportation (New York: John Wiley and Sons, Inc., 1967), Chapter 4, for characteristics of morning arrivals.

11. The work of Gaver and Powell was used for equations of expected round trip time and system capacity. See D.P. Gaver and B.A. Powell, "Variability in Round Trip Times for an Elevator Car During Up-Peak", Transportation Science, 5:2 (May, 1971), pp.169-179.

12. In queuing theory, as the mean arrival rate approaches the mean processing rate (from below), the length of the queue increases.

13. Byrne, E.J. and Landry, R.A. "RAM-30: A Computer Model for Real Estate Investment Analysis." The Arthur Young Journal (Summer, 1971).

14. Nelson, J.R. "The Value of Travel Time." Problems in Public Expenditure Analysis. Edited by Samuel Chase (Washington, D.C.: The Brookings Institution, 1968).
Beesley, M.E. "The Value of Time Spent in Travelling: Some New Evidence." Economica (May, 1965).

15. Brown, J.J. and Kelly, J.J. "Simulation of Elevator System for World's Tallest Buildings." Transportation Science, 2:1 (February, 1968), pp.35-36.

16. For a study of office building population density, see R. Fisher, The Boom in Office Buildings (Washington, D.C.: Urban Land Institute, 1967).

17. Powell, B.A. "Optimal Elevator Banking Under Heavy Up-Traffic." Transportation Science, 5:2 (May, 1971), pp.109-121.

18. Eastman, C.M. and Penz, A.J. "Decision Making in Adaptive Environments", IPP Report 32, Carnegie-Mellon University (November, 1972).

NINE DESIGN LANGUAGES AND METHODS

Chairman: Charles H. Burnette, AIA, Philadelphia, Pa.

Panelists: Juan Pablo Bonta, Architecture, Univ. of
                Buenos Aires, Argentina
            Jon Lang, Urban Studies, Univ. of Penn.,
                Philadelphia
            Marvin Manheim, Civil Engineering, MIT,
                Cambridge, Mass.
            Stuart Rose, American Inst. of Architects,
                Washington, D. C.

Authors: David Anstrand, Richard Dagenhart, John
                Moore, Val Thomas, "Modular
                Programming"
            Charles H. Burnette, Gary T. Moore, Lynn
                Simek, "A Role-Oriented Approach to
                Problem Solving by Groups"
            Patrick J. Quinn and J. MacGregor Smith,
                "Piecemeal Social Engineering: A Case
                Study"
            Pattabi G. Raman, "Synthesis in Design-An
                Interdisciplinary Essay"
            Arie P. Schinnar, "U Graph: A Decision
                Making Aid"
            Michael Kreski and David Rejeski, "Designing
                Off the Street"

DESIGN LANGUAGES AND METHODS: INTRODUCTION   9.0

Charles H. Burnette, Session Chairman
Center for Planning, Design and Construction, Philadelphia, Pa.

As people become increasingly aware that the organization of communication is the organization of control itself, it would seem that the means by which information is organized, expressed and brought to bear on problems of environmental design should be more important as an issue in research and education than it appears to be. In many minds, design method has been discredited by the dogmatic arbitrariness of earlier descriptive formulations, while inherent restrictions, presentation and complexity have rendered statistical decision theory unusable for most designers.

However, time and need have worked a gentle metamorphosis. As the papers to follow suggest, design languages and methods have become more practical, pluralistic and concerned with the designer as an involved, expressive human being in need of usable tools to assist in the organization and conduct of what I have called elsewhere "architectonic communication" (constructive, interdisciplinary, problem solving communication). Games and group techniques such as "The Role Oriented Approach to Creative Group Problem Solving" and decision making aids such as the U Graph have been developed to facilitate this design communication. There is also complimentary evidence of an increased emphasis on ways to more easily represent and handle communication. (Once again, Chris Alexander has anticipated things with his work on the Pattern Language as a means to represent behavioral information for design). The paper on Modular Programming is indicative of the relaxed and practical sophistication with which such tools of organization and display are being developed to aid work. Similarly, as the paper on Piecemeal Social Engineering suggests, problems are being conceptualized in ways which profoundly affect the way they are carried out. Most of these papers share what might be called (with intended irony) a "reactionary" movement toward a more modest and humble conception of the designer. The paper, "Designing Off the Street" manifests this attitude as an attractive nostalgic philosophy and a valuable corrective but/and one which turns back from the complexity and scale of our knowledge and needs.

That the issues of design language and method are diffused and tolerant enough to accept all efforts is guaranteed by other well motivated and informed attempts to rethink the philosophical, psychological and empirical foundations of the design process (as exemplified by the final paper on Synthesis in Design. An increasing involvement with the linguistic aspects of "architectonic communication," while not represented in these proceedings is a vital part of current work. Perhaps, in closing, it is best to acknowledge that design languages and methods will develop through many varied efforts to reconceive, retool and reprogram what we do. Human beings adapt to change....but not all at once or in any one way.

MODULAR PROGRAMMING 9.1

David Anstrand
David A. Crane and Partners, Philadelphia, Pa.

Richard Dagenhart
Wallace, McHarg Roberts and Todd, Philadelphia, Pa.

John Moore
Kling Planning, Philadelphia, Pa.

Val Thomas
David A. Crane and Partners, Philadelphia, Pa.

Abstract
Modular Programming is a methodology for programming large scale urban design projects such as Planned Residential Developments. It is a pragmatic, professionally useful problem solving model designed to package, manipulate and display information for intelligent decision making by designers, clients, users and other participants. The methodooogy utilizes a simple conceptual formulation which expands upon designer's frequent use of "paper blocks" and rules of thumb for programming exercizes. The central concept is the Module, an internally consistent set of data and relationships which can be abstracted from the larger problem environment. Modules are then combined into "basic program mixes" to be (1) tested against threhholds for supplying "external" program provisions through the use of a computer calculator program and (2) related to the specific site and its context through a technique of graphic display. The resulting information is displayed as the basis for final program construction. The significance of Modular Programming lies in (1) forcing the definition of spatial and aspatial relationships early in the programming process, (2) providing a framework for using complex data and relationships, (3) displaying information for various trade-offs from which designers and others may choose and (4) providing a conceptual and methodological connection between analytical work and physical design. This paper presents and evaluates an application of the method for a specific residential problem.

Introduction
In Urban Design the programming of the physical environment deserves particular attention because it addresses the difficult problem of determining what to build. Programming is the linkage between analysis and physical design within the overall design process. Improving this "inception process", as Constance Perin (1) has termed it, can produce higher quality built environments since opportunities are provided for coordinating early multi-disciplinary decisions, meeting the needs of future users, bringing citizen

participation into the "front-end" of the design process, resolving conflicts among participants and constructing achievable objectives.

Unfortunately, urban design programs seldom take advantage of these opportunities; and as a result, design solutions are often constrained by an inadequately formulated program. The following reasons, among others, may account for this situation:
1. Designers often resort to over simplified concepts because dealing with complex data within a professional office is too difficult and time consuming; only a very limited number of alternatives can be defined and evaluated.
2. Sophisticated models, such as computerized models for land allocation, are either too expensive for normal project budgets or inflexible for designer's use. (2)
3. Spatial impact of a program is seldom evaluated before designing actually begins; and, therefore, design and program processes become linked together resulting in program redefinition for the designer's expediency.
4. Participation by the many groups involved--clients, public agencies, future users and surrounding residents--usually occurs only during the design phase; and the program is usually their first concern. (3,4) Earlier participation requires a clear framework for program decisions.

This paper attempts to deal with these programming problems. Although it is certainly not a panacea, we believe that Modular Programming is a framework that can be continuously developed in professional design offices within the very real constraints of a client, budget, time and political feasibility. We developed the method at the University of Pennsylvania during the Spring of 1972 specifically for a low density residential project; it is now being used for programming a Planned Unit Development and a Central Business District project within our respective design offices.

The concepts and the process of Modular Programming are discussed in the second section, and a representative application with sample results are presented in the third section. Finally, in section four the methodology is evaluated in detail.

Basic Concepts in Modular Programming

Modular Programming is a simple conceptual formulation that builds upon techniques that designers frequently use. The Module, the central concept, is analagous to "paper blocks": the pieces of cardboard that are used to arrange spatially the individual elements of a program and to fit them to a site. This set of "internally" defined relationships are related "externally" to the site and its context (with a graphic display technique) and other program elements which depend upon the quantity or nature of demand generated by the particular mix of Modules (with a "threshold" testing

9. DESIGN LANGUAGES AND METHODS / 465

technique). The result of the process is not a definitive program but data displays in a framework for examining trade-offs between facilities and the site, cost and quality, facility and facility and participant preferences from which a final program can be intelligently constructed. The following paragraphs discuss each of the concepts and techniques in detail.

Module
The module is an information block which allows manipulation of sets of data and relationships rather than individual components. It can be adapted spatially to any size area such as a city block, single acre, or larger area depending on the planning problem. Several internal characteristics are described and defined for this area depending on the problem situation. In the case of downtown planning these attributes might be floor area ratios, employee populations, trip attractions and parking relationships related to a use unit such as 1000 square feet of office space; in a natural environment planning problem they might be types of soil, subsurface geology, hydrology, vegetation, and wildlife described for a five acre area. In our use of the method for residential development we constructed one acre planning modules assigning density and dwelling unit types, ratios for parking and open space (derived from FHA land use intensity standards) as well as an assumed population profile. The programming Module can be conceived as an internally consistent block of data and relationships abstracted from a larger environment.

Threshold
The Threshold is a construct describing the nature and scale of demand necessary to support a service facility, piece of environment hardware, construction process, etc. It represents a boundary between two quantative or qualitative levels. For example, it is the point that must be crossed to supply two swimming pools instead of one, or to support a higher level of technology, such as on-site sewage treatment, rather than individual septic tanks. A threshold function has two major determinants:
1. Shape of the curve. This is dependent upon the particular characteristics of the product to be supplied. Some facilities such as swimming pools are "stepped", since a complete product of a fixed size must be constructed. Other facilities may be "continuous"; that is, they can expand in very small increments. Functions may be either linear or non-linear.
2. Nature of demand. This describes the quantity and type of demand necessary to support the facility. This demand may be expressed for residential use as number of families, population subgroup, or population by house type or density. Such quantities act as the independent variables

in threshold testing.

Threshold is an expanded concept for "standards" or "rules of thumb" that planners must use in order to expedite their work. Use of the Threshold concept allows flexibility to avoid over-simplification of the true relationships involved when adapting them to specific problem situations. It is possible to devise Thresholds to test different levels of environmental quality and to provide opportunities of trading off one facility for another to meet the same objectives.

Computer Calculator Program
A simple computer program for calculation purposes was developed to expedite the rapid manipulation of data. The program has two purposes:
1. Aggregation of quantities indexed within the various modules. This involves totaling the information for a given program so that the overall impacts can be evaluated on-site and off-site. For example, the number of school children by age group is of significance for local agencies and citizens groups, yet it is a tedious calculation to do by hand for each alternative; in fact, it is often a tedious task such as this that limits the number of alternatives that can be examined.
2. Performing threshold tests based on the aggregated demand of the combined modules. For example, the total number of school children (aggregated above) is tested to determine the number of schools of a particular type or size called for.

Graphic Technique
This is a method formulated for quick displaying of alternative possibilities and fitting the Modules to the specific constraint of the site. It works in conjunction with the computer calculator program which performs actual threshold tests. The advantages of the technique are the capabilities to make early locational design decisions and to hand manipulate and refine rough program quantities to achieve environmental quality goals. The choice of module types and facilities to be provided can be made "at the board" using this display and the catalogue of construction and maintenance costs on the computer printout. The example problem illustrates the use of the graphic technique.

Module Programming Process
A diagram of the process is illustrated in Figure 1. The first step is to synthesize site and site context information. This information must then be interpreted and formulated into working

assumptions and objectives for programming. A site analysis diagram is prepared as a working base map, and Modules and "Thresholds" are constructed or adjusted from previous experience to the specific problem situation. Initial program mixes are then formulated for detail program formulation.

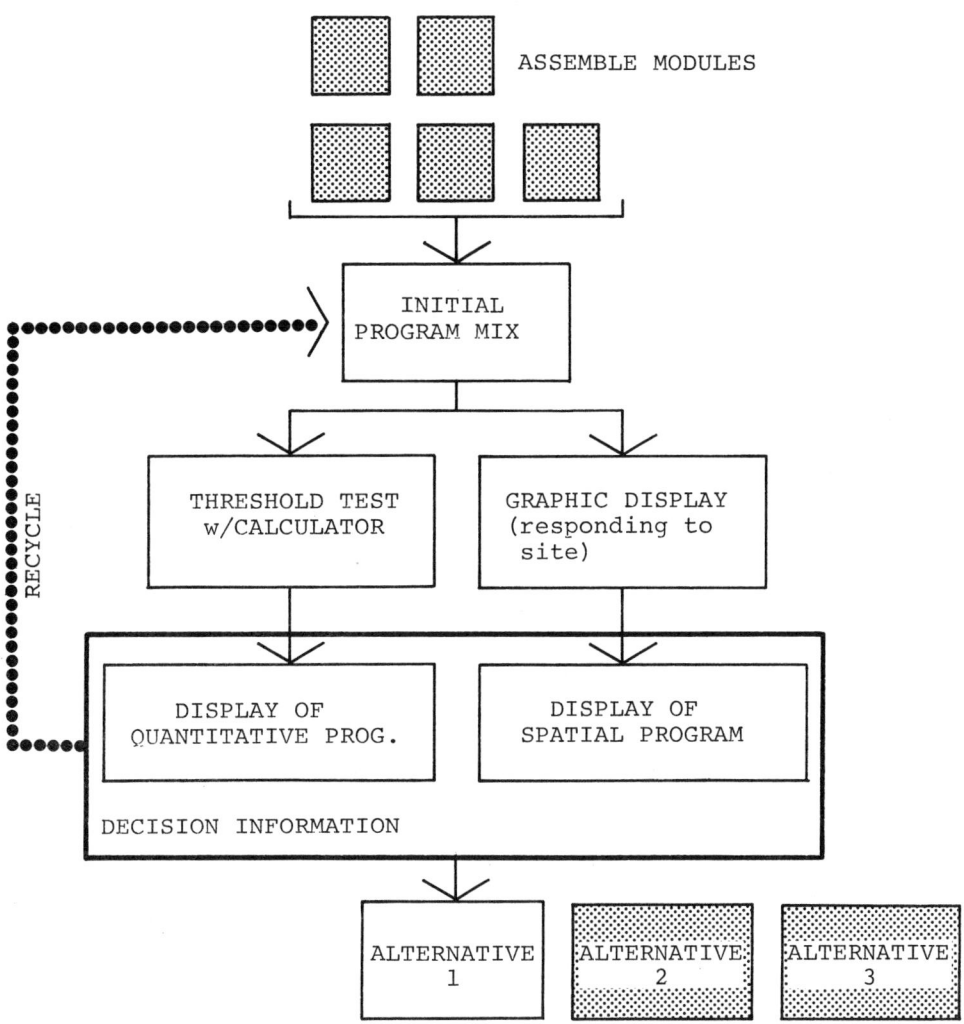

Figure 1. MODULE PROGRAMMING PROCESS

The Modules are then graphically displayed and the first threshold tests run for the initial Module mixes. The programs can then be adjusted graphically in response to site constraints (buildability of land, accessibility, land value, etc.) and to the quantity and quality of the alternative facilities and their spatial distribution. Refinement of the program mix then allows display of refined decision information, and the process of "trading-off" can begin by weighing thresholds in combination, by examining the relationships between Modules and Thresholds and by examining partial aggregations of Modules. The information can also be examined in terms of timing and phasing strategies. This forms the basis for defining clear program alternatives complete with costing data for evaluation by other participants.

Recycling this process can continue well into the design process as constraints become more clearly defined and participants requirements are verbalized. In fact, design issues will be raised during the first graphic display allowing the designer to establish a design concept that does not conflict with the program.

APPLICATION OF METHOD
The Modular Programming method was tested and refined through use on a residential development project. A brief summary of this work provides an introduction to the operational technique of the method and illustrates its benefits and shortcomings.

The site chosen was a 225-acre parcel of undeveloped land within an older suburban area in North Chicago, Illinois. Detailed information was available concerning market analysis, general population profiles, site analysis, service standards for surrounding jurisdictions and cost data. (5) Figure 2, a simple site analysis diagram, illustrates the major features on and surrounding the site and forms the working base map for the process.

From available market data we developed four Modules relating to development density and generic housing types:
1. 4 MODULE - 4 dwelling units/acre:
    (single family detached)
2. 12 MODULE - 12 dwelling units/acre:
    single family attached (townhouses and patio houses)
3. 24 MODULE - 24 dwelling units/acre:
    multi-family walk-up (garden apartments and maisonettes)
4. 60 MODULE - 60 dwelling units/acre:
    multi-family elevator.

9. DESIGN LANGUAGES AND METHODS / 469

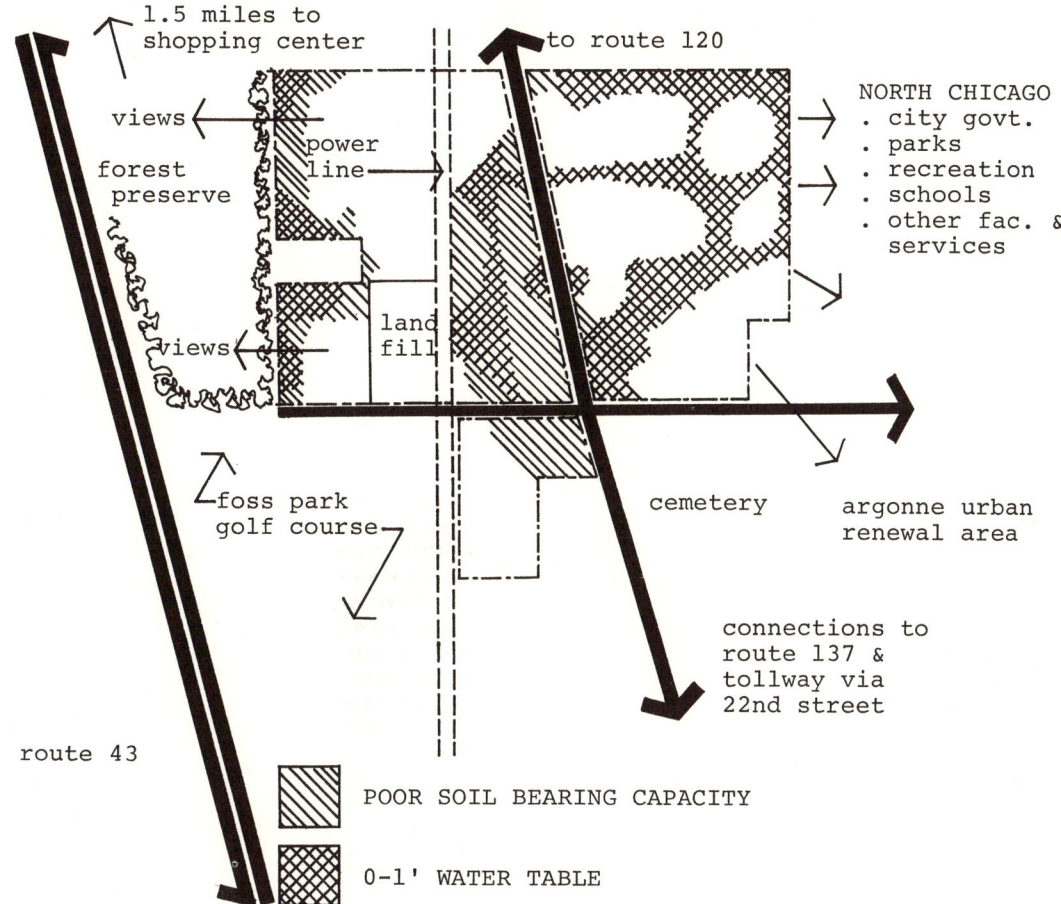

Figure 2. SITE ANALYSIS DIAGRAM

The Modules were constructed with information which is related to dwelling unit size, density, and type. This data is organized in two categories:
 1. The Population Profile was determined by making assumptions about the probable unit occupancy through time. This profile includes:
    A. Household income range
    B. Household type and size distribution
    C. Age group distribution
    D. Dwelling unit size distribution

Only partial data was available in some categories; therefore some data was "manufactured" based upon our past experience and adapted from other projects.

2. Spatial Coefficients include information within the Module that is intimately related to the dwelling type and does not vary significantly with project scale. They contain sufficient flexibility to allow numerous design responses. Within the Module these coefficients are described (6):

ASSUMPTIONS

1. Land Area = 43,560 sq. ft.
2. Land Use Intensity Ratio = 5.0
3. FHA Minimum Property Stds.
4. Residents's household income = $10,300 - $20,000

AGE GROUPS

Age Groups	Pop/DU	Pop/Module
0-4	.30	3.5
5-9	.34	4.1
10-11	.16	1.8
12-13	.22	2.6
14-18	.36	4.3
18-65	2.00	24.0
65+	.15	1.7
TOTAL SCHOOL CHILDREN		12.8
TOTAL POPULATION		42.4

MODULE SPACES

Space Type	Ratio	Space ft.^{2}
Land Area Total	.39 ÷ LA	43,560
Floor Area Total (max.)	.39 ÷ LA	17,000
Open Space Total (min.)	1.8 x FA	30,600
. livability space (min.)	1.1 x FA	18,700
. recreation space (min.)	.13x FA	2,205
. parking space	1.3 x DU (15.6 spaces)	5,460
. resident's parking	1.1 x DU (13.2 spaces)	4,580

HOUSEHOLD TYPES

H.H. Type	H.H. Size	% Dist.
Elderly	2.0	15%
Single	1.0	5%
Couple	2.1	15%
Family 3-4	3.5	40%
Family 5+	5.5	25%

AVERAGE POPULATION/DU 3.50
AVERAGE POPULATION/MODULE 42.40

DU SIZES

No.Bdrms.	Floor Area	% Dist.
0		
1		
2	1150 ft.^{2}	17%
3	1350 ft.^{2}	58%
4+	1500 ft.^{2}	23%

AVERAGE FLOOR AREA/DU 1410 ft.^{2}
AVERAGE FLOOR AREA/MOD 17000 ft.^{2}

VARIATION 1

1 MODULE		1 DU
43,560	Total Area	3,640
18,105	Attached Space	1,510
11,350	Bldg. Coverage@1.0	950
5,460	Parking	455
6,440	Circulation	536
2,205	Recreation	184

VARIATION 2

1 MODULE		1 DU
43,560	Total Area	3,640
16,495	Attached Space	1,340
5,675	Bldg. Coverage@3.0	472
5,460	Parking	455
6,440	Circulation	536
9,490	Recreation	780

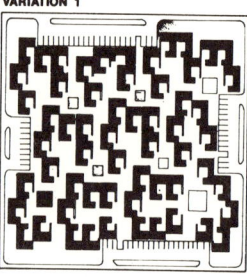
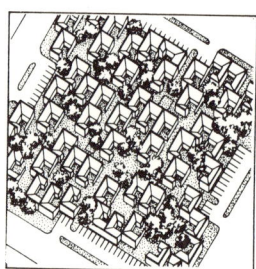
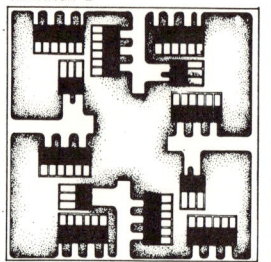
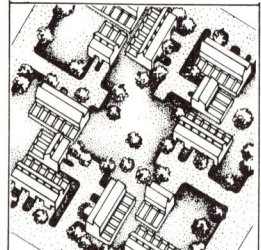

Figure 3. 12 MODULE (12 DU/ACRE-SINGLE FAMILY ATTACHED) (7)

9. DESIGN LANGUAGES AND METHODS / 471

    A. Housing type and density
    B. Floor area ratio
    C. Building coverage ratio
    D. Parking ratio
    E. Ratios of private and public open space
    F. Ratio of recreation space

A typical Module organized for a one acre space grid is illustrated in Figure 3 complete with data developed for the North Chicago site siutation and market. Two possible design variations are presented to demonstrate the internal flexibility of the Module and its flexibility for later detailed physical design.

The thresholds developed for the example problem include only residential services such as: open space, schools, corner stores, etc. Thresholds for infrastructure or other components could have been formulated as well. Services are divided into four components: space, equipment, management personnel, and facilities. This organization responds to the fact that these components have significantly different threshold functions, types of demand, and minimum levels of support quantities. The Thresholds for which functions and values were developed included the following:

1. <u>Spaces</u>
    1. Adjacent Recreation Space
       (for each Module)
    2. Common Recreation Space
       (for each Module)
    3. Community Recreation Space
2. <u>Equipment</u>
    4. Tot Lot Equipment (applied to
       adjacent recreation space)
    5. Older Children and Adult Recreation
       Equipment (applied to adjacent
       recreation space)
    6. Adult and Elderly Recreation Equipment
       (applied to adjacent recreation space)
    7. Playground Equipment (applied to local
       or community recreation space)
    8. Swimming Pool (applied to local or community recreation space)
    9. Softball Field (applied to local or
       community recreation space)
    10. Basketball Courts (applied to local or
        community recreation space)
    11. Tennis Courts (applied to local or
        community recreation space)
3. <u>Management</u>
    12. Recreation Management - Tot Lots
    13. Playground Management (associated with local
        or community recreation)

14. Community Management (associated with school facilities)
4. <u>Facilities</u>
    15. Day Care
    16. K-4 School
    17. K-6 School
    18. K-8 School
    19. Local Store
    20. Supermarket

Thresholds for each service were developed and included the following information. A typical example is illustrated in Figure 4.

DAY CARE (DAY CARE)

THRESHOLD FUNCTION

ASSUMPTIONS

FUNCTION: STEP SLOPE
TYPE: POP. GROUP (0-4)
APPLICATION: 4, 12, 24, 60 MODULES
SUPPORT: 196 POP. (0-4)
INITIAL QUANTITY:
  5,000 SQ FT
  7,000 SQ FT

(48 child capacity, 25% capture at peak hour, 50 sf/child)

CATCHMENT AREA:
LAND COST: $8,000/AC

SWIMMING POOL (SWIM)

THRESHOLD FUNCTION

ASSUMPTIONS

FUNCTION: STEP
TYPE: TOTAL DU
APPLICATION: 4, 12, 24, 60 MODULES
SUPPORT: 600 DU
INITIAL QUANTITY:

CATCHMENT AREA:
LAND COST:

ENVIRONMENTAL LEVEL

	1	2	3
SPACE	5,000 sf $900	7,000 sf $12,600	
IMPROVEMENTS	2,400 sf bldg. 50,000 playground 8,000	2,400 sf bldg. 50,000 playground 15,000	
ANNUAL MAINTENANCE	maint. 2,000 personnel 20,000 misc. 6,000	maint. 3,000 personnel 25,000 misc. 8,000	
TOT DEVEL $	58,900	77,600	
TOT OPER $	28,000	36,000	
DEVEL $/DU			
ANNUAL $/DU			

ENVIRONMENTAL LEVEL

	1	2	3
SPACE			
IMPROVEMENTS	25 yd pool $20,000 lighting 10,000 paving 8,000	25 yd pool $20,000 bathhouse 10,000 lighting 10,000 paving 8,000	50 yd pool $35,000 bathhouse 15,000 lighting 10,000 paving 11,500
ANNUAL MAINTENANCE/OPERATION	personnel 2,500 maint. 2,000	personnel 4,000 maint. 2,500	personnel 7,000 maint. 4,000
TOT DEVEL $	38,000	48,000	71,000
TOT OPER $	5,500	6,500	11,000
DEVEL $/DU	63	80	118
ANNUAL $/DU	9	11	18

Figure 4. DAY CARE AND SWIMMING POOL THRESHOLDS

1. Threshold Function - We utilized only step and step slope curves in the example. The Thresholds were dependent on the following types of demand:
   A. Total dwelling units (demand related to the number of households)
   B. Total dwelling units of a particular housing type or density
   C. Total population
   D. Population by subgroup (such as demand related to the number of children)

9. DESIGN LANGUAGES AND METHODS / 473

2. Application - These assumptions state the limiting factors for applying threshold tests to the above types of demand. They also state the quantity necessary for the initial increment of the service facility.
3. Environmental Level - For each threshold different levels of environmental quality are defined which include combinations of space, equipment, and personnel with associated construction and maintenance costs.

In using the Modular Programming to generate residential programs for the North Chicago site we wished to be able to make three comparisons between alternatives:

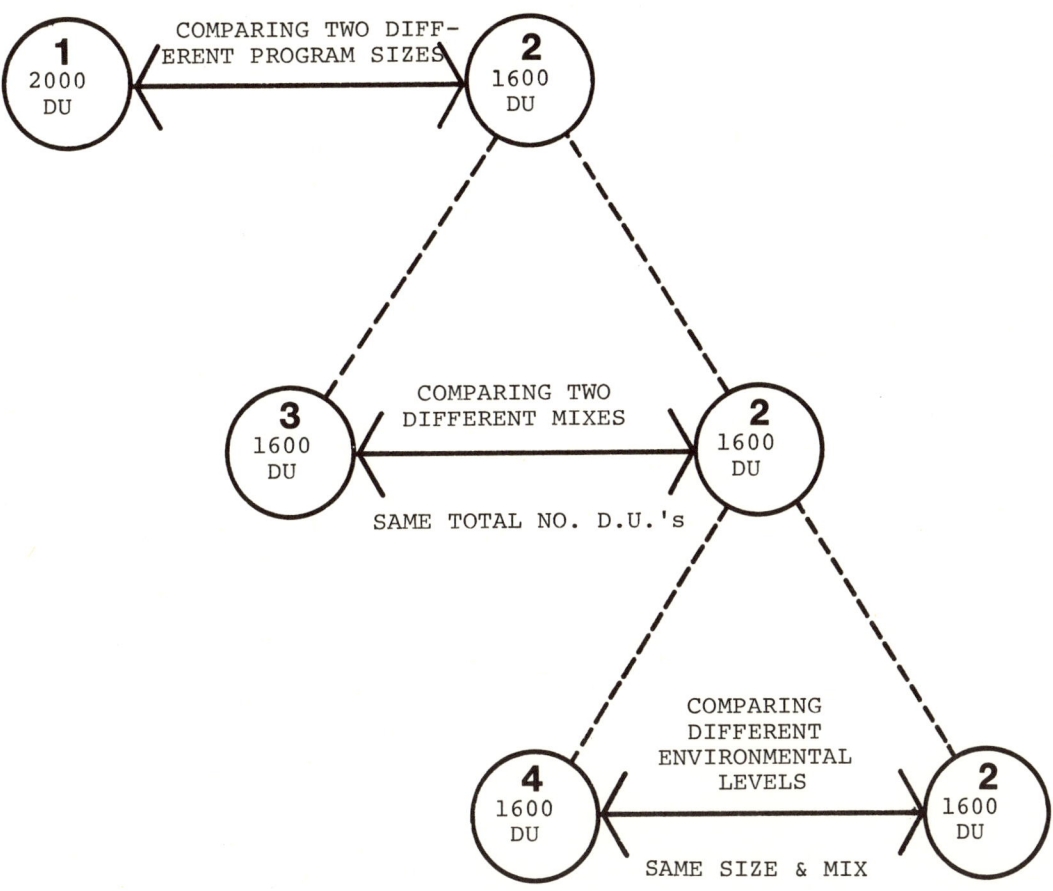

Figure 5. COMPARATIVE PROGRAMS

1. Variation in the total number of units.
2. Variation of the mix of housing types within the same total number of units.
3. Changing the range and quantity of services provided within a given residential mix.

Using the programming Modules and the Thresholds, we generated four basic programs which make these comparisons. The program mixes were generated by hand; the computer was then used to run threshold tests on given program mixes and to output results in three environmental quality levels. Figure 5 illustrates the programs generated and the ones between which particular comparisons are made.

The four alternative base program packages are shown in Figure 6. Program 1 for 2,000 total dwelling units can be compared with Program 2 for 1,600 to observe differences in density and range of services. Program 2 can then be compared with Program 3 which also has 1,600 dwelling units but a different mix of dwelling types. Finally, Program 4 compares with Program 2 to illustrate the results of choosing different combinations of service spaces and facilities for the same mix and number of units. At the bottom can be seen the total land consumed by residential modules and the remaining land available for additional service spaces and facilities on the site.

MODULE SIZE	PROGRAM 1			PROGRAM 2			PROGRAM 3			PROGRAM 4		
	NO. MOD.	NO. D.U.	%TOT D.U.	NO. MOD.	NO. D.U.	%TOT D.U.	NO. MOD.	NO. D.U.	%TOT D.U.	NO. MOD.	NO. D.U.	%TOT D.U.
4	100	400	20%	149	596	37%	120	480	30%	149	596	37%
12	68	815	40	25	300	19	67	793	50	25	300	19
24	16	386	20	22	528	33	14	336	20	22	528	33
60	6.6	400	20	3	180	11	0	0	0	3	180	11
TOTAL	190.6	2001	100%	199	1604	100%	201	1609	100%	199	1604	100%
SITE AREA	225.0			225			225			225		
LAND AVAIL FOR THRESH SERVICES	34AC			26AC			24AC			26AC		

Figure 6. ALTERNATIVE BASE PROGRAMS

In the next step the computer calculator program was used to perform threshold calculations at three environmental levels. Figure 7 illustrates the output at the first level.

ENVIRONMENTAL LEVEL 1

TYPE	THRESHOLD	INITIAL QUANT	SUP QUANT	ABV OR BLW T	INITIAL COST	ANNUAL MAINT	COST/DU	MAINT/DU
ADJREC12	108.00	3000.00	8333.33	0.00	9277.77	833.33	30.93	2.78
ADJREC24	144.00	3000.00	11000.00	0.00	12246.66	1100.00	23.19	2.08
ADJREC60	96.00	2000.00	3750.00	0.00	4706.25	562.50	26.15	3.12
LOCREC4	178.00	1.50	5.02	0.00	54744.94	5357.30	91.85	8.99
LOCREC12	520.00	1.50	0.00	-220.00	0.00	0.00	0.00	0.00
LOCREC24	720.00	1.50	0.00	-192.00	0.00	0.00	0.00	0.00
LOCREC60	1380.00	1.50	0.00	-1200.00	0.00	0.00	0.00	0.00
COMREC	800.00	6.00	12.03	0.00	114284.90	5714.25	71.25	3.56
SWIMPOOL	600.00	0.00	3.00	-196.00	114000.00	16500.00	71.07	10.29
DAYCARE	196.00	5000.00	9978.47	0.00	117546.30	55879.42	73.28	34.84
LOC ST	500.00	5000.00	3.00	104.00	139650.00	0.00	87.06	0.00
PLAYEQPT	500.00	0.00	3.00	-144.73	96000.00	3000.00	59.85	1.87
TOTEQPT	100.00	0.00	4.00	-8.84	19000.00	2000.00	11.97	1.25
ADLTEQPT	400.00	0.00	11.00	60.18	20350.00	2200.00	12.69	1.37
ELD EQPT	75.00	0.00	3.00	-8.28	9300.00	900.00	5.80	0.56
K-4 SCHL	400.00	4.00	4.89	0.00	956369.00	0.00	596.24	0.00
K-6 SCHL	400.00	4.00	7.25	0.00	1507515.00	0.00	939.85	0.00
K-8 SCHL	600.00	4.00	6.33	0.00	1475593.00	0.00	919.95	0.00

Figure 7. THRESHOLD CALCULATION WORK SHEET

From these calculations we see the amount of each service function that is supported by the various types of demand within the Modules. We can then choose particular services to achieve desired environmental levels within the constraints of land available and cost budgets--both for initial construction and yearly maintenance. Based on our objectives for comparison outlined above, we chose the services listed in Figure 8.

Figure 9, illustrating Program 1 on the site, shows an example of the graphic technique. The display uses white one acre modules printed with their densities. For example, a square printed with 12 indicates one 12 module (and thus 12 housing units). Service facilities derived from Threshold testing are represented by white symbols on a black background.

The space consumption in the graphic technique works as follows: Adjacent and local recreation spaces are the accumulated space quantities of the dwelling unit Modules; thus they are overlaid on top of housing Modules, and placed within the type that generated them. (i.e. Those generated by 4 du/acre Modules are placed within a cluster of 4 du Modules and require no additional area of their own). Community recreation and service facilities such as schools, supermarkets, day care, and local stores occupy their own areas independent housing Modules. Equipment and personnel are overlaid on recreation spaces as required. The Modules and service functions can thus be distributed according to design considerations and site opportunities and constraints.

For the example problem we attempted several alternative layouts of each program responding to different possible design rationales.

	PROGRAM 1 2,000 D.U.			PROGRAM 2 1,600 D.U.			PROGRAM 3 1,600 D.U.			PROGRAM 4 1,600 D.U.		
SERVICES	NO. UNITS	SIZE (AC)	TOT AC	NO. UNITS	SIZE (AC)	TOT AC	NO. UNITS	SIZE (AC)	TOT AC	NO. UNITS	SIZE (AC)	TOT AC
ADJ REC 4	5	.18	.90	7	.18	1.26				7	.07	.49
ADJ REC 12	7	.21	1.47	3	.21	.63	7	.07	.49	3	.41	.33
ADJ REC 24	3	.16	.48	3	.16	.48	2	.07	.14	3	.11	.33
ADJ REC 60	4	.09	.36	2	.09	.18				2	.07	.14
TOT ADJ REC			3.21			2.57			.63			1.29
LOC REC 4	2	1.00	2.00	3	1.00	3.00	3	1.50	4.50	3	2.00	6.00
LOC REC 12	2	1.00	2.00	1	1.00	1.00	2	1.50	3.00	1	2.00	2.00
LOC REC 24	1	1.00	1.00	1	1.00	1.00	1	1.50	1.50	1	2.00	2.00
LOC REC 60												
TOT LOC REC			5.00			5.00			9.00			10.00
COM REC	2.5	8.00	20.00	2	8.00	16.00	2	6.00	12.00	2	6.00	12.00
K4 SCH				2	4.00	8.00						
K6 SCH	2	4.00	8.00							2	5.00	10.00
K8 SCH							2	5.00	10.00			
DAY CARE	2	.16	.32	2	.16	.32	2	.16	.36	2	.16	.32
LOC STORE	4	.11	.50	3	.11	.33	3	.22	.66	3	.22	.66
SUPRMKT	1	1.80	1.80	1	1.80	1.80	1	2.0	2.00	1	1.80	1.80
SWIM POOL	3			2			2			1		
SOFTBALL	2			2			2			2		
BSKTBALL	4			3			3			2		
TENNIS	2			2			2			1		
TOT LOT	4			4			4			3		
PLAY EQP	3			2			3			2		
ADLT EQP	5			5			4			3		
ELD EQP	4			3			2			2		
TOT LOT MGT				1			1					
PLAY MGT	1			1			1					
COM FAC MGT	2			1			1			1		
TOTAL ADD. SERVICES AREA			30.60			26.50			24.50			25.00

Figure 8.   SERVICES PROVIDED - PROGRAMS 1-4

9. DESIGN LANGUAGES AND METHODS / 477

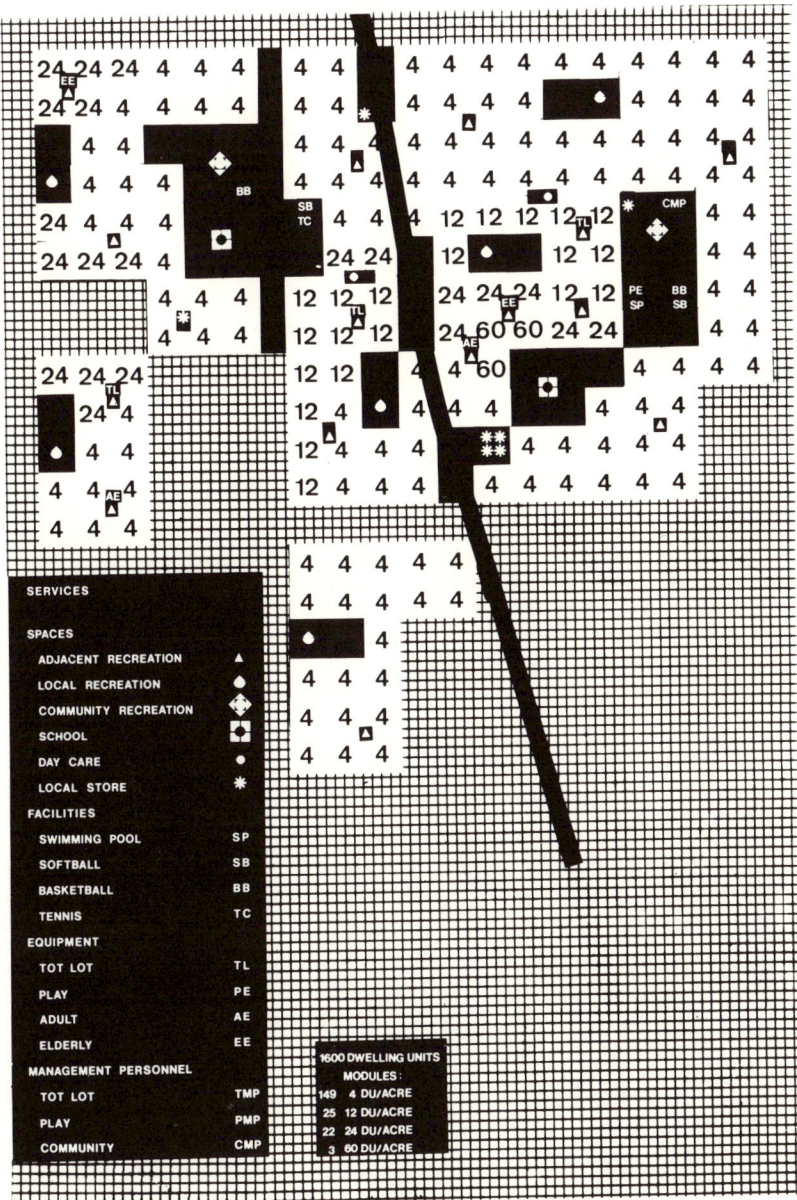

Figure 9. ILLUSTRATIVE SCHEME - 1600 DU

The one illustrated corresponds to a rough site plan leading into the design phase.

EVALUATION OF METHODOLOGY AS APPLIED TO RESIDENTIAL PROGRAMMING

Response to Program Issues

We feel that the graphic representation quickly gives a clear picture of housing mix and the distribution of services. It allows us to see what happens when we change the total number of units on the site or the mix of housing types or alter the levels of services being provided. It is uncomplicated to use to the degree of accuracy generally required in large scale programming. The computer calculator program can rapidly generate mixes of housing types and services based on many different assumptions complete with incremental costing, and the graphic pieces can quickly be assembled to display these alternatives. Then an examination can be made at the distribution, quantity, and quality of the alternatives, establishing a good information base for making decisions for the site as a whole or for sub-areas of development.

Further, the tool can be used effectively throughout the development process. Often, after an initial construction phase, market or site conditions are quite different from those in the initial planning. Only Phase I construction is determined; any number of alternative second phase strategies can be projected for different expectations of the future and services and housing mix in Phase I can be made accordingly flexible. Final decisions on a Phase II program can be made once later conditions, costs, and requirements are known. Likewise for Phase III, etc. Thus, overall goals for housing quality and levels of goals can be set initially and a flexible incremental development process established for actual development through time.

Response to Design Issues

We feel the method proved quite flexible in responding to design constraints and opportunities. Modules were distributed according to site ecological capabilities, fitting the best unit type to the soil conditions in each area. This was quite important due to poor bearing capacity and a high water table in particular areas. Another major locational determinant was vehicular access. Access determinants are more important to certain housing types than others e.g., high rise buildings require better access than single family dwellings where isolation is often desired, and some services such as a supermarket require a location adjacent to a major road. Thus the distribution of Modules and services could be easily related to ecological and access constraints as well as to each other.

It should be emphasized that the residential Modules do not indicate spatial design; they are simply quantitative relationships that aid the design process. Determining the actual physical relationships must still be done by a designer creatively using and interpreting

the Module quantities, service Thresholds, site conditions, and
people to be served. The programming method can be used throughout
the design process to make general locational decisions and to test
the programmatic implications of particular design ideas.

Connection to Other Tools
Several interesting possibilities exist for combining or using the
method in connection with other tools. The threshold testing mechanisms show what and how many services can be supported on the site
or within an area of the site. Where thresholds are not met and
facilities are not provided on-site, an impact on the surrounding
community occurs due to this latent or unfulfilled demand. This
demand could be put into a municipal fiscal impact model for
development. The reverse can work as well; unfulfilled demand already existing in the surrounding area can be added to on-site demand in testing service thresholds.

A vital tool from the developer's point of view is cash flow analysis
which can be used to test the financial feasibility of various alternative programs. Available computerized cash flow models could
be indirectly linked to the Modular Programming method by having
the Threshold calculator program output the cost data in proper
format. This is not an automatic link of course, since we wish to
retain the choice of total program quantities separate from machine
calculations. Data fro the selected modules and thresholds would
be a simple matter of sorting through data cards.

In most development situations, whether publicly or privately
sponsored, funds for providing services are limited. Usually land
is also limited since most developers wish to build as much on their
ground as they feel the market will bear. These two factors indicate
the possibility of linking linear programming to the Modular Programming to be able to choose and allocate between service facilities ovided within the constraints of available land and money.
This would require a means of weighting the im ortance of various
services; a risky exercise due to the lack of satisfactory measures.
However, we feel it would provide further information for making
decisions by displaying the alternative programs which are possible
with capital and operating budgets and quantities of land available.
The importance of this addition to our present method lies 1) in the
ability to perform sensitivity testing and 2) inbbeing able to compare preassembled total site programs at least for the purposes of
defining the limits of the solution space (i.e. the reasonable
range of programs based on alternative objectives).

Increasing Programming Knowledge and Capability
A significant value of Modular Programming when used in a variety
of housing development situations is the ability to assemble a
range of programming information with accurate space and cost data.
Although dwelling unit Modules can be readily adapted to many

different situations, certain ones will often reappear.

Similarly, while threshold scales may be locally adaptable, much of their content will apply to many problem situations. Thus, it is possible to build up a library of useful programming data, cataloging Thresholds and Modules that can be changed and updated as necessary. The purpose of this exercise is not merely to simplify the programmer's and designer's tasks for future projects but to build up an information system that can be used in creating consistently higher quality living environments. For once Modules, Thresholds, and cost data are known, the professional can array many alternative program opportunities against financial possibilities and site constraints so that increasingly better choices can be made to maximize living quality.

Notes

1. Constance Perin, With Man in Mind: An Interdisciplinary Prospectus for Environmental Design, MIT Press Cambridge, Mass., 1970.

2. David Anstrand, Richard Dagenhart, John Moore, Val Thomas "Spatial Allocation Modeling Using IBM-360 Mathematical Programming Systems" (Jersey City Operation Breakthrough Site) Unpublished Paper, University of Pennsylvania, March 1972.

3. Robert Venturi, Denise Scott Brown and Steven Izenour Learning from Las Vegas, MIT Press Cambridge, Mass., 1972.

4. David Anstrand, Richard Dagenhart, Norman Day New Community Design and Development a Research Monograph published by Kling Planning, April 1972.

5. David A. Crane and Partners and Gladstone Associates "Preliminary Development Program Analysis-North Chicago Development Study." A joint memorandum prepared for the Illinois Housing Development Authority, November 1971.

6. Department of Housing and Urban Development "Design Criteria for Multi-family Housing, Washington, D.C., 1969.

7. Princeton University School of Architecture "Prototype Housing Design" 1970.

A ROLE-ORIENTED APPROACH TO PROBLEM SOLVING BY GROUPS   9.2

Charles H. Burnette, Ph.D., AIA
Center for Planning, Design, and Construction, Philadelphia, Pa.

Gary T. Moore
Department of Psychology, Clark University, Worcester, Mass.

Lynn Simek
Clark University, Worcester, Mass.

Abstract

A procedure for solving problems by structuring the efforts of a group according to roles which control particular types of information has been developed and augmented in practice by operational methods for stimulating creativity and principles drawn from the study of group dynamics. The result, after over a dozen applications in different academic and professional settings, is thought to be a highly effective procedure for structuring and guiding the effort of groups engaged in problem solving, for illustrating and analyzing the power, timing, and effects of various types of information and constraints introduced during problem solving, and for studying interpersonal behavior within problem-solving groups. A general description of the procedure is presented.

Introduction

As problems are realized to be more complex and difficult, as problem situations are seen to require knowledge and ideas from many disciplines, and as teams of people from diverse backgrounds, values, and perspectives are brought together to assist in obtaining creative solutions to problems, it becomes increasingly clear that new ways to structure and facilitate such group efforts are needed. To meet this need, new procedures should accomplish at least the following: a) provide a non-threatening situation in which a full range of perspectives, information, and constraints relative to the problem can be brought forward by all interested parties, b) structure the efforts of diverse people -- professionals, clients, users, and consultants -- into a cohesive, cooperative, and efficient team, and c) encourage the free-flow of creative ideas toward problem solution. This paper is a progress report on the development of a group problem-solving procedure to meet these demands. The paper describes the origins of the various techniques proposed and briefly outlines the operational procedure (1).

The proposed method is a blending of a relatively new Role-Defined approach for structuring the efforts of problem-solving teams with methods for stimulating creativity and facilitating the interpersonal dynamics of task-oriented groups. The latter methods are derived in part from the well-known "Synectics" method and from other approaches to problem solving and group dynamics. They focus primarily on the enhancement and cultivation of the psychological components of individual and group creativity. The Role-Defined method, on the other hand, focuses on the structure and handling of substantive information, constraints, and perspectives. The proposed method, therefore, is an attempt to integrate the best of these two approaches and also to include the use of group dynamic principles to assist constructive interpersonal communication and cooperation. Each of these facets of the new approach will be described below in brief.

The Role-Defined method has been developed over the past several years by Burnette and received its first major public exposure at the 1971 Conference of Environmental Design Educators in Key Biscayne, Florida (2). The approach is based on a theoretical structure for a comprehensive information-processing system conceived to complement thought and expression (3). Its application to group problem solving developed from an effort to demonstrate the structure of the system in concrete behavioral terms. The method makes use of seven roles defined in terms of the type of information each controls. These roles and the authority they confer are assumed by members of the group during problem solving. They also function as descriptive categorizations in the study of information flow and problem-solving behavior. Taken together, the role categorizations provide a system for organizing information and focussing effort which can assist the communication and understanding of the group.

This method has been augmented by the incorporation of certain principles and techniques derived from "Synectics", an operational method for stimulating and buiding creativity that has been developed over the past decade by William Gordon, George Prince, and others (4). This method may be seen as consisting of two parts: the use of metaphorical thinking as an aide to creativity, and an overall procedure for generating and testing creative ideas. This method's great value lies in the speed by which ideas and potential solutions can be generated.

These two methods have been augmented with some of the idea-generation techniques of the Osborn-Parnes Buffalo school of creative problem solving (5) and additional principles for effective group dynamics developed by Ruth Cohn and others (6).

The proposed procedure, therefore, is an integration of these various methods based on our experience with them singly and in combination. The new procedure makes use of the following: a) seven information specialized roles divided among the group members to assure a full range of viewpoints, and a ready reference framework for organizing and communicating information and constraints, b) group dynamics principles for helping the team members work cooperatively and effectively, and c) various techniques for liberating and stimulating the creative imagination, and building individual ideas into an integrated solution to the problem at hand.

Operational Procedure

Workshops designed to introduce the procedure have typically involved five phases: introduction and discussion of the group dynamics policies which the group will follow for the rest of the session; introduction, description, and assignment of roles; introduction and exercises in the creative idea generation techniques; the actual problem solving task, and finally a post session discussion of the experience. The sessions to date have all combined learning and application.

Group Dynamics Policies

All participation in the workshop -- role defining, various creativity exercises, and the actual solving of the problem -- is conducted in accordance with three group dynamic policies. These policies are designed to foster cooperation, communication, and group creativity. They are introduced by example, sometimes supplemented with brief exercises. The three policies are:

Constructive Listening Policy: Participants are encouraged to listen, not to interrupt, to ask for clarifications, and to paraphrase in order to better understand each other's points and to support all efforts to contribute to the work of the group.

Positive Selection and Structured Feedback Policy: Participants are encouraged to give credit for another person's contributions, to select the best in each idea, and to itemize specifically those things thought to be of positive importance to the problem at hand, before stating any concerns with the idea.

Synthesis Policy: Participants are encouraged to build on the ideas of others, to correlate and integrate earlier ideas, and to look for syntheses between new and previous ideas.

Said otherwise, these policies mean that each time a team member offers a suggestion, no matter how "trivial" or "wrong" it may initially appear, it is given careful, open-minded consideration by the other members. In order to avoid losing the good points in an idea through too quick criticism, each member is encouraged to give it full, constructive listening, to request clarification and to paraphrase it in order to confirm understanding, and then to state the points which seem to be the best aspects or implications of the idea. After this he may state his concerns about the idea, but should immediately try to offer alternatives to the parts about which he had concern. Finally, he is encouraged to synthesize the best parts of the original idea which these suggestions.

Informational Roles

Next, the seven roles are defined, described, and assigned. Each role allows the person to exercise absolute authority (presumed expertise) over a certain type of information. These roles and the type of information each controls are:

<u>Problem Designator</u>: Defines the problem; controls all directive and interpretive information.

<u>Resource Specifier</u>: Identifies, itemizes, and describes resources which might be marshalled in response to the problem statement; controls all substantive descriptive information.

<u>Resource Organizer</u>: Organizes the specified resources in response to the problem specification; controls all relational information.

<u>Form Giver</u>: Provides physical or communicable forms to manifest the organization of information; controls all representational information.

<u>Actualizer-Doer</u>: Realizes and implements the form or plan; controls all procedural information.

<u>User</u>: Responds to all preceeding roles and information in relationship to his own perceived needs, desires, and reactions; controls all behavioral information.

<u>Evaluator</u>: Evaluates progress in each role and redirects the team; controls all analytic information and structured feedback.

After discussion of the above roles, one person chooses or is assigned each of the roles. A second person may back-up or assist each of the principle role-players. While the roles are described as they are introduced, they are learned primarily through their use. In this regard, the roles and the constraints which they imply relative to one another have proven to be readily recognizable, easily learned, and quickly internalized by participants. Once internalized, they become the normal basis for communication within the group.

Ideally, the role-players sit around a large, circular table next to a blackboard. The Organizer and Form Giver, who will present the work of the group on the board, sit nearest it. The rest of the group sit as shown in Figure 1, the Problem Designator roughly opposite the board, the Resource Specifier to the left of the Problem Designator, then the Organizer, Form Giver, Actualizer, User(s) and the Evaluator. The Workshop leader or Facilitator roves around outside the group, offering procedural clarifications, suggestions, and structured feedback until the procedure is clear and internalized, at which time the group is on its own.

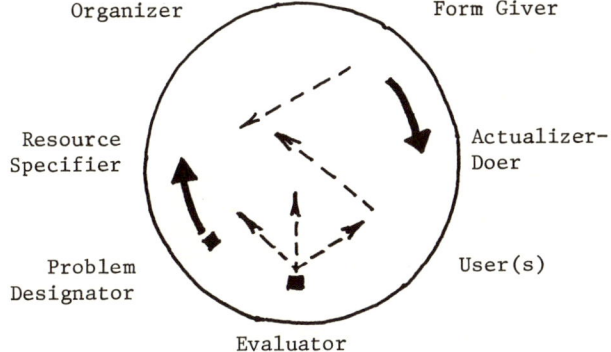

Figure 1.

The first approach to the problem is usually sequential since each role must account for information established and controlled by the preceding role and a logical dependency exists from one role to the next.

In the basic sequence, the Problem Designator briefly states and analyzes the problem, paying special attention to the crux, its symptoms, and the boundaries or initial criteria for the solution, giving examples of suggested solutions and structured feedback as to why they are inadequate. He also states what he as "client" wishes to accomplish during the session, (a general or specific solution; one well articulated solution or a range of alternatives; and so on). Depending on the nature of the problem, the User(s) may aid this process by stating their felt reactions to the situation (as in a community-related problem) or the Resource Specifier may describe the resources available for his solution, as in a product for a manufacturing plant. The Facilitator then asks the other participants to generate specific, realizable goals or objectives based on the problem designation and analysis. The Problem Designator selects for initial consideration a few highly interrelated goals which seem to capture the essense of the problem and which, if solved, will greatly contribute to the solution of the problem as a whole.

The method can then proceed sequentially, clockwise around the group, each role-player in turn responding to the problem as given by the Designator or User, the selected specific goals to be solved, and the responses, information, and ideas of those in preceeding roles.

Once the first cycle of seven roles has been completed, the procedure may be opened to questions in order to clarify roles and involve the group in discussion as a whole. Usually, this results in greater role differentiation and in role focussing, that is, in greater clarification of the roles in general and in the selection, more or less by consensus, of one of the roles as requiring the immediate attention of the group. In this manner, the dialogue between roles continues to evolve with further definition of the problem, the constraints, and the potentials that arise as the needs and contributions of each role and each person become apparent.

For one or two cycles, each person is encouraged to respond quickly and in order. Soon, however, there is dialogue, suggestions toward solution, and idea exchange as the participants recognize the implications of their roles and of the information and ideas contributed by others.

If at any time anyone feels a blockage on creative directions toward solutions, he may call for a break in the basic sequence and suggest using one of the techniques for generating new ideas. The Evaluator is especially on the look-out for these needs of the group. One person, usually the person suggesting the need for new ideas, the Evaluator or Facilitator, then conducts the group on an idea-generation trip away from the immediate, real context of the problem into the world of fantasy and metaphorical thinking.

Although role selections usually remain constant for any one problem session, all participants are exposed to all roles and often assist the person with authority for another role in his exercise of it. Cooperation is often the result of an anticipation of a desire by one role relative to the constraints being set through the playing out of another. Like chess, the role defined approach tends to involve its participants in thinking out and evaluating implications and strategies by providing a framework of authority and power into which to project. Synectics ideas and techniques, on the other hand, offer an effective counterpoint

9. DESIGN LANGUAGES AND METHODS / 487

to the logical framework of roles by offering escape from them through fantasy and metaphor. Together, both methods complement and produce the subject matter for each other.

Creative Idea-Generation Techniques

The seven most-used idea generation techniques are the following, listed from general to specific:

Brief Vacation: The group simply takes a rest from their effort and thinks about something other than the problem in order to gain renewed energy and a fresh perspective.

Brainstorming: A list of ideas are thrust out quickly without any judgment or discussion; these are recorded on the board and subsequently examined to select the best ideas for further consideration.

Free Association: One idea is suggested, as in brainstorming, to which the first person's free association is recorded, then another person's free association to this first association, and so on until someone gets an interesting association which he feels might be helpful in the problem context.

Morphological Synthesis: A matrix is formed comprised of rows and columns representing the formal properties of the idea or object at which the blockage occurred; alternative ideas in relation to each property are generated by brainstorming until the matrix is full, then interesting lines of solution ideas are extrapolated from the matrix and subjected to further development.

Direct Analogy: An example, or analogy, is examined from another context which has characteristics similar to those of the problem or place in the process where the blockage occurred.

Personal Analogy: The participant identifies himself with the object in order to get inside and examine it from an experiential point of view, i.e., the emphasis is on the participants identifying with the object and developing empathic involvement with it.

Essential Paradox: A search for the nexus of the problem is undertaken through examining contradictions and suggesting two-word phrases which capture the essential paradox involved (7).

Each technique offers a particular kind of license to the imagination to escape from the constraints of finding an immediately logical and workable solution to the problem. Each results in a typical kind of product -- a list of alternate ideas from brainstorming, a useful insight from analogy, etc.

These idea generating techniques are used by themselves or in combination as they seem appropriate. For example, the Facilitator may select a key word in one of the goals (say, "selection" in the goal "select a population across income levels to live in an ecologically planned site.") He may then ask for several examples of "selection" from a different conceptual universe (say "advertising "). Then he may ask for discussion of one of the suggested areas (say, "real estate advertising") to suggest some directions toward reaching the original goal (perhaps "self-selection by those who respond to advertisements of the virtues of the site in media normal to their income level.") During such an idea-generation trip, all participants abandon their roles and contribute to the process.

After new ideas or insights are generated, and given that they are usually not full-fledged solutions, the Facilitator must guide the group back into the role-playing situation where the ideas are examined, developed more fully, refined, evaluated, and integrated with the original framework. This process of bringing the results of an idea generating trip back down from the world of fantasy and metaphorical thinking to the reality of the actual problem situation is one of the most difficult and most important jobs of the trip-leader or Facilitator, yet one on which we have the least current understanding. To accomplish his task, the leader must skillfully and gently bring the group back to the problem situation without losing any of the creative nuances and still tenuous metaphorical connections between the ideas and the problem. In this regard, the seven roles offer a variety of "handles" to assist him. Particular ideas may be returned to the most appropriate role and there is a natural predisposition for each role player to assimilate them into his own role view. Whether all or a few roles are invoked, the products of the trip can be returned to a full framework of evaluation understood by the entire group.

After the new ideas are reintroduced to the problem context, they are examined, the best are developed and refined more fully in the context provided by all roles until one or more possible solutions emerge. Structured feedback on the ideas and possible solutions is given by the Problem Designator and User(s) and the Evaluator analyses and strengthens the performance of the team. The process is recycled through the generation or selection of new goals, the generation of new ideas, further assimilation and refinement, etc. The process continues until the Problem Designator and User(s) are satisfied with the development of one or more solutions. If a solution is not found in one session, the group can gather more information and reconvene for another session at a later time. The entire procedure is summarized in figure 2 below.

PROCEDURE

Training:

1. Group dynamics policies
2. Informational roles and their use
3. Creative idea-generation techniques

Problem Session:

    Problem statement by Problem Designator or User(s)

    Role oriented goals or objectives from all participants

    Selection of initial goals by Problem Designator or User(s)

    Sequential introduction of information from all roles

    Idea-Generation trips (as necessary)

    Development of ideas and information in all roles

    Feedback on ideas and possible solutions from Problem Designator and User(s)

    Procedural feedback from Evaluator

    Refining of ideas toward possible solutions

    Recycling through generation of more goals, etc.

Figure 2

Summary

The Role-Oriented approach to creative group problem solving has been tried in a number of different contexts, with variations both in format and in focus, sometimes the focus being on learning and exploring new approaches to problem solving, other times with the focus being on the finding of a solution for a real-world problem. In most cases in which it has been used as a problem-solving tool, it has been found that preliminary solutions were obtained within three hours from initial exposure to the method, in circumstances which were less than ideal, and with problems which were very complex and hard to apprehend (for example, "How to involve the outside world in the affairs of the Martin Luther King Boulevard Project?" -- a Model Cities project in Miami -- or "How to specify the population and density of development in a site to assure ecologically sound development?" -- a project in Landscape Architecture in Seattle, or "How to Renovate An Adandoned Railroad Station to Serve as a Community Center for Several User Groups" -- Department of Architecture, Yale University.)

It is believed that the method has many uses and applications. The role categorizations in particular facilitate its use for the descriptive analysis of communication. For example, the roles make it possible to categorize the information used, to analyse information handling dynamics through an accounting of the amount of information processed by each role and the transactions between roles, thus to analyse characteristic problem structures as well as the strategies employed in their solution. In a similar manner, the roles also provide the structure for studying interpersonal dynamics and creative behavior within groups. The method as a whole can be used for studying and developing personal aptitudes for various roles, as well as for the group as a whole.

We are greatly concerned about many current man-environment problems and are most interested in exploring the method and improving it in these contexts. We believe that the procedure is particularly suited for problems where there are several views or perspectives on the problem which must be reconciled, where there are great quantities of constraints and information (economic, behavioral, etc.) to be introduced and balanced, and where the problem has been particularly impervious to solution by traditional means. All of these are true, of course, for many community and urban planning situations.

The method is still in its infancy. Hopefully it can be constantly improved as we learn more about it through use in a variety of different contexts and through related conceptual and empirical investigations.

Notes

1. This paper is intended as a working communication or preliminary progress report on a group problem-solving procedure presently under development. This brief account is not meant as a final statement -- many conceptual and practical questions remain to be answered. However, the proposed procedure has been seen to be a powerful technique in a number of practical applications and some communication about it, however tentative, seemed appropriate.

9. DESIGN LANGUAGES AND METHODS / 491

2. There is no comprehensive description of this method beyond the present paper. A one page summary description has, however, been quoted in the program and report of the American Collegiate Schools of Architecture/AIA 1971 Teachers' Seminar/ A Case Study: 1st Annual Environmental Educators' Conference/ An Assessment of Strategies, November 11-14, Key Biscayne, Florida, David Clarke, ACSA, 1785 Massachusetts Avenue, N.W., Washington, D.C. 20036. An NSF supported team observed, and reported 100 per cent participant satisfaction and intention to use the method. Schools at which the method has been run include the University of Pennsylvania, Washington, Drexel, Yale, Detroit, Clark, Nova Scotia Tech College, and Philadelphia College of Art.

3. C.H. Burnette, "The Design of Comprehensive Information Systems for Design," EDRA Two: Proceedings of the Second Annual Environmental Design Research Association Conference (J. Archea and C. Eastman, Eds.), Pittsburgh: Carnegie-Mellon University, 1970, pp. 271-279. "Designing to Reinforce the Mental Image: An Infant Learning Environment," EDRA 3/AR8 Conference Proceedings, (William J. Mitchell, Ed.), University of California at Los Angeles, 1972, p. 29-1-1.

4. The formal "Synectics" method is outlined in three books: For metaphorical thinking, see W. J. J. Gordon, "Synectics," New York: Collier, 1961, and Gordon, "The Metaphorical Way," Cambridge, Mass.: Porpoise Books, 1971. For the overall procedure, see G. M. Prince, "The Practice of Creativity," New York: Harper and Tow, 1970. For a discussion of the theoretical notions behind these methods, see L. Simek, "The theory of the creative process underlying the 'Synectics' method, Unpublished seminar paper, Office of Academic Innovation, Clark University, January 1972. It should be pointed out that the term "Synectics" is a trademark for a formal, copyrighted method. Though we have learned much from the method and from our experience with it, no claim is made or implied that we are using the formal method or representing either of the "Synectics" organizations. Rather, we are incorporating certain principles from this method and others, as we have construed them from our experience to create a new, integrated method which we feel is better suited than any one method in isolation for the complexities of large-scale urban and natural environmental problems.

5. See A. F. Osborn, "Applied Imagination," New York: Scribners, 1963; S. J. Parnes and Staff, "Workbook for Creative Problem-Solving Institutes and Courses," Buffalo: Creative Education Foundation, 1966; and other publications of the Creative Education Foundation.

6. R. C. Cohn, The Theme-centered Interactional Method, "Journal of Group Psychoanalysis and Process," 1969-70, 2 (2), 19-36. See also R. F. Bales, "Personality and Interpersonal Behavior," New York: Holt, Rinehart and Winston, 1970; and M. B. Miles, "Learning to Work in Groups," New York: Teacher's College, Columbia University, 1959; both of which analyze the structure and functioning of task-oriented groups.

7. Our use of these techniques is similar, though not identical, to the way they are elaborated in the various publications cited in notes 4 and 5.

PIECEMEAL SOCIAL ENGINEERING: A CASE STUDY

Patrick J. Quinn

Rensselaer Polytechnic
Institute

J. MacGregor Smith

University of Illinois
Urbana

Abstract

This paper discusses the process of piecemeal social engineering. The authors argue that planners must be more modest in their efforts to carry out social and physical change; utopian planning strategies are not workable or realistic. A collection of seminary communities is used as the vehicle for this analysis. The issues and problems encountered by these communities and the recommended social and physical changes are explained. One particular concept for shared studying facilities, the study set, together with its logic, consequences and supporting evidence is examined in detail.

Introduction and Problem Statement

Architects, planners and designers of the environment frequently discuss and argue at great lengths over the concept of community. There is, however, little agreement among these groups on a satisfactory definition for this concept. Perhaps the question "what is community?" is incorrectly construed. Instead of seeking to know the definitive essence of community, conceivably the question should be phrased: "how does a group of persons behave under certain environmental conditions or circumstances?" or "how will one group of persons behave when brought together with different groups of people having diverse life-styles, attitudes, and ideologies?" Utopian planners and often systems analysts in their attempt to describe the particular essence of a social system and then redesign the system as a whole, fail simply because the task is too grand and overly complex. Rather, the worldview underlying the latter questions mentioned above, calls for investigations of a social system or institutions on a more workable, empirical foundation. Certain hypothetical situations or conditions can be proposed and tested in the context where the communities operate. One can then examine the particular consequences of these actions on the groups' behavior. This tact or strategy is termed piecemeal social engineering by Popper and represents the viewpoint adopted in the following analysis. (1)

The process of deriving and testing a series of behavioral changes in the environment for a collection of communities will be examined in the following paragraphs. These behavioral changes were necessary to allow these communities to effectively adapt to a new unfamiliar physical setting. The authors wish to demonstrate the process of piecemeal social engineering and its possible consequences for designing the environment.

The communities examined in this study are theological seminaries which have recently relocated themselves from a basically rural environment to an urban setting. Each theological community represents a distinct subculture with strong ideological orientations, a sustained existence over many years, and fairly well-defined life styles, habits, and values for their constituent members.

Seminaries have traditionally been physically separate from one another because of varied ideological outlooks and the desire to maximize their privacy to practice and study their religion apart from the pressures imposed by secular life. Physical barriers such as wall fortifications, bodies of water, etc. have been utilized in the past to emphasize and reinforce this sense of isolation. Currently, however, ecumenical, economic, social and cultural forces have resulted in the desire for more interfaith collaboration and closer ties with the secular community. Integrating such denominations as Lutherans and Catholics, Presbyterians and Unitarians, where shared classrooms, offices and other facilities occur, creates an entirely different aspatial and spatial environment which these communities must adapt to. How are these communities to maintain their identities, boundaries, and life-styles? What types of shared activities are to be created? What types of new activity settings are necessary to allow the individual subculture to maintain its identity while at the same time establishing a communal identity among the seminaries? These and other questions will be looked at in more detail.

The specific subcultures examined in this study are all a part of the Graduate Theological Union (GTU) in Berkeley, California. The GTU is a corporate body of ten different seminaries with approximately 1000 students and 100 faculty. The seminary constituents are listed below:

1. American Baptist Seminary of the West (ABSW)..........Baptists
2. Church Divinity School of the Pacific (CDSP)..........Episcopal
3. Franciscan School of Theology (FST)...................Catholic
4. Graduate Theological Union (GTU)......................Interfaith
5. Jesuit School of Theology at Berkeley (JSTB)..........Catholic
6. Pacific Lutheran Theological Seminary (PLTS)..........Lutheran
7. Pacific School of Religion (PSR)......................Methodist
8. St. Alberts College (SAC).............................Catholic
9. San Francisco Theological Seminary (SFTS).............Presbyterian
10. Starr King School for the Ministry (SKSM).............Unitarian

The following map illustrates the disposition of buildings owned by the member seminaries of the GTU. The scatteration of facilities and their integration with other buildings in the neighborhood attests to the significant de-emphasis upon physical isolation and separation of the different groups.

Methodology

The problem of integrating these subcultures while at the same time maintaining their identities implied the need for some new activities and activity settings to facilitate the adaptive process of the seminaries in the new environmental context. The planning team was charged with planning the growth and change of the GTU community for the next twenty years. To accomplish this task, it was felt that a

9. DESIGN LANGUAGES AND METHODS / 495

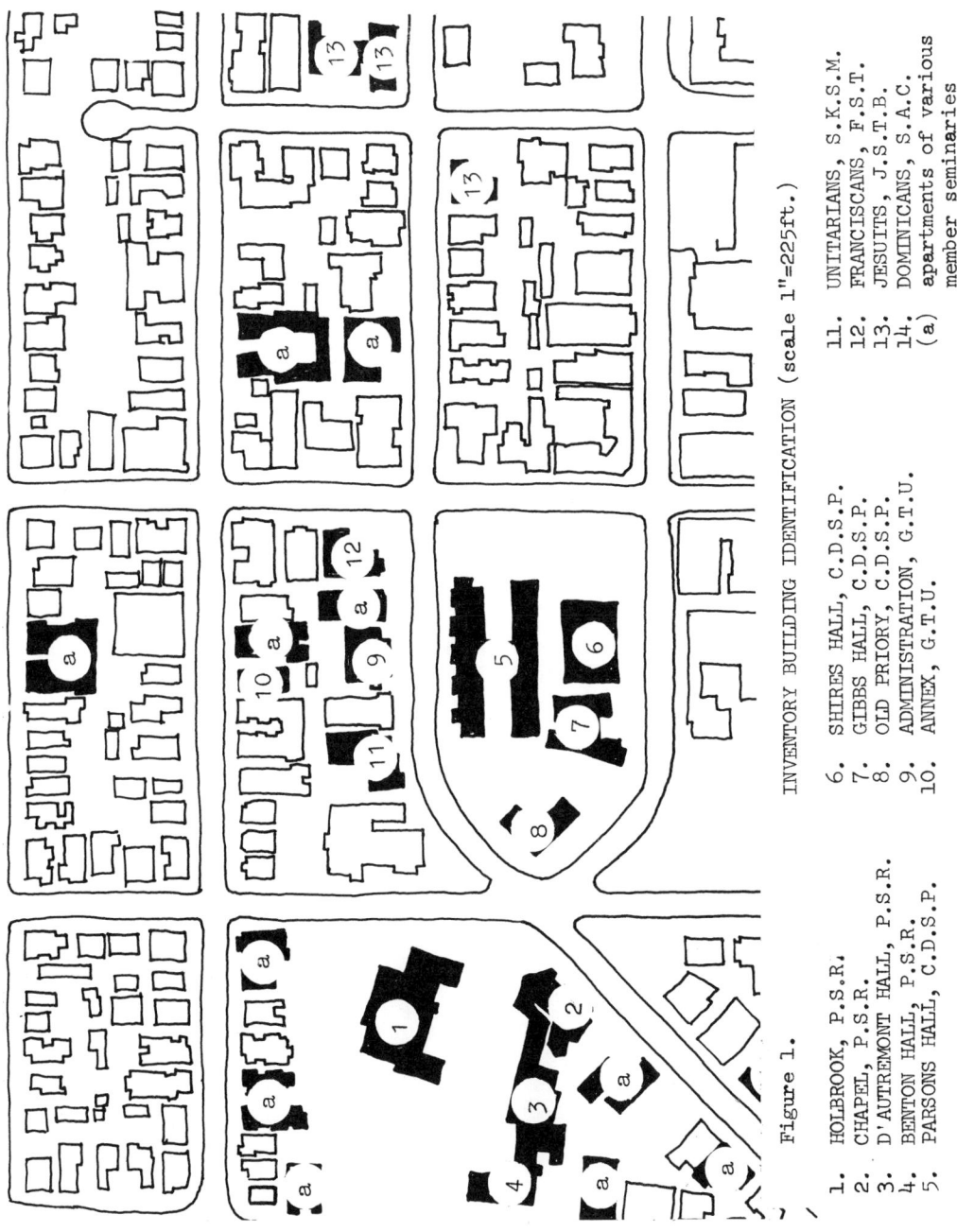

Figure 1.

INVENTORY BUILDING IDENTIFICATION (scale 1"=225ft.)

1. HOLBROOK, P.S.R.
2. CHAPEL, P.S.R.
3. D'AUTREMONT HALL, P.S.R.
4. BENTON HALL, P.S.R.
5. PARSONS HALL, C.D.S.P.
6. SHIRES HALL, C.D.S.P.
7. GIBBS HALL, C.D.S.P.
8. OLD PRIORY, C.D.S.P.
9. ADMINISTRATION, G.T.U.
10. ANNEX, G.T.U.
11. UNITARIANS, S.K.S.M.
12. FRANCISCANS, F.S.T.
13. JESUITS, J.S.T.B.
14. DOMINICANS, S.A.C.
(a) apartments of various member seminaries

thorough examination of the educational and social objectives of these communities would be necessary before any attempt could be made regarding the design of new activity settings both from a social and physical viewpoint. Creating physical settings without first examining the social context was felt to be unwise.

In attacking this problem, a number of strategies were followed. Periodic meetings were held with the various heads of institutions, individually and collectively, in order to establish agreement upon the major issues to be dealt with and to determine the most pressing needs of each individual seminary institution. In addition, a questionnaire was circulated to the students and staff which sampled over 45 percent of both groups. The survey covered educational objectives, searched for programmatic requirements, and gathered data concerning the evaluation of the existing physical plant of the GTU. An inventory of existing buildings and spaces together with a space utilization analysis was also conducted.

Response to the survey questionnaire revealed some of the environmental deficiencies of the present seminary complex, and introduced issues critical to facility planning. Sixty percent of the students and 70 percent of the staff felt that future changes in the physical plant and facilities should be planned to increase the identity of the entire seminary community. Only 28 percent of the students desired increased identity for the individual seminaries; whereas 55 percent of the staff desired it. As can be seen by the following graphs (y-axis represents percentage response; x-axis represents attitudes) students and staff favored physical changes which would substantially increase the interaction between them, both within their own seminary and with other seminaries.

As revealed by the survey, approximately one-half of the population resides at least one mile or more from the GTU area. Thirty-nine percent of the population is married and many are employed in outside jobs. The living arrangement coupled with the marital and job status creates a situation where personal, sustained contact with the GTU community is not easily achieved. Students come to classes and then immediately leave for home or their jobs. Most of them, then, do not have a place to secure their belongings, study between class periods, or utilize during late night hours before term papers are due and examinations occur. In the evaluation of the existing physical environment, a large number of vacated and underutilized spaces were discovered in the buildings owned by the member seminaries. These two factors: 1) the lack of personal space on campus; and 2) the excessive amount of underutilized, vacant space proved to be significant as the analysis unfolded.

It became more apparent that some activity situation had to be designed and implemented that would begin to touch on the above issues so that the goals of increased interaction between and among the students and staff could be realized. Moreover, the new activity setting or settings should begin to resolve the problems brought about by the residential separation of the students and staff, the marital and employment status of most of the students, the lack of personal space at the GTU, and the excessive amount of underutilized building area.

As the analysis progressed, it became apparent that studying was a very important mediating activity between the student's place of residence, the classroom, and the library--the linking pin between the student and his educational resources: <u>i.e.</u>,

9. DESIGN LANGUAGES AND METHODS / 497

TABLE 1
ANALYSIS OF GTU STUDENT AND STAFF ATTITUDES

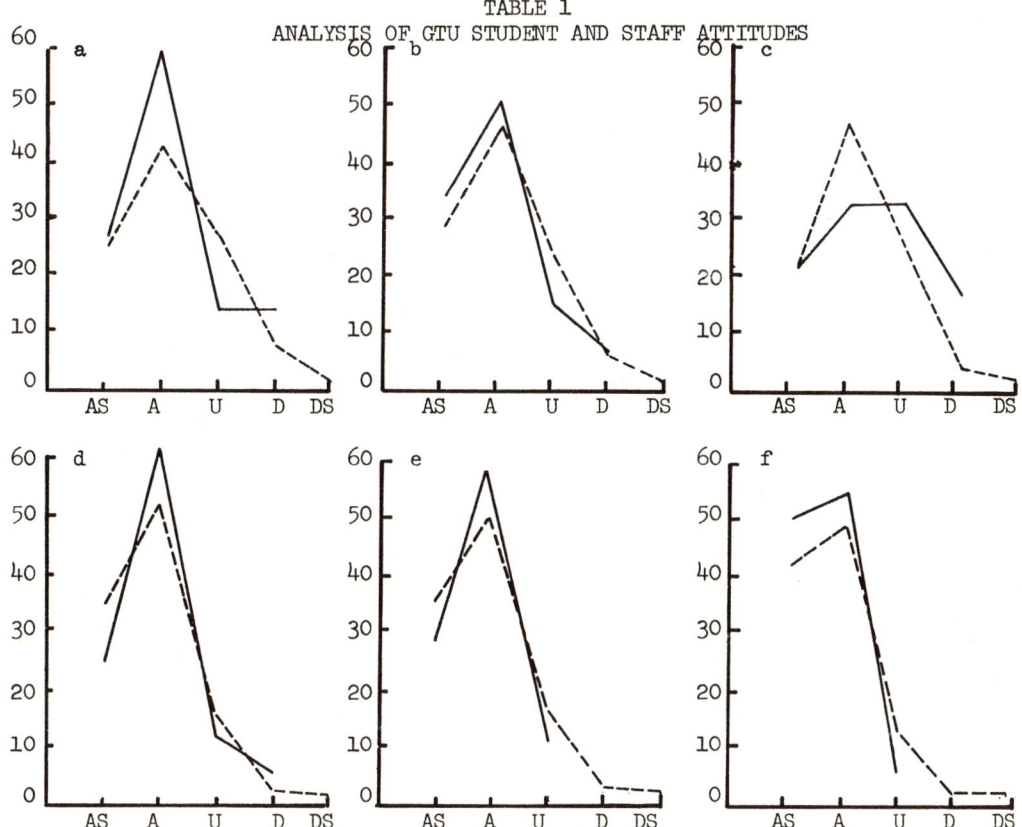

Note.--Solid line represents faculty attitude. Broken line represents student attitude. X-axis represents attitude: AS (agree strongly); A (agree); U (uncertain); D (disagree); DS (disagree strongly). Y-axis represents percentage response of both populations. Following are the various attitudes examined.
a The identity of the entire seminary community should be increased.
b The identity of the GTU administration should be increased.
c The involvement of the neighboring community in the different seminary activities and facilities should be increased.
d Future changes should encourage your interaction with fellow seminarians.
e Future changes should encourage your interaction with seminarians from other seminaries.
f Future changes should encourage your interaction with faculty.

his books, teachers, and colleagues.

The process of studying is really an information exchange process. Transmission of ideas and information occurs between the student and other students, between the student and his teachers, or between the student and the printed document. This information exchange process, if it is to be effective, must rely on a rich assortment of medias and experiences. Informal conversation and browsing are extremely important elements in the exchange process. In all graduate institutions, as in the GTU, creative thinking depends a great deal on the number, frequency, and length of contacts the student maintains with his colleagues and professors. This is especially true for scientific research organizations:

> The variety of ways by which scientists seeks information important in their work is illustrated by answers given by fifty U.S. scientists in a survey made by Glass and Norwood a few years ago: from journals regularly scanned, 30.4 per cent; from citations in other papers, 10.9 per cent; from author reprints, 5.8 per cent; from abstracting or indexing services, 6.4 per cent; from compendia, 4.3 per cent; from casual conversation, 22.6 per cent; by asking colleagues, 8.1 per cent; other methods, 11.5 per cent. Closely similar results have emerged from other inquiries into the problem.... Discussion with colleagues takes up at least as much time as consultation with formal information. (2)

The GTU student already possesses one degree and usually a few years experience working in some field of endeavor. Graduate study is not so much a disciplinary process as it is more an exploratory discovery process where creative thinking is the rule rather than the exception. Therefore, dialogue between the students and his colleagues and staff is just as relevant and important as the personal contact the student carries out with his/her documents.

Once it was decided that the notion of dialogue was an indispensable, significant factor in the information exchange process, it was proposed that group study settings could perhaps become an appropriate mechanism for generating the desired interaction among and within the seminaries. Related to this concept, which will be referred to as study sets for the rest of this analysis, the following corollary conditions were formulated:

1. That each study set should be in an area accommodating at least two persons but no more than four persons.

2. That a collection of study sets should comprise no more than 15 persons. A lounge-seminar discussion area should be provided adjacent to each collection of group study spaces.

3. That the study sets should be near the common library but separate from it. They should be adjacent and interspersed among faculty and staff offices as well as classroom and other meeting spaces.

4. That all the study sets should be located in existing buildings in and around the community and not be concentrated in one single area. The individual

spaces should be personalized and remodeled by the occupants.

The rationale and implications of these statements is as follows:

1. That each study set should be in an area accommodating at least two persons but no more than four persons.

The importance of dialogue in the study process has already been discussed. Individual study spaces would not provide the setting appropriate for such dialogue. Groups perform a number of different functions and offer a number of opportunities and advantages including the following:

  a) An outlet for affiliation needs, that is, needs for friendship, support, and love.

  b) A means of developing, enhancing, or confirming a sense of identity and maintaining self-esteem. Through group membership, a person can develop or confirm some feelings of who he is, can gain some status, and thereby enhance his sense of self-esteem.

  ...One of the commonest findings which comes from the study of groups in organizations--and which, incidentally, is a reason why organizations are so much more complex than traditional organization theory envisioned--is that most groups turn out to have both formal and informal functions; they serve the needs of both the organization and the individual members. Psychological groups, therefore, may well be the key for facilitating the integration of organizational goals and personal needs. (3)

Individual offices are not only less desirable from a sociological and psychological viewpoint, but are also more expensive to provide. With the study set concept, the surplus existing spaces in GTU buildings could become occupied more efficiently. What, then, should be the group size these study sets accommodate?

The two-person group size was chosen as the lower limit for the study set simply because it is the smallest group size possible. Two-person settings, however, are not necessarily the most ideal, especially if the individuals are incompatible with each other--all communication will be at a standstill. Three-person settings are more dynamic than two and in some ways more desirable. Research on scientists in organizations has revealed rather interesting results on the effectiveness of the number of contacts scientists have during their research work. Pelz and Andrews, in their book SCIENTISTS IN ORGANIZATIONS, have noted that scientists seem to have a better performance output in terms of articles written, scientific contribution and overall usefulness when the number of contacts is limited to three sources rather than two or four. (4) The three sources being the scientist himself and two other colleagues. Although three seems to be an optimal number, often in a triad one of the individuals can become isolated with rather undesirable results of exclusion from dialogue with the other two persons. The maximum limit of four persons was suggested so that deviations from the optimum figure could be tolerated. Beyond four persons, the complexity of communications starts to accelerate very rapidly. Hare's formula for the potential number of relationships between members of various sized groups: (5)

$$X = \frac{3^n - 2^{n+1} + 1}{2}$$

yields a sharply ascending curve beyond n=4: (6)

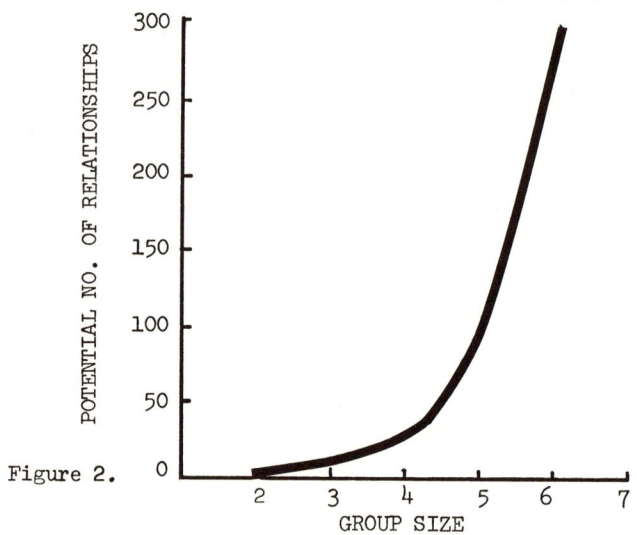

Figure 2.

Although none of the above information is thoroughly conclusive, it seemed safe to assume that a range of 2 - 4 persons in a study set area would not be undesirable.

2. That a collection of study sets should comprise no more than 15 persons. A lounge-seminar discussion area should be provided adjacent to each collection of group study spaces.

Once the study set was established, the next consideration was the maximum size for a grouping of study sets. In reviewing the literature and from discussions with the staff and students, 15 persons was suggested as the upper limit for a collection of these group studies.

For those persons in a collection of study sets, a common area is needed to serve as a lunchroom, coffee lounge, typing area, relaxed study area and as space for other activities which would adversely affect those doing contemplative work. Another important possibility which the common area could allow would be classroom discussions. Many GTU classes were being held in the apartments and houses of the students for less formal, unstructured settings. Therefore, it was important to specify the size which the area would serve so that the collection of study sets would not lose their identity and sense of community. Two significant factors assisted in determining the upper limit.

First, in an informal group discussion, the probable number of persons not speaking up increases as the size of the discussion group increases. In fact, for groups smaller than 24 persons, the size of the gathering n, roughly equals the probability figure p of those not speaking up. (7) If we take k as the number of those not speaking up, then:

$$k = p(n)$$

Thus, if $n = 20$, then $p = 20$ percent, and $k = 4$ persons. Again this seems to hold only for groups under 24 persons. For larger groups, the probability is somewhat less than n. Thus, beyond 15 persons, communication is much less democratic. Besides this communication factor, the physical problems of accommodating groups larger than 15 persons becomes physiologically uncomfortable and physically difficult. (8)

Second, class sizes at the GTU community were diminishing to afford more discussion and fewer lectures. Fifty-eight percent of all class sessions were composed of 15 persons or less, see Table no 2. The spatial utilization study also revealed a deficiency in adequate seminar space. Therefore, increasing the supply of seminar spaces would benefit the seminary educational system.

Thus the upper limit of 15 persons seemed to insure a group size which would effectively maintain the desired interaction between the participants of the study sets while also providing needed seminar space for the GTU. The lounge/seminar space represents one of the shared activity settings to be created for the seminary community.

3. That the study sets should be near the common library, but separate from it. They should be adjacent and interspersed among faculty and staff offices as well as classroom and meeting spaces.

This requirement was felt to be important simply because the library restricts and limits a person in his/her studying habits. The need for personal space here is essential. Many persons like to smoke, drink coffee, or listen to music when writing, thinking, or reading. A library setting usually does not permit such activities. Besides the restriction on studying habits, the hours of use of a library usually do not conform to the students' erratic and unpredictable study patterns. Therefore, it seemed essential that the study sets be separated from the library building.

As the desire for increased interaction between students and staff was requested, it seemed natural that the study sets be interspersed throughout the staff office areas. This proximity relationship would then allow for more sustained, informal interaction between the parties. Perhaps, certain staff members would even desire to share his/her office with the students.

Moving the studying activity outside the library would thus provide more flexi-

TABLE 2.

DISTRIBUTION OF CLASS SIZES[a]

Class size	Total Number	Percentage of Total	Cumulative Percentage
tutorials (1-4 persons)	34	17%	17%
seminars (5-15 persons)	85	41%	58%
small lecture (16-30 persons)	72	34%	92%
medium lecture (31-75 persons)	14	7%	99%
large lecture (76-150 persons)	2	1%	100%
TOTALS:	207 sessions	100%	

[a]source: class schedules of the GTU, Fall 1970.

bility in its use while also increasing the informal encounters between the students and staff.

4. That all the study sets should be located in existing buildings in and around the community and not concentrated in one single area. The individual spaces should be personalized and remodeled by the occupants.

A rather significant issue identified in the space utilization analysis was the above mentioned excessive amount of unused, vacant space scattered throughout the GTU facilities. The study set concept afforded an important means for increasing the space utilization of these excess areas.

In a cost comparison between providing individual carrel space in a new library building and remodeling existing space for the study sets, it was shown that the average costs for space and furnishings in a new library would be somewhere around $1400.00 per student. Remodeled space in existing buildings should cost about $300.00 per student, and this is a rather extravagant estimate. Not only would remodeled space thus reduce capital expenditures, but operating and maintenance costs could also be greatly curtailed. Late night encounters could occur outside of the library, where staffing, air-conditioning, and the flourescent lighting would not have to be in operation all hours of the night.

Remodeled space does not require new furnishings. Old rugs, chairs, desks, and old doors for partitions can more than adequately define accommodations for studying. Finally, acquisition of space in the existing buildings would allow the occupant to personalize the space according to his own tastes and canon of aesthetics. This certainly would not be possible in a brand new library building erected through gifts and donations.

Experimental Results

In order to test out more effectively the notions of the study set, two experimental settings were created by different student groups of the GTU community. Both experiments involved the students in designing their own study environments with available space in the existing buildings of the GTU.

The first group of students were from the American Baptist Seminary of the West (ABSW). They assembled old desks, lounge chairs, rugs, etc. in two adjacent rooms. Each student personalized his own space by defining a small alcove within one of the two rooms. Four students located themselves in one of the rooms while the other two and the lounge area were placed in the adjoining room. All of the students were married and all preferred studying in the group situation even though they all lived within a few blocks from the study set. The study set was adjacent to faculty and staff offices while also being close to their own seminary library. The advantages of being separated from their home environment and also the flexibility of use and personalized space were all dearly appreciated.

Another experiment carried out in the GTU administration building, provided space for 15 students plus one administrator. A separate lounge space was provided adjacent to the study spaces. The cost of converting the basement area into study space (wood studs and gypboard walls, wood doors for tables and partitions, etc.), came to somewhere around $500.00 for the entire group of occupants. The success of this experiment was similar to that of the previous one. Even more encouraging was the request by a young administrator of the GTU to have his office located with the students.

After these experiments were conducted, the student population was canvassed to determine their preference for the study set concept. Twenty-five percent of the population showed an interest in a group study space. Of those showing a preference for personal group studies, 40 percent of the sample desired a two-person office; 30 percent desired a three-person office; and around 30 percent desired a four-person office. Although this preference ordering seemed to indicate a stronger desire for two-person settings, no definite trend could be detected until implementation of the concept throughout the community; and its subsequent utilization, would reveal demand. It was felt that students should self-select themselves into the study sets rather than be assigned according to their seminary affiliation. This self-selection process would allow students from various backgrounds to work together with other students who shared the same interests and values. This would hopefully result in a rich intermixing of students, staff members, and seminary affiliations and serve to create a very healthy attitude and image for the GTU community.

Conclusions

This rather modest demonstration of piecemeal social engineering seemed to be quite effective in resolving some of the issues and problems faced by these communities. Other issues and problems which existed in the context of planning for the GTU are discussed in two other documents. (9)

One of the most important aspects of this study is the viewpoint which it followed: that social and physical change should proceed in a piecemeal manner rather than in a utopian large scale design. To design for change in the manner often advocated by systems analysts and designers is not very practical or realistic. The resources available frequently do not permit an extensive analysis where all the objectives can be identified and assigned weights of importance, all the alternative solutions uncovered, and the optimal solution chosen.

Braybrooke and Lindbloom in their book A Strategy of Decision examine the futility of the systems approach and other utopian engineering schemes. They propose a similar process to piecemeal social engineering called disjointed incrementalism. (10) Proceeding in a piecemeal manner by constructing testable concepts of change and adaptation in the environment for a group of communities can provide a scientific foundation for the creation of a suitable social and physical environment for these communities. This piecemeal process parallels the process in science where testable hypotheses are either shown to be valid or invalid for a particular situation. In the same piecemeal manner, formulating knowledge of how the social and physical environments operate and interrelate can begin to better describe how the social and physical design of communities should proceed.

Notes

1. Karl R. Popper, The Poverty of Historicism (New York and Evanston: Harper & Row, 1964), p. 58.

2. B.C. Vickery, Techniques of Information Retrieval (Hamden, Connecticut: Archon Books, 1970), pp. 7-8.

3. E. H. Schein, Organizational Psychology (Englewood Cliffs, New Jersey: Prentice-Hall, Inc., 1965), p. 70.

4. D.C. Pelz and F. M. Andrews, Scientists in Organizations: Productive Climates for Research and Development (New York: Wiley, 1966), pp. 16-21.

5. P. Hare, Handbook of Small Group Research (New York: The Free Press, 1962), p. 229.

6. Ibid., p. 229.

7. Transaction of discussion with R. Seaton Concerning the data from B. Bass, Organizational Psychology (Boston: Allyn and Bacon, 1965), p. 200.

8. C. Alexander, S. Ishikawa, and M. Silverstein, A Pattern Language Which Generates Multi-Service Centers (Berkeley, California: Center for Environmental Structure, 1968), pp. 265-269.

9. P. Quinn and F. Oda, GTU Master Plan (Berkeley, California: GTU Press, 1971).

10. D. Braybrooke and C. E. Lindbloom, A Strategy of Decision (Glencoe, Illinois: The Free Press, 1963), p. 82.

SYNTHESIS IN DESIGN - AN INTERDISCIPLINARY ESSAY

Pattabi G. Raman

Department of Architecture
University of Edinburgh

Abstract
Methodological works usually view the design process as a problem-solving activity, and this in my view is a simplistic notion. No design problem has only one answer and design synthesis is not as much a function of externally specified input as it is believed by methodologists. It is not the designer's unannounced thoughts which are important in themselves, but what he consciously does with them. This is true of all forms of thought which have a synthetic content. I have attempted to analyse design synthesis, fully recognising the designer's selectivity in receiving and structuring information.

Introduction
This paper began in my own mind as a reaction against methodological literature which relies heavily on what is termed 'systems approach'. The argument for systems approach has always been that there are Gestalten qualities of living organisations that are unlikely to be revealed by the ordinary modes of scientific analysis (1). Methodologists were right in thinking that such a view was needed in the domain of architectural design. My own dissatisfaction with the works that draw heavily from systems theory stems from this fact: systems theory is mostly concerned with analysis of living organisations, whose goals are seen as capable of formulation in a tangible way. But architectural design is concerned with discovering spatial organisation in response to certain needs; it involves problems of prediction and formulation of such needs. Furthermore synthesis, which characterizes the design process, makes it an activity that is different from that encountered in systems theory.(2).

Because of its synthetic content the thinking that leads to the formulation of a theory in pure science would seem closer to design thinking. Although the subject-matters pursued by scientists and designers are quite different, they are similar at least in so far as they both are concerned with ways of dealing with complexity. I have attempted to examine this similarity and to link the ideas emerging from such an examination with psychological notions that explain the information structuring process and with philosophical views on form and content. Wherever possible, I have attempted to illustrate the ideas with real examples of design synthesis. As can be seen the ideas emerging from each field are by no means complete on their own. They tend to overlap considerably, yet taken together they provide a useful lesson for the designer.

9. DESIGN LANGUAGES AND METHODS / 507

A Scientific Analogy

The usual criticism levelled against design methodologists is that they are too preoccupied with the direct application of techniques from other fields. Our concern here is with analogies rather than with direct application of techniques. Analogies help us to understand new ideas by linking them with notions that are already well established. Furthermore, if new ideas in a relatively young field are fragmented, by means of this link with older ones in established fields we are led to see these new ideas in a related way.

Much of recent works on design methodology is reminiscent of Baconian induction. Bacon believed, as Ernest Hutten (3) has suggested, in "putting Nature to the question and to torture her into giving up her secrets". The state of affairs in design methodology is something similar to this. It is quite unnecessary here to introduce the problem of induction. Royston Landau (4) has already considered its relevance to the design process and indeed his observations offer a point of departure. Landau asserts that conceptual approaches to design bear a close likeness to accounts of scientific methods. He then draws an analogy between design in architecture and theory in science. Rhetorical though this observation may seem to some methodologists, it contains a remarkable lesson for the practising designer. The line of thought that leads to the formulation of a scientific theory has received considerable attention by philosophers and scientists and their accounts by analogy offer some help for a better understanding of the thinking involved in design. In three respects such accounts on scientific methods seem relevant to design process.

First is the question of attitude to methodology. As Sir Peter Medawar (5) has pointed out, "if a scientist is more or less successful in the enterprise he is engaged on, he attributes it to having enjoyed more or less of luck or learning or perceptiveness or flair, never to the use or misuse of formal methodology.". It would seem then that the widespread indifference of practising architects to design methodology is not entirely unjustifiable. Secondly there is the plausibility of the idea that one must approach a problem without any preconceived notions. Modern psychology teaches us that not everyone appreciates data in the same way. Again according to Sir Peter Medawar "the initiative for the kind of action that is distinctly scientific is held to come, not from apprehension of 'facts' but from having an imaginative preconception of what might be true." (6). In contrast design teaching at present stresses that preconception must at all costs be avoided. In the context of a reaction against beaux-arts formal tradition this stress is admirable. However since this reaction could now be claimed to have achieved its objectives, the problem is how can we in our design efforts go beyond the irrelevancies of daily life to discover built forms that satisfy needs of ever increasing complexity.

Finally there is the question of how science copes with complexity. In science the formulation of a new theory and the attainment of a higher level of abstraction **seem to go together** (7): we know, for example, that Relativity has brought about a higher level of abstraction in the description of the physical world. But in the domain of architectural design, with one or two exceptions, the tendency recently has been to shy away from the very word abstraction. Abstraction as applied in science and architectural design will be discussed in greater detail later on in this paper. In science more often than not complexity is resolved by idealising a situation (for example, experiments of unifactorial design, omission of friction in analysing certain problems in mechanics), at least initially, in the hope that a formalism capable of describing the situation in greater complexity may later be found (8). The idealised view of design is not altogether unfamiliar; indeed the whole way of working from the particular to the general comes under this category. To give a specific example, systems building are instances where the idealised view is employed. Systems building are situations where at first one concentrates on the ease with which components can be manufactured, transported and assembled on site in minimum time, and deprives the building of other factors, very often those that constitute the environmental quality (9). As a first approximation there is very little wrong with this, and the success or failure of systems building will eventually depend on how well the other factors which were omitted in the first instance are integrated into the building, even if these are considered separately. But problems for systems building begin when the idealised view is allowed to remain a fixed conception of design. In this respect, the nature of scientific idealisation has some clues for the designer. Very few idealisations remain a fixed conception in physical theory. They are formulated as a temporary measure and then discarded as soon as possible (10).

The Problem of Synthesis in Design
Although much of what has been presented so far is critical, my intention is not to suggest that methodology has no place in design activity. Indeed the contribution made by research into design methodology to the teaching of design is well known and it might well be that the purpose of most methodological works is to offer some guidance to the teacher. Furthermore, if we consider the conventional description of the design process - analysis, synthesis and evaluation - methodologists say very little about synthesis and at the most their accounts are full of perplexing phrases like 'creative leaps', 'flashes of insight' 'inspired guesswork', etc.

Here, however, a reference should be made to Alexander's "Pattern Language" which is perhaps the first work that recognizes the problems of synthesis in design (11). Patterns are a collection of already synthesised bits of a design problem which the designer uses to synthesise the total form he requires. In other words the process is a synthesis of ready synthesised bits of a design problem. The question posed by Royston Landau of whether architectural design is simply a summation of facts or whether like theory in science it goes beyond this is relevant here Furthermore, anyone who has had the personal experience of designing an artefact will know that he always operates under conditions where the information available is much less than that which will eventually be contained in the designed artefact. Indeed this is why all designs call for a certain amount of invention.

Discussion of synthetic aspects of design becomes difficult because it involves the introduction of the characteristics of the designer into the design process. Any discussion of synthesis in design calls for a more dynamic view of design where there is a genuine interaction between the designer and the nature of the problem.

Synthesis in Design : a Psychological Viewpoint
Accounts of synthesis usually include the two extremes of a) attempting it directly from the definition of parameters (e.g. Alexander's method) and b) deliberately stimulating the intuition through the use of techniques like synectics (12). Neither of these has found much application in practising designers' offices. Somewhere in between these extremes lie many possible gradations of approaches which in fact most designers use. It is the task of researchers in design methodology to attempt to describe these. There is a need to study the structures of design thought through the dissection of real examples of synthesis and through systematic links with some established theories of perception. The selective and interpretative nature of perception as it exists in design activity must be emphasised and illustrated with examples of synthesis in design.

Psychology of perception demonstrates in various ways that what is perceived depends not only on the objects perceived (13) but also on the mental state of the perceiver, that is his expectations and preconceptions about the objects perceived. The term 'schema' as used by Head and Bartlett is one of the terms which illustrate this fact. Bartlett's work (14) on memory first introduced the notion of 'schema' into psychology. The usefulness of this term is well known. As Oldfield has put it "Perhaps the chief merit of this has been that it enables us to move with cenceptual ease among a great variety of psychological phenomena, many of them very obscure, and reaching far beyond memory itself." (15). In the field of architectural design it does provide us with a conceptual aid for exploring on the one hand the designer's characteristics, that is his personality, and on the other hand the expectations of the user.

Schema can be thought of as "habitual perception" (16) and such a notion aids our understanding of the relationship between old and new information. It helps us to consider how past information predisposes us to act in certain ways rather than in others (17). The way children learn to draw provides a good example of schematization. Studies on babies' eye movements reveal that at the earliest age babies are interested in the human face more than in anything else. Jane Abercrombie has suggested that babies build up schemata of people's faces and use them for interpreting the world (18). Although there seems to be a controversy about how much the early drawings of children are based on their observation of faces, it is clear that children's first drawing of animals are not very different from those of humans. A human schema made with ears on top of the head becomes a rabbit or a bear (19). Figure 1 is an example of this. In a drawing of Red Riding Hood and the wolf by a four-year old (Figure 2) the only difference between them is the number of legs (18). In illustrating children's drawings of houses Jane Abercrombie goes even further (18). She observes that next to humans children like to draw their own houses. These drawings resemble human faces very closely; they tend to be symmetrical, the mouth becomes the door, the windows the eyes (Figure 3).

Figure 1. Humans or Animals ? (Drawing by a four-year old)

Figure 2.

Figure 3.

9. DESIGN LANGUAGES AND METHODS / 511

It is extremely difficult to observe schematization as an identifiable process in architects' works. Obviously, the phenomena involved are much more complex. Again as Oldfield has put it "Bartlett's methods do not lend themselves much to further refinement in the direction of systematic quantitative experimental design. They remain somewhat esoteric, their precise character less important than the conclusions of the man who devised and used them"(20). It hardly seems worthwhile making attempts to pin down the 'schematization' in architects' works. Here certainly is a familiar conflict between relevance and objectivity. It seems that no direct testing by physical manipulation of the outside world is possible for an architect's 'schemata' other than simply discussing them and comparing and contrasting them with with those of other architects. The following is some evidence for the fact that schematization is at work in many architects' works.

The notion of 'schematization' becomes explicit in many of Ponti's works, particularly in his conscious efforts to achieve a **closed form (21)**. When Norman Foster was describing his scheme for the Norwegian shipping company amenity centre at the Royal Institute of British Architects, he explained how the scheme had its origins in his scheme for the school at Newport, which in turn was influenced by Californian SCSD work (22). Giedion's suggestion that Le Corbusier always deliberately sought for experiences which eventually found their way into his design solutions can also be regarded as an illustration of the notion of 'schematization'. Giedion suggests that Le Corbusier "... always looked for the experiences of former times in his travels and he was equally interested in crystalline Greek forms and in the forms of Roman vaults or Islamic and Gothic architecture. His search for inner similarities had nothing to do with art history: it embraced the experiences of the entire architectural development. It is no accident that the tower of Ronchamps has been compared with a primitive cult structure" (23). Aalto and Pietila find the Finnish landscape full of ideas for built form, and it is sometimes suggested that deriving built forms from landscape is done with a view to integrating architecture with the surrounding landscape. Yet Aalto's Finnish pavilion for the New York 1963 Fair, the Dormitory at MIT and Pietila's entry for the Finnish Embassy competition in New Delhi, all with their forms derived from Finnish landscape, demonstrate that this is not a satisfactory explanation. It does however raise the important philosophical question of the relation between a form and its content. This will be dealt with later on in the paper.

It must nonetheless be stressed that schematization is not a rigid, closed process. Analysing a design problem for instance could to some extent affect our schema, yet it is quite possible that our prejudices may be so entrenched that we would never revise it. Mental rigidity in design situations can be thought of as the state of mind which acquires certain schemata and attempts to force reality at all costs into them (24). When we talk about deliberate stimulation of intuitive leaps we simply mean that we give ourselves a chance to make new use of acquired schemata. The above examples are intended only to illustrate the notion of schematization in an architectural context. The purpose is not to launch a new method, less still to suggest that the efforts of those architects are worth imitating. The aim is simply to point out that the designer's mind is not a clean slate in which the nature of problems inscribes their solutions automatically. From the users' point of view, they too build up expectations based on their own past experience. There might be a gap between users' expectations and the designer's schemata. Competent research in design methodology ought to investigate this.

Design Synthesis - a Viewpoint Based on the Relationship between Form and Content

If we consider architectural design as attempts to discover built forms in response to needs, much philosophical discussion on form and content (25) seems relevant. Knowledge has its origin in the recognition that familiar things assume different forms. Water, for instance, exists in the form of ice, snow, and so on. Electricity is another example which takes on a variety of forms. Similarly, the same form may be exemplified by different contents. Consider for instance Picasso's sculpture "Bull's Head". The bycicle seat and the handle bar have nothing in common with a bull's head except the way they are arranged. It is this arrangement that gives the form of a bull's head. Obviously here the meaning of the term 'form' is stretched beyond the usual connotation of shape and implies an orderly arrangement of parts that are found in nature and in artefacts.

Different things that exhibit the same form are analogous. Analogy is a means by which we represent one thing using another which has similar structure. When we draw the floor plan of a building or a street map we draw what can be called the 'logical picture' of something. The logical picture differs from the oridinary picture or photograph in that it need not look in any way like the original object. The relationship between the original object and its 'logical picture' is not that of a copy, but of analogy. Indeed the more like the reflections of the original objects street plans and building plans look, the less useful they are. Floor plans and maps are useful not only beacuse of what they show, but also because of what they leave out. The recognition of a form which several things have in common is termed 'abstraction'.

Generally speaking forms of things are discovered in two ways: (a) by abstraction, that is through the recognition of common forms from instances which nature as well as artefacts provide and (b) by reinterpreting forms we already know. The latter way is usually easier, since our search is confined to forms that are already known. But the power to recognize a common form in a chance collection produces innovative design (and scientific breakthroughs). This is simply another way of saying that inventiveness depends on one's ability to make new use of acquired 'schemata'.

Architecture Science and Methodology: Points of Convergence

As mentioned earlier, in science the search for a new theory and the attainment of a higher level of abstraction seem to go hand in hand. Abstraction is in fact consideration of form apart from content (26). The use of models has played an important part in the construction of abstract theories in science. A model offers certain structural analogies to the situation for which it is a model. In physics, for instance, a variety of phenomena varying from mechanics to quantum mechanics can be seen through the use of the mechanical oscillator as a model (27). The oscillator is simply a mass connected to a spring held firm at the other end. Any periodic process can be conceived with the aid of this model. Quite clearly on no occasion can we ascribe to the model aspects of the actual situation for which it is a model. Again, as was the case with the logical picture, the relation between a model and the actual situation is not that of a copy but of analogy. In fact it is a prerequisite for anything to be a model that it is not the same as the phenomenon for which it is a model.

Thus the model in physics is very much the same thing as Pietila's landscape abstractions (or "hand-made landscapes" as he called them) from which he derives his built forms. When he presented his "Approach to Architecture" at the Royal Institute of British Architects (28) Pietila described his abstractions as follows: "They are prototype landscapes of Finland seen as a national romantic photographer sees his native country. Or, they are imaginary shapes designed to present Finnish landscape. Further, there are abstract forms that one can see as fictitious spaces and at the end there are diagrams or graphs simulating some environmental space characters.". In either case the object is, as Ernest Hutten (29) has put it, to "penetrate beyond the likenesses of daily life to discover patterns - structure and function - that exist in the world and, in particular, those that otherwise lie hidden in our inner 'unconscious' world of feelings.". This is crucial since anyone who has had the practical experience of designing knows how difficult it is to release oneself from the mental block imposed mainly by the weight of information related to the problem. Aalto describes his mental state when he is designing like this:

> "Whenever I have to solve an architectural problem, I am inevitably held up by the thought of its realisation - it is the sort of 'three o'clock in the morning-feeling', probably due to the difficulties caused by the weight of the different elements at the moment when the design is being carried out.
>
> The social, human, technical and economic demands which are found alongside psychological factors and which concern each individual and each group, their rhythm and the effect they have on each other, are so numerous that they form a maze which cannot be worked out by rational methods. The ensuing complexity prevents the basic architectural idea from taking shape.
>
> In such cases I proceed in an irrational way as follows: for a moment I forget all the maze of problems, I erase them from my mind and busy myself with something which can best be described as abstract art. I start drawing, giving free rein to my instinct, and suddenly the basic idea is born, a starting-point which links the numerous, often contradictory elements already mentioned, and brings them into harmony with each other.
>
> While designing the Municipal Library in Viipuri (I had a lot of time at my disposal - five long years), I spent a great deal of time making children's drawings, representing an imaginary mountain, with different shapes on the slopes and a sort of celestial superstructure consisting of several suns, which shed an equal light on the sides of the mountain. In themselves these drawings had nothing to do with architecture, but from these seemingly childish drawings sprang a combination of plans and sections which, although it would be difficult to describe how, were all interwoven. And this became the basic idea for the library which, unfortunately, has now been destroyed. This basic idea consisted in grouping the reading rooms and the lending rooms on different levels, like on the slope of a mountain, around a central control desk uppermost in the

building. Above everything was erected a sort of solar system - the round conical skylights." (30).

Aalto's 'abstract' work has the same purpose as that of a model in physics, it is performed for the sake of finding a physically interpretable form.

Some objections to this point of view must now be considered. It is usually argued that when an architect proceeds in what Aalto calls an "irrational way" certain visual characteristics of space (which are probably very personal) guide his search. Here, as C.H. Waddington has suggested, we need to distinguish between the two faculties that are in action when we perceive. In his study of the relations between painting and the natural sciences in this century Waddington uses Whitehead's account of the way we perceive to explain this distinction: "Every act of perception, he maintained, involves two modes: perception by 'presentational immediacy' and by 'causal efficacy'. By 'presentational immediacy' he referred to what we have been trained to think of as the immediate data presented to us by our senses - patches of colour seen with the eyes, noises of a certain pitch and intensity, sensations of hard, rough, or soft surfaces, and so on. These are the 'sense data' of more old-fashioned philosophers, and were formerly regarded as the basic elementary facts of experience. But Whitehead claims that a more primitive element in perception is awareness, not of sheer sensations, but rather of entities which are perceived as having some potential effectiveness in the world. He argues that it is more elementary to perceive causally efficacious things, such as a chair (something suitable for sitting in) or a table (somethings to put things down on), and that it is a relatively sophisticated business to 'see' such things as mere coloured patches or other pure sensations; in fact, it takes training - ..." (31).

The way we use a graph is a simple example of perception by causal efficacy. When we look at a graph the phenomenon that it represents is more important than the picture itself. It is doubtful whether images of any kind are ever used by architects like Aalto purely for their presentational immediacy. In architectural design abstractions are attempts to strip bare certain aspects in order to emphasize certain others, particularly those that written language fails to transmit. It seems justifiable that architects should use abstraction as a means of giving rein to their instincts. Indeed there are enough examples which illustrate clearly that their abstract constructions do simulate certain environmental characters and are not just images.

Idealisation and abstraction are devices which the scientist and the designer employ for dealing with complexity. Idealisation is used for simplifying the state of affairs to be considered by omitting some of the parameters needed to describe it fully. One of the dangers of idealised view in the design process arises out of a temptation to allow the idealised view to remain a fixed concept. This is partly beacuse parameters of design are not yet capable of being formulated as clearly as their counterparts in science and therefore their omission or reintroduction into the design process at will is that much more difficult. Abstraction is an instrument we use to strip off irrelevancies in order to lay bare more structure (32). Abstraction has its pitfalls as well. As Morris has put it: "One of the dangers of the use of models in science, for instance, arises out of

9. DESIGN LANGUAGES AND METHODS / 515

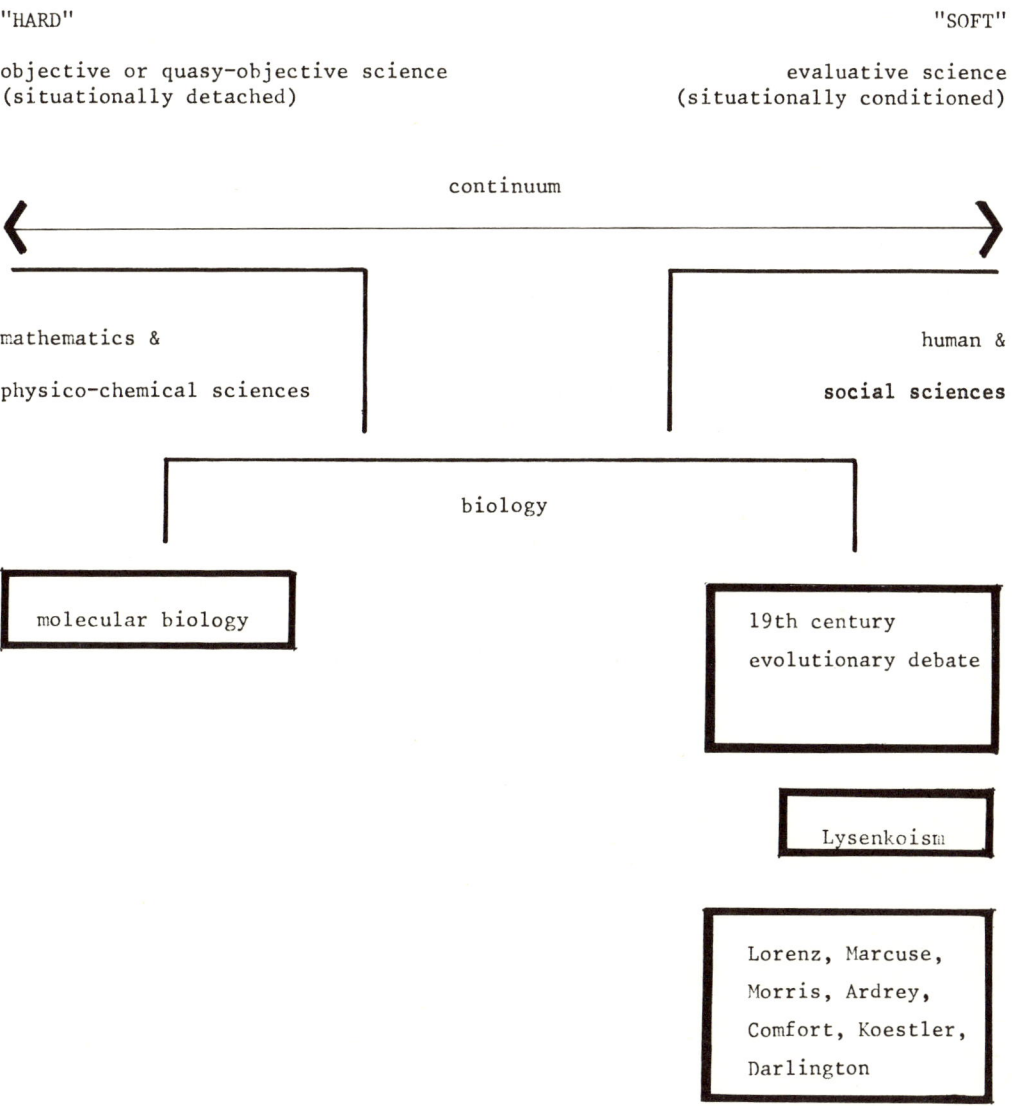

Figure 4.

the temptation to ascribe to the subject matter of a theory properties of the model illustrating the theory which are not involved in the theory itself." (33). Similar difficulties exist in the realm of the design process too. Examples of 'media take-over' in architectural design are well known. In physics models render phenomena picturable; however modern physics does encounter conceptions that are not picturable through the use of models. In such cases the situation is mathematically described (34); and this is made possible by a clear formulation of the parameters involved. It would appear then that the study of design process needs to be two-pronged. Firstly we should search for effective instruments that would enable us to go beyond mere consideration of facts to discover built forms matching our needs which are all the time increasing in their complexity. Secondly we should record a great variety of needs that are relevant to design and attempt to classify them.

Architecture Science and Methodology: Points of Difference
Similarities in approaches, especially in their ways of dealing with complexity, may be better understood if some of the differences are identified. The basic difference between theory in science and design in architecture is obviously the problem of verifiability. Verifiability and closer definition of parameters go hand in hand. It can only be hoped that the present interest in evaluative techniques will lead to an equally vigorous quest for establishing the parameters involved in design. From this major difference others emerge. In physics, for instance, models once formulated become universally accepted as representing certain processes. As Ernest Hutten points out: "The model is a bit of an established, known theory - and so we obtain an equation which is then interpreted in terms of the new theory. The new interpretation is linked to the interpretation in terms of the old theory by the Correspondence Principle." (35). Architectural abstractions are not like this. Also, in the field of design it is difficult to use mathematics for describing the various parameters. These differences do not necessarily reflect the backwardness of this field, but rather the complexity of problems belonging to it.

Figure 4 is a diagram devised by Jonathan Benthall (36) to show how different branches of science fit into a continuum between the objectivity of mathematics at one extreme and all ideologically-based thinking about man and society at the other. If architectural design is to be regarded as a science it should be accepted that its location in this continuum is somewhere near the soft end.

Concluding Remarks
Devising ways to deal with the complexity of issues that arise in design is the subject-matter of all methodological works in architecture. By analogy with how science copes with complexity we have the terms 'idealisation' and 'abstraction'. Idealisation involves simplifying the state of affairs. Abstraction is concerned with accentuating the dominant factors, which obviously involves cutting of information no longer relevant. Both idealisation and abstraction are frequently used in the domain of architectural design.

9. DESIGN LANGUAGES AND METHODS / 517

From the psychological standpoint we have th term 'schematization' which effectively accounts for the designer's selectivity in receiving information. Idealisation, abstraction and schematization are all in a sense terms that are conncetedwith the structuring of information. Schematization involves unconscious selectivity in the structuring of information. Both idealisation and abstraction are attempts where selectivity is made explicit.

When an abstract approach is used, the main aim is to release oneself from the complexity of information about a problem. In idealisation we concentrate initially on a few factors only and consider other factors separately later. In approaches where neither idealisation nor abstraction are used the designer's structuring of information can be said to be entirely unconscious.

We all have our preconceptions which we use all the time for interpreting the world. It may be worthwhile exploring ways of making these preconceptions conscious and explicit.

Notes

1. Emery, F.E., Systems Thinking, Harmondsworth, 1969, p.7.
2. In his article 'Operations Research and Architecture' Churchman argues that O.R. problems are not design problems. See Architectural Design, September 1969.
3. Hutten, Ernest H., The Origins of Science, London, 1962, p.197.
4. Landau, Royston, Towards a Structure for Architectural Ideas, Arena, June 1965.
5. Medawar, Sir Peter, Induction and Intuition in Scientific Thought, London, 1969, pp.8-9.
6. ibid., p.51.
7. Hutten, Ernest H., The Ideas of Physics, Edinburgh, 1967, p.62.
8. ibid., p.77.
9. Broadbent, Geoffrey, Systematic Confusion, The Architects Journal, 6th January, 1971.
10. Hutten, Ernest H., The Ideas of Physics, op.cit., p.80.
11. Alexander, Christopher, et al, Houses Generated by Patterns, Berkeley, 1969.
12. See, for instance, J.C. Jones, Design Methods, London, 1970.
13. Abercrombie, M.L.J., The Anatomy of Judgement, Harmondsworth, 1969.
14. Bartlett, F.C., Remembering, London, 1932.
15. Oldfield, R.C., Experiment in Psychology - A Centenary and an Outlook, in Readings in Psychology, ed. John Cohen, Woking, 1964, p.36.
16. Norberg-Schulz, Christian, Intentions in Architecture, Oslo, 1966.

17. Abercrombie, M.L.J., Anatomy of Judgement, op.cit., pp.30-31.
18. Abercrombie, M.L.J., Perception and Construction, in Design Methods in Architecture, ed. Broadbent, G. and Ward, A., London, 1969.
19. Kellogg, Rhoda, Analysing Children's Art, Palo Alto, 1969, p.114.
20. Oldfield, R.C., op.cit., p.36.
21. Ponti, Gio, Espressione dell'Edificio Pirelli in Costruzione a Milano, Domus, March, 1966.
22. Foster, Norman, Architects Approach to Architecture, RIBA Journal, June, 1970.
23. Giedion, S., Space Time and Architecture, Cambridge, 1967, p.578.
24. Norberg-Schulz, Christian, op.cit., p.41.
25. Much of the discussion on form and content has been abstracted from Susanne K. Langer's An Introduction to Symbolic Logic, New York, 1967, pp.21-43.
26. ibid, p.43.
27. Hutten, Ernest H., Ideas in Physics, op.cit., p.66.
   For an illustration of how a vast range of phenomena can be seen in terms of the model oscillator see pp.69,72,74, illustrations 12,13 and 14.
28. Pietila, R., Approach to Architecture, talk given in 1970 at the Royal Institute of British Architects. Paper not published but privately obtained from the author.
29. **Ernest** H. Hutten, Ideas in Physics, op.cit., p.81. In Hutten's text he in fact suggests that "the model is very much what has occurred in modern art".
30. Aalto, Alvar, Abstract Art and Architecture, in Aalto Synopsis, Basel, 1970.
31. Waddington, C.H., Behind Appearance, Edinburgh, 1969, pp.114-115.
32. Hutten, Ernest H., Ideas in Physics, op.cit., p.81.
33. Morris, Charles, Signs Language and Behaviour, New York, 1949, p.27.
34. For striking examples of unpicturable phenomena in physics see Norwood Russell Hanson, Patterns of Discovery, Cambridge, 1965, pp.123-124.
35. Hutten, Ernest H., Ideas in Physics, op.cit., p.80.
36. Benthall, Jonathan, The Sociology of Knowledge in Studio, January, 1971.

U GRAPH: A DECISION MAKING AID

Arie P. Schinnar

Program Coordinator
School of Architecture and Environmental Design
State University of New York at Buffalo
2917 Main Street, Buffalo, New York 14214

Abstract
This paper presents a graphical tool to aid groups in decision making in the context of problem-solving and program development. It is basically a U-shaped graph which serves as a reference-base on which alternative proposals, competing for the priority of being selected and implemented, are plotted; then discussed; then manipulated by prescribed resolution-strategies; and finally interpreted to establish preference of one over the other. The U graph, as the group proceeds through its various phases, leads the participants into exploring the implications of each alternative decision and in conjunction helps develop the reasoning behind the ultimate choice. The objective of this paper is to acquaint the reader with the process and its outcomes, and is too short to accommodate all of its subtleties.

Introduction
The current trend and emerging practice to increase community awareness and activate community participation in planning community programs, has brought about the use of various decision making theories in the context of group dynamics. The present engagement of the Buffalo Organization for Social and Technological Innovation (BOSTI) in the project PAK (a Planning Aid Kit for mental health program development with community participation) is one such example. Yet, inevitably, along the path of problem solving and program development entailing group decision making, one recurring symptom has been the emergence of intra-group conflicts which in turn have hampered the decision making. In addition, decision making is further weakened by the existing discontinuity in the decision making process itself.

Any decision process can be systematically described by a set of operations that lead the decision-maker from a point of uncertainty, where all alternatives are equally attractive, to a point of certainty or preference. These operations may be clustered into two consecutive major phases:

    a. Reduction of uncertainty(1), and

    b. Expression of preference.

Thus, given that the decision-maker has to choose from among various alternatives, and operating under the premise that at the start all alternatives appear to him equally attractive, the decision-maker finds himself vacillating in a state of uncertainty, the reduction of which is a necessary prerequisite for decision making. Reduction of uncertainty will therefore be achieved by employing a set of operations that help either devaluate or evaluate the relative attractiveness (or effectiveness, or performance, or...) of each alternative, with the final outcome being alternatives rated according to one or several criteria.

While reduction of uncertainty is essentially an assessment procedure, the next phase, which is the expression of preference, is a discrimination operation involving first the exploration of the implications of each decision and then the development of the reasoning behind the particularly preferred decision. Most of the available decision techniques deal very well with the reduction of uncertainty, yet none is suitable for the expression of preference in the context of group problem solving. To close this gap in the continuum of systematic decision making and to facilitate decision-making by groups, the U-graph was developed.

The U-Graph: Definition
The U-graph, i.e., a "U" shaped graph, is a simple and visual aid with a solid mathematical support(2) that makes the decision process and its implications public. The U-graph serves as a reference-base on which alternative proposals, competing for the priority of being selected and implemented, are plotted; then discussed; then manipulated by prescribed resolution-strategies, and finally interpreted to establish preference of one over the other. The U-graph, as the group proceeds through its various phases, leads the participants into exploring the implications of each alternative decision and in conjunction helps develop the reasoning behind the ultimate choice.

The U graph has three axes, two vertical and one horizontal. Each axis is subdivided decimally into ten intervals which in addition are identified by a verbal scale to facilitate discussion and verbal manipulation. (See Figure 1.)

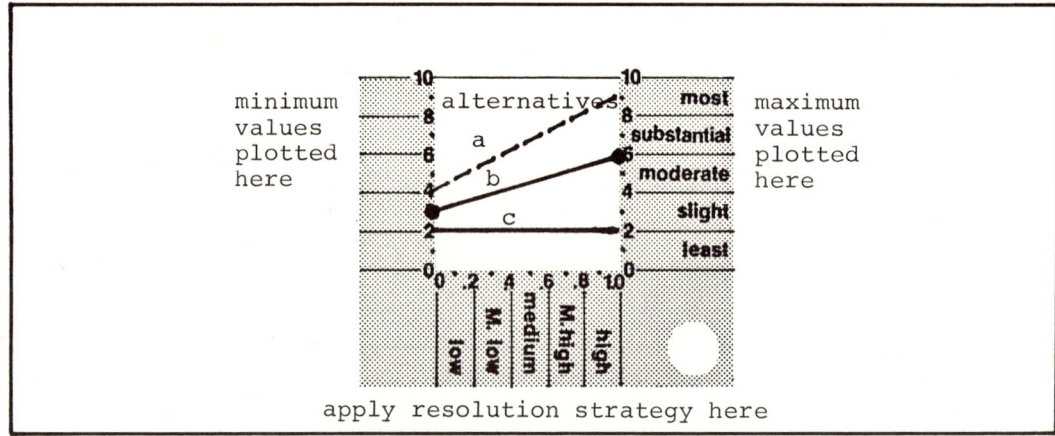

Figure 1.

Every competing alternative is represented by a linear model in the form of a segment of a line that spans between the two vertical axes. The two points formed by the intersection of that line with the left and the right vertical axes represent the respective minimum and maximum possible values of the assessed attractiveness (or performance, or effectiveness, or cost, etc...) of that alternative. Thus, lines represent alternatives, and end points of the lines represent the values assigned to the alternatives via assessment by a criterion. Whereas the vertical axes serve both to plot the various alternatives (i.e. lines) and to ultimately assess the overall attractiveness of each in order to establish the preference of one over another, it is the horizontal axis that enables the final resolution through the application of a proper resolution strategy.

The U-Graph: Operation
Operating the U-graph involves a sequencial five-step process (see Figure 2).

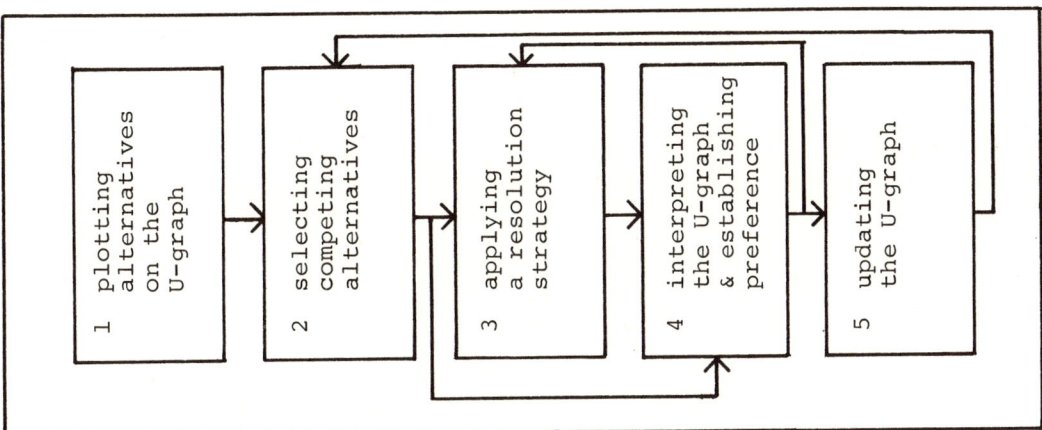

Figure 2.

However, prior to initiating any U graph activity, checking three items for compliance is a prerequisite:

a. Alternatives have to be assessed either quantitatively or qualitatively via one criterion at least.

b. The frequency distribution for each assessed alternative has to be determined by reviewing the voting pattern of the decision making group. Note should be made of the three possible voting patterns that groups regardless of their size may display: NORMATIVE or converged agreement -- a normal distribution that should be represented by one value (mean, mode, or median); EXTREME or converged disagreement -- a bimodal distribution that should be represented by two values (means, modes, or medians of the two modals) and the relative commulative frequency of each modal; and RANGE or diverged disagreement -- an even distribution that should be represented by two values (the boundaries of the range).

c. All numerical values of the assessed alternatives have to be proportionately converted into a decimal scale.

After completing this preparation of the alternatives, the U graph is introduced.(3)

Step 1: Plotting Alternatives on The U-Graph
The first step consists of merely plotting all the alternatives on a 'master' U graph which serves as a record for all the proposed alternatives and is also utilized to determine which of the alternatives are actually competing. Different voting patterns, whether normative, extreme or range, should be distinguished graphically either by different colors or line quality. Also, in plotting the alternatives, all possible maximum values should be plotted on the right axis (maximum-axis), whereas all possible minimum values should show on the left axis (minimum-axis) (4). (as shown in Figure 1)

Step 2: Selecting Competing Alternatives
The second step consists of scanning the master graph to select from among all the plotted alternatives only those alternatives which compete for the same rank order. Determining which alternatives are in fact in competition for the same rank requires:

a. Identification of the potentially most dominant alternative (the one that has the highest maximum value); then

b. Identification of its minimum value and construction of a horizontal line from the minimum value. All those alternatives that intersect the recently constructed horizontal or that lie above it are competitive with the potentially dominant alternative; and finally

c. Reploting these competing alternatives on another U graph. (exemplified in Figures 6 and 7)

However, if no competing alternatives are observed, one is to proceed to Step Four.

Step 3: Applying a Resolution Strategy
The third step consists of choosing the appropriate resolution strategy and applying it to the graph, always to the lower horizontal axis of the graph. A resolution strategy is a means of resolving a preference dilemma among competing alternatives; and even if on occasion the resolution strategy fails to clearly resolve between competing alternatives, it at least defines a solution space in the form of hypothetical 'if...then...' statements, thus helping the decision-makers make at least better educated guesses. Five different strategies have been developed to enable consideration of different circumstances as they present themselves in a group, and they can be used either individually or in alternating combinations

depending on the particular characteristics of the problem (see Figure 3).

```
Resolution Strategies:                    Combinations:
                                  1 2 3 4 5 6 7 8 9 10 etc.
   a    Attitude                  ● ○ ○ ○ ○ ○ ○ ○ ● ○
   b    Frequency Dist.             ●         ○ ○ ○ ●
   c    Independent Var.              ●         ● ● ●
   d    Dependent Var.                  ●       ● ● ●
   e    Relative Weight                   ● ● ● ○ ○ ○

Legend:
   ● ....primary strategy
   ○ ....secondary strategy
```

Figure 3.

Though each strategy requires slightly different procedures, the interpretation process in any of them is basically the same. These resolution strategies are:

a. <u>Attitude Strategy</u> -- The most basic of strategies that does not require any additional information except for the general attitude inclination of the group in regard to a particular criterion used in assessing the alternatives. That is to say that in this case the strategy enables the decision-makers treat 'attitude' as a variable in itself. By plotting this variable on the lower axis and projecting lines that intersect with the linear models of the alternatives, the points of intersection will define the relative attractiveness of each competing alternative (see example in Figure 6-b). This strategy is especially effective for testing the implications of different and divergent attitudes within the group on the ultimate decision.

b. <u>Frequency Distribution Strategy</u> -- is to be applied only in cases where at least one of the competing alternatives manifests an 'extreme' voting pattern (presented bimodally), and only when the decision-making group attributes great significance to the size of the disagreeing subgroups (i.e., size of the subgroups is treated as a variable in itself, and its impact on establishing priorities between alternatives is explored).

c. <u>Independent Variable Strategy</u> -- is applied only when the group considers from the start at least two independent variables or criteria (e.g., performance and feasibility) to determine the overall attractiveness of each competing alternative. Variables are treated two at a time by having one plotted on the vertical axis and the other on the horizontal axis to define the commulative attractiveness of each alternative (as shown in Figure 6).

d. <u>Dependent Variable Strategy</u> -- is employed when a group wants to formulate preference only with respect to one variable or criterion, disregarding in this instance the group's attitude. It does so by reintroducing a part of the variable as the dependent variable on the lower axis. It is that part of the variable that is considered the most important component of it that is extracted and reapplied as a resolution strategy.

   While all previous strategies attribute equal significance (weight) to every variable or criterion, strategy (c) enables a weighted consideration of variables.

   e. <u>Relative Weight Ratio Strategy</u> -- is applicable only in combination with either strategy (c) or (d) in which at least two variables are involved. The application of this strategy entails a readjusting of the U-graph in proportion to the weights given to variables, thus yielding the relative impact that the decision-makers want each variable to have on the ultimate decision (e.g., performance is three times as important as feasibility). (5)

<u>Step 4: Interpreting the U-Graph and Establishing Preference</u>
The fourth step consists of the interpretation of the U-graph and the establishment of preference among the competing alternatives. Preference is established when the decision-makers are able to identify dominance (superiority) of one alternative over the others. A dominant alternative is, therefore, one that even when it assumes its lowest value, it nevertheless appears to be superior to its competitors. If, on the other hand, no dominance can be established through this initial interpretation, an "if-then" sort of game is introduced. Its purpose is to maximize the relative attractiveness of each competing alternative by considering the best and worst circumstances should each of the alternatives be tried out or acted upon, and the implications of such actions. This step requires paying attention to the type of voting distribution of the values assigned to each alternative because it is this factor which determines the upper and lower bounds as well as the range of choices of values to be considered in the "if-then" exercise. If a clear preference has not been established following such as exercise, the decision makers should recycle through a different resolution strategy in Step Three.(6)

<u>Step 5: Updating the U-Graph</u>
The fifth step requires the updating of the 'master' U graph by removing from it the now-established prefered alternative with its assigned rank, and repeating steps Two through Five for the remaining alternatives.

<u>An Experimental Use of the U-Graph</u>
For a semester project in the School of Architecture and Environmental Design, SUNY at Buffalo, a team of students engaged in a comparative analysis between three alternative land development plans:

   a. conventional lot by lot development.

9. DESIGN LANGUAGES AND METHODS / 525

    b. clustered development -- uniform dwelling types.

    c. clustered development -- mixed dwelling types.

The team simulated a meeting between a potential resident, a developer, an ecologist and a representative from the community planning board. Each of the alternatives was analyzed and evaluated with respect to two criteria:

    a. The expected performance of each development plan (taking into consideration such varied factors as the attainment of social goals in terms of privacy, recreation, density, etc... and the meeting of environmental goals in terms of preservation of natural features, availability of open space, etc...).

    b. The feasibility or likelihood of implementing each development plan (taking into consideration such factors as cost of land, utilities and road development, property value, etc... and legal administration of land in terms of ordinance reinforcement, need for ordinance revision, etc...).

At this point the team made use of the U graph. A quantitative and qualitative assessment of the alternatives with respect to the chosen criteria was performed and the appropriate voting distribution was observed and recorded (Figure 4).

Alternatives:	Performance:	Feasibility:
(a) Lot by lot development	least 0.6 (normative)	medium or medium high 4.3 or 6.3 (extreme)
(b) Cluster develop. - Uniform.	moderate 4.8 or 6.0 (extreme)	medium low to medium 3.3 to 5.5 (range)
(c) Cluster develop. - Mixed.	moderate to most 6.0 to 9.0 (range)	medium low 2.2 or 3.3 (extreme)

Figure 4

Reading the content of the table shows, for instance, that alternative a (lot by lot development) received a 'least' or .6 assessment for performance in a normative type vote (converged agreement within the group), while regarding feasibility the group split into an extreme type voting (bimodal), with some assessing feasibility as 'medium' (4.3) and others as 'medium-high' (6.3).

Next, the data compiled for all three alternatives on one variable (performance) was plotted on a U graph, with the data compiled on the second variable (feasibility) left out, to be used later as a resolution strategy (see Figure 5-a) (dotted line representing range type voting pattern; continuous line representing normative;

line with circles on both ends representing extreme voting pattern). For each alternative, the maximum value was plotted on the right axis, and the minimum value on the left axis, and connected by a segment inbetween.

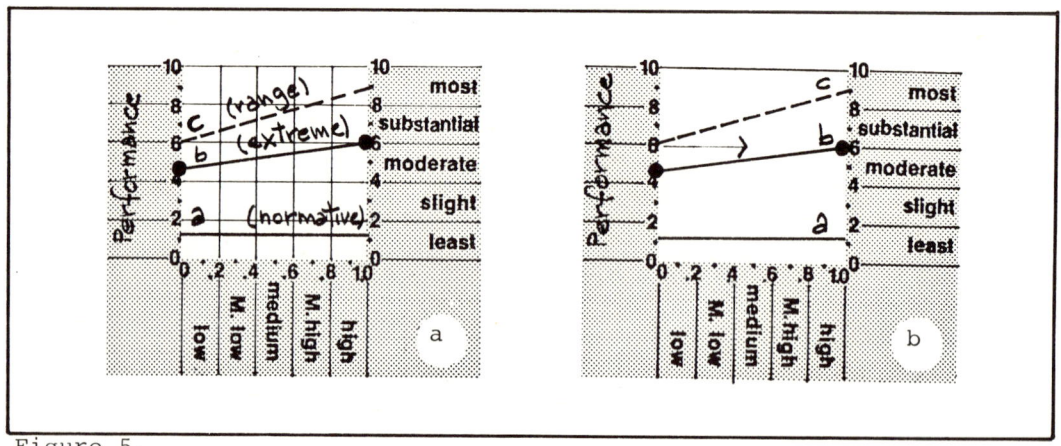

Figure 5.

Alternatives b and c were identified as competing because the construction of a horizontal projection from the minimum value of c intersected the segment representing alternative b, thus manifesting competition (Figure 5-b), while alternative a appeared clearly inferior to the other alternatives in this case. Since feasibility of implementation was judged to be independent of the performance of each alternative development plan, the team proceeded by adopting the Independent Variable Resolution Strategy. (This particular strategy requires special replotting of the alternatives on a new U graph in the following manner: both minimum and maximum values of each alternative are plotted on the right axis alone and connected with lines to the point of origin of the left axis.) (Figure 6-a)

Following this replotting, the strategy was implemented by plotting on the horizontal axis the assessed values of the feasibility of each development plan. Vertical projections then intersected their respective alternatives, with the points of intersection identifying the overall attractiveness of each alternative. For the purpose of clarity, the obtained values of attractiveness were replotted on yet another U graph (Figure 6-b).

Interpretation followed. At first glance, alternatives b and c still appeared to be highly competing even after the application of the resolution strategy. An "if-then" exercise showed that with a highly optimistic assessment of the feasibility of development plan c (i.e., when assigning it 'high' values), that alternative was competing with the moderately-high assessment values for feasibility of development plan b; and for c to top b, the feasibility of b had to lag behind that of c. Yet an inspection of the original data collected on feasibility, revealed that the lowest assessment of b never lagged behind c but rather equalled the highest assessment of c, hence manifesting a superiority for development plan b. Therefore, the team concluded that the clustered development plan with uniform dwelling types was superior to either clustered development of mixed dwelling types or conventional lot by lot development.

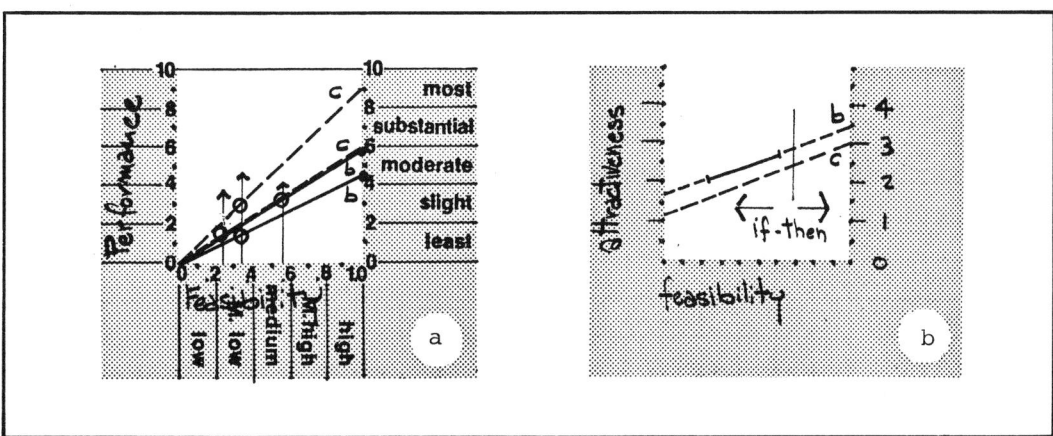

Figure 6.

Conclusion
The advantages of the decision making tool presented in this paper are emphasized when reviewing the shortcomings of presently available decision making procedures. Conflicting opinions arising from any group interaction cannot, in the present decision procedures, be accommodated, weighted and appropriately processed, with the resulting outcome that only neutral alternatives (i.e., alternatives evoking least disagreement and not necessarily good ones) are consistently favored. The U-graph compensates for this and generates a decision-making process which does manifest sensitivity to both extreme and normative attitudes.

Another weakness of most decision mechanisms that operate on a relative high level of sophistication is their nontransparency to the average user. The user provides the input and is expected to accept without any reservations the outcome provided by the process. Again, the U graph attempts to overcome this shortcoming by providing transparency via the visual graph. As such, it displays tradeoffs to be made when more than one aspect is involved. Furthermore, it permits the decision-maker to change his opinions and explore different assessments of an alternative in the midst of processing the data. In addition, it displays the various implications of each decision and thus promotes clarity of reasoning by the decision-maker.

The U graph is favored by yet another asset, its tolerance toward different types of information. In principle, it will tolerate any level of precision in the data; will accommodate both gross and approximate assessments of an alternative, as well as very fine and precise definitions (regardless of whether a criteria is implicit or explicit).

Finally, the possibility of developing the U-graph into an automated (computerized) mode to be used interactively with groups involved in decision making processes, promises the opening of new horizons in such areas as management, policy making, and design.

Acknowledgements

The author is grateful to Rita Schinnar for her extensive help in writing and editing this paper. Thanks are also due to Michael Brill and Bonnie See of BOSTI, Inc., for their encouragement and support during the conceptual development of the U-graph.

Notes and References

1. Papers discussing further the Reduction of Uncertainty are:

   Pleasie Mill, "Decision Making in PAK" (Unpublished BOSTI notes), March 1972.

   Arie Schinnar, "An Outline of a Proposed Decision Making Tool for PAK II" (Unpublished BOSTI notes), July 1972.

2. In its various phases of development the U graph has drawn from several other contributing theories such as: Game Theory, Probability and Statistical Theory, Decision Making and Optimization Techniques.

3. The mechanical details of operating the U graph are available in another paper by the author, yet unpublished.

4. However, only in cases employing the Independent Variable Resolution Strategy the alternatives are replotted differently on the U graph, as shown in Figure 6.

5. Strategies (a), (c), and (e) are recommended as the most effective, and strategies (b) and (d) should be applied only when (a), (c) and (e) have failed to achieve conclusive results.

6. The complexity of the analysis depends on the particular circumstances, ranging from rather simple and deterministic occasions to "iffy" and probabilistic situations, as well as on the level of sophistication of the decision makers.

Peggy Bloch Memorial Reading Room, Interior View.

Michael Kreski and David Rejeski

PIDS photograph

DESIGNING OFF THE STREET.

529

DESIGNING OFF THE STREET   Michael Kreski & David Rejeski   Pids   9.6

Abstract

"Designing Off The Street" is an attempt to synergically combine existing items in our environment to answer human needs. It is a process which involves people directly in the solutions, and allows them to grasp the implications of creatively utilizing what is already available in their environment. In this sense, "Designing Off The Street" is an attempt to teach a systemized form of "make-doism" to people who have been conditioned by our technocracy to expect and accept obsolescense.

It is also a process conceived to purposely limit the designer's sphere of influence to a point where he is unable to try and juggle huge amounts of space, time, money and material; this reduces the negative effect of any conceptual errors or miscalculations he might make. It represents a reversal of trends towards computer-analogs, supersystems, and conceptual grandeur, moving instead in the direction of concise and perspicuous solutions which reflect the charm, humor and intricacies of everyday life.

"There are always going to be some designers who want to design sewing machines and vacuum cleaners, and they will be the millionaires of the world. Then there are going to be the idealists who are interested in pollution and city planning-and they will starve." H. Dreyfuss 1904-1972

Everyone in the design profession today has become keenly aware of this apparent conflict between idealism and profit motivation. Ultimately, design is a matter of conscience for all of us involved in its implementation. We are in a position to affect the environment, and the environment affects people. In light of the new findings in behaviorism, we can begin to grasp the widening implications of our solutions, but are we, as designers, prepared to take on this increased responsibility?

There was a time when as a designer you could make a conceptual mistake, a styling error, or a miscalculation on dimensions without seeing that lapse of judgement multiplied a thousand-fold and poured into the consumer market like so much mercury-laden salmon. The designer of many years ago was able to take his time, to do the job with his own hands, or at least to delegate the problem to a close associate. His sphere of influence was limited to the people he could personally provide for, whereas today, one man is in the position of being able to make decisions that affect millions of lives.

In a time when we are talking about lack of design methodology and thinking processes suitable to meeting the incredible complexities of modern problems, we are failing to consider that the real difficulty may be our ability to actually carry out whatever decision we finally do end up making. We are as a profession devising more and more sophisticated ways to put larger and larger projects into the hands of fewer people. We tell the people of a city that they need a professional to solve their problems, that they can't possibly do it themselves and after they believe that, we come in and bite off a chunk of the problem that we can't possibly chew without some outside mastication assistance, be it in the form of computers, systemology, or whatever.

As projects grow larger, and systems expand to super-systems, there is a tendency to abstract people into generalized groups for easy management. The unfortunate result of this practice is generalized solutions. People are tired of sitting in chairs made for the 50th percentile male, and people are tired of living in housing made to accommodate the "average" middle income family. What we need is a reversal; less systems, and smaller problems.

Since the number of designers needed to implement a plan such as this would be prohibitive, we must get people, our clients, directly involved in the design experience. We must climb down off our pedestals, go out into the street, and be willing to communicate with the people on a level which is commensurate with their vocabulary. If necessary, the designer must become a handyman; the architect, a construction worker; and the graphic designer, a sign painter. We must start making creative use of what is available in the environment, and start working with things on a hand-to-result scale.

It is time we stopped answering to multi-million dollar corporations, bogus city governments, and profit-oriented building contractors, and became sensitive to the public on a one-to-one basis. Today, egotism is replacing altruism, and designers are quickly losing the magnanimity and the humbleness needed to respond to human problems. Our goal is therefore to secularize design; a process we think can best be accomplished by "Designing Off the Street."

We have always been cautious when appling generalized design philosophies to unique problems, but "Designing Off the Street" has proved itself adaptive enough to meet our needs. We attribute this open-endedness to the fact that it relies on people and their environment for solutions. It is a method by which those incongruities and intricacies of life are included in the designed solution. It is an attempt at low-cost design, a way of preventing the modern designer from juggling huge amounts of material, space, and people through the use of technological means. Hopefully, it would prevent finished projects from looking like full-scale versions of simplified problem abstracts.

We also see it as a means of handling some of today's environmental problems by getting people directly involved in a recycling process which produces tangible rewards for their efforts-reinforcement. Recycling was once thought by many people to be our environmental panacea, but it has not been totally effective because it is a non-creative, technologically implemented process. It usually produces changes in form only; not function. So a Coke bottle is recycled to become a milk bottle, and the shell from a discarded Pontiac becomes the body of a new Dodge. In short, trash once again becomes trash in a never ending cycle which remains oblivious to meaningful creative input and direct human involvement.

"Designing Off the Street" is a non-cyclical, people-oriented process which deals with changes in the function of objects without necessarily altering their form. It is an attempt to synergically combine existing items in our environment to answer human needs. Enough raw material already exists on this planet in the form of useless and redundant objects to supply our needs for years to come. By selectively altering the function of various items, we can eventually reach a steady-state condition in which materials are conserved and distributed evenly for our daily lives. Though at first this concept might appear grandiose, we have found it to be a viable alternative to standard design methodology and would like to illustrate its application with the following projects.

We were recently commissioned by the City of Cranston, Rhode Island, to prepare a proposal for the usage of a vacant piece of property in the city's limits as an exposition site to serve as a year-round attraction for tourism. Upon investigation of the site, we discovered it to be serving as the neighborhood dump. It was landscaped with piles of wrecked cars, old refrigerators and other large and small artifacts of the American throwout culture.

The city council desired this exposition to relate to the theme of "Italian Heritage and Brotherhood in American History," and asked us to expound upon this theme in the form of pavilions and possible exhibits. However, we instead took the existing na-

9. DESIGN LANGUAGES AND METHODS / 533

ture of the site as our theme inspiration. We let the site talk to the designers rather than vice-versa. Instead of coming up with paper proposals for exhorbitantly expensive buildings and media presentations, we worked backwards from what was on the site. Normal designers would not do this, as the temptation is very strong to impose one's will upon a vacant piece of land. Having rejected the proposal for a Drive-In Pizzaria among other suggested exhibits, we turned to the inspiration of the site itself and developed a theme based on the idea of "Man and His Garbage."

America is the disposable culture nation; we can trace the history of our industrial and social development through the contents of our junkyards. Junkyards are enjoyable places to browse and to pick up unique objects of special interest. Once easily accessible to children, today's junkyards have become more and more isolated from them. They are hidden behind "beautifying" walls and landscaping, or are placed in the middle of undesirable and dangerous industrial neighborhoods.

Our "Man and His Garbage" ThemePark was to be a living, seething experience in discovery and self-education through trash. Plenty of raw material was available in surrounding towns for the asking; in fact, many individuals and groups paid us to acquire their disposable culture artifacts and materials. Within a few months, the park was beginning to take shape.

In one section, old automobile parts were employed in the creation of a sunning area for senior citizens: old car seats became soft and comfortable no-cost benches, headlamps became luminaires for night lighting, and engine blocks had their pistons removed and the vacant holes were filled with fresh earth and planted with flowering plants. In another area, old wood pieces became playground equipment. Visitors were permitted to browse for unusual discards throughout the site; one of the most popular exhibits featured a wide variety of old kitchen appliances, some dating back as far as thirty years. The contrast between old and new served as as educational demonstration of how the most modern products of technology become obsolete. Special exhibits featured historical discards and waste materials from times as early as those of Cro-Magnon man; old useless weapons, broken pottery and leaky containers all served to draw parallels between today's garbage and that of our ancestors.

In retrospect, the Cranston ThemePark had an impact on its environment far beyond what it might have had had we simply started designing without making use of what was available. The city is cleaner, the residents have a place to meet and enjoy themselves, and the fascinating subject of the American Disposable Culture now has a permanent place of honor. In addition, we were able to produce this ThemePark at a profit; we actually made money in its development. It was a concept well-suited to its site, and able to be participated in by every local resident. It was truly an example of "Designing Off the Street."

We employed this design system again when the Paulsen Library in San Antonio, Texas wanted to expand its facilities to provide a quiet, comfortable reading room. Being extremely low on funds, the cost factor was important to the Library. Several design firms were requested to suggest solutions; some gave up because of the lack of space available. Others decided that the space was sufficient but that the cost of furniture and other installations would be prohibitive. A few suggested custom installations at accordingly high prices.

None of the design teams attempted to make use of what was available. We did.

Our observations indicated that space was indeed available but in the form of oblong and intricate spaces tucked behind cabinets and between air conditioning units. It was decided that these spaces, however irregular in shape, could be converted into a comfortable environment; cost, however, was still a problem.

We scavenged the entire library area and came up with an inventory of unused furniture, seven chairs and two tables. A tour of nearby construction sites yielded several van loads of off-size lumber scraps and leftovers. With the purchase of some nails and building tools, we were ready to go.

Letting the shape of the available wood guide our design, we started to build. If we couldn't find a piece shaped to fit our plans, the plans were changed to fit the piece we had. Some library departments were able to provide us with old bulletin boards which, cut up, were used as paneling. The wood, left unpainted, had a dynamic natural appearance; taking our clue from this, our floor material (which had to be soft underfoot and sound-absorbing) became wood chips from a nearby sawmill, spread to a depth of two inches.

As it developed, the Peggy Bloch Memorial Reading Room became one of the most popular and enjoyable spaces in the library. It was an informal place, not at all cold and antiseptic like many modern libraries. Bare lightbulbs hanging among the wooden beams and furniture produced a warm, comforting lighting effect. People could curl up in the little niches and corners formed by the serpentine shape of the Room (defined by an arrangement of large filing cabinets, pieces of cloth, homasote panels and existing walls) and read for hours in secure isolation.

"It's not so much a room as a trail of cubbyholes hidden throughout our library," commented librarian Charles Gremling, "and it fills our need for isolation of readers from library traffic. The fact that it looks like somebody's cellar is in fact the very quality that makes the Room the most liveable and home-like public space I've seen."

Unfortunately, the Paulsen Library was forced to close some months later due to lack of funds, and the hundreds of people who had come to enjoy the Peggy Bloch Memorial Reading Room in that time were forced to do their future reading elsewhere. The fact that it was designed with available scraps and leftovers and not as the result of a master plan or architectural conception made it what it was, a totally successful reading room.

Since then, we have seen new multi-million dollar libraries whose interiors, while done in what can be called good taste, still leave one cold and repulsed by their scale and efficiency. We think the P.B.M.R.R. was an ideal solution to the problem.

Up to this point, we've been discussing projects which were done at a scale allowing us to hand-build the finished product. The logical next step was for us to try and adapt the designing off the street system to something beyond our personal capacity to construct.

The project: The Warsaw Public Transit Network (PolandTrans).

9. DESIGN LANGUAGES AND METHODS / 535

The city of Warsaw had long been hampered by traffic congestion caused by the continued conflict of motorized and non-motorized vehicles. There had been a movement afoot for many years to exclude the picturesque ox-cart from the public roadways, and to replace it with more conventional modern forms of mass transit.

Seeing this as a chance to preserve a part of the ethnic heritage of the nation of Poland, we began development of systems for improving the efficiency and aesthetics of the ox-cart as a mass transit system.

Our first pressing problem was one of creating in the minds of potential riders an image of the ox-cart as a viable yet familiar and friendly city transportation. A graphics and visuals program was called for.

In choosing a specific typeface for usage on PolandTrans signs, we turned not to a typeface catalog but to the street, in this case the open-air markets of the Northeast Warsaw area. We found the hand-made signs of this area s markets to be of a distinctive typeface; gathering dozens of these signs together, we created a composite of their letterforms and produced a typeface called Medium Condensed Bulbous. In later public opinion tests, this face was proven to be more popular among Warsawers than Helvetica Medium or Copper Plate Bold.

We studied Warsaw s existing transit network and found it to be very confusing. We produced an official Route Map (hand drawn) to demonstrate to citizens what was already available; thus the existing system became better understood and used.

By evaluating numerous aerial photographs of ox-cart traffic, we found that by re-routing existing vehicular traffic we could leave a variety of street networks open to slow-speed traffic. This meant that the ox-cart system could have its own right of way, not competing with other forms of transit

We invited neighborhood store owners to a series of luncheons at which plans for the ox-cart network were discussed. We all agreed on several main points:
1) Oxen are quiet urban transporters.
2) They require less maintanence and upkeep than motor vehicles.
3) They last longer than other transportation vehicles (some ox-carts have been in service in excess of 85 years)
4) Ox-cart routes are desirable for Urban Peoplemovers.

The problem then was one of implementation. We were faced with a lot of criteria and very conflicting needs. How could we successfully re-orient Warsaw to the ox-cart idea as opposed to more ordinary forms of transit. For our solution we turned to the street and it didn't let us down.

In talking to many street vendors along the proposed ox-cart routes, we learned that vehicular traffic often interrupted business, causing pedestrians to crowd uncomfortably together to allow cars or buses to pass. The solution taken by many vendors was to park their own cars in such positions as to make it virtually impossible for traffic to enter certain streets. While illegal, the system worked and was of no disadvantage to anyone, since alternative traffic routes were always available. From this bit of street knowledge came the idea we needed.

In order to make ox-cart travel a more successful form of transit, we proposed to the Warsaw Traffic Ministry a campaign of street signage and directioning which invariably would lead to confusion on behalf of those persons driving Warsaw s streets. As a result, fewer and fewer people would be willing to drive in the areas of downtown Warsaw in which ox-cart routes were planned, because driving there would be so difficult. In this way we would increase ox-cart traffic by decreasing other vehicular traffic; unwilling to drive their cars in these areas, Warsawers would take readily to the ox-carts.

What might have been another typically dull experiment in transit busing or taxicab improvement might instead become a refreshingly unhurried way to browse the charming streets of downtown Warsaw.

An ambitious project now under consideration by the Traffic Ministry and Sklovnet Vlos & cie. of Warsaw is the creation of the Teodoro Czaplicki Open Air Shopping Mall on one major thoroughfare of the city. Its parallel rows of open air stalls and vending stands would form routes on which continuing streams of ox-cart shuttles and expresses could run in city-circling loops.

In short, we didn't come into Warsaw and start designing major traffic systems; we worked a little at a time, not fighting against the forces of the city's life but absorbing and using its energy to propose better transit that is of the street, not in spite of it. The project continues; we cannot yet tell if it will be fully implemented. In the meantime.........

We are continuing to fight against boredom. We are trying to bring a sense of non-systemized "make-doism" to all kinds of design situations and letting the problems and limitations almost provide their own solutions. We are not saying that this is a unique talent of ours; it is a matter of attitude more than ability, and we are presenting this paper in the hopes that our form of attitude may serve other designers and their clients well.

We are indeed tired of seeing things done "right". The time is approaching when "Designing Off the Street" may be the only thing that stands between our lives and a prefabricated world of crisp clean solutions that solve problems but don't have the ability (or indeed even the need) to charm, to amuse, to influence.

"Designing Off the Street": no huge master plans, just an interesting way of surrounding ourselves with a friendly environment.

APPENDIX

LIST OF REVIEWERS OF ABSTRACTS (A) AND PAPERS (P)

Irwin Altman, (A,P), Dept. of Psychology, Univ. of Utah, Salt Lake City
Howard Andrews, (A,P), Dept. of Geography, Univ. of Toronto, Canada
John Archea, (A), Div. of Man-Environment Relations, University Park, Pennsylvania
Martin Baker, (P), Demov, Morris, Levin & Shein, New York, N.Y.
Vladimir Bazjanac, (A), Dept. of Architecture, Univ. of California, Berkeley
Robert B. Bechtel, (A,P), Environmental Research & Devel. Foundation, Kansas City
Fred Binding, (A,P), Waterloo, Ontario, Canada
Michael Brill, (A), School of Arch., State University of New York, Buffalo
Charles Burnette, (A,P), American Inst. of Architects, Philadelphia, Pa.
Daniel H. Carson, (A,P), School of Arch., Univ. of Wisconsin, Milwaukee
Kenneth H. Craik, (A,P), Inst. of Person. Assessment, Univ. of California, Berkeley
Michael Cunningham, (P), College of Arch., V.P.I. & S.U., Blacksburg, Virginia
Robert David, (P), Dept. of Ind. Design, Kansas City Art Inst., Kansas City
Charles Davis, (P), Skidmore, Owings & Merrill, New York, N.Y.
Gerald Davis, (A), The Environmental Analysis Group, Vancouver, Canada
Al DeLong, (A,P), School of Arch., Univ. of Texas at Austin
John Dickey, (P), College of Arch., V.P.I. & S.U., Blacksburg, Virginia
G. Day Ding, (A), Dept. of Arch.,Miami University, Oxford, Ohio
Jerry Durlak, (A), Div. of Social Science, York Univ., Downsview, Canada
Charles Eastman, (A,P), School of Urban & Public Affairs, Pittsburgh, Pa.
John Eberhard, (A,P), School of Arch., State Univ. of New York, Buffalo
Peter Eisenman, (A,P), Inst. for Arch. & Urban Studies, New York, N.Y.
Andrew Euston, (A), Department of HUD, Washington, D.C.
Jerry Finrow, (A), Center for Environ. Research, Univ. of Oregon, Eugene
Robert Frew, (A), New Haven, Connecticut
Philip Gary, (P), College of Arch., V.P.I. & S.U., Blacksburg, Virginia
Lawrence Good, (A,P), Lawrence R. Good & Associates, Lawrence, Kansas
Donald Grant, (A,P), School of Arch. & Environ. Design, Cal. Tech, San Luis Obispo
Robert Gutman, (A,P), Dept. of Sociology, Rutgers University, New Brunswick, N.J.
Robert G. Hershberger, (A,P), Dept. of Arch., Arizona State University, Tempe
Michael Kennedy, (A,P), School of Arch., Univ. of Kentucky, Lexington
Jon Lang, (A,P), Dept. of Urban Studies, Univ. of Pa., Philadelphia, Pa.
Powell Lawton, (P), Philadelphia, Pa., Geriatric Center
Theodore Lundy, (A,P), Lake Oswego, Oregon
William Michelson, (A,P), Urban Affairs, Ottawa, Canada
William Mitchell, (A), School of Arch., Univ. of California, Los Angeles
Gary T. Moore, (A,P), Dept. of Psychology, Clark Univ., Worcester, Mass.
Alex Murray, (A,P), Fac. of Enviornmental Studies, York Univ., Downsview, Canada
Len Olson, (A,P), College of Arch., V.P.I. & S.U., Blacksburg, Virginia
Henry Sanoff, (A,P), School of Design, N. C. State University, Raleigh
Leonard Simutis, (P), College of Arch., V.P.I. & S.U., Blacksburg, Virginia
Robert Stuart, (P), College of Arch., V.P.I. & S.U., Blacksburg, Virginia
Raymond G. Studer, (A,P), Div. of Man-Environ. Relations, University Park, Pa.
Philip Thiel, (A,P), College of Arch. & Urb. Planning, Univ. of Washington, Seattle
Thomas L. Thomson, (A,P), School of Arch., Washington Univ., Saint Louis, Missouri
Alan Wicker, (A,P), Claremont Grad. School, Dept. of Psychology, Claremont, Cal.
Gary Winkel, (A,P), The City Univ. of New York, Grad. Center, New York, N.Y.

AUTHOR INDEX

Aalto, A., 513,514,518
Abercrombie, M.L.J., 509,510,517,518
Acking, C.A., 64,74,83
Adler, R., 448
Alexander, C., 63,284,290,372,374
   375,381,382,403,405,429,501,508,517
American Institute of Architects, 396
American Jewish Committee, 155
American Society of Mechanical Engineering, 375,381
Anderson, J., 64
Anderson, M., 21
Anstrand, D., 464
Armour, G.C., 419
Arnold, D., 150
Appleyard, D., 279,375,381
Asch, S.E., 313
Austin, W.M., 8
Avery, M.C., 329

Bales, R.F., 482
Bannister, D., 243,251,285
Barker, R., 35,147,191,205,358,379,382
Barnes, R.M., 375,381
Bartlett, F.C., 509,511,517
Bass, B., 501
Bauer, R., 151
Bayes, K., 191,192
Bechtel, R.B., 145,146
Beck. H., 9
Beery, 4,13
Beesley, M.E., 449
Benthall, J., 515,516,518
Berlyne, D.E., 84
Bernstein, L., 375,381
Bonarius, J., 246
Birdwhistell, R., 375,381
Birkmayer, W., 83
Birren, J.E., 389
Bishop, R.L., 166
Blalock, H., 315
Bliss, C., 375,381
Borgen, F.H., 332
Bornstein, 375,381
Bosowitz, H., 48
Boyd, R.R., 205
Braybrooke, D., 504
Bridge, A., 249
Brillouin, L., 334
Broadbent, D.E., 49

Broadbent, G., 508,517
Brock University Planning Department, 325,329
Brown, J.J., 452
Brown, P.J., 166
Brotchie, J., 416
Brunswik, E., 85
Building Construction Cost Data, 444
Bullock,
Bultena, G.L., 164,165
Burnette, C.H., 482
Byrne, E.J., 453

Calhoun, J.B., 17,38,124,204
Campbell, D.T., 226
Campbell, S., 355
Canter, D.V., 246
Carp, F.M., 231
Carson, D.H., 47,444
Cassel J., 21
Cassier, E., 183
Cassirer, E., 183
Cattell, R.B., 218,288
Catton, W.R., 165
Census of Population,338
Chapanis, A., 389
Chapin, F.S., 52,138
Chapman, W.P., 389
Chermayeff, S., 372,381
Chomsky, N., 429
Chown, S.M., 218
Churchman, C.W., 443,446,506,517
Clark, R.N., 170
Clowes, M., 429
Coates, G., 313
Cohn, R.C., 482
Community Mental Health Centers Act, 36
County and City Data Book, 22,30,153 154
Cox, K., 24,153
Craik, K.H., 164,165,265,267,281
Crane, D.A., 468
Craun, R., 313
Cronbach, L.J., 220,226
Crouch, D.C., 389
Cullen, G., 272
Cumming, E., 204,219
Cypra, K., 162

543

Darling, 38
Dattner, R., 166
Davis, F.A., 205
deCharms, R., 281
Department of Health Education and Welfare, 151
Department of Housing and Urban Development, 470
Desor, J.A., 145,147
DeVore, I., 43
Diaiso, R.J., 321
Dineen, F., 375,382
Diringer, D., 375,381
Dixon, W.J., 315
DeLong, A.J., 5,8
Deutsch, P.S., 287
Dohrenwend, B.P., 48,50
Douek, E.E., 389
Douglas, P., 33,34,155
Downs, R.M., 276,313
Dreyfuss, H, 375,381
Driver, B.L., 162,167,168
Dubos, R., 41
Duncan, O., 23,153

Eastman, C., 398,400,401,402,405,415, 428,429,459
Emery, F.E., 506,517
Engelbart, D.C., 382
Erikson, E.H., 43,204
Esser, A.H., 8,16,124,125,339,340
Estes, M., 410
Evans, T., 408

Farrell, P.B., 448
Farr, L.E., 389
Feder, J., 429
Federal Electric Corporation, 375,381
Federal Housing Administration, 302
Fischer, F., 19
Fishburn, P.C., 441,443
Fisher, R., 390,458
Flachsbart, P.G., 100
Forstall, R., 26,153
Foster, N., 511,518
Frank. A.J., 398
Fraser, D., 38
Fried, M., 50
Fromboluti, C., 400

Fruchter, B., 315

Gans, H.J., 14,145,292
Garling, T., 74
Gaver, D.P., 453
Geldard, F.A., 390
Gesell, A., 191
Giedion, S, 511,518
Gilbert, C.G., 165,170
Gilbert, J., 390
Glaser, D., 145,148
Gordon, W.J.J., 4,482
Gregory, R., 64
Grether, W.F., 162,163
Grigsby, W., 155
Glazer, N., 14
Goulomb, S., 375,381
Grant, D., 416
Grant, E., 366
Grason, J., 415
Gutman, G.M., 218

Hadden, J., 3,9,10,28,150,151,153
Hall, E.T., 14,54,98,99,375,381
Halprin, L., 375,381
Hamilton, E., 43
Hamovitch, M.B., 231
Hare, P., 5,500
Harman, H.H., 304,315
Harris, C., 2, 150
Harris, H., 398
Harrison, J., 285
Hart, R.A., 192,256
Hase, J.D., 221
Hawley, A.N., 22
Heacock, F., 375,381
Hebb, D.O., 278
Hediger, H., 33,37,44
Helson, H., 26
Hendee, J.C., 165
Henderson, A., 112
Hershberger, R., 63,284,289,295
Hickey, T., 204
Hilger, J.A., 390
Hillier, W., 125,131,185
Hinkle, D.N., 244
Hofstetter, H.W., 390
Hole, V., 50
Hollingshead, A., 287
Honikman, B., 244,285

Hood, E., 32,155
Horst, P., 315
Howard, A., 48
Howard, R.B., 261
Huang, K., 334
Hubbard, H.V., 274
Humber College of Applied Arts and Technology, 4,350
Hutchinson, A., 375,381
Hutten, E.H., 507,508,512,513,514,516, 517,518

IBM, 382
Internal Revenue Service, 364
Ittleson, W.H., 148,194,200,206

Jacobs, J., 21
Janis, I.L., 48
Jansma, J.D., 444
Jensen, C.R., 162,165
Johnsen, J., 375,382
Johnson, T., 398,416,429
Johnson, S.C., 114
Joiner, D., 251
Jolly, A., 38
Jones, J.C., 509,517
Jonouskova, K., 390

Kaplan, R., 267,281
Kaplan, S., 219,256,265,268,276,280
Kasmar, J., 375,382
Kates, R.W., 164
Kazmierczak, H., 398
Kellogg, R., 509,510,518
Kelly, G., 84,243,244,251,285
Kerlinger, F.N., 314
Kira, A., 145,440
King, J.A., 37
Klee, P., 375,381
Kleemeier, R.W., 390
Kluckhohn, C., 220
Knapper, C., 312
Kuhlen, R.G., 191,219,268
Knapper, C., 312
Kuhlen, R.G., 191,218,268
Kuller, R., 83
Kuper, L., 50
Kusgen, H., 364

Ladd, F.C., 256
Lamanna, R.A., 99
Landau, R., 507,508,517
Langer, S.K., 512,518
Lansing, J.B., 99,212
Lawton, M.P., 24,25,219,232
Lazarus, R.S., 48,50
Lederberg, J., 410
Lee, R.B., 400,417
Lee, T.F., 276
Leff, H., 63
Lefferts, R., 205
Levin, P.H., 123
Levinson, 375,381
Lew, 419
Lewin, K., 52,54,218
Leyhausen, P., 23
Lime, D.W., 169
Lingoes, J.L., 267
Lipman, A., 205,206
Lockwood, A., 375,381
Lorenz, K.Z., 15
Lucas, R.C., 165,168,169,170
Lynch, K., 266,279

McBride, G., 124
McClelland, D.C., 26
McCormick, E.J., 163
McKechnie, G., 85,165
MacLean, P., 8,16
MacMillan, C., 442
Maloney, J., 11,19,20,25,151,152,153
Mangum, W., 25
Mannheim, M., 382
Markman, R., 85
Martens, R., 166
Maslow, A.H., 50
Mead, M., 322
Mechanic, D., 48
Medawar, P., 507,517
Mercer, D., 166
Meresko, R., 220
Mesolella, V., 390
Michelson, W., 99,206
Miles, M.B., 6,482
Milgram, S., 147
Milne, M., 382
Mitchell, W., 415,417
Moncrief, R.W., 390
Moore, G., 403

Morgan, M.W., 390
Morris, C., 514,516,518
Morrissett, L.E., 390
Murray, H.A., 26
Myer, T., 410

National Academy of Sciences, 151
National Science Borad, 6,151
Negroponte, N., 407
Nelson, I., 27,153
Nelson, J.R., 456
Neugarten, B.L., 218
Neulinger, J., 166
Neumann, E.S., 166
Neutrath, O., 374,381
Newell, A., 400
Nilsson, N.J., 401
Norbert-Schulz, 509,511,517,518
Norwood, R.H., 516,518
Novak, M., 14
Nugent, C., 406
Nunnally, J., 218,220,222,255

Obi, S., 390
Ogden, C., 375,382
Oklahoma Employment Security Commission, 111
Oldfield, R. C., 509,511,517,518
Olver, R., 85
Ontario Department of Education, 349, 350,354
Ore, O., 375,381
Osborn, A.F., 5,482
Osgood, C.E., 72,73,75,314
Owen, C., 376,382

Parnes, S.J., 5,482
Pastalan, L.A., 219,390
Pelz, D.C., 499
Penz, A.J., 450
Perin, C., 53,355,375,381,463
Peterson, G.L., 99,165,166,168, 170,313
Pfeffercorn, C., 401,429
Piaget, J., 8,191,201
Pietila, R., 513,518
Plant, J., 50
Pribram, K., 8
Prince, G.M., 4,482
Princeton University School of Architecture, 470

Proshansky, H.M., 124,164
Pollowy, A.M., 192
Ponti, G., 511,518
Popper, K., 493
Powell, B.A., 459

Quinn, P., 504

Rainwater, L., 99
Rapoport, A., 219,265,281,375,381
Reichard, S., 204
Reynolds, I., 50
RIBA, Research Committee, 133
Rivlin, L.G., 196
Robinson, A., 285
Rohn, J., 419
Ronge H., 390
Rose, S., 65
Rosenthan, J., 21
Rosow, I., 24,205
Ruder, E.m 375,381
Rusch, C., 375,382
Rush, C., 374,381
Rummel, R.J., 315
Rupp, C.W., 170
Rutherford, W., 165

Saltz, R., 205
Sammet, J., 382
Sanoff, H., 99,313
Schaller, G., 38
Schein, E., 499
Schenk, H., 390
Schorr, A.L., 54
Schrag, P.,
Scott, D.H., 351
Selye, H., 48
Shafer, E.J., 165,169
Shannon, C., 372,381
Shapiro, J., 231
Shaw, A., 382
Sherman, S., 231
Shevky, E., 152
Shipley, W., 287
Siegel, S., 288
Simek, L., 482
Simmons, L.W., 204,205
Simon, H., 374,381
Sivadon, P., 30
Skinner, B.F., 133
Slagle, J., 402

Slataper, F.J., 391
Slater, P., 244
Sokol, R., 152,325,328
Sommer, R., 14,54,145
Spearman, C., 315
Spitz, R., 8,44
Stankowsky, A., 375,381
Stanky, G.H., 165
Steinitz, V., 84
Stevenson, A., 50
Stone, G.D., 165
Strakosch, G., 449
Stringer, P., 243
Survey Research Center, 205
Sussman, M., 212
Sutherland, I.E., 405
Suttles, G.D., 22
Swartz, W.W., 449

Thiel, P., 375,381
Thurstone, L.L., 304
Tibbetts, P., 8
Tiffin, J., 391
Tolman, E.C., 49,276
Tryon, A., 14,16,21,152,153,329

Ullman, J.,
U.S. Department of Housing and Urban
  Development, 302
U.S. President's Commission on
  Population Growth, 14

Vaschide, N., 391
Venturi, R., 464
Vickery, B.C., 498

Waddington, D.H., 514,518
Wagar, J.A., 169
Walton, W.G., 391
Ward, W.S., 416
Watson, A., 15
Weilgart, J., 375,381
Weinzapfel, G., 403
Wesley, E., 220
West, D.C., 166,168
Westin, A., 219
White, L.E., 50
White. R.H., 53
White, R.W., 278
Whitehead, B., 401,417

Wilkie, R.W., 22
Wilson, R.L., 98
Winer, B.J., 259
Wingo, L., 168
Wohlwill, J.F., 24,164,166,265,281
Wolfe, M., 191,196
Wong, A.C.K., 405
Wurman, R., 350
Wynne-Edwards, V.C., 15

Yancy, W.L., 21
Yessios, C., 404,408,429,430

Zeisel, J., 113
Zuckerman, 85

SUBJECT INDEX

Activities, distribution of, 195,201
  number involved in, 127,130,196,197,
    199-200,316,318,320,494
Activity, place structure, 127,128-131,
    139
  patterns and locations of, 191,196,197,
    198,201
  rate of, 299
  social, 123,126-129,130,133,191-195
Adaptation to environment, 24-32
Affinity interaction matrix, 417
Aged, alternatives of the, 229
  attitudes in the, 204,205,218
  characteristics of, 204,209,212,231,
    383,384,387,533
  cross generation contact, 205,206,
    210,213,214,231
  environmental preferences, 192,201,
    207,218
  housing for, 229-230,231-232
  and human competence, 219
  income groups, 207
  physical mobility of, 209
  privacy and proximity, 207,210,212,
    214,231
  stimulation of, 231
Algorithms, 414,415,416
American Indians, 111-121
Animistic situation, 179
Architect, 184,231,230,284,290
  design process, 511,516-517
  forms, 124,127-129,131-133,178,364,
    511-516
Archetypal place, 33,37,38
Area, quantum of, 336,337,342
Assessment, 5,12,275,278-281,519,520
  as measures of validity, 254,255,
    265-267,269,272,275,278-281,414
  of behavior, 229,493
  of buildings, 324
  of building economy, 361
  of elevator performance, 448,450,454,
    457
  of financial structure, 359,363-369,
    448-450,455-457,479
  of physical characteristics, 266,496,503
  of satisfaction, 448,454-459
  theoretical assumptions, 217
Assessment criteria, 88,125,133,246,
  521-528

Assessment, techniques of, 464-468
  constraint graph, 403
  generated trips, 127,131,132,133
  reliability coefficients, 259,262,263
  T-tests, 315
  U-graphs, 519,528
Aesthetic system, 34
Attitude, group's, 524,527
  preferences, 326
Attitudes, behavioral disposition, 128

Bathroom, 440,441,443,444,445
Behavior, aged, 191,210,212
  determinants of, 125
  developmental phases of, 35,42
  group, 251,481-490,493,498,499
  interaction, 496,501
  modification of, 125,134,135,139
  motivation, 124,125
  patterns, 34,62,123,191,230,292
  response categories, 127
  self-regulating, 142
  setting deprivation, 35-37
  social systems, 335
  solitary, 197
  stress, 142,266,289,290,296
  user, 332
Behaviorism, 531
Behavioral dependence, 25
  setting, 35,37-41,44,380
Binomial tests, 326,328,330
Bi-polar adjectives, 314
Boolean test (& graphs), 402,403
Brose and Einstein, 335
Boundaries, visual, 386
Buffalo organization for social and
  technological innovation (BOSTI),
  519,528
Building characteristics, 329
Building costs, 354
  prototype, 349

Change, implementation of, 123,124,135
Chicago, residential development project, 468-480
Children, cognitive capacities, 201
  color preferences, 192
  conception of places, 178

drawings by, 398,509-510
movements, 509
living units of, 198,200
playgrounds, 193
public areas, 198-200
Circulation, in office buildings, 11-12, 406,407
Citizen participation, 463-464
Classrooms, unstructured, 201
Clients, 312-313,440,441,446,531-532
Clustering effect, 334,336
Cognition, 7,8,201,251,314
Cognitive gap, 62,64,284
mapping, 254,255,275-279
structures, 255,312,321
College, campus, 123-135,349
life styles, 128
management and planning of, 349-358
student enrollment, 350,355-354
systems, 349-358
timetables, 126
of applied arts and technology, 1-2, 4,6,349-350,352,354
Color, 62,386,388
Color preferences, of children, 192
Communications, Shannon's model of, 372,373
systems of, 5-6,371-380,498
Community, sense of, 501
Communities, theological, 493-494,496, 504-505
Complexity, 265,271,272,406,507-508, 511,513-516
cognitive, 246
environmental, 5-9,243,280-281
Computer, aided design, 396,400,403, 405-406,407-408,409-411
Computers, data systems, 396-397, 408-411,463-465
Computer drafting language, 398,400,402, 405,408,411
Computer languages, use of, 127
Computer programs, 466
basic programming, 463
computer implemented site planning system, 428,429
CORELAP, 417
economic base, 363-365
Computers in site planning, 438
Computing time comparison, 415,416,419, 423

Construct eliciting (Kelly), 244
Constructs, 243,285,287,289,290,292
laddered, 247
physical characteristic, 247
Construct repertory test, 285
systems, 249,251
Cost analysis, 363-365,448-449,454, 456-458
Creativity, architectural, 178
in group behavior, 487-488
thinking, 498
Cultural disposition, 125
Cultural heritage, 184
Culturally specific archetypal places, 38
Culture, 37
containers of, 41
modification, 125,133-134
throwaway, 532,533

Decision making, 45,463,519-528,520-521, 524
Decision model, 440,441
Delinquent behavior, 36
Demodynamics, 1,4,6,10,335,338,340,344
theorem of, 336
characteristics, 312-314,321,336,337
clustering, 342
disorder, 336,337
equilibrium, 334,342
system, 1,3,4,6,8,335,337,338,340, 342
micro state of, 337,338
Deprivation, social, 98
Density, 210,214
Design, alternatives, 230,359,361,369, 370,522,528
computer aided, 396-398,400,402-411
constraints, 361,441,445,512
decisions, 230,587
education, 183-184
for children, 202
goals, 351-359
information, 1-16,396,408-410
intelligence, 397,405,410-411
languages, 178,183-184
methodology, 123,315,317,338,340, 416,418,506-517
parameters, 123,128,132-133

SUBJECT INDEX / 553

    patterns, 508,513
    people oriented, 532
    presentation, 63
    process, 184,312,372,373,414,423,508, 516
    profession, 531
    requirements, 376-377
    specifications, 37
    strategies, 33,134,396,400,403-407,516
    synthesis in, 33,42-45,506,508-509,512
    systematic generation, 360
    teaching, 507,508
Designed artifact, 508,512
Designer, 5,12,265,270,284,289,312,321, 423,426,440,506,512,514
Developmental, life cycle, 33
    tasks, 45
Deviation, applicability gap, 133
Disorder, entropy, 334,338,342,344
Distances, perception of, 254,256,259,261
Diversity, 22, 124
Dormitories, 147
Draftsman, 397-398,405,407
Dreyfuss, H., 431

Ecology, 24,38,45
Elderly, see Aged
Elevator system design, 448-459
Entropy, 334,335,337,338,344
Environment, built, 84,125,134,273,332
    cues and codes, 5,6,12,386,388
    perception of, 41, 72-83,98,107,275, 277,280,384
    pollution of, 316
    variety of, 275,278,279,280,281
Environmental, assessment, 243,249,530, 532,533
    cognition, 246,284,285,295
    components and parameters, 124,125, 219-220,222-226,243,246,257
    constraints, 440,441,444
    constructs, 267,268,275,278-280, 312,316,321
    cost factors, 366
    design, 123,124,192,370
    disposition and qualities, 24,26,51 62,105,107,217-218,219-220,222- 226,289,313,536
    "magnet" effect, 196,201

    management, 135
    preferences, 265,267,269,270,271, 272,275,280,536
    professionals, 265,284,289,290,292, 294,296
    professional planner, 284,289,312, 312,313,321
    qualities, 287,289,292,294,536
    stress, 48
Ethology, 14
Evolution, 34

Factor analysis, 116,151,243,285,296, 312,315,321
Feedback cycle, 355
Filter theory, (housing), 49
Formula financing (see also Assessment, 351,353-354
Foster Grandparent Program, The, 205, 212
Freedom of choice, 125,133
Function in Design, polarization of, 37,124,128,133,182

Gerontology, 24
Graduate Theological Union, 494-498, 500-503
Graphic, communication, 396,405,408
    displays and representation, 178, 269,415,466
Group, attitudes, 524,527
    behavior, 194,481-490
    communication, 483
    dynamics, 482-490,499,501,519-528
Guttman-Lingoes Smallest Space Analysis III, 267

Habitats, animal, 37
Heuristics, in concept uncovering, 360,363
Hierarchical taxonomic algorithm, 328
Hinkle's construction laddering technique, 244
Home range, 38
Housing, 47,77,98-109,111,121
    attributes of, 87,98
    conditions, 111-121, 150,160

SUBJECT INDEX

density, 313,315-316,318,320-321
evaluation of, 313,316
form, 89,106
physical patterns, 319
privacy, 5,7,9,10,316,318,320-321
safety, 312, 316-318,320-321
scale, 100
turnkey III, 113
Human, behavior (see Behavior)
interaction, patterns of, 124,133
motivations, 24, 123
needs, 275,277
satisfaction, 280

Information, aspectual integration of codes, 5,8,9
hierarchial organization of codes, 5,6,8,9
needs, 229
perception of 506,513-517
processing, 271,278,279
structuring, 507-517
systems, computer retrieval, 396-397, 408-411
systems, design, 396,408-410
theory, 335
Interest cue, 63

Judgement, of environments (see also Environmental Assessment), 73
Junkyards, and design, 533
Juvenile, 138

Kendall's Coefficient of Concordance, 444
Key concepts, common measurings of, 314
Kindergarten, physical separation, 194
Knowledge, cognitive domains of, 276

Land distribution, patterns of, 334,335, 336
Landscapes, visual characteristics of, 166
Landuse and Built Form Studies, Centre of, 134

Language, diagrammatic systems of, 374-379
primitives and grammar of, 375-379
of design, 372,380
Linguistic structures, 438
Lawton, docility hypothesis, 219
Lay person, 284,287,290,292,294,295,296
Library design, 534
Life cycle, schema, 42-43
Life space, Lenin, 218
Life styles, college, 128
Living units, children, 198,200

MacMillan's 0-1 Linear Programming, 440, 442
Man-environment relations, 277
Man-environment systems, properties of, 123
Mapping, 256,258
behavioral, 194
Measurement, criteria of, 355
instruments of, 257
of behavioral dispositions, 217
Measuring techniques, 254
obtrusiveness, 134
reliability of, 220,224
validity, 220,224,226
verbal/visual, 220-221
Memory, 509
reconstruction process, 264
Mental structures, logic, 183
Methods, absolute judgement, 257,258
abstraction, 506,507,512,514-514, 531
analogy, 507,512-513,516-517
assignment techniques, 415,416
behavior mapping, 138
creative-idea generation, 487,488
cashflow modeling, 448-450,454,458
cluster analysis, 152,267,314,324, 329,331
clustering algorithm, 2,6,325,329
computerized simulation, 350-358, 448-449,458-459
computerized u-graph, 528
computerized relative allocation of facilities technique (CRAFT) 419,423

computerized higher education, 350-358
construction laddering (Hinkle), 244
content analysis, 209,511-516
cost-benefit analysis, 448-449,454, 456-458
decoding, 6,12
for plotting alternatives, 522,527
grid, 285,287,295,296
idealisation, 508,514,517,531
index of relatedness, 325
issues, programming, 3-5, 351-353
linear link diagrams, 247,250
linear programming, 441
nuclear growth, 418,423
optimization, 440
overlay techniques, 416
parameters, 509
perceptual, 260,262,263
performance criteria, 525-526
performance hierarchy, 230
photography techniques, 84, 101
polyonimo assembly, 417,423
projective location generation, 400
random generating approach, 415
random switch, 418
ratio estimation, 257,258
reduction, 519-520
repertory grid techniques, 244,250
survey and questionnaires, 127,356
trial-and-error-solution, 423
Modeling, abstract theories, 512-514,516
architectural design, 359-369
environmental, 276
methods of, 64,256,258
urban design, 4630480
Modular programming, 463,480
Mystery,265,267,269,270,271,272,280

Naturism, 292
Natural laws, 335
Natural settings, 273
New York Urban Development Corporation, 230

Observation methods, 356
Odors, 387
Olfactory loss, 385
Organized space, for orientation, 388

Outdoor recreation, 161
Over-crowding, 14
Ox-cart traffic, 535

Parks, 169
Perception, 386,574
  distortion, 264
  environmental, 278,285,287
  loss of, 380,384-387
  of distances, 254,256,259,261
  of glare, 386
  psychology of, 62,509
  visual recovery of, 386-387
Perceptual capacities, 277
Personal construct theory, 242,243,249, 250,251,285
Personal identify, 493,495,499,501
Personal space, 42,251,342
Physical environment, aged preferences for, 207
  attributes of, 316
  use of, 135
Physical factors, in cross generation contact, 206,210,213
Physical mobility, of aged, 209
  paths of, 254
Physical space, 315,320
Place, 124-128,178-184
Places, conceptions of, 179,180-182,184
Planning considerations, 351,396,403-407
Planning, participation in, 351,519
  strategies, 214,493
  target population for, 212
Playgrounds for children, 193
Playground equipment, 533
Pollution, social, 16
Population density, 312,313,315,316,321, 340
  distribution of, 338
  levels of, 340,342
Poverty, 37
Prediction, of environmental preferences, 265,266,267,269,270,271,272
  of environmental features, 279
Privacy, 125,312,494
  aged,210
  design of, 535
Problem composition, 429,432
Problem-solving, group techniques of, 481-490

Professional socialization, 289,290
Professionalization, 289,290
Proximity, of aged to children, 207
Psychology, types of, 162
Psychological factors, 355-356
Public areas and children, 198-200

Quantum of individual area, 334,336,337, 342
Questionnaires, in assessment, 356

Race, American Indian, 115-121
  housing, integration, 316,317,**318**,320, 321
  housing of blacks, 98-109
Random generating routine, 413,417,423
Random switch method, 418
Range of convenience, 243
Ratio estimation, method of, 257,258
Recycling of Resources, 532
Region, psychological, 53
Reliability coefficients, 259,262,263
Repertory grid techniques, 244,250
Residential development project, 468
Resistance to change, Hinkle's technique, 244
Resource modification, 125,127,134,292
Role-players, 484-487

Satisfaction, 454-459
Scaling methods, 256
Schema, 509,511,512,517
Science, theory in, 506,507,512,516
Scoring algorithm, 417
Security, 534
Shopping mall, 536
Semantic differential, 62,135,284,285, 295,312,313,315,321
Sense data, 514
Sense organs, 384-387
Sensitivity analysis, 366-368,450,456-457
Sensory activity, 388
Sensory deprivation, 386,387,388
Sensory loss, 383-389
Setting, activity, 493,494,496
Setting deprivation, 33,35

Settings, group study, 498-504
Similarity coefficient, 325
Simulation, 62,384,449,453-454
Single switch method, 418
Site plan, 429,430,431
Site planning, residential, 403,404
Social analyst, 5,12
Social behavior, mutation of, 41,45
  pathology, 344
Social engineering, 493,505
Social organization, 18
Social relationships, 251
Social role, 181
Social setting, 184,412
Social structure, 96
Social systems, 493
Sounds, 387
Space, allocation techniques, 414
  characteristics of, 250
  conceptual, 18
  global, 312-315,428
  limbic, 19
  personalization, 503
  use of, 7,191-194,496,503
Space diagram, 429
Space grammars, 428,436-438
Space need, 354
Space planning, 415,416,419,423,428,429, 438,503
Space time, logic-operational models, 179,182-183
Spatial environment, layout of the, 256, 277
Spatial organization, concepts of, 125, 129
State of Equilibrium demographic systems, 340,342
Statistical distribution, 521-522,524-525
Statistical parameters, predictive validity, 4,8,10,220,224,226
Statistical techniques, chi-square analysis, 195,127-129
  measures of validity, 220,224,226
  multiple regression analysis, 70,263
  multivariate statistical techniques, 115
  poisson process, 452-453
  significance tests, 312
  thermodynamics, 335,336

Street, urban, 62
Stress, 37, 82
Structural linguistics, 6
Study facilities, "study set(s)", 493-504
Studying habits, restriction on, 501
Subculture(s), integration, 494
Subject characteristics, domain of, 325
Survey methods, diary questionnaire, 127
    open ended questions, 209
Synectics, 482, 509, 530, 532
Synthesis, 5, 11-12
Super-systems, 531
Systems analysis, steady-state condition, 532
Systems approach, reaction against, 506
Systems building, 508

Tactile loss, 385, 388
Territory, 14, 31, 42, 125
Therapy, of juvenile delinquents, 140
Thought, elements of human, 374
Thresholds, functional types, 471-474
Time, 30, 33, 124, 249, 449, 453, 454, 456-457
    circadian, 43
    distance, pedestrian, 6, 11, 128, 133
Traffic congestion, 535-538
Traffic, elevator patterns simulated, 448-449, 451-453, 456-458
Transgression, in user interactions, 125

Unconscious world, 513, 517
Urban design, 62, 463-464
Urban design tool, 71
Urbanism, 294
Urban Planning, 334
Urban topologies, 150-160
Utility, 292, 416
    theory, 441
User attributes, 3, 5, 6, 7, 125, 127-129, 191, 329
    behavior, 230, 332
    expectations of, 509, 511-512
    needs, diversity of, 1, 9, 324, 332
    preferences, 125
    satisfaction, 328, 440

Validity, cross cultural, 340
Value system, 370
Variables, 1, 3, 4, 7, 9, 312, 314, 315, 318, 320
    dependent, 123, 128, 132-133, 330
    independent, 123, 128, 132-133, 330, 331
Verbal construct systems, 1, 2, 4, 7, 11, 12, 285, 287, 289, 290, 294, 295
Verifiability, 516
Vegetation, 107
Visual detail, 387, 388
Visual interest, 63
Visual preception, loss of, 384, 386-387, 388

Working drawings, production of, 396
Wilderness, 164
    reaction to, 169

Youth, 84, 97

Zoning, 170